Handbook in
Monte Carlo Simulation

Wiley Handbooks in
FINANCIAL ENGINEERING AND ECONOMETRICS

Advisory Editor
Ruey S. Tsay
The University of Chicago Booth School of Business, USA

A complete list of the titles in this series appears at the end of this volume.

Handbook in Monte Carlo Simulation

Applications in Financial Engineering, Risk Management, and Economics

PAOLO BRANDIMARTE
Department of Mathematical Sciences
Politecnico di Torino
Torino, Italy

WILEY

Library of Congress Cataloging-in-Publication Data:

Brandimarte, Paolo.
  Handbook in Monte Carlo simulation : applications in financial engineering, risk management, and economics / Paolo Brandimarte.
     pages cm
  Includes bibliographical references and index.
  ISBN 978-0-470-53111-2 (cloth)
  1. Finance—Mathematical models. 2. Economics—Mathematical models. 3. Monte Carlo method. I. Title.
  HG106.B735 2014
  330.01'518282—dc23   2013047832

Contents

Preface *xiii*

Part I
Overview and Motivation

1 Introduction to Monte Carlo Methods *3*
  1.1 Historical origin of Monte Carlo simulation *4*
  1.2 Monte Carlo simulation vs. Monte Carlo sampling *7*
  1.3 System dynamics and the mechanics of Monte Carlo simulation *10*
    1.3.1 Discrete-time models *10*
    1.3.2 Continuous-time models *13*
    1.3.3 Discrete-event models *16*
  1.4 Simulation and optimization *22*
    1.4.1 Nonconvex optimization *23*
    1.4.2 Stochastic optimization *26*
    1.4.3 Stochastic dynamic programming *28*
  1.5 Pitfalls in Monte Carlo simulation *30*
    1.5.1 Technical issues *31*
    1.5.2 Philosophical issues *33*
  1.6 Software tools for Monte Carlo simulation *35*
  1.7 Prerequisites *37*
    1.7.1 Mathematical background *37*
    1.7.2 Financial background *38*
    1.7.3 Technical background *38*
  For further reading *39*
  References *39*

2 Numerical Integration Methods *41*
  2.1 Classical quadrature formulas *43*
    2.1.1 The rectangle rule *44*
    2.1.2 Interpolatory quadrature formulas *45*
    2.1.3 An alternative derivation *46*
  2.2 Gaussian quadrature *48*
    2.2.1 Theory of Gaussian quadrature: The role of orthogonal polynomials *49*
    2.2.2 Gaussian quadrature in R *51*
  2.3 Extension to higher dimensions: Product rules *53*

v

2.4   Alternative approaches for high-dimensional integration   *55*
    2.4.1   Monte Carlo integration   *56*
    2.4.2   Low-discrepancy sequences   *59*
    2.4.3   Lattice methods   *61*
2.5   Relationship with moment matching   *67*
    2.5.1   Binomial lattices   *67*
    2.5.2   Scenario generation in stochastic programming   *69*
2.6   Numerical integration in R   *69*
For further reading   *71*
References   *71*

Part II
Input Analysis: Modeling and Estimation

3   Stochastic Modeling in Finance and Economics   *75*
  3.1   Introductory examples   *77*
    3.1.1   Single-period portfolio optimization and modeling returns   *78*
    3.1.2   Consumption–saving with uncertain labor income   *81*
    3.1.3   Continuous-time models for asset prices and interest rates   *83*
  3.2   Some common probability distributions   *86*
    3.2.1   Bernoulli, binomial, and geometric variables   *88*
    3.2.2   Exponential and Poisson distributions   *92*
    3.2.3   Normal and related distributions   *98*
    3.2.4   Beta distribution   *105*
    3.2.5   Gamma distribution   *107*
    3.2.6   Empirical distributions   *108*
  3.3   Multivariate distributions: Covariance and correlation   *112*
    3.3.1   Multivariate distributions   *113*
    3.3.2   Covariance and Pearson's correlation   *118*
    3.3.3   R functions for covariance and correlation   *122*
    3.3.4   Some typical multivariate distributions   *124*
  3.4   Modeling dependence with copulas   *128*
    3.4.1   Kendall's tau and Spearman's rho   *134*
    3.4.2   Tail dependence   *136*
  3.5   Linear regression models: A probabilistic view   *137*
  3.6   Time series models   *138*
    3.6.1   Moving-average processes   *142*
    3.6.2   Autoregressive processes   *147*
    3.6.3   ARMA and ARIMA processes   *151*
    3.6.4   Vector autoregressive models   *155*
    3.6.5   Modeling stochastic volatility   *157*
  3.7   Stochastic differential equations   *159*
    3.7.1   From discrete to continuous time   *160*

        3.7.2   Standard Wiener process   *163*
        3.7.3   Stochastic integration and Itô's lemma   *167*
        3.7.4   Geometric Brownian motion   *173*
        3.7.5   Generalizations   *176*
   3.8   Dimensionality reduction   *178*
        3.8.1   Principal component analysis (PCA)   *179*
        3.8.2   Factor models   *189*
   3.9   Risk-neutral derivative pricing   *192*
        3.9.1   Option pricing in the binomial model   *193*
        3.9.2   A continuous-time model for option pricing: The
                 Black–Scholes–Merton formula   *196*
        3.9.3   Option pricing in incomplete markets   *203*
   For further reading   *205*
   References   *206*

4   **Estimation and Fitting**   *209*
   4.1   Basic inferential statistics in R   *211*
        4.1.1   Confidence intervals   *211*
        4.1.2   Hypothesis testing   *214*
        4.1.3   Correlation testing   *218*
   4.2   Parameter estimation   *219*
        4.2.1   Features of point estimators   *221*
        4.2.2   The method of moments   *222*
        4.2.3   The method of maximum likelihood   *223*
        4.2.4   Distribution fitting in R   *227*
   4.3   Checking the fit of hypothetical distributions   *228*
        4.3.1   The chi-square test   *229*
        4.3.2   The Kolmogorov–Smirnov test   *231*
        4.3.3   Testing normality   *232*
   4.4   Estimation of linear regression models by ordinary least squares   *233*
   4.5   Fitting time series models   *237*
   4.6   Subjective probability: The Bayesian view   *239*
        4.6.1   Bayesian estimation   *241*
        4.6.2   Bayesian learning and coin flipping   *243*
   For further reading   *248*
   References   *249*

Part III
Sampling and Path Generation

5   **Random Variate Generation**   *253*
   5.1   The structure of a Monte Carlo simulation   *254*
   5.2   Generating pseudorandom numbers   *256*
        5.2.1   Linear congruential generators   *256*
        5.2.2   Desirable properties of random number generators   *260*

      5.2.3   General structure of random number generators   *264*

      5.2.4   Random number generators in R   *266*

  5.3   The inverse transform method   *267*

  5.4   The acceptance–rejection method   *269*

  5.5   Generating normal variates   *276*

      5.5.1   Sampling the standard normal distribution   *276*

      5.5.2   Sampling a multivariate normal distribution   *278*

  5.6   Other ad hoc methods   *282*

  5.7   Sampling from copulas   *283*

  For further reading   *286*

  References   *287*

6 **Sample Path Generation for Continuous-Time Models**   *289*

  6.1   Issues in path generation   *290*

      6.1.1   Euler vs. Milstein schemes   *293*

      6.1.2   Predictor-corrector methods   *295*

  6.2   Simulating geometric Brownian motion   *297*

      6.2.1   An application: Pricing a vanilla call option   *299*

      6.2.2   Multidimensional GBM   *301*

      6.2.3   The Brownian bridge   *304*

  6.3   Sample paths of short-term interest rates   *308*

      6.3.1   The Vasicek short-rate model   *310*

      6.3.2   The Cox–Ingersoll–Ross short-rate model   *312*

  6.4   Dealing with stochastic volatility   *315*

  6.5   Dealing with jumps   *316*

  For further reading   *319*

  References   *320*

Part IV
Output Analysis and Efficiency Improvement

7 **Output Analysis**   *325*

  7.1   Pitfalls in output analysis   *327*

      7.1.1   Bias and dependence issues: A financial example   *330*

  7.2   Setting the number of replications   *334*

  7.3   A world beyond averages   *335*

  7.4   Good and bad news   *337*

  For further reading   *338*

  References   *338*

8 **Variance Reduction Methods**   *341*

  8.1   Antithetic sampling   *342*

  8.2   Common random numbers   *348*

  8.3   Control variates   *349*

  8.4   Conditional Monte Carlo   *353*

8.5   Stratified sampling   *357*

8.6   Importance sampling   *364*

      8.6.1   Importance sampling and rare events   *370*

      8.6.2   A digression: Moment and cumulant generating functions   *373*

      8.6.3   Exponential tilting   *374*

For further reading   *376*

References   *377*

9   Low-Discrepancy Sequences   *379*

9.1   Low-discrepancy sequences   *380*

9.2   Halton sequences   *382*

9.3   Sobol low-discrepancy sequences   *387*

      9.3.1   Sobol sequences and the algebra of polynomials   *389*

9.4   Randomized and scrambled low-discrepancy sequences   *393*

9.5   Sample path generation with low-discrepancy sequences   *395*

For further reading   *399*

References   *400*

Part V
Miscellaneous Applications

10 Optimization   *405*

10.1   Classification of optimization problems   *407*

10.2   Optimization model building   *421*

      10.2.1   Mean–variance portfolio optimization   *421*

      10.2.2   Modeling with logical decision variables: Optimal portfolio tracking   *422*

      10.2.3   A scenario-based model for the newsvendor problem   *425*

      10.2.4   Fixed-mix asset allocation   *426*

      10.2.5   Asset pricing   *427*

      10.2.6   Parameter estimation and model calibration   *430*

10.3   Monte Carlo methods for global optimization   *432*

      10.3.1   Local search and other metaheuristics   *433*

      10.3.2   Simulated annealing   *435*

      10.3.3   Genetic algorithms   *439*

      10.3.4   Particle swarm optimization   *441*

10.4   Direct search and simulation-based optimization methods   *444*

      10.4.1   Simplex search   *445*

      10.4.2   Metamodeling   *446*

10.5   Stochastic programming models   *448*

      10.5.1   Two-stage stochastic linear programming with recourse   *448*

      10.5.2   A multistage model for portfolio management   *452*

      10.5.3   Scenario generation and stability in stochastic programming   *457*

10.6   Stochastic dynamic programming   *469*

        10.6.1 The shortest path problem *470*
        10.6.2 The functional equation of dynamic programming *473*
        10.6.3 Infinite-horizon stochastic optimization *477*
        10.6.4 Stochastic programming with recourse vs. dynamic
                   programming *478*
   10.7 Numerical dynamic programming *480*
        10.7.1 Approximating the value function: A deterministic
                   example *480*
        10.7.2 Value iteration for infinite-horizon problems *484*
        10.7.3 A numerical approach to consumption–saving *493*
   10.8 Approximate dynamic programming *506*
        10.8.1 A basic version of ADP *507*
        10.8.2 Post-decision state variables in ADP *510*
        10.8.3 *Q*-learning for a simple MDP *513*
   For further reading *519*
   References *520*

11 Option Pricing *525*
   11.1 European-style multidimensional options in the BSM world *526*
   11.2 European-style path-dependent options in the BSM world *532*
        11.2.1 Pricing a barrier option *532*
        11.2.2 Pricing an arithmetic average Asian option *539*
   11.3 Pricing options with early exercise features *546*
        11.3.1 Sources of bias in pricing options with early exercise
                   features *548*
        11.3.2 The scenario tree approach *549*
        11.3.3 The regression-based approach *552*
   11.4 A look outside the BSM world: Equity options under the
            Heston model *560*
   11.5 Pricing interest rate derivatives *563*
        11.5.1 Pricing bonds and bond options under the Vasicek
                   model *565*
        11.5.2 Pricing a zero-coupon bond under the CIR model *567*
   For further reading *569*
   References *569*

12 Sensitivity Estimation *573*
   12.1 Estimating option greeks by finite differences *575*
   12.2 Estimating option greeks by pathwise derivatives *581*
   12.3 Estimating option greeks by the likelihood ratio method *585*
   For further reading *589*
   References *590*

13 Risk Measurement and Management *591*
   13.1 What is a risk measure? *593*
   13.2 Quantile-based risk measures: Value-at-risk *595*

13.3 Issues in Monte Carlo estimation of V@R   *601*

13.4 Variance reduction methods for V@R   *607*

13.5 Mean–risk models in stochastic programming   *613*

13.6 Simulating delta hedging strategies   *619*

13.7 The interplay of financial and nonfinancial risks   *625*

For further reading   *626*

References   *627*

14 Markov Chain Monte Carlo and Bayesian Statistics   *629*

14.1 Acceptance–rejection sampling in Bayesian statistics   *630*

14.2 An introduction to Markov chains   *631*

14.3 The Metropolis–Hastings algorithm   *636*

    14.3.1 The Gibbs sampler   *640*

14.4 A re-examination of simulated annealing   *643*

For further reading   *646*

References   *647*

Index   *649*

Preface

The aim of this book is to provide a wide class of readers with a low- to intermediate-level treatment of Monte Carlo methods for applications in finance and economics. The target audience consists of students and junior practitioners with a quantitative background, and it includes not only students in economics and finance, but also in mathematics, statistics, and engineering. In fact, this is the kind of audience I typically deal with in my courses. Not all of these readers have a strong background in either statistics, financial economics, or econometrics, which is why I have also included some basic material on stochastic modeling in the early chapters, which is typically skipped in higher level books. Clearly, this is not meant as a substitute for a proper treatment, which can be found in the references listed at the end of each chapter. Some level of mathematical maturity is assumed, but the prerequisites are rather low and boil down to the essentials of probability and statistics, as well as some basic programming skills.[1] Advanced readers may skip the introductory chapters on modeling and estimation, which are also included as a reminder that no Monte Carlo method, however sophisticated, will yield useful results if the input model is flawed. Indeed, the power and flexibility of such methods may lure us into a false sense of security, making us forget some of their inherent limitations.

Option pricing is certainly a relevant application domain for the techniques we discuss in the book, but this is not meant to be a book on financial engineering. I have also included a significant amount of material on optimization in its many guises, as well as a chapter related to computational Bayesian statistics. I have favored a wide scope over a deeply technical treatment, for which there are already some really excellent and more demanding books. Many of them, however, do not quite help the reader to really "feel" what she is learning, as no ready-to-use code is offered. In order to allow anyone to run the code, play with it, and hopefully come up with some variations on the theme, I have chosen to develop code in R. Readers familiar with my previous book written in MATLAB might wonder whether I have changed my mind. I did not: I never use R in research or consulting, but I use it a lot for teaching. When I started writing the book, I was less than impressed by the lack of an adequate development environment, and some design choices of the language itself left me a bit puzzled. As an example, the $*$ operator in MATLAB multiplies matrices row by column; whenever you want to work elementwise, you use the . operator, which has a clear and uniform meaning when applied to other operators. On the

---

[1] In case of need, the mathematical prerequisites are covered in my other book: *Quantitative Methods: An Introduction for Business Management*. Wiley, 2011.

xiii

contrary, the operator $*$ works elementwise in R, and row-by-column matrix product is accomplished by the somewhat baroque operator $\%*\%$. Furthermore, having to desperately google every time you have to understand a command, because documentation is a bit poor and you have to make your way in a mess of packages, may be quite frustrating at times. I have also found that some optimization functions are less accurate and less clever in dealing with limit cases than the corresponding MATLAB functions. Having said that, while working on the book, I have started to appreciate R much more. Also my teaching experience with R has certainly been fun and rewarding. A free tool with such a potential as R is certainly most welcome, and R developers must be praised for offering all of this. Hopefully, the reader will find R code useful as a starting point for further experimentation. I did *not* assemble R code into a package, as this would be extremely misleading: I had no plan to develop an integrated and reliable set of functions. I just use R code to illustrate ideas in concrete terms and to encourage active learning. When appropriate, I have pointed out some programming practices that may help in reducing the computational burden, but as a general rule I have tried to emphasize clarity over efficiency. I have also avoided writing an introduction to R programming, as there are many freely available tutorials (and a few good books[2]). A reader with some programming experience in any language should be able to make her way through the code, which has been commented on when necessary. My assumption is that a reader, when stumbling upon an unknown function, will take advantage of the online help and the example I provide in order to understand its use and potentiality. Typically, R library functions are equipped with optional parameters that can be put to good use, but for the sake of conciseness I have refrained from a full description of function inputs.

Book structure

The book is organized in five parts.

1. Part I, Overview and Motivation, consists of two chapters. Chapter 1 provides an introduction to Monte Carlo methods and applications. The different classes of dynamic models that are encountered in simulation are outlined, and due emphasis is placed on pitfalls and limitations of Monte Carlo methods. Chapter 2 deals with numerical integration methods. Numerical integration is quite relevant, as it provided most of the historical motivation to develop Monte Carlo methods; furthermore, there are cases in which one is much better off using good quadrature formulas than throwing random numbers around. Finally, framing Monte Carlo methods within numerical integration provides the necessary discipline to understand and properly use low-discrepancy sequences, sometimes referred to as quasi–Monte Carlo methods.

---

[2]Unfortunately, I have also run into very bad books using R; hopefully, this one will not contribute to the list.

2. Part II, Input Analysis: Modeling and Estimation, is specifically aimed at students and newcomers, as it includes two introductory chapters dealing with stochastic model building (Chapter 3) and model fitting (Chapter 4). Essentially, in this part of the book we are concerned with the modeling of inputs of a Monte Carlo simulation. Many advanced books on Monte Carlo methods for finance skip and take for granted these concepts. I have preferred to offer a limited treatment for the sake of unfamiliar readers, such as students in engineering or practitioners without an econometrics background. Needless to say, space does not allow for a deep treatment, but I believe that it is important to build at least a framework for further study. In order to make this part useful to intermediate readers, too, I have taken each topic as an excuse for a further illustration of R functionalities. Furthermore, some more advanced sections may be useful to students in economics and finance as well, such as those on stochastic calculus, copulas, and Bayesian statistics.

3. Part III, Sampling and Path Generation, is more technical and consists of two chapters. In Chapter 5 we deal with pseudorandom number and variate generation. While it is certainly true that in common practice one takes advantage of reliable generators provided by software tools like R, and there is no need for an overly deep treatment, some basic knowledge is needed in order to select generators and to manage simulation runs properly. We also outline scenario generation using copulas. In Chapter 6 we deal with sample path generation for continuous-time models based on stochastic differential equations. This is an essential tool for any financial engineer and is at the heart of many derivative pricing methods. It is important to point out that this is also relevant for risk managers, insurers, and some economists as well.

4. Part IV, Output Analysis and Efficiency Improvement, looks at the final step of the simulation process. Monte Carlo methods are extremely powerful and flexible; yet, their output may not be quite reliable, and an unreasonable computational effort may be called for, unless suitable countermeasures are taken. Chapter 7 covers very simple, and possibly overlooked, concepts related to confidence intervals. Counterexamples are used to point out the danger of forgetting some underlying assumptions. Chapter 8 deals with variance reduction strategies that are essential in many financial engineering and risk management applications; indeed, the techniques illustrated here are applied in later chapters, too. Chapter 9 deals with low-discrepancy sequences, which are sometimes gathered under the quasi–Monte Carlo nickname. Actually, there is nothing stochastic in low-discrepancy sequences, which should be regarded as deterministic numerical integration strategies. For a certain range of problem dimensionality, they are a good alternative to pseudorandom sampling.

5. Part IV, Miscellaneous Applications, includes five more or less interrelated chapters dealing with:

- The interplay between optimization and Monte Carlo methods, including stochastic methods for deterministic global optimization, scenario generation for stochastic programming with recourse, and stochastic dynamic programming (Chapter 10)

- Option pricing, with an emphasis on variance reduction methods (Chapter 11)

- Obtaining sensitivity of performance measures with Monte Carlo simulation (Chapter 12)

- Risk measurement and management, with an emphasis on value-at-risk and related risk measures for financial risk management (Chapter 13)

- Markov chain Monte Carlo (MCMC) methods, which are relevant for different applications, most notably computational Bayesian statistics (Chapter 14)

There are some logical precedences among these final chapters, but they need not be read in a strict sequential order. Chapter 14 is independent of the others, and the only link is represented by the possibility of regarding simulated annealing, a stochastic approach for both global and combinatorial optimization, as an MCMC method. Stochastic dynamic programming, dealt with in Chapter 10, is needed to understand American-style option pricing in Chapter 11. Measuring option price sensitivity is used as a motivating example in Chapter 12, but the methods outlined there have a much more general applicability, as they can also be used within optimization procedures. Finally, there is certainly a natural link between option pricing and the risk management examples discussed in Chapter 13.

Supplements and R code

The code has been organized in two files per chapter. The first one contains all of the function definitions and the reference to required packages, if any; this file should be sourced before running the scripts, which are included as chunks of code in a second file. An archive including all of the R code will be posted on a webpage. My current URL is

- http://staff.polito.it/paolo.brandimarte/

A hopefully short list of errata will be posted there as well. One of the many corollaries of Murphy's law states that my URL is going to change shortly after publication of the book. An up-to-date link will be maintained on the Wiley webpage:

- http://www.wiley.com/

For comments, suggestions, and criticisms, all of which are quite welcome, my e-mail address is

- paolo.brandimarte@polito.it

PAOLO BRANDIMARTE

Turin, February 2014

Part One

Overview and Motivation

## Chapter One

Introduction to Monte Carlo Methods

The term *Monte Carlo* is typically associated with the process of modeling and simulating a system affected by randomness: Several random scenarios are generated, and relevant statistics are gathered in order to assess, e.g., the performance of a decision policy or the value of an asset. Stated as such, it sounds like a fairly easy task from a conceptual point of view, even though some programming craft might be needed. Although it is certainly true that Monte Carlo methods are extremely flexible and valuable tools, quite often the last resort approach for overly complicated problems impervious to a more mathematically elegant treatment, it is also true that running a *bad* Monte Carlo simulation is very easy as well. There are several reasons why this may happen:

- We are using a wrong model of uncertainty:
    - Because we are using an unrealistic probability distribution
    - Or because we are missing some link among the underlying risk factors
    - Or because some unknown parameters have been poorly estimated
    - Or because the very nature of uncertainty in our problem does not lend itself to a stochastic representation
- The output estimates are not reliable enough, i.e., the estimator variance is so large that a much larger sample size is required.
- There is a systematic error in the estimates, which could be low or high biased.
- The way we generate sample paths, possibly discretizing a continuous-time model, induces a non-negligible error.
- We are using poor random variate generators.
- There is some possibly subtle bug in the computer program implementing the method.

Some of these issues are technical and can be addressed by the techniques that we will explore in this book, but others are more conceptual in nature and point out a few intrinsic and inescapable limitations of Monte Carlo methods; it is wise not to forget them, while exploring the richness and power of the approach.

The best countermeasure, in order to avoid the aforementioned pitfalls, is to build reasonably strong theoretical foundations and to gain a deep understanding of the Monte Carlo approach and its variants. To that end, a good first step is to frame Monte Carlo methods as a numerical integration tool. Indeed, while the term *simulation* sounds more exciting, the term Monte Carlo *sampling* is often used. The latter is more appropriate when we deal with Monte Carlo sampling as a tool for numerical integration or statistical computing. Granted, the idea of simulating financial markets is somewhat more appealing than the idea of computing a multidimensional integral. However, a more conceptual framework helps in understanding powerful methods for reducing, or avoiding altogether, the difficulties related to the variance of random estimators. Some of these methods, such as low-discrepancy sequences and Gaussian quadrature, are actually deterministic in nature, but their understanding needs a view integrating numerical integration and statistical sampling.

In this introductory chapter, we consider first the historical roots of Monte Carlo; we will see in Section 1.1 that some early Monte Carlo methods were actually aimed at solving deterministic problems. Then, in Section 1.2 we compare Monte Carlo sampling and Monte Carlo simulation, showing their deep relationship. Typical simulations deal with dynamic systems evolving in time, and there are three essential kinds of dynamic models:

- Continuous-time models
- Discrete-time models
- Discrete-event models

These model classes are introduced in Section 1.3, where we also illustrate how their nature affects the mechanics of Monte Carlo simulation. In this book, a rather relevant role is played by applications involving optimization. This may sound odd to readers who associate simulation with performance evaluation; on the contrary, there is a multiway interaction between optimization and Monte Carlo methods, which is outlined in Section 1.4.

In this book we illustrate a rather wide range of applications, which may suggest the idea that Monte Carlo methods are almost a panacea. Unfortunately, this power may hide many pitfalls and dangers. In Section 1.5 we aim at making the reader aware of some of these traps. Finally, in Section 1.6 we list a few software tools that are commonly used to implement Monte Carlo simulations, justifying the choice of R as the language of this book, and in Section 1.7 we list prerequisites and references for readers who may need a refresher on some background material.

1.1   Historical origin of Monte Carlo simulation

Monte Carlo methods involve random sampling, but the actual aim is to estimate a deterministic quantity. Indeed, a well-known and early use of Monte Carlo–

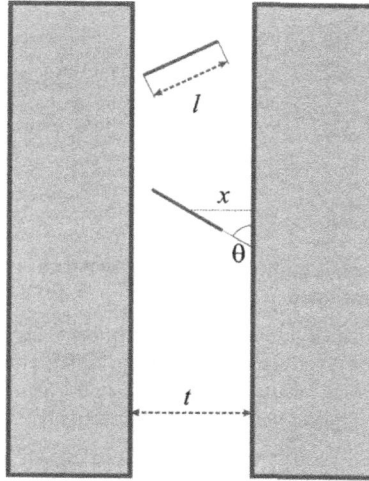

FIGURE 1.1 **An illustration of Buffon's needle.**

like methods is Buffon's needle approach to estimate π.[1] The idea, illustrated in Fig. 1.1, is to randomly throw n times a needle of length l on a floor consisting of wood strips of width $t > l$, and to observe the number of times h that the needle crosses the border between two strips. Let X be the distance from the center of the needle to the closest border; X is a uniform random variable on the range $[0, t/2]$, and its probability density function is $2/t$ on that interval and 0 outside.[2] By a similar token, let θ be the acute angle between the needle and that border; this is another uniformly distributed variable, on the range $[0, \pi/2]$, with density $2/\pi$ on that support. Using elementary trigonometry, it is easy to see that the needle will cross a border when

$$X \leq \frac{l}{2} \sin \theta.$$

The probability of this event is found by integrating the joint density of the random variables X and θ over the relevant domain. Assuming independence, the joint density is just the product of the two marginals. Therefore, we have to compute a double integral as follows:

$$\int_0^{\pi/2} \int_0^{(l/2)\sin\theta} \frac{4}{t\pi} \, dx \, d\theta = \frac{2l}{t\pi} \int_0^{\pi/2} \sin\theta \, d\theta = \frac{2l}{t\pi}.$$

If we want to estimate π, we can use a sample estimate of this probability, i.e., the ratio of h over n:

$$\frac{h}{n} \approx \frac{2l}{t\pi} \quad \Rightarrow \quad \hat{\pi} = \frac{2ln}{th}.$$

---

[1] The estimation of π by Monte Carlo methods will also be used as an illustration of variance reduction strategies in Chapter 8.

[2] In Chapter 3 we offer a refresher on common probability distributions, among other things.

```
estPi <- function(N){
  x <- 2*runif(N)-1
  y <- 2*runif(N)-1
  h <- sum(x^2+y^2<=1)
  return(4*h/N)
}
```

R programming notes:

1. A function is created by the `function` keyword, followed by the input arguments, and assigned to a variable.

2. `runif` is a typical R function, featuring the `r` prefix, for generating a random sample of given size from a selected distribution. In the following, we will meet similar functions, like `rexp` and `rnorm`, to sample from an exponential and a normal distribution, respectively.

3. The expression `sum(x^2+y^2<=1)` illustrates an interesting feature of R to get rid of `for` loops: It returns a vector of elements with values `FALSE` or `TRUE`, which are interpreted in the sum as 0/1, respectively.

FIGURE 1.2   **Using R to estimate π.**

We see that the original problem is purely deterministic, but we use Monte Carlo sampling as an estimation tool.

A slightly more practical approach, taking advantage of R, is illustrated in the code of Fig. 1.2. The idea is shooting n random bullets on the square $[-1, 1] \times [-1, 1]$ and evaluating the fraction h/n of points falling inside the unit circle $x^2 + y^2 = 1$. Since the ratio between the area of the circle and the area of the square is $\pi/4$, we have

$$\hat{\pi} = \frac{4h}{n}.$$

Running the code,[3] we obtain

```
> set.seed(55555)
> estPi(1000)
[1] 3.132
> estPi(1000)
[1] 3.14
> estPi(1000)
[1] 3.084
> estPi(1000000)
[1] 3.141388
```

---

[3]Later, we will understand in more detail what is accomplished by the command `set.seed(55555)`. For now, it is enough to say that we always use this function to set the state of pseudorandom number generators to a given value, so that the reader can replicate the experiments and obtain the same results. There is nothing special in the argument 55555, apart from the fact that 5 is the author's favorite number.

```
> pi
[1] 3.141593
```

Clearly, a huge sample size is needed to obtain a fairly accurate estimate, and this is certainly not the smartest way to estimate π.

In more recent times, in the period between 1930 and 1950, Monte Carlo methods were used by the likes of Enrico Fermi, Stanislaw Ulam, and John von Neumann to solve problems related to physics. Indeed, such methods were used in the 1950s when the hydrogen bomb was developed at the Los Alamos laboratories. It was then that the term *Monte Carlo* was coined after the well-known gambling house. Then, luckily enough, the idea moved to other domains, including operations research and economics. At present, Monte Carlo methods are extremely widespread in many domains, including risk management and financial engineering.

1.2 Monte Carlo simulation vs. Monte Carlo sampling

The terms *simulation* and *sampling* are both used with reference to Monte Carlo methods. Indeed, there is no 100% sharp line separating the two concepts, and a couple of examples are the best way to get the point.

Example 1.1 Shortfall probability in wealth management

Consider an investor who allocates her wealth W_0 to n risky financial assets, such as stock, bonds, derivatives, and whatnot. At time $t = 0$, the asset price of asset i, $i = 1, \ldots, n$, is P_i^0, and the (integer) number of shares of assets i in the portfolio is h_i. Therefore, the initial wealth is

$$W_0 = \sum_{i=1}^n h_i P_i^0.$$

The portfolio is held up to time $t = T$, and the price of asset i at the end of this investment horizon depends on a set of underlying risk factors, such as the term structure of interest rates, spreads due to credit risk, inflation, oil price, the overall state of the economy, etc. Let $\mathbf{X}$ be a vector random variable, with joint density $f_{\mathbf{X}}(\mathbf{x})$, representing these factors. If we assume that the price of each asset at time T is given by a function of the underlying factors,

$$P_i^T(\mathbf{X}),$$

then the terminal wealth is a function of $\mathbf{X}$,

$$W_T = \sum_{i=1}^n h_i P_i^T(\mathbf{X}),$$

which is itself a random variable. On the one hand, we are interested in the expected value of W_T. However, since risk is a quite relevant issue, we might also be interested in the shortfall probability with respect to a target wealth H:

$$P\{W_T < H\}.$$

Whatever the case, we are lead to the expected value of some function $g(\mathbf{X})$:

$$\int g(\mathbf{x}) f_{\mathbf{X}}(\mathbf{x}) \, d\mathbf{x}. \tag{1.1}$$

If we choose $g(\mathbf{X}) = W_T$ we obtain the expected value of future wealth; if we choose the indicator function of the event $\{W_T < H\}$, i.e.,

$$g(\mathbf{X}) = \mathbf{1}_{\{W_T < H\}} \equiv \begin{cases} 1, & \text{if } W_T < H, \\ 0, & \text{otherwise,} \end{cases}$$

then we obtain the shortfall probability. Other risk measures can be selected and will be discussed in Chapter 13. Evaluating the multidimensional integral in Eq. (1.1) is quite difficult, even when we know the analytical form of the involved functions. As we shall see, by using Monte Carlo we may draw a sample of m observations $\mathbf{X}_k$, $k = 1, \ldots, m$, and estimate the quantities of interest. We are using random sampling as a tool for numerical integration.

In the example above, no simulation is involved at all, assuming that we are able to sample directly the random variable $\mathbf{X}$ given its density. In other cases, we may have to cope with a more complicated dynamic model that describes the dynamics of the underlying risk factors over the time interval $[0, T]$. In such a case, we may not be able to sample risk factors directly, and we have to generate a set of sample paths. Dynamic models may be represented by continuous-time stochastic differential equations[4] or by relatively simple discrete-time processes as in the following example.

▣ Example 1.2 **An autoregressive process**

Time series models are quite often used in financial econometrics to describe the evolution of a quantity of interest over time. A simple

---

[4]Stochastic differential equations are introduced in Section 3.7, and approaches for sample path generation are described in Chapter 6.

example of this class of models is an autoregressive (AR) process of order 1 like

$$X_t = X_{t-1} + \epsilon_t, \tag{1.2}$$

where X_0 is given and ϵ_t is the driving stochastic process, i.e., a sequence of random variables ϵ_t, $t = 1, 2, 3, \ldots$. In the simplest models, this noise term consists of a sequence of mutually independent and identically distributed normal random variables. We use the notation $X \sim \mathsf{N}(\mu, \sigma^2)$ to state that X is a normal random variable with expected value μ and variance σ^2. Unfolding the recursion, we immediately see that

$$X_t = X_0 + \sum_{\tau=1}^{t} \epsilon_\tau,$$

which implies

$$X_t \sim \mathsf{N}(X_0, t\sigma^2).$$

The process described by Eq. (1.2) is very easy to analyze, but it is not the most general AR process of order 1. A more general example is

$$X_t = \mu + \alpha X_{t-1} + \epsilon_t,$$

whose properties critically depend on the coefficient α (see Example 1.4). In other cases, we may have a system of such equations, with mutually correlated random driving factors and a more complicated structure, preventing us from obtaining the probability distribution of the variable of interest in an explicit form. In such a case, we may have to simulate the dynamical system.

On the basis of the above examples, we may argue that the term *sampling* could be reserved to those cases in which there is no dynamics to be simulated over time, whereas *simulation* entails the generation of sample paths. We may need to generate sample paths because the dynamical model is complicated and prevents an analytical solution, or because it is important to check the underlying variables over time. For instance, when pricing a European-style option,[5] we just need to sample the underlying variable at the option maturity. In the easy case of geometric Brownian motion (GBM for short) this boils down to sampling a lognormal random variable, but we have to resort to sample path generation when we step outside that safe and comforting domain, and we enter the realm populated by stochastic volatilities and price jumps. On the contrary, when we are pricing an American-style option, featuring early exercise opportunities, we need a whole sample path, whatever price model we adopt.

---

[5] See Section 3.9 and Chapter 11.

As is usually the case, also the above distinction is not so clear-cut. In both settings we are actually evaluating a possibly high-dimensional integral of a function, in order to estimate a probability or an expectation. The function may be given in analytical form, or it may be defined by a possibly complicated black box, requiring several lines of code to be evaluated. Conceptually, there is no difference. Indeed, numerical integration plays such a pivotal role for a full appreciation of Monte Carlo methods that we will devote the whole of Chapter 2 to it.

1.3 System dynamics and the mechanics of Monte Carlo simulation

When we have to describe the dynamic evolution of a system over time, we typically resort to one of the following model classes:

- Discrete-time models
- Continuous-time models
- Discrete-event models

From a technical point of view, as usual, the distinction is not sharp. For instance, even if we choose a continuous-time model, we have to discretize it in some way for the sake of computational feasibility. This implies that we actually simulate a discrete-time approximation, but given the issues involved in sample path generation, which involves nontrivial concepts related to the numerical solution of stochastic differential equations, it is better to stick to the classification above. It is also possible to create somewhat hybrid models, such as stochastic processes including both a diffusion component, which is continuous, and jumps, which are related to discrete-event dynamics.

1.3.1 DISCRETE-TIME MODELS

In discrete-time models we assume that the time horizon $[0, T]$ is discretized in time intervals (time buckets) of width δt. In principle, nothing forbids nonuniform time steps, which may actually be required when dealing with certain financial derivatives; nevertheless, we usually stick to the uniform case for the sake of simplicity. The system is actually observed only at time instants of the form

$$k\,\delta t, \qquad k = 0, 1, \ldots, M,$$

where $T = M\,\delta t$. What happens between two discrete-time instants is not relevant. Typically, we forget about the time bucket size δt, which could be a day, a week, or a month, and we use discrete-time subscripts, like $t = 0, 1, 2, \ldots$, directly.

⬛ Example 1.3 **Cash on hand**

Consider the cash management problem for a retail bank. During each day t, we observe cash inflows w_t^+ and cash outflows w_t^-, corresponding to cash deposits and withdrawals by customers, respectively. Hence, if we denote by H_t the cash holding at the end of day t, we have the dynamic equation

$$H_t = H_{t-1} + w_t^+ - w_t^-.$$

Such a model makes sense if we are only interested in the balance at the end of each day.

The AR model of Example 1.2 is a very basic example of time series models. From a technical point of view, simulating a time series model is usually no big deal.

⬛ Example 1.4 **Simulating a simple AR process**

Let us consider the following autoregressive process:

$$X_t = \mu + aX_{t-1} + \epsilon_t,$$

where $\epsilon_t \sim N(0, \sigma^2)$. The R function in Fig. 1.3 can be used for its simulation. In the specific case

$$X_t = 8 + 0.8X_{t-1} + \epsilon_t, \qquad \epsilon_t \sim N(0, 1), \qquad (1.3)$$

we run the function as follows:

```
> set.seed(55555)
> AR(40,.8,8,1,50)
```

obtaining the plot in Fig. 1.4.

In the above examples we had the purely random evolution of a system. In a more general setting, we may have the possibility of influencing, i.e., of partially controlling its dynamics. For instance, in discrete-time stochastic optimization models, we deal with a system modeled by the state transition equation

$$\mathbf{s}_{t+1} = \mathbf{g}_t(\mathbf{s}_t, \mathbf{x}_t, \epsilon_{t+1}), \quad t = 0, 1, 2, \ldots, \qquad (1.4)$$

where

- $\mathbf{s}_t$ is a vector of state variables, observed at the end of time period t.

```
AR <- function(X0,a,mu,sigma,T){
  X <- rep(0,T+1)
  X[1] <- X0
  for (t in 1:T){
    X[t+1] <- mu+a*X[t]+rnorm(1,0,sigma)
  }
  plot(0:T,X,pch=15)
  lines(0:T,X)
}
```

R programming notes:

1. In order to allocate memory to an array, we use `rep(0,T+1)` here; as an alternative, we may also use `numeric(T+1)`.

2. The syntax of `for` loops is self-explanatory. Braces are used to enclose multiple statements; in this case, since there is only one statement inside the loop, they could be omitted.

3. The function `plot` is used here to lay down dots, with a character specified by the optional `pch` parameter; the function `lines` connects the dots with lines, without opening another window.

FIGURE 1.3 **R code to simulate a simple autoregressive process.**

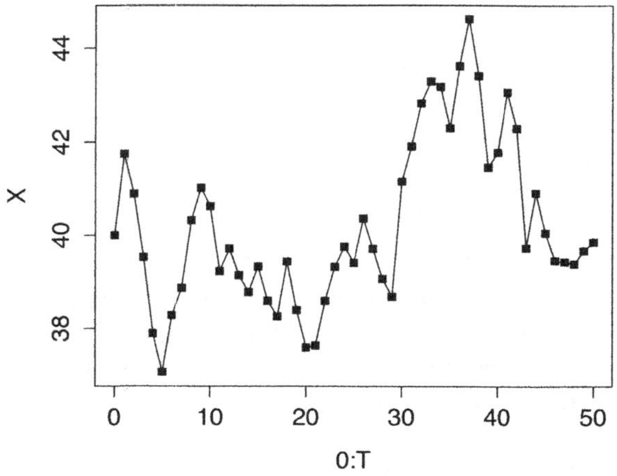

FIGURE 1.4 **Sample path of the AR process of Eq. (1.3).**

- $\mathbf{x}_t$ is a vector of control variables, i.e., decisions made *after* observing the state $\mathbf{s}_t$.
- ϵ_{t+1} is a vector of disturbances, realized *after* we apply the control variable $\mathbf{x}_t$.

If the sequence of disturbances ϵ_t is intertemporally independent, we obtain a Markov process. Markov processes are relatively easy to represent, simulate, and control. We should also observe that the state need not be a continuous variable. We may have a system with discrete states, resulting in a discrete-time Markov chain.[6] As an example, we may consider a Markov chain modeling the migration in the credit rating of a bond issue. Sometimes, we use a discrete-state Markov chain to approximate a continuous-state system by aggregation: We partition the state space with a mutually exclusive and collectively exhaustive collection of subsets and identify each one with a single representative value.

1.3.2 CONTINUOUS-TIME MODELS

The typical representation of a continuous-time model relies on differential equations. We are all familiar with the differential equation $F = ma$ from elementary physics, where F is force, m is mass, and a is acceleration, i.e., the second-order derivative of position with respect to time. As a more financially motivated example, imagine that you deposit an amount of cash $B(0) = B_0$, at time $t = 0$, in a safe bank account. If r is the *continuously compounded* interest rate on your deposit,[7] the evolution in time of wealth over time can be represented as

$$\frac{dB(t)}{dt} = rB(t), \qquad (1.5)$$

whose solution, given the initial condition $B(0) = B_0$, is

$$B(t) = B_0 e^{rt}.$$

If the interest rate is not constant, and it is given by $r(t)$, we obtain

$$B(t) = B(0) \exp\left\{ \int_0^t r(\tau)\, d\tau \right\}. \qquad (1.6)$$

This solution is easily obtained by rewriting Eq. (1.5) in the differential form

$$\frac{dB}{B} = r(t)\, dt,$$

which, by recalling the derivative of the natural logarithm, can be transformed to

$$d\log B = r(t)\, dt.$$

---

[6]The Markov property is essential in dynamic programming, as we shall see in Chapter 10; we will cover some material on Markov chains in Chapter 14.

[7]See Section 3.1.3 for further details and a motivation of such an interest rate.

Integrating on the interval $[0, t]$ and taking the exponential to get rid of the logarithm immediately yields Eq. (1.6). In less lucky cases, we have to resort to numerical methods, which typically yield a discrete-time approximation of the continuous-time system.

Monte Carlo methods come into play when randomness is introduced into the differential equation, yielding a stochastic differential equation. A well-known model in this vein is the GBM

$$dS(t) = \mu S(t)\, dt + \sigma S(t)\, dW(t), \tag{1.7}$$

which is arguably the most elementary model used to represent the random evolution of stock prices. With respect to the previous case, we have an additional term $W(t)$, which is a Wiener process. We will deal later with the underpinnings of this kind of stochastic process,[8] but for now suffice to say that the increment of the Wiener process over a finite time interval $[t, t + \delta t]$ is a normal random variable:

$$\delta W(t) = W(t + \delta t) - W(t) \sim \mathsf{N}(0, \delta t).$$

Loosely speaking, the differential $dW(t)$ is the limit of this increment for $\delta t \to 0$. Thus, we might consider simulating random sample paths of GBM by a straightforward discretization scheme:

$$\delta S(t) = S(t + \delta t) - S(t) = \mu S(t)\, \delta t + \sigma S(t)\, [W(t + \delta t) - W(t)].$$

This is called the Euler discretization scheme, and it immediately leads to the discrete-time process

$$S(t + \delta t) = S(t)(1 + \mu\, \delta t) + \sigma S(t)\sqrt{\delta t}\, \epsilon(t + \delta t),$$

where $\epsilon(t + \delta t) \sim \mathsf{N}(0, 1)$. This kind of process, in terms of simulation, is not unlike an AR process and is easy to deal with. Unfortunately, this rather naive discretization suffers from a significant inconvenience: We obtain the wrong probability distribution. Indeed, it is easy to see that we generate normally distributed prices; therefore, at least in principle, we may generate a negative price, which does not make sense for limited liability assets like stock shares.

Using tools from stochastic calculus, we will see that we may solve the stochastic differential equation exactly and find an exact discretization:

$$S(t + \delta t) = S(t) \exp\left[\left(\mu - \frac{\sigma^2}{2} \right) \delta t + \sigma\sqrt{\delta t}\, \epsilon(t + \delta t) \right]. \tag{1.8}$$

This results in a lognormal distribution, ruling out negative prices. We also notice that μ is related to the drift of the process, i.e., its average rate of change over time, whereas σ plays the role of a volatility.

---

[8]The Wiener process is introduced in Chapter 3, and its sample path generation is discussed in Chapter 6.

```
simGBM <- function (S0,mu,sigma,T,numSteps,numRepl){
  dt <- T/numSteps
  muT <- (mu-sigma^2/2)*dt
  sigmaT <- sqrt(dt)*sigma
  pathMatrix <- matrix(nrow=numRepl,ncol=numSteps+1)
  pathMatrix[,1] <- S0
  for (i in 1:numRepl){
    for (j in 2:(numSteps+1)){
      pathMatrix[i,j] <-
          pathMatrix[i,j-1]*exp(rnorm(1,muT,sigmaT))
    }
  }
  return(pathMatrix)
}
```

R programming notes:

1. In the first few lines we precompute invariant quantities; it is always a good rule to avoid useless recalculations in `for` loops.

2. Note how the function `matrix` is used to preallocate a matrix with a given number or rows and columns; in this book we always store sample paths in rows, rather than columns, and include the initial value in the first column.

3. The `rnorm` function is used to generate a sample from a normal distribution with given parameters.

4. This function is not optimized in the sense that it is not vectorized for efficiency. We will see an alternative implementation in Chapter 6.

FIGURE 1.5 **R code to generate sample paths of geometric Brownian motion.**

Example 1.5 Simulating geometric Brownian motion

The R code in Fig. 1.5 can be used to generate sample paths of GBM. The function returns a matrix whose rows correspond to `numRepl` sample paths starting from initial state `S0`, over `numSteps` time steps from 0 to T, for a GBM with parameters μ and σ. The R snapshot of Fig. 1.6 yields the plots in Fig. 1.7. We immediately appreciate the role of the volatility σ.

This solution is exact in the sense that there is no discretization error in the sample path generation; we are left with sampling errors that are typical of Monte Carlo methods. In other cases, things are not that easy: Even though, in the end, continuous-time models may boil down to discrete-time models, they do have some peculiarities, which may be hard to deal with.

```
set.seed(55555)
S0 <- 40
mu <- 0.25
sigma1 <- 0.1
sigma2 <- 0.4
T <- 1
numSteps <- 50
numRepl <- 10
Paths1 <- simGBM(S0,mu,sigma1,T,numSteps,numRepl)
Paths2 <- simGBM(S0,mu,sigma2,T,numSteps,numRepl)
maxVal <- max(Paths1,Paths2)
minVal <- min(Paths1,Paths2)
labx <- bquote(italic(t))    # how to generate plot labels
laby <- bquote(italic(S(t))) # with italic fonts
par(mfrow = c(1,2))
plot(Paths1[1,],type="l",ylab=laby,xlab=labx,
     main="Vol. 10%",ylim=c(minVal,maxVal))
for (k in 2:10) lines(Paths1[k,])
plot(Paths2[1,],type="l",ylab=laby,xlab=labx,
     main="Vol. 40%",ylim=c(minVal,maxVal))
for (k in 2:10) lines(Paths2[k,])
par(mfrow = c(1,1))
```

R programming notes:

1. Note the use of variables `minVal` and `maxVal` to set an appropriate vertical range for the plot, which is then controlled by the `ylim` parameter of `plot`. You may also use the `range` function to gather the smallest and largest element in an array.

2. The function `plot` opens a new window, whereas `lines` draws a line on the same window.

3. In order to control the layout of several plots on the same window, you may use `par(mfrow = c(1,2))`; it is advisable to reset the layout to the default value after plotting.

FIGURE 1.6   **R script to plot GBM sample paths for different volatilities.**

1.3.3   DISCRETE-EVENT MODELS

The GBM sample paths depicted in Fig. 1.7 look noisy enough, but continuous. Indeed, it can be shown that they are continuous, in a stochastic sense, but nondifferentiable everywhere. The sample paths of a discrete-event dynamic system are quite different in nature. It is important to avoid a possible confusion between *discrete event* and *discrete time*. Discrete-event systems are continuous-time models and, unlike discrete-time models, there is no constant time step δt. Like discrete-time systems, the state changes only at some time

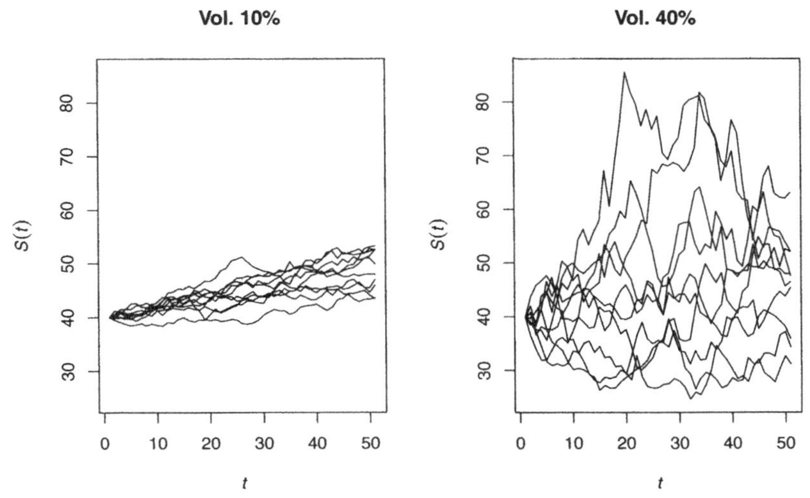

Vol. 10%   **Vol. 40%**

FIGURE 1.7 **Sample paths of geometric Brownian motion for different levels of volatility.**

instants, corresponding to the occurrence of some event, but the time between events is in general a continuous random variable.

Perhaps, the simplest discrete-event model, to be further discussed in Section 3.2.2, is the Poisson process. This is an instance of a counting process: The value $N(t)$ counts the number of events that occurred during the time interval $[0, t]$, where the times X_k elapsing between events $k - 1$ and k, $k = 1, 2, 3, \ldots$, are a sequence of independent and identically distributed exponential variables. By convention, X_1 is the time at which the first event occurs after the start time $t = 0$. A sample path of a Poisson process is illustrated in Fig. 1.8; we see that the process "jumps" whenever an event occurs, so that sample paths are piecewise constant. The underlying exponential distribution is characterized by a single parameter λ that can be interpreted as a "rate," i.e., the average number of events occurring per unit time.[9] This basic Poisson process can be generalized by allowing for a nonconstant rate $\lambda(t)$, which yields an inhomogeneous Poisson process. We may also allow for arbitrary jump sizes. If we associate a continuous probability distribution with the jump size, we obtain a continuous-state system, whereas unit jumps yield a discrete state space corresponding to integer numbers. This second generalization is called compound Poisson process, which is a discrete-event, continuous-state model.

Financially motivated examples of such models are:

---

[9]In Section 3.2.2 we shall see that $1/\lambda$ is the corresponding expected value of the exponential random variable.

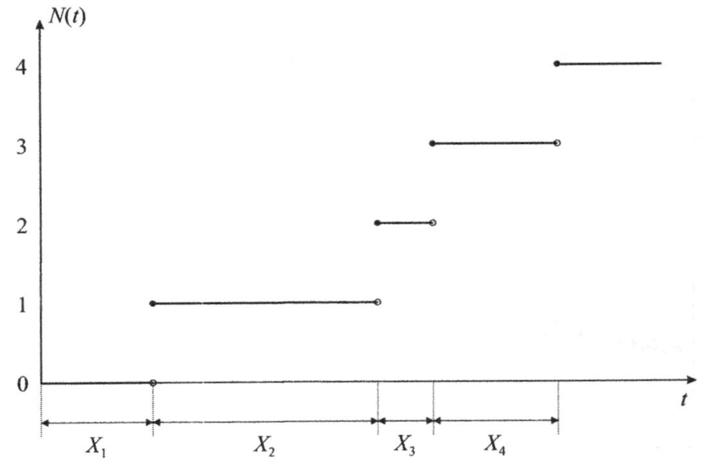

FIGURE 1.8 **Sample path of the Poisson process.**

- Models of stock prices allowing for jumps. Pure-jump models have been proposed, but we may also integrate diffusion processes, like GBM, with jumps. Stochastic processes in this more general form are known as Lévy processes.
- Models of credit rating migration, where the rating of a bond issue changes across a finite set of states at random times. This is a discrete-event, discrete-state model. Continuous-time Markov chains are an example in this family.

Discrete-event models are the rule in other application domains, like the simulation of manufacturing systems and logistic networks.[10] These kinds of simulation problems are, in a sense, dual to the simulation problems faced in economics and finance: The mathematics involved is generally trivial, but there is an incredible complexity in data bookkeeping, resulting in huge computer programs, unless one resorts to a special-purpose simulation tool. On the contrary, when dealing with a stochastic differential equation, the program often consists of just a few lines of code, possibly hiding a quite sophisticated mathematical complexity. In the next section we give a simple example of a discrete-event system, so that the reader can appreciate the mechanics involved.

1.3.3.1 Simulation of discrete-event systems: a single queue

Let us consider a very simple system consisting of a single server (e.g., a bank teller), which must satisfy a set of incoming requests for service (e.g., customers entering a bank). The time between arrivals of consecutive customers is ran-

---

[10]The curious reader might refer, e.g., to [4].

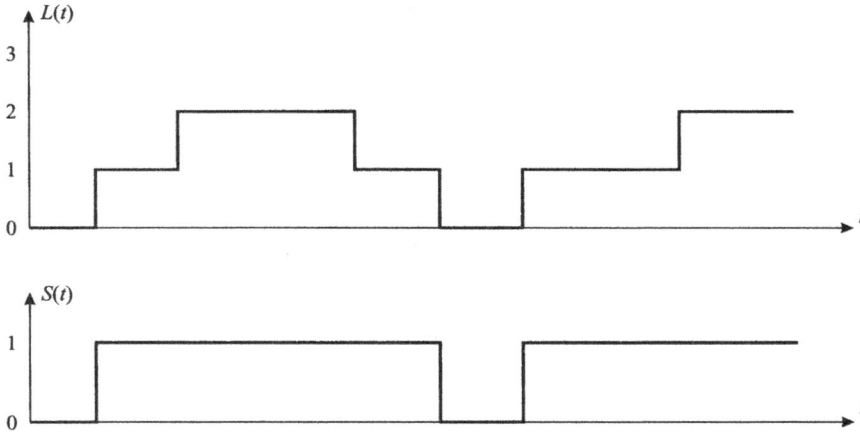

FIGURE 1.9 **Sample path of queuing system consisting of a single server.**

dom; we might assume that it is a plain Poisson process or some time-varying process. The time needed to serve each request is random as well. The system dynamics is not difficult to describe:

- If the arriving customer finds the server idle, it starts her service immediately and will free the resource after service completion; the server state changes from idle to busy.
- If the server is idle, the new customer joins a queue of requests waiting for service. The simplest queuing discipline is first-in, first-out (FIFO), i.e., requests are handled according to the order of their arrivals.
- When the server terminates a service, it starts servicing the first customer in the queue, if any; otherwise, its state changes from busy to idle.

In order to describe the state of this system, we would certainly use $L(t) = 0, 1, 2, \ldots$, the number of customers in the system, which includes the one being served, if any, and the state of the server, $S(t) \in \{0, 1\}$, where we assume that 1 corresponds to busy and 0 to idle. We also need some information about the future departure of the currently served customer, if any. A sample path of this single-server queuing system is depicted in Fig. 1.9. Statistics that we might be interested to collect are the average queue length or the average server utilization, which are essentially the integrals of the sample paths in Fig. 1.9 divided by the simulation time horizon. We immediately see that some bookkeeping is required, as these are not the familiar sample means, but rather integrals over time. A performance measure that can be evaluated as a sample mean is the average customer waiting time.

In some specific cases, there is no need to resort to simulation, as the system can be analyzed mathematically. For instance, if the service and the interarrival times are both exponentially distributed, the system is essentially a continuous-time Markov chain, thanks to the memoryless property of the exponential distribution.[11] Such a queue is denoted by $M/M/1$, where the M refers to the memoryless nature of the two stochastic processes involved, customer arrivals and customer service, and the 1 refers to a system consisting of a single server. If we denote the arrival and service rates by λ and μ, it can be shown that

$$\overline{W} = \frac{\rho/\mu}{1-\rho}, \tag{1.9}$$

where $\rho = \lambda/\mu$ is the utilization rate; clearly, stability requires $\mu > \lambda$, i.e., the service rate, i.e., the average number of requests served per unit time, must be larger than the arrival rate. In more complicated cases, typically involving network of servers, or more difficult distributions, or different queuing disciplines, one may be forced to resort to simulation.

Let us denote by A_j, S_j, and W_j the arrival time, service time, and waiting time, respectively, for customer j. It is possible to find a recursive equation for the waiting time of customer j. Customer j has to wait if it arrives before the completion time of customer $j-1$. Denoting the latter quantity by C_{j-1}, we have

$$W_j = \max\{0, C_{j-1} - A_j\}. \tag{1.10}$$

However, the completion time of customer $j-1$ is

$$C_{j-1} = A_{j-1} + W_{j-1} + S_{j-1}.$$

Plugging this into Eq. (1.10), we find

$$\begin{aligned} W_j &= \max\{0, A_{j-1} + W_{j-1} + S_{j-1} - A_j\} \\ &= \max\{0, W_{j-1} + S_{j-1} - I_j\}, \end{aligned} \tag{1.11}$$

where $I_j \equiv A_j - A_{j-1}$ is the interarrival time between customers $(j-1)$ and j. We note that, in a more general setting possibly involving priorities, balking, or multiple servers, complicated data bookkeeping would be needed to track customer waiting times, requiring suitable data structures and a corresponding programming effort. In this lucky case, it is easy to implement the recursion in R, as shown in Fig. 1.10. In Fig. 1.11 we also show an R script running the simulator in order to check the results against the exact formula of Eq. (1.9). With $\rho = 90.90\%$, which is a fairly high value for the utilization rate, we observe a lot of variability around the correct value:

```
> rho/mu/(1-rho)
[1] 9.090909
> MM1_Queue(lambda,mu,100000)
[1] 9.026053
> MM1_Queue(lambda,mu,100000)
[1] 10.34844
> MM1_Queue(lambda,mu,100000)
[1] 8.143615
```

---

[11]See Section 3.2.2.1.

```
MM1_Queue <- function(lambda=1, mu=1.1, howmany=10000){
    W <- numeric(howmany)
    for (j in 1:howmany){
        intTime <- rexp(1,rate=lambda)
        servTime <- rexp(1,rate=mu)
        W[j] <- max(0, W[j-1] + servTime - intTime)
    }
    return(mean(W))
}
```

R programming notes:

1. As usual, the array `W` is preallocated, using `numeric` in this case.

2. Note how R offers the possibility of setting default values for input arguments. Hence, the function could be called using just `MM1_Queue()`; if you want to assign a different value to the second parameter, you may circumvent parameter positions by naming them when calling the function, as in `MM1_Queue(mu=1.2)`.

3. The `rexp` function is used to sample from an exponential distribution. Note that the input argument to this function is the average *rate* at which events occur, rather than the expected time elapsing between consecutive events.

FIGURE 1.10 **Implementation of Lindley's recursion for a $M/M/1$ queue.**

```
set.seed(55555)    # ensure repeatability
lambda = 1; mu = 1.1
rho = lambda/mu    # utilization is 90.90%
rho/mu/(1-rho)     # exact value
MM1_Queue(lambda,mu,100000)
MM1_Queue(lambda,mu,100000)
MM1_Queue(lambda,mu,100000)
# change parameters
lambda = 1
mu = 2
rho = lambda/mu    # utilization is 50%
rho/mu/(1-rho)     #exact value
MM1_Queue(lambda,mu,100000)
MM1_Queue(lambda,mu,100000)
MM1_Queue(lambda,mu,100000)
```

FIGURE 1.11 **Script to check the function** `MM1_Queue()`.

For a low utilization rate, such as $\rho = 50\%$, we get more reliable results:

```
> rho/mu/(1-rho)
[1] 0.5
> MM1_Queue(lambda,mu,100000)
[1] 0.497959
> MM1_Queue(lambda,mu,100000)
[1] 0.5044687
> MM1_Queue(lambda,mu,100000)
[1] 0.4961641
```

We need to quantify the reliability of the above estimates, possibly by confidence intervals, as we show in Chapter 7 on simulation output analysis.

1.4   Simulation and optimization

There is a considerable interplay between Monte Carlo methods and optimization problems. A generic finite-dimensional optimization problem can be stated as

$$\min_{\mathbf{x} \in S} f(\mathbf{x}),\tag{1.12}$$

where $\mathbf{x} \in \mathbb{R}^n$ is an n-dimensional vector collecting the decision variables, $S \subseteq \mathbb{R}^n$ is the feasible set, and f is the objective function. The feasible set S is usually represented by a collection of equalities and inequalities. In recent years, there has been an astonishing progress in the speed and reliability of commercial software tools to solve wide classes of optimization problems. As a general rule that we shall illustrate in more detail in Chapter 10, convex optimization problems, whereby the function f and the set S are both convex, can be efficiently solved. In particular, large-scale linear programming problems can be solved in a matter of seconds or minutes. Other families of convex problems, such as second-order cone programming models, can be solved rather efficiently by relatively recent interior-point methods.

The joint progress in algorithmic sophistication and hardware speed makes optimization modeling a more and more relevant tool for many practical problems, including those motivated by economical or financial applications. Nevertheless, there are classes of problems that still are a very hard nut to crack. They include the following:

1. Nonconvex problems, i.e., either problems with a nonconvex objective function featuring many local optima, or problems with a nonconvex feasible set, like integer programming and combinatorial optimization problems

2. Stochastic optimization problems involving uncertainty in the problem data

3. Dynamic decision problems affected by uncertainty, which preclude the determination of an optimal static solution, but on the contrary require a dynamic strategy that adapts decisions to the unfolding of uncertainty

We will see much more on this kind of models, and the related solution algorithms, in Chapter 10. Here we just want to point out that Monte Carlo methods play a key role in tackling these problems, which we also illustrate in the next subsections.

1.4.1 NONCONVEX OPTIMIZATION

In a minimization problem with no constraints on the decision variables, if the objective function is convex, then a locally optimal solution is also a global one. This property considerably helps in devising solution algorithms. Equivalently, maximizing a concave function is relatively easy. This is good news for an economist wishing to solve a decision problem featuring a concave utility function or a financial engineer minimizing a convex risk measure. However, there are cases in which such a nice condition is not met, and the need for global optimization methods arises. The following example illustrates a typical function featuring a large number of local optima.

⬛ Example 1.6 **The Rastrigin function**

> The Rastrigin function is a sort of benchmark to test global optimization algorithms, as well as to illustrate how bad an objective function can be. In the general n-dimensional case, the Rastrigin function is defined as
>
> $$R_n(\mathbf{x}) = An + \sum_{i=1}^{n} \left[x_i^2 - A\cos(2\pi x_i) \right],$$
>
> where A is some constant. In a bidimensional case, if we choose $A = 10$, we may boil down the function to
>
> $$R_2(\mathbf{x}) = 20 + x^2 - 10\cos 2\pi x + y^2 - 10\cos 2\pi y,$$
>
> depending on the two variables x and y. A surface plot of the Rastrigin function is displayed in Fig. 1.12, where its nonconvexity can be appreciated. Another view is offered by the contour plot of Fig. 1.13. These two plots have been obtained by the R script in Fig. 1.14.

The Rastrigin function does illustrate why nonconvexities make an optimizer's life difficult, but it definitely looks artificial. To see how nonconvexities may arise in a practical context, let us consider model calibration in asset pricing. Suppose that we have a model for the price of n assets indexed by k, $k =$

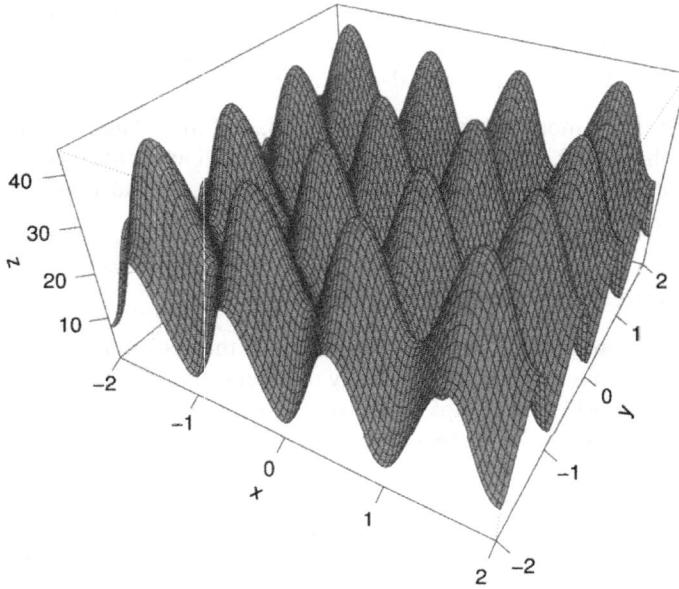

FIGURE 1.12 **Surface plot of Rastrigin function.**

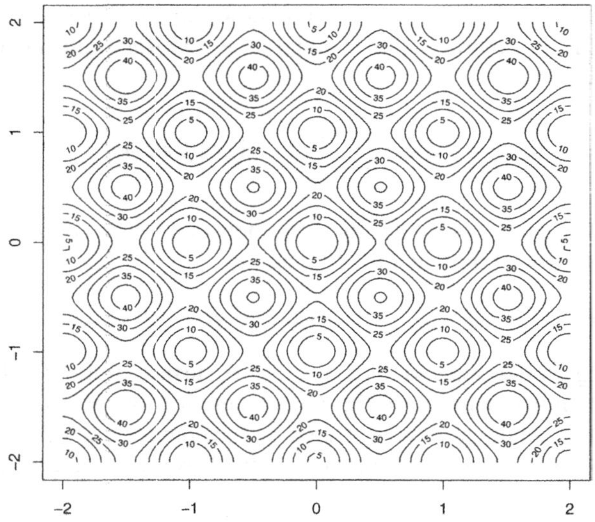

FIGURE 1.13 **Contour plot of Rastrigin function.**

```
rastrigin <- function(x,y)  20+x^2-10*cos(2*pi*x)
    + y^2-10*cos(2*pi*y)
x <- seq(-2,2,length=100)
y <- x
z <- outer(x,y,rastrigin)
persp(x,y,z,theta=30, phi=35, expand=0.6,
    col='lightgreen', shade=0.1, ltheta=120,
    ticktype='detailed')
contour(x,y,z)
```

   R programming notes:

   1. In this function we illustrate one possible use of `seq` to partition an interval using a given number of points.

   2. The functions `persp` and `contour` can be used to draw surface and contour plots, respectively. Their output can be customized with plenty of optional parameters.

   3. To create those plots, we need to combine the coordinates of the two input vectors x and y in any possible way, and then apply the function `rastrigin` on them to create a matrix. This is accomplished by the function `outer`; try `outer(1:4,1:4)` to figure out how it works.

FIGURE 1.14 **Script to produce the plots of the Rastrigin function.**

$1, \ldots, n$. The pricing function

$$\hat{P}_k = F_k(\beta_1, \beta_2, \ldots, \beta_m)$$

predicts the price $\hat{P}_k$ depending on some specific characteristics of the asset k itself, and a set of general parameters β_j, $j = 1, \ldots, m$, common to several assets, where $m << n$. Model calibration requires finding a set of parameters that fit as much as possible a set of observed prices P_k^o. After model calibration, we can move on and price other assets in an internally consistent way. To be concrete, an option pricing model is calibrated on the basis of quoted prices of exchange-traded derivatives, and it can then be used to price an over-the-counter derivative for which a market quote is not available. The model can be used, e.g.,

   • By the client of an investment bank to check if the price proposed by the bank for a specifically tailored contract makes sense

   • By an arbitrageur to spot price inconsistencies paving the way for arbitrage opportunities

   • By a risk manager to predict the impact of changes in the underlying factors on the value of a portfolio

A standard way to tackle the problem is by minimizing the average squared distance between the observed price P_k^o and the theoretical price $\hat{P}_k$ predicted by the model, which leads to the nonlinear least-squares problem:

$$\min \quad \sum_{k=1}^{n} [P_k^o - F_k(\beta_1, \beta_2, \ldots, \beta_m)]^2 \tag{1.13}$$

$$\text{s.t.} \quad \beta \in S, \tag{1.14}$$

where β is a vector collecting the parameters and S is a feasible set representing restrictions on parameters, such as non-negativity. While ordinary least-squares problems are trivial to solve, in this case the pricing functions may be nonlinear; hence, there is no guarantee in general that the resulting problem is convex, and a local solver may well get trapped in a locally optimal solution yielding a set of unreasonable parameters. A similar consideration applies in parameter estimation, when a complicated likelihood function is maximized.[12]

Several global optimization algorithms have been proposed, and some of them do guarantee the global optimality of the reported solution. Unfortunately, these algorithms are typically restricted to a specific class of functions, and the optimality guarantee may require demanding computations. As an alternative, if we are not able to exploit specific structural properties of the problem, and a near-optimal solution is enough, possibly the optimal one but with no proof of its status, we may resort to general-purpose stochastic search algorithms. Among the proposed classes of methods, we mention:

- Pure random sampling of points in the feasible domain
- Using a local optimizer along with random multistart strategies
- Using strategies to escape from local optima, like simulated annealing
- Using strategies based on the evolution of a population of solutions, like genetic algorithms and particle swarm optimization

These methods may also be applied when the complementary kind of nonconvexity arises, i.e., we have to deal with a nonconvex set of feasible solutions, as is the case with combinatorial optimization problems, where we have to explore a finite, but huge discrete set of solutions. The above methods come in a possibly confusing assortment of variants, also including hybrid strategies. The effectiveness of each approach may actually depend on some problem features, most notably the presence or absence of difficult constraints. However, all of them rely on some random exploration behavior, based on Monte Carlo sampling.

1.4.2   STOCHASTIC OPTIMIZATION

Most financial problems involve some form of uncertainty. Hence, we may have to tackle an optimization problem where some data defining the objective function or the feasible set are uncertain. We face the difficult task of choosing an

---

[12] See Section 4.2.3.

optimal solution $\mathbf{x}^*$ in S before knowing the value of a vector random variable $\boldsymbol{\xi}$ taking values on Ξ. If so, two questions arise:

- Is $\mathbf{x}^*$ robust with respect to optimality? In other words, what can we say about the quality of the solution across possible realizations of $\boldsymbol{\xi}$?
- Is $\mathbf{x}^*$ robust with respect to feasibility? This is an even thornier issue, as the solution we choose might be infeasible for some particularly unlucky scenarios. For instance, we may have solvency issues for a pension fund.

For now, let us focus only on the first issue and consider a problem of the form

$$\min_{\mathbf{x} \in S} F(\mathbf{x}) \equiv \mathrm{E}_{\boldsymbol{\xi}}[f(\mathbf{x}, \boldsymbol{\xi})]. \tag{1.15}$$

The objective function $f(\mathbf{x}, \boldsymbol{\xi})$ depends on both decisions $\mathbf{x}$ and random factors $\boldsymbol{\xi}$, but by taking an expectation with respect to $\boldsymbol{\xi}$ we obtain a deterministic function $F(\mathbf{x})$. Assuming that the components of $\boldsymbol{\xi}$ are jointly continuous and distributed according to a density function $g_{\boldsymbol{\xi}}(\mathbf{y})$, with support Ξ, the objective function can be written as

$$F(\mathbf{x}) = \int_{\mathbf{y} \in \Xi} f(\mathbf{x}, \mathbf{y}) g_{\boldsymbol{\xi}}(\mathbf{y}) \, d\mathbf{y}.$$

As we point out in Chapter 2, such a multidimensional integral can be extremely difficult to evaluate for a given $\mathbf{x}$, let alone to optimize in general, even when we are able to prove that $F(\mathbf{x})$ enjoys such nice properties as continuity, differentiability, or even convexity.

A possible approach is to boil down the integral to a sum, by sampling K scenarios characterized by realizations $\boldsymbol{\xi}^k$ and probabilities π^k, $k = 1, \ldots, K$. Then, we solve

$$\min_{\mathbf{x} \in S} \sum_{k=1}^{K} \pi^k f(\mathbf{x}, \boldsymbol{\xi}^k), \tag{1.16}$$

which is a more familiar optimization problem. We note that Monte Carlo sampling may play a key role in optimization as a scenario generation tool. However, *crude* sampling is typically not a viable solution and the following issues must be tackled:

- How can we generate a limited number of scenarios, in order to capture uncertainty without having to solve a huge and intractable problem?
- How can we prove convergence, i.e., that when the sample size goes to infinity, the solution of problem (1.16) converges to the true optimal solution of problem (1.15)?
- The previous question is actually theoretical in nature and definitely beyond the scope of this book. A more practical, but obviously related question is: How can we assess the quality of the solution for a finite sample size?

- How can we assess the robustness of the solution with respect to sampling variability and, possibly, with respect to higher level uncertainty on the parameters defining the uncertainty model itself?

Some of these hard questions will be addressed in this book, when we deal with variance reduction strategies, numerical integration by Gaussian quadrature, and deterministic sampling based on low-discrepancy sequences. If all this does not sound challenging enough, recall that we may also have feasibility issues. What can we do if a solution proves to be infeasible after we observe the true value of the uncertain parameters? More generally, what if the optimization problem is dynamic and we have to make a *sequence* of decisions while we gather information step by step? One approach to cope with these multistage stochastic problems relies on multistage stochastic programming with recourse, discussed in Chapter 10; sometimes, they can also be tackled by stochastic dynamic programming.

1.4.3 STOCHASTIC DYNAMIC PROGRAMMING

In the previous section we have considered how uncertainty issues may affect an optimization problem, but we disregarded a fundamental point: the dynamics of uncertainty over time. In real life, when making decisions under uncertainty, we do start with a plan, but then we adjust our decisions whenever new pieces of information are progressively discovered. Rather than coming up with a plan to be rigidly followed over time, we pursue a dynamic strategy in which decisions are adapted on the basis of contingencies. This kind of consideration suggests the opportunity of formulating multistage, dynamic stochastic optimization models, whereby we have to find a sequence of decisions x_t, $t = 0, 1, \ldots, T$, which are not deterministic but depend on the realized future states. As we shall see, there are different approaches to tackle these quite challenging problems, depending on their nature and on our purpose.

- We may be interested in the first decision x_0, because this is the decision that we have to implement here and now. The next decisions enter into the optimization model in order to avoid an overly myopic behavior, but when time advances, we will solve the model again according to a rolling horizon strategy. Note that this requires the solution of a sequence of multistage models. This approach is common in management science and in some engineering problems, and it is exemplified by multistage stochastic programming models with recourse.

- In other cases, we want to find a *policy* in feedback form, i.e., a rule mapping the current state s_t, which evolves according to the state transition equations (1.4), to an optimal decision. The mapping $x_t^* = F_t(s_t)$ can be found by following another approach, based on stochastic dynamic programming.

To see an example in which the second kind of approach may be preferable, consider pricing American-style options.[13]. In that case, we are interested in the expected *value* of the objective function, resulting from the application of an optimal exercise policy in the future. This requires a strategy, possibly a suboptimal one, to come up with lower bounds on the option value, to be complemented by upper bounds.

Another common kind of problem whereby a dynamic programming approach may be preferable can be found in economics. To illustrate the idea, let us consider a stylized consumption–saving problem.[14] A decision maker must decide, at each discrete-time instant, the fraction of current wealth that she consumes immediately, deriving a corresponding utility, and the fraction of current wealth that is saved and invested. The return on the investment can be random and depending on the asset allocation decisions. Even if we assume that return is deterministic, we should consider uncertainty in labor income, which determines the level of wealth available for consumption/saving. A possible model is the following:

$$
\max \quad \mathrm{E}_t \left[\sum_{\tau=t}^{T} \beta^{\tau-t} U(C_\tau) \right]
$$
$$
\text{s.t.} \quad S_\tau = X_\tau - C_\tau,
$$
$$
X_\tau = R_\tau S_{\tau-1} + Y_\tau,
$$
$$
Y_\tau = P_\tau \epsilon_\tau,
$$

where

- $\mathrm{E}_t[\cdot]$ denotes a *conditional* expectation given the current state at time t.
- $U(\cdot)$ is a utility function related to individual preferences.
- $\beta \in (0,1)$ is a subjective time discount factor.
- X_τ is the cash on hand (wealth) at time τ, before making the consumption/saving decision.
- C_τ is the amount of wealth that is consumed.
- S_τ is the amount of wealth that is saved.
- R_τ is the return from period $\tau - 1$ to τ; note that this return is usually random and that R_τ applies to $S_{\tau-1}$, not S_τ.
- P_τ is the nominal labor income, which may vary over time according to a deterministic or stochastic trajectory.
- Y_τ is the noncapital income, i.e., the income from labor that depends on the nominal labor income and multiplicative shocks ϵ_τ.

---

[13]See Example 1.8. We discuss how to price American-style options with Monte Carlo methods in Chapter 11.

[14]For an extensive discussion of such models see, e.g., [5].

As we shall see in Chapter 10, we may solve such a problem by dynamic programming, which is based on a recursive functional equation for the value function

$$V_t(X_t) = \max_{C_t \geq 0} \{U(C_t) + \beta E_t [V_{t+1}(X_{t+1})]\}, \qquad (1.17)$$

where $V_t(X_t)$ is the expected utility if we start the decision process at time t, from a state with current wealth X_t. In the above model, wealth is the only state variable and randomness is due to random financial return and shocks on nominal income. However, if we model income in a more sophisticated way, including permanent and transitory shocks, we may need additional state variables. From the above equation we see that the problem is decomposed stage by stage, and if we are able to find the sequence of value functions, we have just to solve a static optimization problem to find the optimal decision at each step, given the current state. The value function is the tool to trade off immediate utility against future benefits.

Also note that we make our decision after observing the actually realized state, which does not depend only on our decisions, since it is also affected by external random factors. The policy is implicitly defined by the set of value functions $V_t(\cdot)$ for each time instant. This is definitely handy if we have to carry out an extensive set of simulations in order to investigate the impact of different economic assumptions on the decision maker's behavior, and it is an advantage of dynamic programming over stochastic programming models with recourse. Nevertheless, dynamic programming has some definite limitations on its own, as it relies on a decomposition that is valid only if some Markovian properties hold for the system states, and it is plagued by the aptly called "curse of dimensionality." This refers to the possibly high dimensionality of the state space. If we have only one state variable, it is easy to approximate value functions numerically on a grid, but if there are many state variables, doing so is practically impossible. Actually, other curses of dimensionality may refer to the need to discretize the expectation in Eq. (1.17) and the computational complexity of the optimization required at each step. Nevertheless, there are approximate dynamic programming approaches that do not guarantee optimality, but typically provide us with very good decisions. The exact approach should be chosen case by case, but what they have in common is the role of Monte Carlo methods:

- To generate scenarios
- To learn an optimal policy
- To evaluate the performance of a policy by simulation

1.5 Pitfalls in Monte Carlo simulation

Monte Carlo methods are extremely flexible and powerful tools; furthermore, they are conceptually very simple, at least in their naive and crude form. All of these nice features are counterbalanced by a significant shortcoming: They can be very inefficient, when each single simulation run is computationally expensive and/or a large sample size is needed to obtain reliable estimates. These computational difficulties are technical in nature and can be (at least partially) overcome by clever strategies. However, there are deeper issues that should not be disregarded and are sometimes related to the very meaning of uncertainty.

1.5.1 TECHNICAL ISSUES

Each Monte Carlo replication yields an observation of a random variable, say X_k, for $k = 1, \ldots, N$, which can be thought of as a realization of a random variable X, whose expected value $\mu = \mathrm{E}[X]$ we want to estimate. Basic inferential statistics suggests using the sample mean $\overline{X}$ and sample standard deviation S to build a confidence interval with confidence level $1 - \alpha$:

$$\overline{X} \pm z_{1-\alpha/2} \frac{S}{\sqrt{N}}, \tag{1.18}$$

where we assume that the sample size N is large enough to warrant the use of quantiles $z_{1-\alpha/2}$ from the standard normal distribution.[15] This form of a confidence interval is so deceptively simple that we may forget the conditions for its validity.

> **Example 1.7 Estimating average waiting time**
>
> Consider the queuing system of Section 1.3.3.1, but now assume that there are multiple servers, possibly bank tellers. We might be interested in investigating the trade-off between the number of servers, which has a significant impact on cost, and the average waiting time in the queue, which is a measure of service quality. Using Monte Carlo, we may simulate a fairly large number N of customers and tally the waiting time W_k, $k = 1, \ldots, N$, for each of them. Based on this sample, we may estimate the average waiting time by the sample mean $\overline{W}$. Now can we use Eq. (1.18) to build a confidence interval for the average waiting time?
>
> By doing so, we would commit a few mistakes:
>
> 1. The confidence interval is exact for a normal population, i.e., when observations are normally distributed. Clearly, there is no reason to believe that waiting times are normally distributed.
>
> 2. What is the waiting time of the first customer entering the system? Clearly, this is zero, as the first customer is so lucky that she finds all of the servers idle. In such a system, we must account for

---

[15]The quantile $z_{1-\alpha/2}$ is a number such that $\mathrm{P}\{Z \leq z_{1-\alpha/2}\} = 1 - \alpha/2$, where $Z \sim \mathsf{N}(0,1)$ is standard normal. Output analysis is treated in Chapter 7.

a transient phase, and we should collect statistics only when the system is in steady state.

3. Last but not least, the common way to estimate standard deviation assumes independence among the observations. However, successive waiting times are obviously positively correlated (the degree of correlation depends on the load on the system; for high utilization levels, the correlation is rather strong).

We may find a remedy to overcome these issues, based on a batching strategy, as we shall see in Chapter 7.

The last two mistakes underlined in Example 1.7 remind us that standard inferential statistics rely on a sequence of i.i.d. (independent and identically distributed) random variables. In nontrivial Monte Carlo methods this condition may not apply and due care must be exerted. Another issue that must not be disregarded is bias: Is it always the case that $E[\overline{X}] = \theta$, i.e., that the expected value of the sample mean we are taking is the parameter in which we are interested? This is certainly so if we are sampling a population and the parameter is actually the population mean μ. In general, estimators can be biased.

Example 1.8 Bias in option pricing

Monte Carlo methods are commonly used to price options, i.e., financial derivatives written on an underlying asset. Consider, for instance, an American-style put option, written on a stock that pays no dividends during the option life. This option gives its holder the right to sell the underlying stock at the strike price K, up to option maturity T. Whenever the stock price $S(t)$ at time t, for $t \in [0, T]$, is smaller than K, the option is "in-the-money," and the holder can exercise her right immediately, earning a payoff $K - S(t)$; to see why this is the payoff, consider that the holder can buy the stock share at the current (spot) price $S(t)$ and sell it immediately at the strike price $K > S(t)$. This is a somewhat idealized view, as it neglects transaction costs and the fact that market prices move continuously. However, for some derivatives that are settled in cash, like index options, this is really the payoff. However, when is it really optimal to exercise? Should we wait for better opportunities? Finding the answer requires the solution of a stochastic dynamic optimization problem, whereby we seek to maximize the payoff, whose output is an optimal exercise policy. The fair option value is related to the expected payoff resulting from the application of this optimal policy. When the problem is challenging,

> we may settle for an approximate policy. Unfortunately, a suboptimal policy introduces a low bias, since we fall short of achieving the truly optimal payoff.

Leaving the aforementioned issues aside, the form of the confidence interval in Eq. (1.18) shows the role of estimator's variance:

$$\text{Var}(\overline{X}) = \frac{\sigma}{N},$$

where σ is the (usually unknown) variance of each observation. This implies that the half-length of the confidence interval is given by $z_{1-\alpha/2}\,\sigma/\sqrt{N}$, which is both good and bad news:

- The nice fact is that when the sample size N is increased, the confidence interval shrinks in a way that does not depend on problem dimensionality.[16] This is why Monte Carlo methods are naturally suitable to high-dimensional problems.
- The not so nice fact is that this convergence is related to the square root of N. To get the message loud and clear, this implies that to gain one order of precision, i.e., to divide the length of the confidence interval by a factor of 10, we should multiply N by a factor of 100.

The last observation explains why a huge number of replications may be needed in order to obtain reliable estimates. This brute force approach may be feasible thanks to another nice feature of Monte Carlo methods: They lend themselves to parallelization, and the availability of multicore machines at cheap prices is certainly helpful. However, a smarter approach does not rely on brute computational force. Using proper variance reduction techniques, discussed in Chapter 8, we may be able to reduce σ without introducing any bias in the estimator. Even this improvement may not be enough when dealing with complex multistage optimization problems, though.

1.5.2 PHILOSOPHICAL ISSUES

The issues that we have discussed in the previous section are technical in nature, and can be at least partially overcome by suitable tricks of the trade. However, there are some deeper issues that one must be aware of when tackling a problem by Monte Carlo, especially in risk management.[17] While in other cases we might be interested in an average performance measure, in risk management

---

[16]Well, we should say that there is no *explicit* dependence on dimensionality, even though this may affect variability, not to mention computational effort.

[17]We shall deal with risk measurement and management in Chapter 13.

we are interested in the "bad tail" of a distribution, where extreme events may occur, such as huge losses in a financial portfolio.

▣ Example 1.9 **Value-at-risk**

> One of the most common (and most criticized) financial risk measures is value-at-risk (V@R). Technically, V@R is the quantile of the probability distribution of loss on a portfolio of assets, over a specified time horizon, with a given confidence level. In layman's terms, if the one-day V@R at 99% level is $1000, it means that we are "99% sure" that we will not lose more than $1000 the next day. If we know the probability density of loss, finding the quantile is a rather easy matter. In general, however, we may hold a portfolio with widely different asset classes, including stock shares, bonds, and perhaps quite sophisticated derivatives, whose value depends on a large number of underlying factors. Monte Carlo methods can be used to sample a suitable set of scenarios in terms of factors that are mapped to scenarios in terms of portfolio value; then, we may estimate the quantile in which we are interested. It stands to reason that estimating V@R at 99.9% will take a larger computational effort than estimating V@R at 90%, since we have to deal with extreme events. Using brute force, or clever variance reduction, we might be able to tackle such a task. However, there are more serious difficulties:
>
> - Can we really trust our probabilistic model when we deal with very rare, but potentially disastrous events?
> - What about the reliability of the estimates of small probabilities?
> - What happens to correlations under extreme conditions?
> - Did we include every risk factors in our model? Assuming we did so in financial terms, what about the consequence of an earthquake?

Sometimes, a line is drawn between *uncertainty* and *risk*. The latter term is reserved to cases in which the rules of the game are clear and perfectly known, as in the case of dice throwing and coin flipping. Uncertainty, in this framework, is a more general concept that is not really reduced to the uncertainty in the realization of random variables. To get the picture, let us consider the scenario tree depicted in Fig. 1.15. The tree can be interpreted in terms of a discrete probability distribution: Future scenarios S_k are associated with probabilities π_k. Risk has to do with the case in which both possible scenarios and their probabilities are known. However, we may have another level of uncertainty when the probabilities are not really known, i.e., we have uncertainty about the uncertainty. Indeed, when dealing with extreme risks, the very notion of

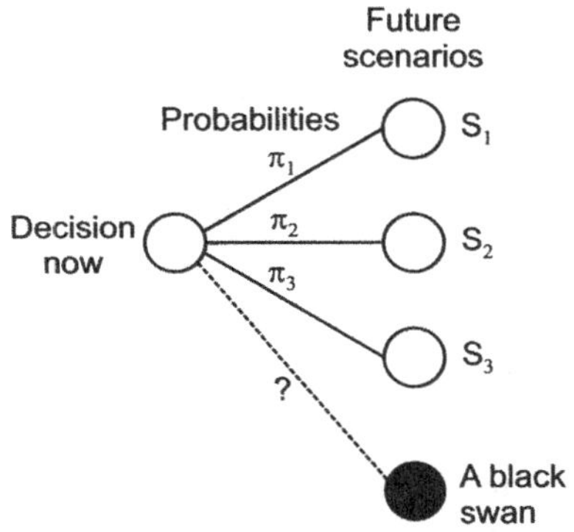

FIGURE 1.15 **Schematic illustration of different kinds of uncertainty.**

probability may be questioned: Are we talking about frequentist probabilities, or subjective/Bayesian views? We may even fail to consider all of the scenarios. The black scenario in the figure is something that we have failed to consider among the plausible contingencies. The term *black swan* is commonly used, as well as *unk-unks*, i.e., unknown unknowns. To add some complexity to the picture, in most stochastic models uncertainty is exogenous. In social systems, however, uncertainty is at least partially endogenous. To see concrete examples, think of the effect of herding behavior in finance and trading in thin and illiquid markets.

To see the issue within a more general setting, imagine a mathematical model of a physical system described by a difficult partial differential equation. Can we obtain brand new knowledge by solving the equation using accurate numerical methods? Arguably, the knowledge itself is encoded in the equation. Solving it may be useful in understanding the consequence of that knowledge, possibly invalidating the model when results do not match observed reality. By a similar token, Monte Carlo methods are numerical methods applied to a stochastic model. There is high value in flexible and powerful methods that allow tackling complex models that defy analytical solution. But if the very assumptions of the model are flawed, the results will be flawed as well. If we analyze risks using sophisticated Monte Carlo strategies, we may fall into the trap of a false sense of security, and pay the consequence if some relevant source of uncertainty has not been included in the model or if probabilities have been poorly assessed. Monte Carlo is an excellent tool, but it is just that: a tool.

1.6 Software tools for Monte Carlo simulation

In general, Monte Carlo simulations require the following:

1. The possibility of describing the system behavior using an imperative programming language or a connection of graphical blocks

2. Random number generators

3. Statistical tools to collect and analyze results

4. Visualization/animation facilities

There is a wide array of software systems that can be used to meet the above requirements, which may be more or less relevant, depending on the underlying kind of model:

- General-purpose programming languages like C++, Visual Basic, or Java.
- Spreadsheet add-ins like @Risk[18]
- Software environments like R and MATLAB,[19] which also allow for classical imperative programming, in a relatively high-level language
- Graphical editors for the description of a system in terms of interconnected building blocks like Simulink, which is an outgrowth of MATLAB, and Arena[20]

All of these options have advantages and disadvantages in terms of cost, ease of use, flexibility, readability, and ease of maintenance of the models. Clearly, the maximum power is associated with general-purpose languages. By using proper object-oriented methods, we may also build reusable libraries, and compiled general-purpose languages are definitely the best option in terms of execution speed. However, they require a non-negligible coding effort, particularly if some essential modeling infrastructure must be built from scratch. Furthermore, with respect to more powerful programming languages like MATLAB, they require many more lines of code. This is due to the need of declaring variables, which allows for good type checking and efficient compiled code, but it is definitely bad for rapid prototyping. Many lines of code may also obscure the underlying model. Furthermore, visualization and random number generation require a significant coding effort as well, unless one relies on external libraries.

Spreadsheets provide the user with a familiar interface, and this can be exploited both by writing code, e.g., in VBA for Microsoft Excel, or by using add-ins, which automate some tasks and typically provide the user with much better quality random number generators. However, if VBA code is written, we essentially fall back on general-purpose programming languages; if a ready-to-use tool is adopted, we may lack flexibility.

On the opposite side of the spectrum we find tools relying on graphical editors to connect building blocks. This is really essential when dealing with complex discrete-event systems. When modeling manufacturing systems, building everything from scratch is a daunting and error-prone effort. A similar consideration applies when simulating nonlinear systems with complicated dynamics. Indeed, this is a domain where such tools like Arena or Simulink are quite help-

---

[18]See http://www.palisade.com and the copyright notices there.

[19]See http://www.mathworks.com and the copyright notices there.

[20]See http://www.arenasimulation.com and the copyright notices there.

ful. Unfortunately, these tools are not really aimed at problems in finance and economics, and are rather expensive.

The kind of simulation models we are concerned with in this book are somewhere in between, as they do not involve overly complex event bookkeeping, but may require careful discretization of stochastic differential equations. They do not involve the level of sophistication needed to simulate complicate engineering systems featuring nonlinearities, but we do face the additional issues related with uncertainty. Hence, we need a set of functions enabling us to sample a variety of probability distributions, as well as to statistically analyze the output. Therefore, in most financial and some economic applications, a good compromise between flexibility and power is represented by numerical and statistical computing environments like MATLAB and R. They offer a library of statistical functions, a rich programming language, and powerful visualization tools. They are both excellent tools for teaching, and even if they are not best suited in terms of efficiency, they can be used to validate a model and to prototype code; the code can be then translated to a more efficient compiled language. In this book, we will adopt R, which has the very welcome feature of being free.[21] All the code displayed in this book is available on my website:

`http://staff.polito.it/paolo.brandimarte`

1.7   Prerequisites

The prerequisites for reading this book involve:

- Mathematical background
- Financial/economical background
- Technical/programming background

Generally speaking, the requirements are rather mild and any professional or advanced undergraduate student in mathematics, statistics, economics, finance, and engineering should be able to take advantage of this book.

1.7.1   MATHEMATICAL BACKGROUND

An obvious prerequisite for a book on Monte Carlo methods is represented by the basics of probability theory and inferential statistics, including:

- Events and conditional probabilities

---

[21] I have made my point about my personal preference between MATLAB and R in the Preface, and I will not repeat myself here. For more information on the use of MATLAB for financial applications, see `http://www.mathworks.com/financial-services/`, and for an illustration of using MATLAB in Monte Carlo financial applications see [2].

- Discrete and continuous random variables (including distribution functions, quantiles, moments)
- A bit of stochastic processes
- A modicum of multivariate distributions, with emphasis on concepts related to covariance and correlation
- Statistical inference (confidence intervals and linear regression)

All of the rest will be explained when needed, including some basic concepts about stochastic integration and stochastic differential equations, as well as copula theory. Note that we will never use measure-theoretic probability. In the chapters involving optimization, we also assume some familiarity with linear and nonlinear programming. The required background is typically covered in undergraduate courses, and any basic book could be used for a brush-up. In my book on quantitative methods [3], all of the required background is given, both for probability/statistics and optimization applications.

1.7.2 FINANCIAL BACKGROUND

This book does touch some topics that may be useful for an economist, but the main focus is on finance. To keep this book to a reasonable size, I will not provide much background on, say, portfolio management and option pricing theory. When illustrating examples, we will just state which kind of mathematical problem we have to face, then we will proceed and solve it. To be specific, when illustrating Monte Carlo methods for option pricing, I will just recapitulate what options are, but I will not provide any deep explanation of risk-neutral pricing principles; a quick outline is provided in Section 3.9. The reader interested in that background could consult, e.g., [1] or [10].

1.7.3 TECHNICAL BACKGROUND

I assume that the reader has a working R installation and knows how to run a simple session, and how to source functions and scripts. There are plenty of online tutorials addressing the needs of a newcomer. Therefore, rather than adding a dry appendix quickly reviewing R programming, I will provide R notes along the way, when discussing concrete examples. Certainly, some familiarity with the basic concepts of any programming language is essential. However, we will not need complicated data structures, and we will also do without sophisticated object-oriented principles. The code in this book is provided only to help the reader in understanding Monte Carlo methods, but it is far from being industrial strength, reusable, and efficient code.

For further reading

- In this book, we will deal with Monte Carlo methods at an intermediate level. Complementary references, including some more advanced material, are [7, 13, 15]. In particular, [13] is a comprehensive treatment, offering pieces of MATLAB code. See also [18] for a treatment with emphasis on Bayesian statistics. By the way, a quite readable book on the important topic of Bayesian statistics, including Monte Carlo methods in R, is [14].

- The role of Monte Carlo methods for optimization is discussed in [8] and [19]; see [17] for an extensive treatment of dynamic stochastic optimization based on Monte Carlo simulation.

- Stochastic programming theory and models are treated, e.g., in [11].

- Probably, most readers of this book are interested in financial engineering. An excellent treatment of Monte Carlo methods for financial engineering is given in [9]; see also [12] or, at a more elementary level, [2], which is MATLAB based.

- On the technical side, good readings on R programming are [6] and [16].

References

1 Z. Bodie, A. Kane, and A. Marcus. *Investments* (9th ed.). McGraw-Hill, New York, 2010.

2 P. Brandimarte. *Numerical Methods in Finance and Economics: A MATLAB-Based Introduction* (2nd ed.). Wiley, Hoboken, NJ, 2006.

3 P. Brandimarte. *Quantitative Methods: An Introduction for Business Management*. Wiley, Hoboken, NJ, 2011.

4 P. Brandimarte and G. Zotteri. *Introduction to Distribution Logistics*. Wiley, Hoboken, NJ, 2007.

5 J.Y. Campbell and L.M. Viceira. *Strategic Asset Allocation*. Oxford University Press, Oxford, 2002.

6 J.M. Chambers. *Software for Data Analysis: Programming with R*. Springer, New York, 2008.

7 G.S. Fishman. *Monte Carlo: Concepts, Algorithms, and Applications*. Springer, Berlin, 1996.

8 M.C. Fu. Optimization by simulation: A review. *Annals of Operations Research*, 53:199–247, 1994.

9 P. Glasserman. *Monte Carlo Methods in Financial Engineering*. Springer, New York, 2004.

10 J.C. Hull. *Options, Futures, and Other Derivatives* (8th ed.). Prentice Hall, Upper Saddle River, NJ, 2011.

11 P. Kall and S.W. Wallace. *Stochastic Programming*. Wiley, Chichester, 1994.

12 R. Korn, E. Korn, and G. Kroisandt. *Monte Carlo Methods and Models in Finance and Insurance*. CRC Press, Boca Raton, FL, 2010.

13 D.P. Kroese, T. Taimre, and Z.I. Botev. *Handbook of Monte Carlo Methods*. Wiley, Hoboken, NJ, 2011.

14 J.K. Kruschke. *Doing Bayesian Data Analysis: A Tutorial with R and BUGS*. Academic Press, Burlington, MA, 2011.

15 C. Lemieux. *Monte Carlo and Quasi–Monte Carlo Sampling*. Springer, New York, 2009.

16 N.S. Matloff. *The Art of R Programming: A Tour of Statistical Software Design*. No Starch Press, San Francisco, 2011.

17 W.B. Powell. *Approximate Dynamic Programming: Solving the Curses of Dimensionality* (2nd ed.). Wiley, Hoboken, NJ, 2011.

18 C.P. Robert and G. Casella. *Introducing Monte Carlo Methods with R*. Springer, New York, 2011.

19 J.C. Spall. *Introduction to Stochastic Search and Optimization: Estimation, Simulation, and Control*. Wiley, Hoboken, NJ, 2003.

## Chapter Two

Numerical Integration Methods

Numerical integration is a standard topic in numerical analysis, and in the previous chapter we have hinted at the link between integration and Monte Carlo methods. In this chapter we have a twofold objective:

- On the one hand, we want to insist on the link between numerical integration and Monte Carlo methods, as this provides us with the correct framework to understand some variance reduction methods, such as importance sampling, as well as alternative approaches based on low-discrepancy sequences.

- On the other hand, we want to outline classical and less classical approaches to numerical integration, which are deterministic in nature, to stress the fact that there are sometimes valuable alternatives to crude Monte Carlo; actually, stochastic and deterministic approaches to numerical integration should both be included in our bag of tricks and can sometimes be integrated (no pun intended).

In many financial problems we are interested in the expected value of a function of random variables. For instance, the fair price of a European-style option may be evaluated as the discounted expected value of its payoff under a risk-neutral probability measure. In the one-dimensional case, the expected value of a function $g(\cdot)$ of a single random variable X with probability density $f_X(x)$ is given by the following integral:

$$\mathrm{E}[g(X)] = \int_{-\infty}^{+\infty} g(x) f_X(x) \, d\ddot{x}.$$

This is just an ordinary and quite familiar integral. We also recall that when we need to estimate the probability of an event A, which may occur or not depending on the realization of a random variable X, we are back to the case above:

$$\mathrm{P}(A) = \int_{-\infty}^{+\infty} \mathbf{1}_A(x) f_X(x) \, dx,$$

where $\mathbf{1}_A(x)$ is the indicator function for event A (taking the value 1 if event A occurs when $X = x$, 0 otherwise).

In lucky cases, an integral may be evaluated in analytical or semianalytical form. To illustrate the latter case, consider the Black–Scholes–Merton (BSM) formula for the price of a call option written on a non-dividend-paying stock share at time t:

$$C(t) = S(t)\Phi(d_1) - Ke^{-r(T-t)}\Phi(d_2).$$

In the formula, $S(t)$ is the current spot price of the underlying asset, K is the strike price, T is the option maturity, i.e., when it can be exercised, and

$$d_1 = \frac{\log[S(t)/K] + (r + \sigma^2/2)(T-t)}{\sigma\sqrt{T-t}}, \qquad d_2 = d_1 - \sigma\sqrt{T-t},$$

where r is the continuously compounded risk-free rate, and σ is the annualized volatility of the underlying asset. Both r and σ are assumed constant in this model. Finally, $\Phi(z)$ is the cumulative distribution function of a standard normal variable:

$$\Phi(z) \equiv \frac{1}{\sqrt{2\pi}} \int_{-\infty}^{z} \exp\left(-\frac{y^2}{2}\right) dy.$$

The BSM formula is typically labeled as analytical, even though, strictly speaking, it is semianalytical, since there is no analytical expression for $\Phi(z)$; nevertheless, there are quite efficient and accurate ways to approximate this function, and we need not resort to general-purpose numerical integration. In other cases, we do need a numerical approach, typically based on quadrature formulas. In this chapter we will only give the basics of such numerical strategies, but quite sophisticated methods are available. Indeed, there is no need for Monte Carlo methods in low-dimensional problems. The difficulty arises when we deal with a vector random variable $\mathbf{X}$, with support Ξ, and the corresponding multidimensional integral

$$\mathrm{E}[g(\mathbf{X})] = \int_{\Xi} g(\mathbf{x}) f_{\mathbf{X}}(\mathbf{x})\,d\mathbf{x}.$$

Product quadrature formulas quickly grow out of hand, in the sense that they require a huge number of points, and this is where Monte Carlo comes into play. Indeed, even sophisticated numerical computing environments only offer limited support for multidimensional integration via classical quadrature formulas.

However, we cannot afford to dismiss deterministic approaches to numerical integration. To see why, consider a stochastic optimization problem, whereby we need the expected value of a function $h(\cdot,\cdot)$ depending on both a vector of control variables $\mathbf{x}$, modeling our decisions, and a vector of random variables $\boldsymbol{\xi}$ with joint density $f_{\boldsymbol{\xi}}(\cdot)$, modeling what we cannot control:

$$H(\mathbf{x}) \equiv \mathrm{E}_{\boldsymbol{\xi}}[h(\mathbf{x},\boldsymbol{\xi})] = \int_{\Xi} h(\mathbf{x},\mathbf{z}) f_{\boldsymbol{\xi}}(\mathbf{z})\,d\mathbf{z}.$$

In such a case, we do need just a single number, i.e., an expected value estimated by a sample mean, but a way to approximate the function $H(\mathbf{x})$, possibly obtained by selecting S scenarios characterized by realizations ξ_s and probabilities π_s, $s = 1, \ldots, S$:

$$H(\mathbf{x}) \approx \sum_{s=1}^{S} \pi_s h(\mathbf{x}, \xi_s).$$

As we have already pointed out, Monte Carlo methods may need a large sample size to achieve an acceptable degree of accuracy, and we may not afford that when solving a difficult optimization problem. In this setting, alternative approaches may be valuable. Some of them are deterministic in nature and are best understood with reference to a view integrating numerical integration and random sampling.

We start in Section 2.1 with a brief overview of quadrature formulas, emphasizing the role of interpolation based on polynomials. Then, we outline more sophisticated Gaussian quadrature formulas in Section 2.2. These approaches are developed for one-dimensional integration but, at least in principle, they can be extended to the multidimensional case by product formulas; this idea and its limitations are discussed in Section 2.3. Alternative approaches for high-dimensional problems, such as good lattices, Monte Carlo, and quasi–Monte Carlo, are introduced in Section 2.4. In Section 2.5 we pause to illustrate some important links between statistics and numerical integration, by considering moment matching methods. We close the chapter by outlining some R functions for numerical integration in Section 2.6.

2.1   Classical quadrature formulas

Consider the problem of approximating the value of the one-dimensional definite integral

$$I[f] = \int_a^b f(x)\, dx,$$

involving a function $f(\cdot)$ of a single variable, over a bounded interval $[a, b]$. Since the integration is a linear operator, i.e.,

$$I[\alpha f + \beta g] = \alpha I[f] + \beta I[g],$$

for all functions f, g and real numbers α, β, it is natural to look for an approximation preserving linearity. A quadrature formula for the interval $[a, b]$ is defined by two ingredients:

   1. A set of $n + 1$ *nodes* x_j, $j = 0, 1, \ldots, n$, such that

$$a = x_0 < x_1 < \cdots < x_N = b.$$

   2. A corresponding set of *weights* w_j, $j = 0, 1, \ldots, n$.

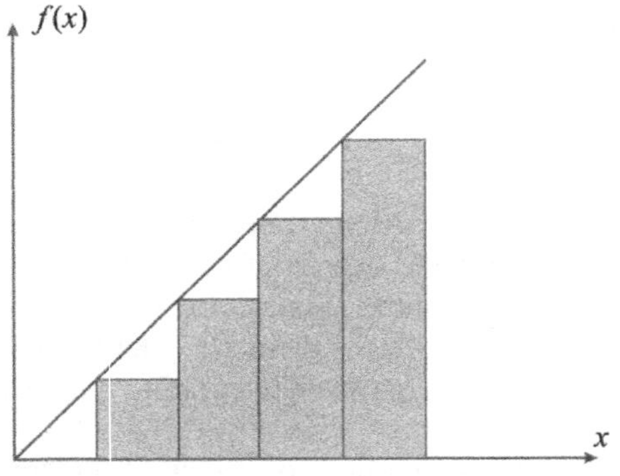

FIGURE 2.1 **Illustration of the rectangle rule.**

Then, we approximate $I[f]$ by

$$Q[f] = \sum_{j=0}^{n} w_j f(x_j).$$

To be precise, a quadrature formula like the one we are describing is called a *closed* formula, since the set of nodes includes the extreme points a and b of the interval. Open formulas are used when the function is not well-behaved near a or b, or when we are integrating on an infinite interval. In this book we will only consider closed formulas, as in most cases of interest for Monte Carlo simulation, as we shall see later, the integration domain is the unit interval $[0,1]$ or the unit hypercube $[0,1]^p$.

2.1.1 THE RECTANGLE RULE

One intuitive way to choose nodes and weights is the rectangle rule. The idea is illustrated in Fig. 2.1, where the underlying idea is clearly to regard the integral as an area and to approximate it by the union of rectangles. If we are integrating over the unit interval $[0,1]$, the rectangle rule may be expressed as

$$R_n[f] = \frac{1}{n} \sum_{j=0}^{n-1} f\left(\frac{j}{n}\right). \tag{2.1}$$

Note that, in such a case, we divide the unit interval in n slices, and we take the function value at the left end of each subinterval as the representative value for that slice; if we sum for j in the range from 1 to n, we are considering right end points.

For the case in the figure, i.e., $f(x) = x$, it is fairly easy to see that

$$R_n[f] = \frac{1}{n^2} \sum_{j=0}^{n-1} j = \frac{1}{n^2} \times \frac{n(n-1)}{2} = \frac{1}{2} - \frac{1}{2n}.$$

Hence, the absolute integration error is

$$\left| R_n[f] - I[f] \right| = \frac{1}{2n}.$$

As expected, the error goes to zero when n goes to infinity, but perhaps we may do better. By "better" we do not only mean faster convergence. One would expect that, at least for such an easy function like $f(x) = x$, a quadrature formula would be able to hand us the exact result. We shall see next how to accomplish this, but we should mention that the simple rectangle formula can be generalized to much more sophisticated rules based on good lattices, as we shall see later.

2.1.2   INTERPOLATORY QUADRATURE FORMULAS

Polynomials are arguably the easiest function to integrate. We also know, from one of the many Weierstrass theorems, that a continuous function can be approximated on an interval by a suitably high-order polynomial with arbitrary precision. So, it is natural to look for quadrature formulas such that the integration error

$$E = I[f] - Q[f]$$

is zero for a wide class of polynomials.

DEFINITION 2.1 (Order of a quadrature formula) *We say that a quadrature formula Q is of order m if the integration error is zero for all the polynomials of degree m or less, but there is a polynomial of degree $m + 1$ such that the error is not zero.*

Now, how can we approximate a function by a polynomial? One possibility is represented by Lagrange polynomials, which we outline very briefly. Given a function $f(\cdot)$, say that we want to interpolate it at a given set of points (x_j, y_j), $j = 0, 1, \ldots, n$, where $y_j = f(x_j)$, and $x_j \neq x_k$ for $j \neq k$. It is easy to find a polynomial P_n of degree (at most) n such that $P_n(x_j) = y_j$ for all j. The Lagrange polynomial $L_j(x)$ is defined as

$$L_j(x) = \prod_{\substack{k=0 \\ k \neq j}}^{n} \frac{x - x_k}{x_j - x_k}. \tag{2.2}$$

Note that this is a polynomial of degree n and that

$$L_j(x_k) = \begin{cases} 1, & \text{if } j = k, \\ 0, & \text{otherwise.} \end{cases}$$

Then, an interpolating polynomial can be easily written as

$$P_n(x) = \sum_{j=0}^{n} y_j L_j(x).$$

Now, in order to define a quadrature formula, we have to define its nodes. The seemingly obvious way is to consider equally spaced nodes:

$$x_j = a + jh, \qquad j = 0, 1, 2, \ldots, n,$$

where $h = (b - a)/n$; also let $f_j = f(x_j)$. As it turns out, this choice need not be the best one, but it is a natural starting point. Selecting equally spaced nodes yields the set of Newton–Cotes quadrature formulas. Given these $n + 1$ nodes and the corresponding Lagrange polynomial $P_n(x)$ of degree n, we compute the quadrature weights as follows:

$$\int_a^b f(x)\, dx \approx \int_a^b P_n(x)\, dx = \int_a^b \left[\sum_{j=0}^n f_j L_j(x) \right] dx$$

$$= \sum_{j=0}^n f_j \left[\int_a^b L_j(x)\, dx \right],$$

which yields

$$w_j = \int_a^b L_j(x)\, dx. \qquad (2.3)$$

Consider the case of two nodes only, $x_0 = a$ and $x_1 = b$, with step $h = x_1 - x_0$. Here we are just interpolating f by a straight line:

$$P_1(x) = \frac{x - x_1}{x_0 - x_1} f_0 + \frac{x - x_0}{x_1 - x_0} f_1.$$

A straightforward calculation yields

$$\int_{x_0}^{x_1} P_1(x)\, dx = \frac{f_1 + f_0}{2} h.$$

A closer look shows that this formula approximates the area below the function using a trapeze. We may improve the integration error by applying the formula on subintervals, thus approximating the overall area by trapezoidal elements, as depicted in Fig. 2.2. Thus, we find the trapezoidal rule:

$$Q[f] = h \left[\frac{1}{2} f_0 + \sum_{j=1}^{n-1} f_j + \frac{1}{2} f_n \right].$$

This idea is general: Given any quadrature formula for an interval, we can build a *composite* formula by applying the same pattern to small subintervals of the integration domain.

2.1.3   AN ALTERNATIVE DERIVATION

A quadrature formula based on $n + 1$ nodes and the corresponding weights obtained by Lagrange polynomials is by construction exact for polynomials of

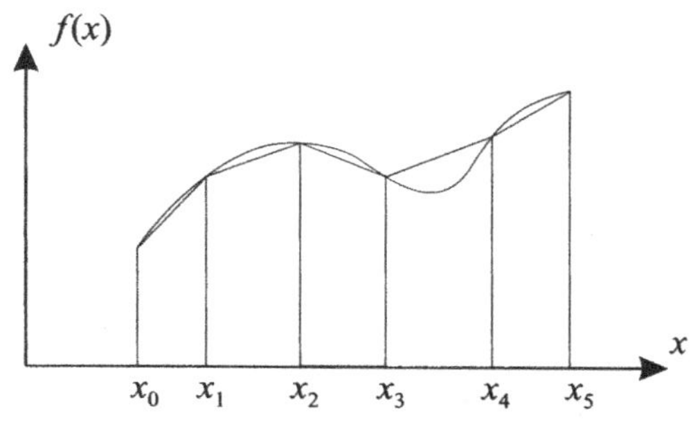

FIGURE 2.2 **Illustrating the trapezoidal quadrature formula.**

degree $\leq n$. We may go the other way around, and find weights of a formula by requiring that it features a given order. For the sake of simplicity, let us consider an integral on the unit interval, and the three quadrature nodes $0, 0.5, 1$:

$$\int_0^1 f(x)\,dx \approx w_0 f(0) + w_1 f(0.5) + w_2 f(1).$$

Since we have three nodes, we should be able to find a formula that is exact for polynomials of degree ≤ 2. Given linearity of integration, we may enforce this requirement by making sure that the formula is exact for the following monomials:

$$q_0(x) = 1, \qquad q_1(x) = x, \qquad q_2(x) = x^2.$$

The corresponding weights can be found by solving the following system of linear equations:

$$1 = \int_0^1 dx = w_0 + w_1 + w_2,$$

$$\frac{1}{2} = \int_0^1 x\,dx = \frac{1}{2}w_1 + w_2, \qquad (2.4)$$

$$\frac{1}{3} = \int_0^1 x^2\,dx = \frac{1}{4}w_1 + w_2,$$

which yields $w_0 = \frac{1}{6}$, $w_1 = \frac{2}{3}$, $w_2 = \frac{1}{6}$. Applying the same idea on the more general interval $[a, b]$, we obtain the well-known Simpson's rule:

$$\int_a^b f(x)\,dx \approx \frac{b-a}{6}\left[f(a) + 4f\left(\frac{b+a}{2}\right) + f(b)\right].$$

It is fairly easy to see that, somewhat surprisingly, this formula is actually exact for polynomials of degree ≤ 3. In fact, we have

$$\int_a^b x^3 dx = \frac{b^4 - a^4}{4}.$$

Applying Simpson's rule we have, by some straightforward algebra:

$$\frac{b-a}{6}\left[a^3 + 4\left(\frac{b+a}{2}\right)^3 + b^3\right]$$

$$= \frac{b-a}{6}\left[a^3 + \frac{1}{2}\left(a^3 + 3a^2b + 3ab^2 + b^3\right) + b^3\right] = \frac{b^4 - a^4}{4}.$$

In order to improve precision, Simpson's rule may be applied to subintervals of (a, b), resulting in a composite formula.

What we have discussed only scratches the surface of numerical integration. We would also need:

- A way to estimate the integration error in order to decide if a more refined partition of the interval is needed.
- A way to adapt partitions to the variability of the function in different integration subintervals.
- A non-naive way to choose nodes.

In the next section we give a partial answer to the last question. What is fundamental, however, is to start seeing the connection between the system of equations (2.4), which is motivated by a classical problem in numerical analysis, and the statistical concept of moment matching, to be discussed in Section 2.5.

2.2 Gaussian quadrature

In Newton–Cotes formulas, nodes are fixed and suitable weights are sought in order to squeeze the most in terms of quadrature order. With $n + 1$ nodes, we may find a quadrature formula with order n. However, we might improve the order by finding nodes and weights jointly, as this essentially means doubling the degrees of freedom. In fact, this is the rationale behind Gaussian quadrature. Gaussian quadrature may yield high-quality numerical integration, and it is particularly relevant for some problems in finance, statistics, and stochastic optimization, as it may be interpreted in terms of discretization of a continuous probability distribution. There are different Gaussian quadrature formulas, associated with a non-negative weight function $w(x)$ and an integration interval:

$$\int_a^b w(x)f(x)\,dx \approx \sum_{i=1}^n w_i f(x_i), \tag{2.5}$$

for a weight function $w(x)$ and an integration interval $[a, b]$. In this section, for the sake of convenience, we consider n nodes x_i, $i = 1, \ldots, n$. The integration interval can also be $(-\infty, +\infty)$, in which case we need a weight function $w(x)$ that goes to zero quickly enough to guarantee, as far as possible, the existence of the integral. This is quite relevant for our purposes, as we will relate $w(x)$ to a probability density, such as the density of a normal random variable.[1] Indeed, the integration interval is $(-\infty, +\infty)$, and a weight function doing the job is $w(x) = e^{-x^2}$. With such a choice of the weight function, we talk of Gauss–Hermite quadrature formulas. Actually, this weight function is not exactly the density of a normal variable, but it is easy to adjust the formula by a change of variable. Let Y be a random variable with normal distribution $N(\mu, \sigma^2)$. Then, recalling the familiar density of a normal variable:

$$E[f(Y)] = \int_{-\infty}^{+\infty} \frac{1}{\sqrt{2\pi}\,\sigma} \exp\left\{ -\frac{1}{2} \left(\frac{y-\mu}{\sigma} \right)^2 \right\} f(y)\,dy.$$

In order to use weights and nodes from a Gauss–Hermite formula, we need the following change of variable:

$$-x^2 = -\frac{1}{2} \left(\frac{y-\mu}{\sigma} \right)^2 \quad \Rightarrow \quad y = \sqrt{2}\sigma x + \mu \quad \Rightarrow \quad \frac{dy}{\sqrt{2}\sigma} = dx.$$

Hence, we find

$$E[f(Y)] \approx \frac{1}{\sqrt{\pi}} \sum_{i=1}^{n} w_i f(\sqrt{2}\sigma x_i + \mu). \qquad (2.6)$$

In practice, we may regard Gauss–Hermite formulas as a way to approximate a normal distribution by a discrete random variable with realizations x_i and probabilities w_i, $i = 1, \ldots, n$. In the next two sections we provide the interested reader with the theoretical background behind Gaussian quadrature, showing the role of orthogonal polynomials, and some hints about computational issues. However, the practically oriented reader may safely skip these sections and proceed directly to Section 2.2.2, where we show how to use the corresponding R package.

2.2.1 THEORY OF GAUSSIAN QUADRATURE: THE ROLE OF ORTHOGONAL POLYNOMIALS

There is a rich theory behind Gaussian quadrature formulas, and the key concept is related to families of orthogonal polynomials. In fact, a quadrature formula with maximum order is found by choosing nodes nodes as the n roots of a polynomial of order n, selected within a family of orthogonal polynomials with respect to the inner product

$$<f, g> = \int_a^b w(x)f(x)g(x)\,dx,$$

where $w(\cdot)$ is a weight function and the integration interval can be bounded or, with a suitable choice of the weight function, unbounded. The following properties, among many others, can be proved:[2]

---

[1] In case of need, the reader may find a refresher on the normal distribution in Section 3.2.3.
[2] See. e.g., [2, Chapter 2].

- A polynomial of degree n within such a family has n distinct real roots, all located within the interval (a, b).
- These roots are interleaved, in the sense that each of the $n - 1$ roots of the polynomial of degree $n - 1$ lies in an interval defined by a pair of consecutive roots of the polynomial of degree n.

Using this choice of nodes, along with a proper choice of weights, yields a quadrature formula with order $2n - 1$. To see this, let Π_n denote the set of polynomials of degree n, and let us consider a polynomial $q \in \Pi_n$ that is orthogonal to all polynomials in Π_{n-1}, i.e.,

$$\int_a^b w(x)q(x)p(x)\,dx = 0, \qquad \forall p \in \Pi_{n-1}.$$

One might wonder whether such a polynomial can always be found. The answer lies in how families of orthogonal polynomials can be built, on the basis of recursive formulas that we hint at later. We obtain a sequence of polynomials of increasing degree and, since orthogonal polynomials of degree up to $n - 1$ form a basis for Π_{n-1}, the next polynomial in the sequence is of degree n and orthogonal to all polynomials in Π_{n-1}.

Now, let us consider a function $f \in \Pi_{2n-1}$, i.e., a polynomial of order $2n - 1$. Any such polynomial can be divided by the aforementioned q, obtaining a quotient polynomial p and a remainder polynomial r:

$$f = qp + r,$$

where $p, r \in \Pi_{n-1}$. Now let us integrate the (ordinary) function product wf by a quadrature formula based on n nodes x_i, $i = 1, \ldots, n$, chosen as the zeros of q:

$$\int_a^b w(x)f(x)\,dx$$

$$= \int_a^b w(x)p(x)q(x)\,dx + \int_a^b w(x)r(x)\,dx \qquad \text{(division)}$$

$$= 0 + \int_a^b w(x)r(x)\,dx \qquad (q \text{ is orthogonal to } p)$$

$$= \sum_{i=1}^n w_i r(x_i) \qquad \text{(quadrature is exact for } r \in \Pi_{n-1})$$

$$= \sum_{i=1}^n w_i f(x_i) \qquad (x_i \text{ is a zero of } q).$$

This shows that indeed the formula is exact for polynomials of degree $2n - 1$. Using this property, it is possible to prove the following essential facts:

- The sum of weights is 1.

▪ Weights are positive.

These are essential facts in light of our probabilistic applications. In fact, different formulas can be adapted to different probability distributions, like we have shown for Gauss–Hermite formulas. The specific family of orthogonal polynomials to be used depends on the weight function and the integration interval. Other examples are:

- Gauss–Legendre formulas, where the integration domain is $(-1, 1)$ and the weight function is $w(x) = 1$.
- Gauss–Laguerre formulas, where the integration domain is $(0, +\infty)$ and the weight function is $w(x) = x^\alpha e^{-x}$.

After clarifying the link between orthogonal polynomials and Gaussian quadrature, how can we actually find nodes and weights? There are different approaches.

1. We may build the relevant sequence of orthogonal polynomials by a recursion of the form

$$p_{j+1}(x) = (x - a_j)p_j(x) - b_j p_{j-1}(x).$$

We omit further details, such as the form of the coefficients a_j and b_j, referring the reader to the references. We just stress two facts:

- Since roots of successive polynomials in the sequence are interleaved, the roots of the previous polynomials are useful to initialize fast root-finding algorithms.
- After determining the nodes, we may find the corresponding weights, e.g., by solving a system of linear equations or by using a set of formulas involving the underlying orthogonal polynomials and their derivatives; see, e.g., [10, Chapter 4].

2. Another kind of approach has the same starting point as the previous one, but the recursive formulas are used to build a matrix (tridiagonal and symmetric), whose eigenvalues are, in fact, the roots we need. Indeed, the R functions we describe in the next section follow an approach in this vein; see [4].

2.2.2 GAUSSIAN QUADRATURE IN R

Gaussian quadrature in R is accomplished by the package `statmod`, which offers the following two functions:

- `gauss.quad(n, kind="legendre")`, whose main input arguments are the number `n` of nodes and weights, and the kind of quadrature, which can be of many types, including `"hermite"`; `"legendre"` is the default setting of the parameter. If we choose this function, we have to use the transformations of Eq. (2.6) to deal with the normal distribution.

▪ If we want a more direct approach, we may use

```
gauss.quad.prob(n,dist="uniform",l=0,u=1,mu=0,sigma=1,
                alpha=1,beta=1),
```

which allows to select a distribution `"uniform"`, `"normal"`, `"beta"`, or `"gamma"`, along with the relevant parameters.[3]

In both cases, the output is a list including weights and nodes. To illustrate, let us check a Gauss–Hermite quadrature with 5 nodes, for a normal random variable with parameters $\mu = 50$ and $\sigma = 20$:

```
> mu <- 50
> sigma <- 20
> N <- 5
> out <- gauss.quad(N, kind="hermite")
> w1 <- out$weights/sqrt(pi)
> x1 <- sqrt(2)*sigma*out$nodes + mu
> out <- gauss.quad.prob(N,dist="normal",mu=mu,sigma=sigma)
> w2 <- out$weights
> x2 <- out$nodes
> rbind(w1,w2)
           [,1]       [,2]      [,3]      [,4]       [,5]
w1 0.01125741 0.2220759 0.5333333 0.2220759 0.01125741
w2 0.01125741 0.2220759 0.5333333 0.2220759 0.01125741
> rbind(x1,x2)
       [,1]     [,2] [,3]     [,4]      [,5]
x1 -7.1394 22.88748   50 77.11252 107.1394
x2 -7.1394 22.88748   50 77.11252 107.1394
```

We immediately verify that the same results are obtained by using the two functions, provided that we apply the correct variable transformations. The nodes, as expected, are symmetrically centered around the expected value. Furthermore, we may also check that probabilities add up to one and we get the correct first- and second-order moments:

```
> sum(w1)
[1] 1
> sum(w1*x1)
[1] 50
> sqrt(sum(w1*x1^2)-sum(w1*x1)^2)
[1] 20
```

Readers familiar with the concepts of skewness and kurtosis, will also appreciate the following results:[4]

---

[3]The reader who needs a refresher on such probability distributions may have a look at Chapter 3.

[4]Skewness and kurtosis for a random variable X with expected value μ and standard deviation σ are defined as $E[(X-\mu)^3]/\sigma^3$ and $E[(X-\mu)^4]/\sigma^4$, respectively. Skewness measures the lack of symmetry and is zero for a normal distribution, which is symmetric; kurtosis is related

```
> sum(w1*(x1-50)^3)/20^3
[1] -8.526513e-16
> sum(w1*(x1-50)^4)/20^4
[1] 3
```

Within numerical accuracy, the above results are correct for a normal distribution: We are able to match moments of order 3 and 4 as well, with only five nodes. We will come back to this in Section 2.5. All of the results above are related to the quadrature order of Gauss–Hermite formulas when applied to polynomials. With a more complicated function, say, an exponential, we cannot expect exact results; so, let us compute the expected value of an exponential function of a normal random variable. In this particular case we can even afford an instructive comparison with the exact result. Indeed, given the properties of the lognormal distribution,[5] we know that if $X \sim \mathsf{N}(\mu, \sigma^2)$, then

$$\mathrm{E}\left[e^X\right] = e^{\mu + \sigma^2/2}.$$

The script displayed in Fig. 2.3 computes the percentage error for $\mu = 4$ and $\sigma = 2$, when 5, 10, 15, and 20 nodes are used. Running the script, we observe that a remarkable precision is achieved with a fairly modest number of nodes:

```
N= 5    True= 403.4288   Approx= 398.6568   %error 1.182865
N= 10   True= 403.4288   Approx= 403.4286   %error 5.537709e-05
N= 15   True= 403.4288   Approx= 403.4288   %error 1.893283e-10
N= 20   True= 403.4288   Approx= 403.4288   %error 9.863052e-14
```

These results explain why Gaussian quadrature is used, e.g., when solving stochastic dynamic optimization problems by dynamic programming. However, there remains an issue: How does all of this scale to higher dimensions? We investigate the issue next; unfortunately, the answer is not quite positive.

2.3 Extension to higher dimensions: Product rules

Let us consider a function of p variables

$$f(\mathbf{x}) = f(x_1, x_2, \ldots, x_p).$$

How can we integrate this function over a domain Ω? We know from Fubini theorem that a multidimensional integral can be transformed into a sequence of nested one-dimensional integrals. If we consider the case in which the integration region is just the Cartesian product of subsets,

$$\mathbf{\Omega} = \Omega_1 \times \Omega_2 \times \cdots \times \Omega_p,$$

---

to the tails of the distribution and is 3 for any normal distribution. Distributions with fatter tails, which play a key role in risk management, have a larger kurtosis.

[5] See Section 3.2.3.

```
require(statmod)
mu <- 4
sigma <- 2
numNodes <- c(5,10,15,20)
trueValue <- exp(mu + sigma^2/2)
for (i in 1:length(numNodes)){
  out <- gauss.quad(numNodes[i], kind="hermite")
  w <- out$weights/sqrt(pi)
  x <- sqrt(2)*sigma*out$nodes + mu
  approxValue <- sum(w*exp(x))
  percError <- 100*abs(trueValue-approxValue)/trueValue
  cat("N=",numNodes[i]," True=",trueValue, " Approx=",
      approxValue, " %error", percError,"\n")
}
```

FIGURE 2.3 **Script to check Gauss–Hermite quadrature.**

we have

$$\int_{\boldsymbol{\Omega}} f(\mathbf{x})\, d\mathbf{x} = \int_{\Omega_1} \int_{\Omega_2} \cdots \int_{\Omega_p} f(x_1, x_2, \ldots, x_p)\, dx_p \cdots dx_2\, dx_1.$$

It is worth mentioning that this is indeed the relevant case for Monte Carlo simulation, where $\boldsymbol{\Omega} = [0,1]^p = [0,1] \times [0,1] \times \cdots \times [0,1]$.

Therefore, it seems quite natural to extend quadrature formulas to multi-dimensional integration by using a product rule reflecting the above structure. More precisely, let us assume that we have weights and nodes for a Newton–Cotes quadrature formula along each dimension: For dimension k, $k = 1, \ldots, p$, we have weights w_i^k and nodes x_i^k, $i = 1, \ldots, m_k$. A product rule approximates the integral as

$$\sum_{i_1=1}^{m_1} \sum_{i_2=1}^{m_2} \cdots \sum_{i_p=1}^{m_p} w_{i_1}^1 w_{i_2}^2 \cdots w_{i_p}^p\, f\left(x_{i_1}^1, x_{i_2}^2, \ldots, x_{i_p}^p\right).$$

The idea behind a product rule is quite intuitive, and it is illustrated in Fig. 2.4. From the figure, we may immediately appreciate the drawbacks of the approach:

- If N nodes per dimension are required to achieve a satisfactory precision, a total of N^p nodes are required; this exponential growth makes the approach impractical for high-dimensional integrals.
- A subtler issue is that if we want to add a few points to improve precision, there is no obvious way to do it; adding a single point would disrupt the overall structure.
- Furthermore, we also observe that the same coordinates are used repeatedly for each dimension: The grid of Fig. 2.4 consists of $11 \times 11 = 121$

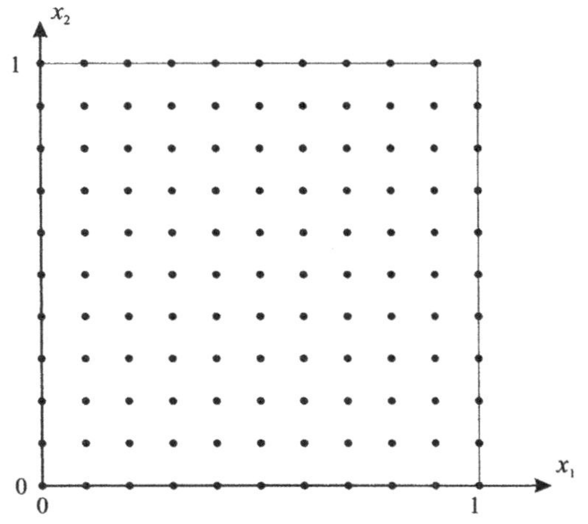

FIGURE 2.4 **An illustration of product rules.**

points in the plane, but if we project them on each coordinate axis, we only find 11 points. One might wonder if there is a better way to gather information about the function.

Since such regular grid is going to be impractical for large p, we have to find alternative approaches, either deterministic or stochastic.

2.4 Alternative approaches for high-dimensional integration

In order to overcome the limitations of classical quadrature, alternative approaches have been developed. Needless to say, random sampling strategies collectively known as Monte Carlo methods play a prominent role in this respect, but we should not take for granted that random methods are the only alternative. In fact, there are deterministic strategies that work well in medium-sized problems:

- Lattice methods
- Low-discrepancy sequences

Here we just introduce Monte Carlo methods within the framework of numerical integration, leaving further development to later chapters. Low-discrepancy sequences are also just outlined here, as they are the subject of Chapter 9. We discuss lattice methods in some more detail, as they will not be dealt with elsewhere in this book.

2.4.1   MONTE CARLO INTEGRATION

To see how random sampling may be used to compute a definite integral, which is a deterministic quantity, let us consider a one-dimensional integral on the unit interval $[0, 1]$:

$$I[g] = \int_0^1 g(x)\,dx = \int_0^1 g(u) \cdot 1\,du.$$

The last rewriting, where we multiply the function $g(u)$ by 1, may not look much of an achievement. However, let us recall that the density of a uniform random variable U on the unit interval $[0, 1]$ is, in fact,

$$f_U(u) = \begin{cases} 1, & \text{if } 0 \le u \le 1, \\ 0, & \text{otherwise.} \end{cases}$$

Therefore, we may think of this integral as the expected value

$$\mathrm{E}[g(U)] = \int_0^1 g(u) f_U(u)\,du$$

for $U \sim \mathsf{U}(0, 1)$. Now it is natural to estimate the expected value by a sample mean. What we have to do is to generate a sequence $\{U_i\}$ of *independent* random realizations from the uniform distribution, and then to estimate I as

$$\hat{I}_m[g] = \frac{1}{m} \sum_{i=1}^m g(U_i).$$

The strong law of large numbers implies that, with probability 1,

$$\lim_{m \to \infty} \hat{I}_m[g] = I[g].$$

Genuine random sampling is not really feasible with a computer, but we can produce a sequence of *pseudo*random numbers using generators provided by most programming languages and environments, including R, as we shall see in Chapter 5.

Example 2.1 A first Monte Carlo integration

Consider the trivial case

$$\int_0^1 e^x\,dx = e - 1 \approx 1.718282.$$

To sample the uniform distribution, we rely on the `runif` function, whose first argument is the sample size, setting the seed of the generator in order to allow the replication of the experiment:

```
> set.seed(55555)
> mean(exp(runif(10)))
[1] 1.657509
> mean(exp(runif(10)))
[1] 1.860796
> mean(exp(runif(10)))
[1] 1.468096
> mean(exp(runif(100)))
[1] 1.701009
> mean(exp(runif(100)))
[1] 1.762114
> mean(exp(runif(100)))
[1] 1.706034
> mean(exp(runif(1000000)))
[1] 1.718733
```

Clearly, a sample size $m = 10$ does not yield a reliable estimate, and $m = 100$ is just a bit better. A large sample size is needed to get closer to the actual result. Monte Carlo looks extremely straightforward and flexible, but not necessarily quite efficient. It is useful to compare the above results with those obtained by Gaussian quadrature:

```
> require(statmod)
Loading required package: statmod
> out=gauss.quad.prob(10,dist="uniform",l=0,u=1)
> sum(out$weight*exp(out$nodes))
[1] 1.718282
```

Needless to say, we do not know the exact result in practice, and we should wonder how to qualify the reliability of the estimate. This can be done by computing a confidence interval as follows:

```
> set.seed(55555)
> out <- t.test(exp(runif(100)))
> out$estimate
mean of x
 1.717044
> out$conf.int
[1] 1.621329 1.812759
attr(,"conf.level")
[1] 0.95
```

Note how a confidence interval for a mean is obtained in R by what may seem a weird route, using the t.test function, which returns a data structure including the estimate and, by default, a 95% confidence interval, among other things. This is due to the link between testing hypotheses about the mean of a normal population and con-

> fidence intervals for the mean. Even though the result may not look
> bad, the 95% confidence interval points out the issues with Monte
> Carlo. A large sample size is needed, unless suitable variance reduc-
> tion strategies are employed, as we discuss at length in Chapter 8.

We will learn later why integrating on the unit interval is so relevant, but we may
estimate the expected value with respect to other distributions. For instance, we
have already considered the expected value of the exponential function of a
normal random variable. In Section 2.2.2 we have seen that $\mathrm{E}[e^X] = 403.4288$
for $X \sim \mathsf{N}(4, 2^2)$. Here we sample the normal distribution using `rnorm`:[6]

```
> set.seed(55555)
> X=exp(rnorm(100,mean=4,sd=2))
> t.test(X)

        One Sample t-test

data:  X
t = 4.1016, df = 99, p-value = 8.426e-05
alternative hypothesis: true mean is not equal to 0
95 percent confidence interval:
 153.2929 440.5990
sample estimates:
mean of x
 296.9459
```

Once again, the results are disappointing. Indeed, for one-dimensional integration, Monte Carlo is hardly competitive with deterministic quadrature, but
when computing a multidimensional integral it may be the only viable option.
In general, if we have an integral like

$$I[g] = \int_{\Omega} g(\mathbf{x})\, d\mathbf{x}, \tag{2.7}$$

where $\Omega \subset \mathbb{R}^p$, we may estimate $I[g]$ by randomly sampling a sequence of
points $\mathbf{x}^i \in \Omega$, $i = 1, \dots, m$, and building the estimator

$$\hat{I}_m[g] = \frac{\mathrm{vol}(\Omega)}{m} \sum_{i=1}^{m} g(\mathbf{x}_i), \tag{2.8}$$

where $\mathrm{vol}(\Omega)$ denotes the volume of the region Ω. To understand the formula,
we should think that the ratio $(1/m)\sum_{i=1}^{m} g(\mathbf{x}_i)$ estimates the average value of

---

[6]Unlike the last example, we let R print the whole output from `t.test`; clearly, we are only
interested in the estimate, and the test p-value is irrelevant here.

the function, which must be multiplied by the volume of the integration region in order to find the integral.

We will see that in practice we need only to integrate over the unit hypercube, i.e.,

$$\Omega = [0, 1]^p,$$

hence $\text{vol}(\Omega) = 1$. More generally, if we consider a vector random variable

$$\mathbf{X} = \begin{bmatrix} X_1 \\ X_2 \\ \vdots \\ X_p \end{bmatrix},$$

with joint density function $f_{\mathbf{X}}(x_1, \ldots, x_p)$, we have to sample $\mathbf{X}$ to estimate:

$$\text{E}[g(\mathbf{X})] = \int \int \cdots \int g(x_1, \ldots, x_p) f(x_1, \ldots, x_p) \, dx_p \cdots dx_1.$$

The good news is that, as we know from elementary inferential statistics, the half-length of a standard confidence interval with confidence level $1 - \alpha$ is

$$H = z_{1-\alpha/2} \frac{\sigma}{\sqrt{n}},$$

assuming that the sample size n is large enough to warrant use of quantiles $z_{1-\alpha/2}$ from the standard distribution. Here σ is the standard deviation of the estimator, and we see that H does not depend on the problem dimensionality[7] p, whereas product rules based on classical quadrature are highly impractical for large p. The bad news is in the denominator, $\sqrt{n}$, which does not grow quickly enough. In fact, to reduce H by a factor of 10, we have to increase n by a factor of 100, which shows that Monte Carlo estimates may converge slowly in practice.

2.4.2 LOW-DISCREPANCY SEQUENCES

As we have seen in the previous sections, the performance of naive Monte Carlo is far from impressive for small sample sizes. We may also get a further intuition by plotting a bidimensional random sample on the unit square. The following R commands

```
> set.seed(55555)
> plot(runif(100),runif(100),pch=20)
> grid(nx=10)
```

---

[7]As we shall also remark later, this statement should be somewhat tempered. It is true that p does not play an explicit role, but it may be expected that a large p will influence σ implicitly, not to mention the computational effort.

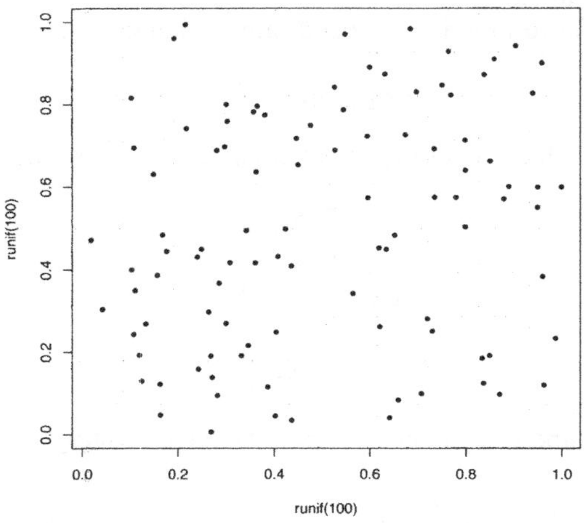

FIGURE 2.5 **A uniform random sample within the unit square.**

produce the plot in Fig. 2.5. Here we are sampling a bidimensional uniform distribution, with independent marginals. We observe that the coverage is far from accurate and uniform. Using the `grid(nx=10)` function, we plot a grid of 10×10 small squares, many of which do not contain any point (ideally, each subsquare should contain exactly one). One possibility, of course, is to increase the sample size. In Chapter 9 we will see that we may define an objective measure of lack of uniform coverage of the unit hypercube, namely, the star discrepancy. Good set of points should have low discrepancy, and this leads to the concept of low-discrepancy sequences. They are deterministic sequences but, unlike grids produced by product quadrature rules, they do not display any apparent regularity. The Halton sequence is arguably the simplest low-discrepancy sequence, which is included in the library `randtoolbox` along with other sequences, such as the Sobol sequence. To illustrate the idea, let us generate and plot a two-dimensional sequence of 100 points using the Halton sequence:

```
> library(randtoolbox)
> x <- halton(100,dim=2)
> plot(x[,1], x[,2], pch=20)
> grid(nx=10)
```

The resulting sequence, displayed in Fig. 2.6, looks more evenly distributed than the random one of Fig. 2.5. More formally, while Monte Carlo has an error of order $O(1/\sqrt{n})$, low-discrepancy sequences have an error of order $O(\log n^p/n)$. This is bad news on the one hand, since the problem dimensionality p does play a role; on the other hand, we get rid of the square root, which makes Monte Carlo not quite efficient. In practice, it turns out that low-discrepancy sequences are efficient for problems with a moderate dimensionality. Another point worth

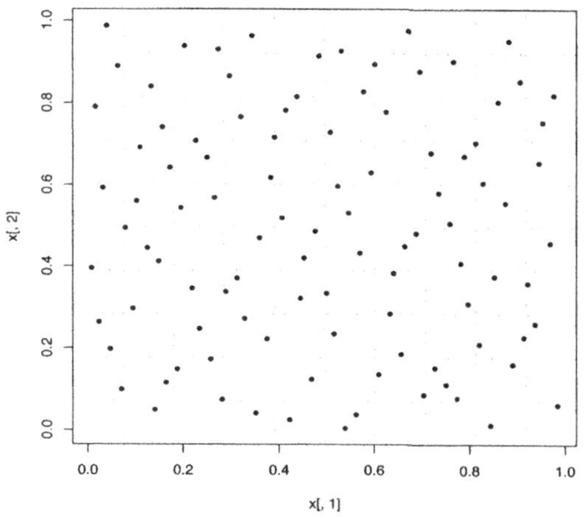

FIGURE 2.6 **A bidimensional low-discrepancy sequence of points.**

mentioning is that if we want to add one point to a low-discrepancy, we can do so without disrupting its structure, just like with random sequences; this is not true for regular grids. We should also note that deterministic and stochastic methods are not mutually exclusive, as scrambled low-discrepancy sequences can also be used, as we shall see later in Chapter 9.

2.4.3 LATTICE METHODS

Another class of deterministic integration methods that is often discussed in conjunction with low-discrepancy sequences is based on lattices. Since the term "lattice" is used in diverse ways with a different meaning, it is best to state clearly what is meant in this context.

DEFINITION 2.2 (Lattice) *A p-dimensional lattice is a discrete subset of $\mathbb{R}^p$ that is closed under addition and subtraction.*

A familiar lattice is depicted in Fig. 2.4. Indeed, a regular grid associated with product rules is a lattice or, maybe more precisely, a lattice point set. Lattices may be associated with a *basis*, i.e., a linearly independent set of vectors $\{\mathbf{w}_1, \ldots, \mathbf{w}_p\}$, called generators, such that the whole lattice is obtained by taking suitable linear combinations of the generators:

$$L_p = \{v_1\mathbf{w}_1 + \cdots + v_p\mathbf{w}_p, \ \mathbf{v} \in \mathbb{Z}^p\}, \tag{2.9}$$

where $\mathbb{Z}^p$ is the set of p-dimensional vectors with integer coordinates. Actually, different bases may generate the same lattice.

🖳 Example 2.2 A regular grid as a lattice

Consider the following basis:

$$\mathbf{w}_1 = \left[\frac{1}{5}, 0\right]^{\mathsf{T}}, \quad \mathbf{w}_2 = \left[0, \frac{1}{5}\right]^{\mathsf{T}},$$

where the superscript ^{T} denotes vector and matrix transposition. This basis generates a grid of points with vertical and horizontal spacing given by 0.2. If we take the intersection of this grid with the half-open square $[0, 1)^2$, we find points with coordinates in the set

$$\{0, 0.2, 0.4, 0.6, 0.8\}.$$

Clearly, there are $N = 25$ points in the set. Now, let us consider the matrix formed by lining up the vectors of the basis in columns:

$$\mathbf{W} = \begin{bmatrix} \dfrac{1}{5} & 0 \\ 0 & \dfrac{1}{5} \end{bmatrix}.$$

Clearly, $1/\det(\mathbf{W}) = 25$, i.e., the determinant of $\mathbf{W}$ is related to the number N of points in the intersection between the lattice and the half-open unit square. As we shall see, this is no coincidence. More generally, if we consider a basis formed by vectors

$$\frac{1}{n_j}\mathbf{e}_j, \qquad j = 1, \ldots, p,$$

where $\mathbf{e}_j$ is the jth unit vector in $\mathbb{R}^p$ and n_j is an integer number associated with dimension j, we obtain a grid with

$$N = \prod_{j=1}^{p} n_j$$

points. If we adopt the same spacing along each dimension, i.e., we consider a single value n, we obtain a grid with $N = n^p$ points.

In the following, we will be interested in integrating a function on the half-open unit hypercube $[0, 1)^p$. One might wonder why we should consider $[0, 1)^p$ rather than $[0, 1]^p$. From a theoretical point of view, lattice rules are designed for fairly smooth and periodic functions,[8] such that $f(\mathbf{x}) = f(\mathbf{x} + \mathbf{z})$ for any $\mathbf{x} \in \mathbb{R}^p$ and $\mathbf{z} \in \mathbb{Z}^p$. Thus, in lattice rules, the North–East border is ruled

---

[8]Periodicity is required for the validity of certain arguments showing the precision of lattice rules, which rely on Fourier series. See [11] for details.

out as there the function is supposed to take the same values as in the opposite border, and we do not want to double-count them. When applying lattice rules to more general functions, in principle, nothing changes since we are ruling out a set of measure zero. From a numerical point of view, we should also note that many pseudorandom number generators employed in Monte Carlo integration actually produce numbers in the interval $[0, 1)$.

Using Eq. (2.9), an integration lattice L_p is generated; the lattice point set is just the intersection

$$P_N = L_p \cap [0, 1)^p,$$

which includes N points $\mathbf{u}_i$, $i = 1, \ldots, N$. The integral of a function $f(\mathbf{x})$ on the unit hypercube is approximated as

$$\frac{1}{N} \sum_{i=1}^{N} f(\mathbf{u}_i).$$

Note that weights are not related to quadrature formulas, and that this rule can be associated with sampling. We may also consider lattice rules as a generalization of the rectangle rule of Section 2.1.1. The key feature of lattices is that they can be generated in a more general way than a rectangular grid, improving the accuracy of numerical integration. To this aim, a proper choice of the basis is in order.

To get a better understanding of lattices, it is useful to associate a "unit cell" with each choice of generators. The unit cell is generated by taking linear combinations as follows:

$$\{\lambda_1 \mathbf{w}_1 + \cdots + \lambda_p \mathbf{w}_p : 0 \le \lambda_i \le 0, \ i = 1, \ldots, p\}.$$

It is easy to see that this is parallelepiped. To illustrate, let us consider the basis

$$\mathbf{w}_1 = \left[\frac{1}{5}, \frac{2}{5}\right]^{\mathsf{T}}, \quad \mathbf{w}_2 = \left[\frac{2}{5}, -\frac{1}{5}\right]^{\mathsf{T}}. \tag{2.10}$$

The corresponding lattice is displayed in Fig. 2.7, along with an arbitrarily selected shaded unit cell. We note the following:

- The unit shaded cell can be shifted repeatedly to generate the lattice.
- The area of the parallelepiped is just the determinant of the matrix $\mathbf{W}$ whose columns are the generators.
- This area is easily seen to be $\det(\mathbf{W}) = \frac{1}{5}$.
- If we imagine tiling the space by shifting the unit cell, we see that the average number of lattice points per unit volume is just $1/\det(\mathbf{W}) = 5$.
- Indeed, in the unit square $[0, 1)^2$, we have five points, the vertices of the shaded cell plus the origin.

When choosing the basis, we have to fix the number N of points that we want to include in the point set, and this in turn is influenced by the choice of generators.

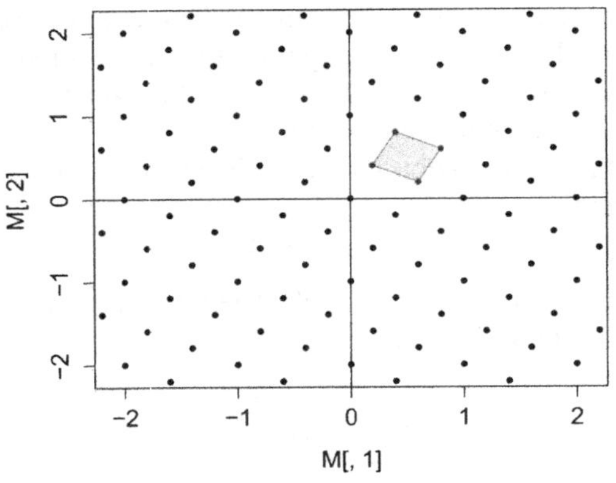

FIGURE 2.7 **Lattice generated by the basis in Eq. (2.10).**

Actually, we need not select p generators. We may simplify the choice and the related analysis by selecting a number $r < p$ of generators. This results in rank-r lattice rules, which may be rewritten as

$$P_N = \left\{ \left(\frac{i_1}{n_1} \mathbf{w}_1 + \cdots + \frac{i_r}{n_r} \mathbf{w}_r \right) \bmod 1, \quad 0 \le i_k < n_k, \ k = 1, \ldots, r \right\}.$$

Note that by taking a number x modulo 1 we obtain its fractional part. The simplest rank-r rule is obtained for $r = 1$, and an example of rank-1 rule is the Korobov point set, which is generated as follows:

$$P_N = \left\{ \frac{i}{N} [1, a, a^2, \ldots, a^{p-1}]^\mathsf{T} \bmod 1, \quad i = 1, \ldots, N - 1 \right\}.$$

This point set can be also generated by the basis

$$\mathbf{w}_1 = \frac{1}{N} [1, a, a^2, \ldots, a^{p-1}]^\mathsf{T},$$

$$\mathbf{w}_2 = [0, 1, 0, \ldots, 0]^\mathsf{T},$$

$$\vdots$$

$$\mathbf{w}_p = [0, 0, 0, \ldots, 1]^\mathsf{T}.$$

The determinant is clearly $1/N$, and in fact the point set, with a proper choice of parameters, includes N points. An R function to generate a Korobov point set is illustrated in Fig. 2.8. In the code we take the increasing powers of a immediately modulo N, rather than computing possibly large powers of a and then applying the modulo operator; this may avoid issues with overflow or loss of precision. The following snapshot generates the point set displayed in Fig. 2.9.

```
korobov <- function(a,d,N){
  z <- numeric(d)
  z[1] <- 1
  for (i in 2:d){
    z[i] <- (z[i-1]*a) %% N
  }
  Z <- matrix(rep(z,N),N,d,byrow=TRUE)
  B <- matrix(rep(1:N),N,d)
  return( (B*Z/N) %% 1)
}
```

R programming notes:

1. The %% operator implements mod in R.

2. In order to avoid for loops, we use matrix to build matrices consisting of replicated copies of vectors. To replicate a vector, we use rep, which yields another vector. Then matrix uses the vector to build the matrix by tiling vectors by columns. The byrow parameter is used when you want to override the default behavior and tile vectors by rows.

FIGURE 2.8 **Function to generate a Korobov point set.**

```
> a <- 3
> d <- 2
> N <- 10
> X <- korobov(a,d,N)
> X
        [,1] [,2]
 [1,]   0.1  0.3
 [2,]   0.2  0.6
 [3,]   0.3  0.9
 [4,]   0.4  0.2
 [5,]   0.5  0.5
 [6,]   0.6  0.8
 [7,]   0.7  0.1
 [8,]   0.8  0.4
 [9,]   0.9  0.7
[10,]   0.0  0.0
> plot(X[,1],X[,2],pch=20)
```

This choice of parameters is good in the sense that it is *projection regular*. A point set consisting of N points is projection regular if the projection of its points on each coordinate axis consists of N distinct points:

```
> sort(X[,1])
 [1] 0.0 0.1 0.2 0.3 0.4 0.5 0.6 0.7 0.8 0.9
> sort(X[,2])
 [1] 0.0 0.1 0.2 0.3 0.4 0.5 0.6 0.7 0.8 0.9
```

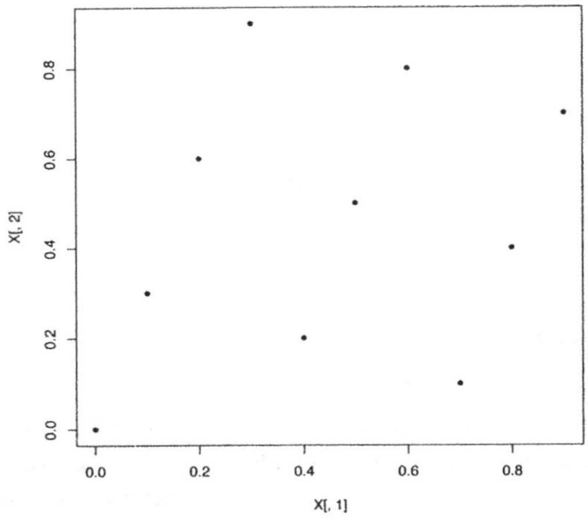

FIGURE 2.9 **Korobov point set for** $a = 3, d = 2, N = 10.$

We observe that a regular grid does not enjoy such a property, as it is clear in Fig. 2.4. For instance, if we build a 5×5 grid, we have 25 points on the grid, but only 5 along each projection. This "shadowing" effect may have the effect that we do not collect rich enough information about the function.[9] Full projection is not to be taken for granted. For instance, if we take $N = 12$ points in the above sequence, the point set is less satisfactory:

```
> N <- 12
> X <- korobov(a,d,N)
> sort(X[,1])
 [1]  0.00000000 0.08333333 0.16666667 0.25000000
      0.33333333 0.41666667 0.50000000 0.58333333
 [9]  0.66666667 0.75000000 0.83333333 0.91666667
> sort(X[,2])
 [1]  0.00 0.00 0.00 0.25 0.25 0.25 0.50 0.50 0.50
      0.75 0.75 0.75
```

In this case, we do not have 12 distinct points along the second dimension. There are theoretical conditions ensuring projection regularity, as well as tables suggesting parameters in order to generate good lattices.

Another important feature of lattice rules is that they are associated with a fixed number N of points. Unlike pseudorandom or low-discrepancy sequences, there is no obvious way to add a single point. However, extensible lattice rules have been proposed in the literature.

---

[9]This is not only relevant in numerical integration but also in global optimization.

2.5   Relationship with moment matching

In Section 2.1.3, when developing quadrature formulas, we have considered matching the integral of a monomial exactly:

$$\int_0^1 x^k \, dx = \sum_{j=0}^m w_j x_j^k, \quad k = 0, 1, \ldots, m.$$

If we interpret integrals as expectations and quadrature formulas as discretizations of continuous probability distributions, with realizations x_j and probabilities w_j, we may wonder if there is a link with matching moments $\mathrm{E}[X^k]$ of the continuous random variable with the corresponding moments of the discrete random variable approximating X. Indeed, the link is pretty evident for Gaussian quadrature, as we have observed in Section 2.2.2. In the following two sections we further illustrate the idea with reference to scenario generation. Again, it is important to understand that plain Monte Carlo scenario generation need not be the best option.

2.5.1   BINOMIAL LATTICES

A widely used approach to price financial options relies on a binomial lattice[10] discretizing the continuous-time, continuous-state stochastic process followed by the underlying variable. The underlying variable could be the price $S(t)$ of a stock share. Let S_0 be the price of the asset at time $t = 0$, and consider the new price $S_{\delta t}$ after a time period δt. We discretize the process with respect to both time and price, assuming that the new price can only take two values:

$$S_{\delta t} = \begin{cases} u S_0, & \text{with probability } p, \\ d S_0, & \text{with probability } 1 - p. \end{cases}$$

Here u is a multiplicative shock corresponding to an "up" move, whereas d corresponds to a "down" move. The advantage of using multiplicative shocks is that the lattice recombines: Since $S_0 u d = S_0 d u$, an up-down sequence leads to the same price as a down-up sequence. The resulting lattice is illustrated in Fig. 2.10. Clearly, the size of a recombining lattice grows linearly in time, whereas a non-recombining tree explodes exponentially. If we assume that the parameters u, d, and p apply to the whole lattice, how can we choose them? Since we have three degrees of freedom, we may match two moments of the random variable $S_{\delta t}$. Let

$$\nu = \mathrm{E}[S_{\delta t}], \qquad \xi^2 = \mathrm{Var}(S_{\delta t}).$$

We will see in Chapter 3 that if we assume a geometric Brownian motion, the resulting distribution is lognormal, and its expected value and variance are related to the drift and the volatility of the stochastic process; for now, let us just

---

[10]Needless to say, this kind of lattice has nothing to do with lattices discussed in Section 2.4.3.

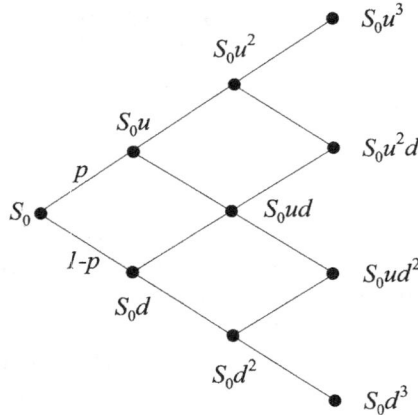

FIGURE 2.10 **A recombining binomial lattice.**

take the above values as given. The moment matching conditions are

$$puS_0 + (1-p)dS_0 = \nu,$$
$$pu^2 S_0^2 + (1-p)d^2 S_0^2 = \sigma^2 + \nu^2.$$

We are left with a residual degree of freedom that can be used to set additional conditions. One possible choice is $p = 0.5$; another widely used choice is $u = 1/d$. In general, we find a system of possibly nonlinear equations that can be solved exactly or approximately, depending on its complexity. For instance, if $S(t)$ follows a geometric Brownian motion with drift μ and volatility σ, a common calibration of the lattice yields[11]

$$u = e^{\sigma\sqrt{\delta t}}, \quad d = e^{-\sigma\sqrt{\delta t}}, \quad p = \frac{e^{\mu\,\delta t} - d}{u - d}.$$

For small values of δt, this discretized lattice does approximate very well the continuous-time, continuous-state stochastic process, allowing for efficient pricing of options, including those featuring early exercise opportunities. However, the approach can only be applied to low-dimensional problems. A subtler issue is path dependency: Some options, like Asian-style options, have a payoff depending on the whole path of the underlying variables. The recombining binomial lattice has a Markovian feature preventing its direct use in such a case. For high-dimensional and path-dependent options, Monte Carlo is the method of choice.

---

[11] See, e.g., [1] for a derivation and an illustration of possible computational advantages of this choice.

2.5.2   SCENARIO GENERATION IN STOCHASTIC PROGRAMMING

We have introduced stochastic programming models in Section 1.4.2. One of the main issues when formulating such problems is scenario generation. Given, e.g., a multivariate normal variable $\mathbf{Y}$, modeling the random data, with expected value $\boldsymbol{\mu}$ and covariance matrix $\boldsymbol{\Sigma}$, we might use Monte Carlo sampling to generate a discrete set of equiprobable scenarios. As an alternative, we may look for a set of scenarios matching moments and, possibly, a few more desirable properties. Since we are dealing with a multivariate normal, we know that for each marginal Y_i skewness should be zero and kurtosis should be 3:

$$\gamma_i = \mathrm{E}\left[\frac{(Y_i - \mu_i)^3}{\sigma_i^3}\right] = 0, \qquad \kappa_i = \mathrm{E}\left[\frac{(Y_i - \mu_i)^4}{\sigma_i^4}\right] = 3.$$

Say that we want to generate a scenario fan (the single stage of a tree) of size S and that each realization has probability $1/S$, for the sake of simplicity. Let us denote by y_i^s the value of the random variable Y_i in scenario s. Then, we should have

$$\frac{1}{S}\sum_s y_i^s \approx \mu_i, \quad \forall i, \qquad \frac{1}{S}\sum_s (y_i^s - \mu_i)(y_j^s - \mu_j) \approx \sigma_{ij}, \quad \forall i,j,$$

$$\frac{1}{S}\sum_s \frac{(y_i^s - \mu_i)^3}{\sigma_i^3} \approx 0, \quad \forall i, \qquad \frac{1}{S}\sum_s \frac{(y_i^s - \mu_i)^4}{\sigma_i^4} \approx 3, \quad \forall i.$$

Approximate moment matching is obtained by solving the problem:

$$\min \quad w_1 \sum_i \left[\frac{1}{S}\sum_s y_i^s - \mu_i\right]^2 + w_2 \sum_{i,j} \left[\frac{1}{S}\sum_s (y_i^s - \mu_i)(y_j^s - \mu_j) - \sigma_{ij}\right]^2$$

$$+ w_3 \sum_i \left[\frac{1}{S}\sum_s \left(\frac{y_i^s - \mu_i}{\sigma_i}\right)^3\right]^2 + w_4 \sum_i \left[\frac{1}{S}\sum_s \left(\frac{y_i^s - \mu_i}{\sigma_i}\right)^4 - 3\right]^2.$$

The objective function includes four weights w_k, which may be used to fine tune performance. Again, this is a deterministic approach that should be compared against random sampling. We will consider scenario generation for stochastic programming models in Section 10.5.3.

2.6   Numerical integration in R

Available functions to compute integrals in R are, e.g., `integrate` for one-dimensional integrals and `adaptIntegrate` for multidimensional integrals on hypercubes. Consider the integral

$$I = \int_0^{2\pi} e^{-x} \sin(10x)\, dx.$$

Integration by parts yields the exact result, which we may use to check the accuracy of numerical solutions:

$$I = -\frac{1}{101}e^{-x}\left[\sin(10x) + 10\cos(10x)\right]\Big|_0^{2\pi} \approx 0.0988.$$

The following R snapshot illustrates how to do the job:

```
> f <- function(x) exp(-x)*sin(10*x)
> integrate(f, lower=0, upper=2*pi)
0.09882501 with absolute error < 5.9e-08
```

It is worth noting that, because of its internal working, integrate needs a vectorized function, i.e., a function that can receive a vector input and return the corresponding vector output. The following example, involving a constant function, illustrates the issue:

```
> g <- function(x) return(1)
> integrate(g,0,10)
Error in integrate(g, 0, 10) :
  evaluation of function gave a result of wrong length
```

One possibility to overcome the obstacle is to vectorize the function manually:

```
> g1 <- function(x) rep(1,length(x))
> g(1:5)
[1] 1
> g1(1:5)
[1] 1 1 1 1 1
> integrate(g1,0,10)
10 with absolute error < 1.1e-13
```

Alternatively, we may use the higher level function Vectorize:

```
> g2 <- Vectorize(g)
> g2(1:5)
[1] 1 1 1 1 1
> integrate(g2,0,10)
10 with absolute error < 1.1e-13
```

The integrate function can deal with integrals on unbounded domains, and it is always suggested to take advantage of that feature, rather than using very small or very large values for lower or upper integration limits, respectively. Let us try integrating the density of a standard normal distribution to see the point:

```
> integrate(dnorm,0,2)
0.4772499 with absolute error < 5.3e-15
> integrate(dnorm,0,2000)
0.5 with absolute error < 4.4e-06
> integrate(dnorm,0,20000000)
```

```
0 with absolute error < 0
> integrate(dnorm,0,Inf)
0.5 with absolute error < 4.7e-05
```

To deal with multidimensional integrals, we may use `adaptIntegrate` from the package `cubature`. As an example, let us consider the bidimensional integral:

$$\int_0^1 \int_0^1 e^{-xy} \left(\sin 6\pi x + \cos 8\pi y\right) \, dx \, dy.$$

```
> library(cubature)
> h <- function(x) exp(-x[1]*x[2])*(sin(6*pi*x[1])+
  cos(8*pi*x[2]))
> adaptIntegrate(h,lower=c(0,0),upper=c(1,1))$integral
[1] 0.01986377
```

For further reading

- Numerical integration methods are covered in any textbook on numerical analysis. One such general reference is [7]; for numerical methods in economics see [6].

- A more specific reference for numerical integration, with emphasis on Gaussian quadrature, is [2].

- Computational approaches for Gaussian quadrature are discussed, e.g., in [10], where approaches based on finding roots of orthogonal polynomials are illustrated, and in [4], which deals with approaches based on matrix eigenvalues. A related interesting reading is [3], where the role of matrix computations is emphasized, as well as the relationship with moments, which is also interesting from a probabilistic viewpoint.

- A deep treatment on lattice methods for numerical integration is [11]. You may also see [9] for a treatment related to Monte Carlo and quasi–Monte Carlo sampling, including randomized versions of these methods.

- Scenario generation for stochastic programming is discussed, e.g., in [5] and [8].

References

1 P. Brandimarte. *Numerical Methods in Finance and Economics: A MATLAB-Based Introduction* (2nd ed.). Wiley, Hoboken, NJ, 2006.

2 P.J. Davis and P. Rabinowitz. *Methods of Numerical Integration* (2nd ed.). Academic Press, Orlando, FL, 1984.

3 G.H. Golub and G. Meurant. *Matrices, Moments, and Quadrature with Applications*. Princeton University Press, Princeton, NJ, 2010.

4 G.H. Golub and J.H. Welsch. Calculation of Gaussian quadrature rules. *Mathematics of Computation*, 23:221–230, 1969.

5 K. Hoyland and S.W. Wallace. Generating scenario trees for multistage decision problems. *Management Science*, 47:296–307, 2001.

6 K.L. Judd. *Numerical Methods in Economics*. MIT Press, Cambridge, MA, 1998.

7 D. Kincaid and W. Cheney. *Numerical Analysis: Mathematics of Scientific Computing*. Brooks/Cole Publishing Company, Pacific Grove, CA, 1991.

8 A.J. King and S.W. Wallace. *Modeling with Stochastic Programming*. Springer, Berlin, 2012.

9 C. Lemieux. *Monte Carlo and Quasi–Monte Carlo Sampling*. Springer, New York, 2009.

10 W.H. Press, S.A. Teukolsky, W.T. Vetterling, and B.P. Flannery. *Numerical Recipes in C* (2nd ed.). Cambridge University Press, Cambridge, 1992.

11 I.H. Sloan and S. Joe. *Lattice Methods for Multiple Integration*. Oxford University Press, New York, 1994.

## Part Two

Input Analysis: Modeling and Estimation

## Chapter Three

Stochastic Modeling in Finance and Economics

This chapter is included for the sake of readers who are not familiar with econometrics and the modeling techniques that are commonly used in finance and financial economics. Needless to say, the following treatment is not meant as a substitute of one of the several excellent books on the subject. Our limited aims are:

1. To offer a concise introduction for readers with a background, e.g., in mathematics or engineering.

2. To provide readers with a short refresher, and possibly an introduction to some specific topics with which they might be not quite familiar.

3. At the very least, we use the subject of this chapter as an excuse to introduce some useful R functions to deal with probability distributions and some multivariate analysis methods.

Here we only deal with modeling; issues related to estimation are deferred to Chapter 4, which has similar objectives and limitations. In modeling, one typically works with parameters, expected values, variances, etc. In estimation, one typically works with their sample counterparts, like sample mean and sample variance. However, since it is often useful to illustrate ideas by simple but concrete examples, we will occasionally use sample counterparts of probabilistic concepts. In doing so, we will rely on readers' intuition and, possibly, some background in elementary inferential statistics, as well as the availability of easy to use R functions; more advanced concepts and issues are dealt with in the next chapter.

Within this framework, the choice of topics is admittedly open to criticism. We cover some rather basic issues related to probability distributions, together with some more advanced ones, such as copulas and stochastic differential equations. However, the coverage of some standard tools in econometrics, such as multiple linear regression models, will leave much to be desired. Our choice is deliberate and related to what one needs in order to appreciate Monte Carlo techniques.

In Section 3.1 we use elementary examples to motivate our study and to further illustrate the differences between static, discrete-time, and continuous-time models. Even these deceptively simple examples raise a number of issues that may be addressed by rather sophisticated tools. Section 3.2 deals with rather basic topics, as we outline a few essential (univariate) probability distributions that are commonly used in finance. Probably, many readers are quite familiar with the content of this section, which may be skipped and referred back when needed. In order to make the section as useful as possible, we take advantage of it by introducing some R functions to deal with probability distributions, including random sample generation; the technicalities involved in the generation of pseudorandom samples from these distributions is deferred to Chapter 5. We move on to deal with multivariate distributions and dependence in Sections 3.3 and 3.4. In the former section we outline multivariate distributions, including a brief review of standard concepts related to covariance and Pearson's correlation. We also cover multivariate extensions of normal and t distributions, but the main aim of the section is to point out the limitations of these standard approaches, paving the way for more advanced ideas revolving around copula theory, which is the subject of Section 3.4. This section is a bit more advanced and also covers alternative correlation measures, such as Spearman's rho and Kendall's tau; we also outline such concepts as tail dependence and independence, which are relevant for risk management. Section 3.5 is a bit of an interlude, where we cover linear regression models within a probabilistic framework. Usually, linear regression is associated with the statistical estimation of linear models;[1] here we insist on the probabilistic view of regression as an orthogonal projection, laying down the foundations for variance reduction by control variates[2] and model complexity reduction by factor models. Regression models are related and contrasted with time series models, which are described in Section 3.6, within the framework of discrete-time models. We should note that a great deal of work has been carried out on representing, estimating, and simulating time series models in R, which also includes a specific time series object. Continuous-time models, based on stochastic differential equations, are introduced in Section 3.7. These models are most useful in financial engineering applications, but they require some background in stochastic calculus and stochastic integration. Here we introduce the basic concepts, most notably Itô's lemma, without paying too much attention to mathematical rigor. In Section 3.8 we outline two approaches that can be used to streamline a model, namely, principal component analysis and factor models. These multivariate analysis tools may be useful from both a model estimation and a simulation perspective, since they reduce data and computational requirements. Finally, in Section 3.9 we rely on Itô's lemma to develop the powerful tools of risk-neutral option pricing. This section may be obviously skipped by the familiar reader, but also by readers who are not interested in option pricing.

---

[1] See Section 4.4.
[2] See Section 8.3.

3.1 Introductory examples

In Chapter 1 we considered the difference between Monte Carlo sampling and simulation. In the former case, we are essentially approximating an integral, which is related to an expectation; in the latter case, we have to simulate dynamics over time. We have also seen that, in principle, any dynamic simulation can be considered as the estimation of the integral of a possibly quite complicated function. Hence, the line between sampling and simulation is not so sharp at all, from a Monte Carlo viewpoint. From a modeling viewpoint, however, the issues involved may be quite different. In this section we outline three related financial modeling problems:

1. A static model for portfolio optimization
2. A discrete-time consumption/saving model
3. A continuous-time model to describe asset price dynamics

The best known static portfolio optimization model is certainly the mean–variance model of Markowitz. In the basic version of this model, time is disregarded and issues related to rebalancing are neglected. To build the model, we need to characterize the joint distribution of asset returns over the investment time horizon, which involves subtle issues, particularly from a risk management perspective.

More realistic models, possibly involving asset and liabilities, transaction costs, etc., require the introduction of time. We may accomplish the task within a discrete- or a continuous-time framework. The former approach typically requires the machinery of time series analysis, whereas the latter one relies on stochastic calculus concepts. When time enters the stage, we may face the task of representing intertemporal dependence. Actually, if we believe the efficient market hypothesis (EMH), we may avoid the task. EMH can be stated in different, more or less stringent forms. The bottom line is that it is impossible to beat the market by stock picking or buy/sell timing. From our point of view, the essential implication is that there is no intertemporal dependence in returns over different time periods. On the contrary, if we have a different view or we are modeling something else and the EMH is not relevant, we are faced with the task of modeling some form of dependence over time.

In econometrics, to categorize the issues a modeler faces when dealing with dependence, the following model classification is used:

- We have a cross-sectional model, when we are representing different variables at the same time.
- We have a longitudinal model, when we observe a single variable at several time instants.
- Finally, we have a panel model, when both dimensions are relevant.

One of the main tasks of this chapter is to illustrate a few of these models, pointing out the related issues.

3.1.1 SINGLE-PERIOD PORTFOLIO OPTIMIZATION AND MODELING RETURNS

To build a single-period portfolio optimization problem, we need the following ingredients:

- A universe of n assets, indexed by $i = 1, \ldots, n$.
- An initial wealth W_0 that we have to allocate among the assets, taking into account some representation of investor's preferences in terms of risk aversion.
- A model to represent the uncertainty in asset prices or returns.

In principle, the microeconomics of uncertainty offers a way to represent investor's preferences, in the form of a utility function $u(\cdot)$ mapping a monetary amount into a real number. Utility theory posits that we should maximize the expected utility of wealth W_T, a random variable, at the end of the holding period T. As we did in Example 1.1, we denote the initial asset price by P_i^0 and the amount of shares of asset i held in the portfolio by h_i; for the sake of simplicity, here we assume that the number of shares is large enough to be approximated by a continuous variable and avoid integrality restrictions. If we neglect transaction costs and tax issues, the problem can be written as

$$
\max \quad \mathrm{E}\left[u\left(\sum_{i=1}^{n} h_i P_i^T \right) \right]
$$
$$
\text{s.t.} \quad \sum_{i=1}^{n} h_i P_i^0 = W_0, \tag{3.1}
$$

where P_i^T is the asset price at the end of the holding period. This models looks easy enough, but there are two serious difficulties:

- How can we find a suitable utility function?
- How can we specify a joint distribution of prices and then compute the expectation?

Computing the expectation is a technical problem, which may be tackled by the methods described in the rest of this book. However, defining a utility function is a much more serious issue. In fact, despite their popularity in most economic literature, utility functions are not easily applicable to real-life financial optimization. On the one hand, even a sophisticated investor would find it hard to assess her utility function; on the other hand, which utility function should a professional fund manager use, when reporting to her many clients? Should she consider her own attitude toward risk or that of her wealthiest client? A common approach is to resort to a risk measure, related to uncertainty in future wealth. We will have more to say about risk measures later, but since they should capture uncertainty, a seemingly natural measure is standard deviation. Furthermore, it may also be convenient to get rid of wealth W_0 and deal with a sort of adimensional model in terms of portfolio weights, i.e., fraction of total wealth allocated to each asset, and asset returns.[3] To be more specific, if the assets we are considering are stock shares, we should also consider any dividend that is paid within the planning horizon. If asset i pays a dividend D_i during the holding period, its (rate of) return can be written as

$$R_i = \frac{P_i^T + D_i - P_i^0}{P_i^0}.$$

Note that the return can be negative, but it cannot be less than -1, since stock shares are limited liability assets and the worst price that we may observe in the future is $P_i^T = 0$. To be precise, this definition does not account for the time value of money; when the dividend is paid exactly does make a difference if the time period is long enough and we consider the possibility of reinvesting the dividend. For the sake of simplicity, let us neglect this issue.

Now, what about the joint distribution of returns? If we assume that the investor only cares about expected value and standard deviation of portfolio return, we may streamline our task considerably. Let:

- $\mu \in \mathbb{R}^n$ be the vector of expected returns, with components $\mu_i = \mathrm{E}[R_i]$.
- $\Sigma \in \mathbb{R}^{n,n}$ be the return covariance matrix,[4] collecting entries $\sigma_{ij} = \mathrm{Cov}(R_i, R_j)$.

If we denote the wealth fraction allocated to asset i by w_i, so that $\sum_i w_i = 1$, we immediately find that the portfolio return R_p is

$$R_p = \sum_{i=1}^n w_i R_i,$$

with expected value

$$\mathrm{E}[R_p] = \sum_{i=1}^n w_i \mu_i = \mathbf{w}^\mathsf{T} \boldsymbol{\mu},$$

where $\mathbf{w}$ is the column vector collecting portfolio weights, which is transposed by the operator $^\mathsf{T}$. The variance of portfolio return is given by

$$\mathrm{Var}(R_p) = \sum_{i=1}^n \sum_{j=1}^n w_i \sigma_{ij} w_j = \mathbf{w}^\mathsf{T} \Sigma \mathbf{w}.$$

In the classical Markowitz portfolio optimization model, a mean–variance efficient portfolio is found by minimizing risk subject to a constraint on expected portfolio return:

$$\begin{aligned}
\min \quad & \mathbf{w}^\mathsf{T} \Sigma \mathbf{w} \\
\text{s.t.} \quad & \mathbf{w}^\mathsf{T} \boldsymbol{\mu} \geq \alpha_T, \\
& \mathbf{w}^\mathsf{T} \mathbf{e} = 1,
\end{aligned}$$

---

[3]What we use here as return is referred to as *rate* of return in some treatments, but we sometimes prefer to streamline terminology a bit.

[4]For the unfamiliar reader, we formally introduce covariance later in Section 3.3.

where α_T is a minimal target for expected return and $\mathbf{e} = [1, 1, \ldots, 1]^\mathsf{T}$; non-negativity on portfolio weights may be added if short sales are forbidden. Note that we minimize variance, rather than standard deviation, for the sake of computational convenience. Doing so, the solution of the optimization problem does not change, but we have to solve a (convex) quadratic programming problem, for which extremely efficient algorithms are widely available.

In fact this new model looks remarkably simple, and there is no need for Monte Carlo methods for its solution. Furthermore, there is no need to specify a full-fledged joint return distribution, as first- and second-order moments are all we need. However, severe difficulties have just been swept under the rug. To begin with, how can we estimate the problem data? The covariance matrix consists of many entries, and there is no hope to estimate them all in a reliable way. To see why, assume that we have to deal with $n = 1000$ assets. Then, the covariance matrix consists of one million entries. Actually, given symmetry, "only"

$$n^2 + \frac{n(n-1)}{2} = 500{,}500$$

entries need be specified. This is a formidable task, as we would need a very long history of returns to obtain reliable estimates. Unfortunately, even if we collect many years of data, assuming that they are available for all of the assets that we are considering for inclusion in our portfolio, few of them are actually relevant since market conditions have changed over the years. This is why we need to resort to some data reduction strategies, as described in Section 3.8.2.

Even if we assume that we have perfect knowledge of the involved parameters, tough issues are to be considered:

- Measuring risk by variance or standard deviation is appropriate for symmetric distributions like the multivariate normal. However, returns are not symmetric, especially when financial derivatives are involved.

- We may need alternative risk measures, such as value-at-risk or conditional value-at-risk, which call for a fuller picture of the joint distribution; see Chapter 13. The behavior on the bad tails of the distribution, where we lose money, is relevant, but capturing this is not easy.

- Covariances and correlations may fail to capture true dependency. We stress this point in Section 3.4, where we motivate the introduction of copulas. In particular, can we believe that correlations stay the same in good and bad times, when extreme events occur?

- Last but not least, standard statistical methods are by their very nature backward looking. However, in finance we should look forward.[5] For instance, assume that yesterday was very lucky for a certain stock share that earned a stellar 150% daily return, which may have a sensible impact

---

[5]To put it bluntly, relying on past returns is akin to driving by just looking at the rear-view mirror.

on return statistics. Is this a good reason to invest all of our wealth in that stock?

We see that even specifying a single-period, static model is not trivial at all. No approach based on Monte Carlo sampling will yield useful results, if the underlying uncertainty model is not credible.

3.1.2 CONSUMPTION–SAVING WITH UNCERTAIN LABOR INCOME

In many applications a static and single-period model is inappropriate, as we need to represent a sequence of decisions to be made in the face of possibly complex dynamics of stochastic risk factors. Here we illustrate a discrete-time framework for a stylized model extending the consumption–saving problem of Section 1.4.3. The task is to choose the fraction of current wealth that is consumed, and how the remaining amount saved is allocated between a risky and a risk-free asset.[6] For strategic decision problems like this one, an appropriate time bucket could be one year; in other cases much higher frequencies are needed. Whatever time period we choose, in any discrete-time model it is essential to specify the sequence of events very clearly, as this defines the information that is actually available when making a decision. We consider *time instants* and *time buckets* (or periods):

- The system state is observed and decisions are made only at discrete time instants $t = 0, 1, 2, \ldots$.
- However, the state evolves during time buckets. To avoid excessive notation, we associate time bucket $t = 1, 2, \ldots$ with the pair of time instants $(t - 1, t)$. It is important to clarify if we refer to something happening at the beginning, at the end, or during a time bucket.

The rules of the game in our problem are:

- At time instant $t = 0, 1, 2, \ldots, T$, i.e., at the beginning of the corresponding time bucket, the agent owns a current wealth denoted by W_t, resulting from the previous saving and investment decisions.
- Labor income L_t is collected; actually, this is income earned during the previous time bucket $t - 1$.
- The total cash on hand $W_t + L_t$ is split between saving S_t and consumption C_t during the next time bucket.
- The saved amount is allocated between a risk-free asset, with deterministic rate of return r_f, and a risky asset, with random rate of return R_{t+1}; note how the subscript $t + 1$ emphasizes the fact that this return will be known only at the next time instant $t + 1$, or, if you prefer, at the *end* of

---

[6]This is a streamlined version of the model described in [4, Chapter 7]. See also Chapter 10 to see how such problems may be tackled by stochastic dynamic programming.

the next time bucket, whereas the allocation decision must be made now, at the *beginning* of the time bucket; let us denote by the fraction of saving that is allocated to the risky asset by $\alpha_t \in [0, 1]$.

Note that we are considering two sources of uncertainty here:

1. The uncertain return of the risky asset
2. The uncertainty in the labor income stream

The first source of uncertainty can be represented using the concepts from the previous section, if we do not want to consider any predictability in asset returns. However, it is likely that the labor income L_t is not quite independent from L_{t-1}; therefore, we must choose some way to represent both predictable and unpredictable (shock) components, as we show below.

Given a decision α_t, the rate of return of the investment portfolio over the next time bucket is

$$\alpha_t R_{t+1} + (1 - \alpha_t)r_f = \alpha_t(R_{t+1} - r_f) + r_f,$$

where $R_{t+1} - r_f$ is typically referred to as *excess return*. Therefore, wealth at time instant $t + 1$ is

$$W_{t+1} = \left[1 + \alpha_t(R_{t+1} - r_f) + r_f\right](W_t + L_t - C_t). \tag{3.2}$$

We emphasize again that the risky asset return R_{t+1} is not known when making the asset allocation decision, whereas the risk-free rate is known. Actually the risk-free return should also be modeled as a stochastic process, as the level of interest rates does change over time because of changing economic conditions and the corresponding monetary policy of central banks. Hence, we should denote this rate as r_{ft}. The main difference with respect to the risky return R_{t+1} is that, at the beginning of a time bucket, we do know the risk-free rate for that time bucket; however, we do not know the interest rates for the subsequent time buckets. So, r_{ft} is a predictable stochastic process. For the sake of simplicity, let us assume a constant interest rate r_f. We might also wish to model correlations between the two returns. After all, we know that when central banks change the level of interest rates, stock exchanges are indeed affected.

To model labor income, we really must be more specific and decompose L_t in components. A model proposed, e.g., in [4, Chapter 7], is

$$L_t = f(t, Z_t) + \nu_t + \epsilon_t, \tag{3.3}$$

where

- $f(t, Z_t)$ is a deterministic component, depending on time/age and other individual characteristics like the level of education (the more you invest in education, the more, hopefully, your wage should increase over time).
- ν_t is a random *permanent* shock.
- ϵ_t is a random *transitory* shock.

The difference between the two random components is that the transitory random term should have an impact in one year, and then it should leave no effect, whereas permanent shocks have to be cumulated, or integrated, if you prefer, over time. This can be represented as a random walk:

$$\nu_t = \nu_{t-1} + \eta_t,$$

where, for instance, $\eta_t \sim \mathrm{N}(0, \sigma_\eta^2)$, i.e., η_t is a normal random variable with variance σ_η^2 and expected value zero. By a similar token, one could assume $\epsilon_t \sim \mathrm{N}(0, \sigma_\epsilon^2)$. Note that the expected value of a shock is zero, since any predictable component should be included in the deterministic component. In simple models, η_t and ϵ_t are assumed independent over time.

Now, we should see a problem with an additive decomposition like the one in Eq. (3.3). If shocks are normally distributed, in principle, income could get negative. Furthermore, using the same volatilities (standard deviations) for all levels of income is questionable. One way to overcome these issues is to resort to a multiplicative decomposition. We might just reinterpret equation Eq. (3.3) in terms of log-income, defined as $l_t \equiv \log L_t$.[7] In other words, we posit

$$L_t = e^{l_t} = \exp\left\{f(t, Z_t) + \nu_t + \epsilon_t\right\}. \tag{3.4}$$

Since we take an exponential of a normal, a negative shock will be transformed into a multiplicative shock less than 1, i.e., a reduction in income. We will also see, in Section 3.2.3, that when we take the exponential of a normal random variable, we obtain a lognormal random variable.

Finally, if we associate a utility function $u(\cdot)$ with consumption, the overall objective could be

$$\max \mathrm{E}\left[\sum_{t=0}^{T} \beta^t u(C_t)\right],$$

where $\beta \in (0, 1)$ is a subjective discount factor. Thus, we face a dynamic, stochastic optimization problem, which cannot be solved analytically in general; Monte Carlo and related methods play a key role in the solution of these models, as well as in the simulation of decisions over time to analyze the resulting policy. In Section 3.6 we will analyze a few more discrete-time models that can be used to represent dependency over time, changes in volatility, etc.

3.1.3 CONTINUOUS-TIME MODELS FOR ASSET PRICES AND INTEREST RATES

Time discretization may look like an approximation of an intrinsically continuous time. This need not be the case, actually: If decisions are made periodically, then a discretized time bucket corresponding to decision periods is all we need. In other cases, higher frequencies must be accounted for. We

---

[7]Throughout this book, we use log rather than ln, but we always mean *natural* logarithms.

may then take smaller and smaller time periods in order to approximate continuous time. Needless to say, this increases the computational burden considerably and a trade-off must be assessed between modeling needs and computational tractability. A possibly surprising fact is that if we take the process to the limit and consider continuous-time models, we may actually ease many issues and come up with tractable models. Furthermore, when we consider trading on exchanges and the actual flow of information, we have to accept the fact that a more faithful representation does require continuous time. This can help in building a theoretically sound framework, even if numerical methods, most notably Monte Carlo methods, are to be employed when models cannot be solved analytically. Indeed, continuous-time models rely on differential equations, which may look tougher than discrete-time difference equations, especially since we have to cope with *stochastic* differential equations. Stochastic models are in fact needed to represent uncertainty in stock prices, interest rates, foreign exchange rates, etc. Nevertheless, it seems wise to introduce the concept in the simplest deterministic case of some practical relevance, the investment in a safe bank account. We follow here the same path that we have followed in Section 1.3.2, providing some more motivation and details.

Consider a risk-free bank account. If we invest an initial amount $B(0)$ at time $t = 0$, the annual interest rate is r_f, and interest is earned once per year, our wealth after one year will be

$$B(1) = B(0)(1 + r_f). \tag{3.5}$$

However, if semiannual interest $r_f/2$ is paid twice a year and is immediately reinvested, we get

$$B(1) = B(0) \left(1 + \frac{r_f}{2}\right)^2.$$

This new amount, if the rate r_f is the same, is a bit larger than the value in Eq. (3.5), since we earn interest on interest. In such a case, we speak of semiannual compounding. More generally, if interest is paid m times per year, we have

$$B(1) = B(0) \left(1 + \frac{r_f}{m}\right)^m,$$

and it is easy to see that if m goes to infinity, we end up with the limit

$$B(1) = B(0)e^{r_f}. \tag{3.6}$$

Such a continuously compounded interest does not really exist in financial markets, even though daily compounding is essentially the same as continuous-time compounding. Nevertheless, it simplifies considerably several financial calculations and it paves the way for continuous-time models.

It is useful to describe all of the above using a dynamic model. So, let us assume that interest is continuously earned and instantaneously reinvested. Over a short time interval $(t, t + \delta t)$, we have the following change in the bank account:

$$\delta B(t) \equiv B(t + \delta t) - B(t) = rB(t)\,\delta t.$$

If we let $\delta t \to 0$, we obtain the differential equation

$$dB(t) = rB(t)\,dt. \tag{3.7}$$

Many readers may be more familiar with the form

$$\frac{dB}{dt} = rB. \tag{3.8}$$

There are a couple of reasons why we might prefer the differential form of Eq. (3.7). First, a slight manipulation shows a way to solve it. If we rewrite the equation as

$$\frac{dB}{B} = r\,dt,$$

and we recall the form of the derivative of a composite function, we may recognize that a logarithm is involved:

$$\frac{dB}{B} = d\log B = r\,dt.$$

Now we may integrate the differentials over the time interval $[0, T]$:

$$\int_0^T d\log B = r \int_0^T dt \quad \Rightarrow \quad \log B(T) - \log B(0) = rT, \tag{3.9}$$

which can be easily rewritten as

$$B(T) = B(0)e^{rT},$$

which is consistent with Eq. (3.6).

There is another reason why the form (3.7) is preferred to (3.8): It can be extended to cope with stochastic factors. Suppose that we are investing in a risky asset, rather than a risk-free bank account. We need a model for the asset price, accounting for both drift (related to expected return) and volatility (related to random variability in the price). The drift component is essentially the same as in Eq. (3.7), with r_f replaced by some coefficient, say, μ. However, we need to add a suitable source of randomness. A common building block is the Wiener process, denoted by $W(t)$. This is a continuous-time stochastic process, which essentially means that for every value of the continuous parameter t, we have a random variable $W(t)$: In general we may define a stochastic process as a time-indexed collection of random variables $X(t)$ or X_t, where time can be continuous or discrete. To include this noise term in the equation, we should first consider increments in the stochastic process over a small time interval $(t, t + \delta t)$,

$$\delta W(t) = W(t + \delta t) - W(t).$$

The Wiener process is a Gaussian process, in the sense that $\delta W(t) \sim \mathsf{N}(0, \delta t)$.[8] Please note that the variance of $\delta W(t)$ is δt, and thus its standard deviation is

---

[8] Actually, the precise definition of Gaussian processes does not only involve normal marginals but also jointly normal distributions.

$\sqrt{\delta t}$. Then we take the limit for $\delta t \to 0$, which yields the infinitesimal increment $dW(t)$. Adding this term, multiplied by a volatility factor σ, to Eq. (3.7) yields the following stochastic differential equation:

$$dS(t) = \mu S(t)\, dt + \sigma S(t)\, dW(t). \tag{3.10}$$

Later, we will see that this defines an important process called geometric Brownian motion. Actually, we have skipped over some challenging issues, since "taking the limit" for $\delta t \to 0$ is quite delicate when random variables are involved, as one must refer to some stochastic convergence concept. Furthermore, the Wiener process is a very "irregular" and noisy process, with sample paths that are not differentiable. We will deal with sample path generation later, but you may have a peek at Fig. 3.39 to see what we mean. Hence, taking the usual derivative $dW(t)/dt$ is not possible, which justifies the use of this form. In Section 3.7 we will outline the fundamental tools of stochastic calculus, and we will learn that Eq. (3.10) is actually a shorthand for an integral equation:

$$\int_0^T dS(t) = S(T) - S(0) = \mu \int_0^T S(t)\, dt + \sigma \int_0^T S(t)\, dW(t). \tag{3.11}$$

This looks superficially like the integral in Eq. (3.9), but it relies on the concept of Itô stochastic integral, to be discussed in Section 3.7. The differential form, however, is more intuitive and paves the way for the development of several interesting processes.

For instance, let us consider a model for stochastic interest rates. Can we model an interest rate $r(t)$ by geometric Brownian motion? The answer is no, and to see why, let us wonder about the solution of Eq. (3.10) when $\sigma = 0$. When there is no volatility, the equation boils down to Eq. (3.7), whose solution is an increasing exponential. If there is a volatility component, intuition suggests that the solution is an exponentially increasing trend, with some superimposed noise. Financial common sense suggests that interest rates do not grow without bound, and tend to oscillate around a long-term average level $\bar{r}$. To model such a mean reversion, we may use the following equation:

$$dr(t) = \gamma \left(\bar{r} - r(t) \right) + \sigma W(t), \tag{3.12}$$

which is an example of the Ornstein–Uhlenbeck process, to be discussed later in Section 3.7.5. The intuition is that the drift can be positive or negative, pulling back $r(t)$ toward its long-term average level. It is worth mentioning that, in practice, continuous-time models are the rule in financial engineering.

3.2   Some common probability distributions

Most readers are quite familiar with the distributions that we outline in the following and can safely skip the section, possibly referring back when needed. The treatment is limited and just meant to be a refresher, but in order to make the section as useful as possible to the widest readership, we also take the review as an excuse to introduce R functions to deal with probability distributions; we will also use functions to produce some simple plots, and R scripts to illustrate both R programming and some specific properties of distributions. As a general rule, four functions are associated with each standard distribution in R. Given a distribution `distr`, R provides us with:

1. `pdistr` to compute the cumulative distribution function (CDF), defined as $F_X(x) \equiv \mathrm{P}\{X \le x\}$
2. `ddistr` to compute the probability density function (PDF) $f_X(x)$, for a continuous random variable, or the probability mass function (PMF), i.e., $p_X(x) \equiv \mathrm{P}\{X = x\}$, which makes sense only for a discrete random variable
3. `qdistr` to compute a quantile, i.e., to invert the CDF
4. `rdistr` to generate a random sample

Note how the prefixes `p`, `d`, `q`, and `r` characterize each function. The `p` is somewhat confusing, as "probability mass" could be associated with discrete distributions, whereas in R the `p` only refers to cumulative probabilities; the choice of the prefix `d` reminds a PDF and could possibly reflect some bias toward continuous distributions.

We may illustrate these four functions with reference to a continuous uniform variable on the interval $[a, b]$, which features a constant density function on its support:

$$f_X(x) \equiv \begin{cases} 1/(b-a), & \text{if } x \in [a, b], \\ 0, & \text{otherwise.} \end{cases}$$

A commonly used notation to state that a random variable X has this distribution is $X \sim \mathsf{U}[a, b]$, or $X \sim \mathsf{U}(a, b)$. We also recall that

$$\mathrm{E}[X] = \frac{b+a}{2}, \qquad \mathrm{Var}(X) = \frac{(b-a)^2}{12}.$$

The following R snapshots illustrate the related R functions:

```
> a=5
> b=15
> dunif(8,min=a,max=b)
[1] 0.1
> punif(8,min=a,max=b)
[1] 0.3
> dunif(3,min=a,max=b)
[1] 0
> punif(3,min=a,max=b)
[1] 0
> dunif(18,min=a,max=b)
[1] 0
> punif(18,min=a,max=b)
[1] 1
```

Note how `qunif` and `punif` are inverse of each other:

```
> punif(9,min=a,max=b)
[1] 0.4
> qunif(0.4,min=a,max=b)
[1] 9
```

The following snapshot will probably give different results on your computer, as random samples depend on the state of the (pseudo)random number generator:

```
> runif(5,min=a,max=b)
[1]   8.434857  5.190168 10.679197  5.220115 10.036124
```

This is why, throughout this book, we will use the command `set.seed(55555)` to set the state of the generator to a specific state, thus allowing replication of results. We defer to Chapter 5 the technicalities involved in the generation of pseudorandom numbers and variates. By the way, while the uniform distribution does not look too interesting in itself, it is fundamental in Monte Carlo simulation as all distributions are sampled by suitable transformations of $U(0,1)$ variables.

In the following sections we illustrate a few common *univariate* distributions, both discrete and continuous, without any claim of exhaustiveness. Some of the distributions that we describe here have a multivariate counterpart, to be discussed in Section 3.3.

3.2.1 BERNOULLI, BINOMIAL, AND GEOMETRIC VARIABLES

The binomial distribution is a discrete distribution built on top of a Bernoulli random variable. The mechanism behind a Bernoulli variable is carrying out a random experiment, which can result in either a success, with probability p, or a failure, with probability $1 - p$. If we assign to the random variable X the value 1 in case of a success, and 0 in case of a failure, we obtain the PMF:

$$p_X(0) \equiv P\{X = 0\} = 1 - p,$$
$$p_X(1) \equiv P\{X = 1\} = p.$$

For a Bernoulli variable, denoted by $X \sim \text{Ber}(p)$, we can easily calculate the expected value,

$$E[X] = 1 \cdot p + 0 \cdot (1 - p) = p,$$

and the variance,

$$\text{Var}(X) = E[X^2] - E^2[X] = \left[1^2 \cdot p + 0^2 \cdot (1 - p)\right] - p^2 = p(1 - p).$$

The formula for variance makes intuitive sense: The variance is zero for $p = 1$ and $p = 0$ (there is no uncertainty on the outcome of the experiment), and it is maximized for $p = \frac{1}{2}$, where we have the largest uncertainty about the outcome.

When needed, we may use any other value, rather than 0 and 1. A financial motivation for a Bernoulli variable, quite relevant in option pricing, is modeling random asset prices $S(t)$ in simplified form, rather than using stochastic

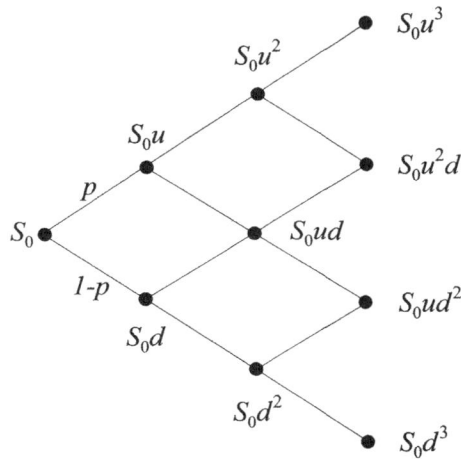

FIGURE 3.1 **A binomial lattice model for a stock price.**

differential equations. According to this simplified model, given the current asset price $S(0) = S_0$, the price after a time step δt, denoted by $S(\delta t)$, can be either $S_u = uS_0$ or $S_d = dS_0$, with probabilities p_u and p_d, respectively. The subscripts may be interpreted as "up" and "down," if we choose multiplicative shocks $d < u$. Now, imagine that the process $S(t)$ is observed up to time $T = n \cdot \delta t$ and that up and down jumps are independent over time. Since there are n steps, the largest value that $S(T)$ can attain is $S_0 u^n$, and the smallest one is $S_0 d^n$. Intermediate values are of the form $S_0 u^m d^{(n-m)}$, where m is the number of up jumps occurred, and $n - m$ is the number of down jumps. Note that the exact sequence of up and down jumps is not relevant to determine the terminal price, as multiplicative shocks commute: $S_0 ud = S_0 du$. A simple model like this is referred to as *binomial model* and can be depicted as the binomial lattice in Fig. 3.1. Such a model is quite convenient for simple, low-dimensional options,[9] as recombination keeps complexity low, whereas the size of a generic tree explodes exponentially.[10] Now, we might wonder what is the probability that $S(T) = S_0 u^m d^{(n-m)}$. Indeed, this is an example of a binomial variable: To build a binomial variable, we carry out a given number n of independent Bernoulli experiments and count the number m of successes. Hence, a binomial variable is just the sum of n independent Bernoulli variables. This distribution has finite support $\{0, 1, \ldots, n\}$ and depends on the two parameters p and n; thus, it may be denoted by $X \sim \text{Bin}(n, p)$. By using properties of sums

---

[9]Needless to say, Monte Carlo methods play a key role in more complicated cases; see Chapter 11.

[10]We have outlined how a simple binomial model may be calibrated by moment matching in Section 2.5.1.

of independent random variables, it is easy to show that

$$E[X] = np, \qquad \text{Var}(X) = np(1-p).$$

The PMF of a binomial variable is not completely trivial:

$$P\{X = m\} = \binom{n}{m} p^m (1-p)^{n-m}, \qquad (3.13)$$

where we use the binomial coefficient[11]

$$\binom{n}{m} \equiv \frac{n!}{(n-m)!m!}.$$

To see why this coefficient is needed, consider that there are many ways in which we may have r successes and $n - r$ failures; which specific experiments succeed or fail is irrelevant. In fact, we have $n!$ permutations of the n individual experiments, $(n-r)!$ permutations of failures, and $r!$ permutations of successes. Since the specific order of each permutation is irrelevant, we divide $n!$ by $(n-r)!$ and $r!$.

Figure 3.2 shows the PMF for $n = 30$, and $p = 0.2$, $p = 0.4$. The plot has been obtained by the following commands:

```
> par(mfrow=c(2,1))
> plot(dbinom(0:30,size=30,p=0.2),type='h',lwd=5)
> plot(dbinom(0:30,size=30,p=0.4),type='h',lwd=5)
```

The first command sets `mfrow` in order to draw plots according to a layout consisting of two rows and one column, thus lining the plots vertically. The `plot` function uses the parameter `lwd` to set the line width. Also check the following calculations:

```
> dbinom(10,size=30,p=0.4)
[1] 0.1151854
> factorial(30)/factorial(20)/factorial(10)*0.4^10*0.6^20
[1] 0.1151854
```

The second way of computing binomial probabilities is a straightforward application of Eq. (3.13), but it is not quite smart, as by taking the factorial of integer numbers there is a danger of overflow, i.e., of producing a number so large that it cannot be represented on a computer. When in need, it is better to leave the task to an appropriately coded function:

```
> choose(30,10)*0.4^10*0.6^20
[1] 0.1151854
```

Another discrete distribution that can be built on top of the Bernoulli is the geometric. This is defined by repeating identical and independent Bernoulli

---

[11] We recall the definition of the factorial of an integer number, $n! \equiv n \cdot (n-1) \cdot (n-2) \cdots 2 \cdot 1$; by convention, $0! = 1$.

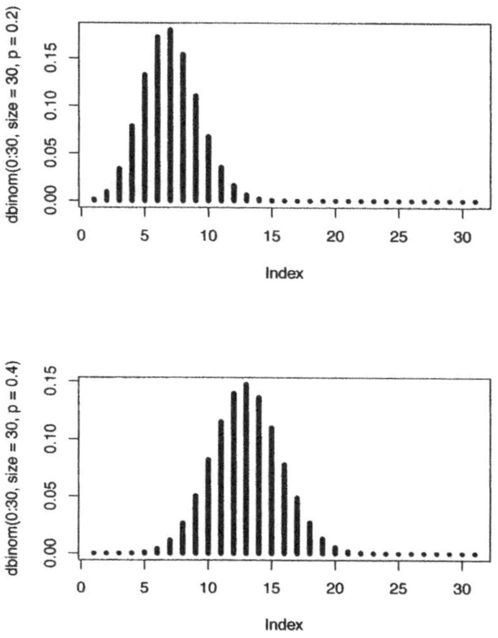

FIGURE 3.2 **Probability mass function of two binomial random variables.**

trials with success probability p, until the first success is obtained and counting the number of experiments.[12] The PMF of the geometric distribution is

$$P\{X = m\} = (1 - p)^{m-1}p.$$

The expected value $E[X]$ can be easily found by conditioning on the result of the first experiment, rather than by applying the definition. If the first experiment is a success, which occurs with probability p, the conditional expected value of the number of experiments is just $E[X \mid \text{success}] = 1$; otherwise, since there is no memory of past outcomes, it is $E[X \mid \text{failure}] = 1 + E[X]$, i.e., we count the first failed experiment and then we are back to square one. Thus:

$$E[X] = p \cdot 1 + (1 - p) \cdot (1 + E[X]) \quad \Rightarrow \quad E[X] = \frac{1}{p}. \tag{3.14}$$

R offers functions for the geometric distribution, too, but there is a little peculiarity:

```
> set.seed(55555)
```

---

[12]Such a mechanism is found, e.g., in rejection sampling, as we shall see in Section 5.4.

```
> rgeom(10,prob=0.5)
 [1] 1 2 3 1 0 2 0 0 0 0
```

We notice that the random variable can take the value 0, because in R the convention for the geometric distribution is to count failures, rather than experiments.

3.2.2 EXPONENTIAL AND POISSON DISTRIBUTIONS

The exponential and the Poisson distributions, despite their different nature, are tightly intertwined: The former is continuous and may be used to model the time elapsing between two events of some kind, whereas the latter is discrete and represents the corresponding number of events occurred within a given time interval. Examples of relevant events in finance are jumps in stock prices or changes in the credit rating of a bond issuer.

3.2.2.1 Exponential random variables

An exponential random variable owes its name to the functional form of its PDF:

$$f_X(x) = \begin{cases} \lambda e^{-\lambda x}, & \text{if } x \geq 0, \\ 0, & \text{if } x < 0, \end{cases}$$

where $\lambda > 0$ is a given parameter. Note that the support of the exponential distribution is $[0, +\infty)$, which makes sense since it is often used to model the length of time intervals. The corresponding CDF is

$$F_X(x) = \int_0^x \lambda e^{-\lambda t}\, dt = 1 - e^{-\lambda x}. \tag{3.15}$$

It is fairly easy to show that

$$E[X] = \frac{1}{\lambda} \quad \text{and} \quad \text{Var}(X) = \frac{1}{\lambda^2}.$$

It is worth noting that the expected value of an exponential random variable is quite different from the mode, which is zero, and that this distribution is right-skewed. If we interpret X as the time elapsing between consecutive events, $1/\lambda$ is the average time between them, and λ is a rate, i.e., the average number of events occurring per unit time. The notation $X \sim \exp(\lambda)$ is typically used to denote an exponential random variable.

A weird but remarkable property of exponential random variables is *lack of memory*. From the cumulative distribution function (3.15) we see that

$$P\{X > t\} = e^{-\lambda t}.$$

Intuitively, this means that, if we are at time zero, the probability that we have to wait more than t for the next event to occur is a decreasing function of t, which makes sense. What is not so intuitive is that if a time period of length t has already elapsed from the last event, this has no influence on the time we still have to wait for the next one. This is clearly false, e.g., for a uniform random variable $U(a, b)$; if a time $b - \epsilon$ has elapsed from the last event, where ϵ is small, we should get ready for the next one. On the contrary, the exponential distribution is memoryless, which can be stated formally as follows:

$$
\begin{aligned}
P\{X > t + s \mid X > t\} &= \frac{P\{(X > t + s) \cap (X > t)\}}{P\{X > t\}} \\
&= \frac{P\{X > t + s\}}{P\{X > t\}} \\
&= \frac{e^{-\lambda(t+s)}}{e^{-\lambda t}} \\
&= e^{-\lambda s} \\
&= P\{X > s\}.
\end{aligned}
$$

This characteristic is important when making modeling choices: If lack of memory is not compatible with the phenomenon we are representing, we should not use an exponential distribution.

Example 3.1 **Checking the memoryless property**

R offers the `exp` family of functions, which we may use to illustrate lack of memory empirically using Monte Carlo sampling as an R programming exercise. The function in Fig. 3.3 generates a sample of `numRepl` exponential random variables with rate `lambda`. Then, the probabilities $P\{X > s\}$, $P\{X > s\}$, and $P\{X > t + s \mid X > t\}$ are estimated empirically. Note how we restrict the sample conditionally on the event $\{X > t\}$, when extracting the subvector `times_t`. Here is the result, based on one million replications:

```
> set.seed(55555)
> CheckMemory(1,1,0.5,1000000)
P(T>t)  =   0.367936
P(T>s)  =   0.606819
P(T>t+s|T>t)  =   0.6064832
```

We see that, despite a small discrepancy due to sampling errors, the estimate of $P\{X > t + s \mid X > t\}$ is fairly close to $P\{X > s\}$. One could wonder whether one million replications are enough. The following exact evaluation is not too different from the above estimates, suggesting that the selected number of replications should be indeed large enough:

```
> 1-pexp(1,rate=1)
```

```
CheckMemory <- function(lambda,t,s,numRepl){
  times <- rexp(numRepl, rate=lambda)
  num_t <- sum(times>t)
  cat('P(T>t) = ', num_t/numRepl,'\n')
  cat('P(T>s) = ', sum(times>s)/numRepl,'\n')
  times_t = times[times>t]
  cat('P(T>t+s|T>t) = ', sum(times_t>(t+s))/num_t,'\n')
}
```

R programming notes:

1. Note how $\texttt{sum}$ is used to count the replications in which some condition occurs: The expression $\texttt{times>t}$ returns a vector with elements TRUE or FALSE, which are interpreted as 1 and 0, respectively, by $\texttt{sum}$.

2. To concatenate and print strings, $\texttt{cat}$ is used.

FIGURE 3.3 **Checking the memoryless property of exponential distribution empirically.**

```
[1] 0.3678794
> 1-pexp(0.5,rate=1)
[1] 0.6065307
```

However, this could just be sheer luck, and in real life we cannot afford the luxury of a comparison with exact numbers (otherwise, we would not be using Monte Carlo methods in the first place). We will address similar issues in Chapter 7, when dealing with simulation output analysis.

3.2.2.2 Erlang random variables

The exponential distribution is the building block of other random variables. By summing n independent exponentials with common rate λ we obtain the Erlang distribution, denoted by $\text{Erl}(n, \lambda)$. Erlang random variables may be used to model time elapsed as a sequence of exponential stages, in order to overcome the memoryless property of a single exponential. The PDF of a few Erlang variables, with a different number of stages, can be plotted using the R script in Fig. 3.4, which produces the plot of Fig. 3.5. From a mathematical point of view we immediately notice that by summing exponential variables we get a rather different distribution. In general, the sum of random variables has a distribution that is quite different from the single terms in the sum. The normal distribution is the most notable exception, as the sum of normals is still normal. We also notice that, in order to plot the PDF of an Erlang variable in R, we have

```
# Erlang.R
lambda <- 1
x <- seq(0,10,0.01)
nValues <- c(1,2,5,10)
data <- matrix(0,nrow=length(nValues),ncol=length(x))
for (i in 1:length(nValues))
  data[i,] <- dgamma(x, shape=nValues[i], rate = lambda)
plot(x,data[1,],ylim=c(0,max(data)),ylab="",type='l')
for (i in 2:length(nValues))
  lines(x,data[i,])
```

R programming notes:

1. To superimpose different line plots, we have to use `plot` first to open the window, and then `lines`.

2. To set the vertical axis properly, we have to set `ylim` to the largest value in the matrix `max(data)`.

FIGURE 3.4 **Plotting the PDF of different Erlang distributions.**

to resort to the `dgamma` function. As we shall see in Section 3.2.5, the gamma distribution generalizes other distributions, including the Erlang.

3.2.2.3 Poisson random variables and the Poisson stochastic process

We have already mentioned in Section 1.3.3 a peculiar stochastic process, the Poisson process, to motivate models based on discrete-event systems. Here we further investigate its properties and link it to a discrete distribution sharing the same name. The process is obtained by counting the number of events occurred in the time interval $[0, t]$, denoted by $N(t)$. Hence, this is a counting process, and Fig. 3.6 illustrates a sample path. Clearly, the sample path is discontinuous, but piecewise constant, as there is a jump corresponding to each event. Let X_k, $k = 1, 2, 3, 4, \ldots$, be the time elapsed between jump $k - 1$ and jump k; by convention, X_1 is the time at which the first jump occurs, after starting the system at time $t = 0$. We obtain a Poisson process if we assume that variables X_k are mutually independent and all exponentially distributed with parameter λ. You might even stumble on exoteric jargon like a *càdlàg function* when dealing with certain stochastic processes in finance. This is just a French acronym for "continue à droite, limitée à gauche," since the sample path is continuous from the right, and is limited (or bounded, i.e., it does not go to infinity) from the left.

If we count the number of events in the $[0, 1]$ time interval, we obtain a discrete random variable, the Poisson random variable. The Poisson distribution is characterized by the same parameter λ as the exponential, and it may take

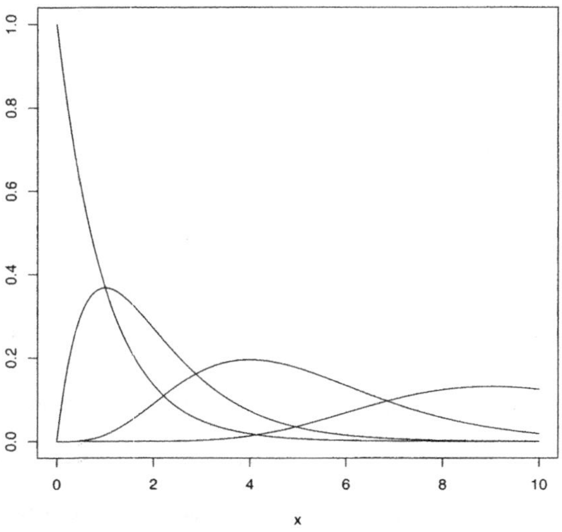

FIGURE 3.5 **Erlang PDFs.**

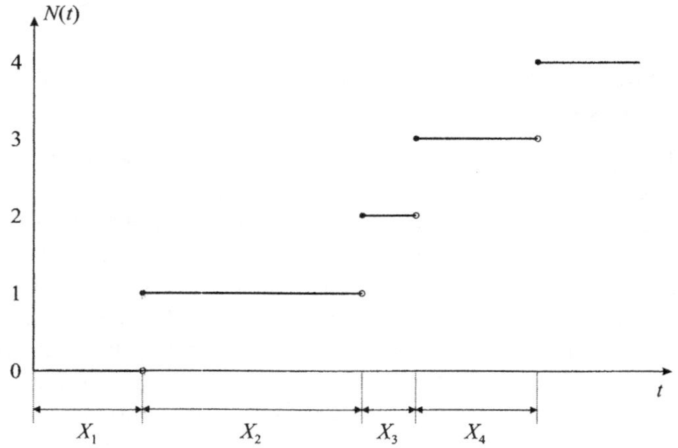

FIGURE 3.6 **Sample path of the Poisson process.**

values on the set $\{0, 1, 2, 3, \ldots\}$. Its PMF is

$$p_i = e^{-\lambda}\frac{\lambda^i}{i!}, \qquad i = 0, 1, 2, \ldots.$$

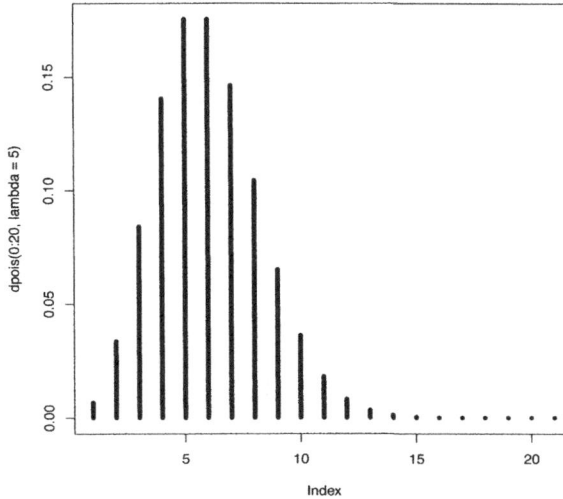

FIGURE 3.7 **Probability mass function of a Poisson random variable.**

Despite the apparent complexity of this function, it is straightforward to check that it meets the fundamental requirement of a PMF:

$$\sum_{i=0}^{\infty} p_i = e^{-\lambda} \sum_{i=0}^{\infty} \frac{\lambda^i}{i!} = e^{-\lambda} e^{\lambda} = 1.$$

By a similar token, we may also find its expected value and variance:

$$E[X] = \lambda, \quad \text{Var}(X) = \lambda.$$

Note that λ is, in fact, the event rate, which characterizes this distribution, denoted by $X \sim \text{Pois}(\lambda)$. If we want to count events occurring in the more general interval $[0, t]$, we have just to consider $X \sim \text{Pois}(\lambda t)$. The R family for the Poisson distribution is `pois`. Figure 3.7 shows the plot of the PMF of a Poisson random variable for $\lambda = 5$, obtained by the following command:

```
> plot(dpois(0:30,lambda=5,p=0.2),type='h',lwd=5)
```

We may state the connection between the Poisson random variable and the Poisson process in more general terms as follows. If we consider a time interval $[t_1, t_2]$, with $t_1 < t_2$, then the number of events occurred in this interval, i.e., $N(t_2) - N(t_1)$, has Poisson distribution with parameter $\lambda(t_2 - t_1)$. Furthermore, if we consider another time interval $[t_3, t_4]$, where $t_3 < t_4$, which is disjoint from the previous one, i.e., $(t_2 < t_3)$, then the random variables $N(t_2) - N(t_1)$

and $N(t_4) - N(t_3)$ are independent. We say that the Poisson process has *stationary* and *independent* increments.

The Poisson process model can be generalized to better fit reality, when needed. If we introduce a time-varying rate $\lambda(t)$, we get the so-called **inhomogeneous** Poisson process. Furthermore, when modeling price jumps, we would like to introduce a random variable modeling the height of the jump. The cumulative sum of jump heights in the time interval $[0, t]$ is another stochastic process, which is known as **compound** Poisson process. We will see how sample paths for such processes can be generated in Chapter 6.

3.2.3 NORMAL AND RELATED DISTRIBUTIONS

It is safe to say that the normal or Gaussian distribution, with its bell shape, is the best-known and most used (and misused) distribution. A normal random variable $X \sim N(\mu, \sigma^2)$ is characterized by the following PDF:

$$f_X(x) = \frac{1}{\sqrt{2\pi}\,\sigma} \exp\left\{ -\frac{1}{2}\left(\frac{x - \mu}{\sigma}\right)^2 \right\}, \qquad -\infty < x < +\infty,$$

which clearly suggests symmetry with respect to the point $x = \mu$. The parameters μ and σ^2 have a clear interpretation,[13] as a few calculations show that

$$E[X] = \mu, \qquad \mathrm{Var}[X] = \sigma^2.$$

R offers the `norm` family of functions. As an illustration, in Fig. 3.8 we show the PDF for two normal distributions with $\mu = 0$ and $\sigma = 1, 2$, respectively; the plot is obtained as follows:

```
> x <- seq(-6.5,6.5,by=0.01)
> plot(x,dnorm(x,mean=0,sd=1),type="l",lty="dashed")
> lines(x,dnorm(x,mean=0,sd=2),type="l",lty="dotdash")
```

From the plot, we also verify that, for a normal distribution, mode, median, and expected value are the same.

In the applications, a very special role is played by the *standard* normal distribution, characterized by parameters $\mu = 0$ and $\sigma = 1$. The reason of its importance is that, if we are able to work with a standard normal in terms of quantiles and distribution function, then we are able to work with a generic normal variable. In fact the CDF is the integral of the bell-shaped function above, for which no analytical formula is known. Although the general CDF for a normal variable is not known analytically, efficient numerical approximations are available for the standard normal case $Z \sim N(0, 1)$:

$$\Phi(x) \equiv P\{Z \le x\} = \frac{1}{\sqrt{2\pi}} \int_{-\infty}^{x} e^{-z^2/2}\, dz.$$

---

[13]One could wonder why we typically use $X \sim N(\mu, \sigma^2)$, referring to variance, and not $X \sim N(\mu, \sigma)$, referring to standard deviation. The reason is that this notation easily generalizes to the multivariate case $\mathbf{X} \sim N(\boldsymbol{\mu}, \boldsymbol{\Sigma})$, where $\boldsymbol{\Sigma}$ is the covariance matrix; see Section 3.3.

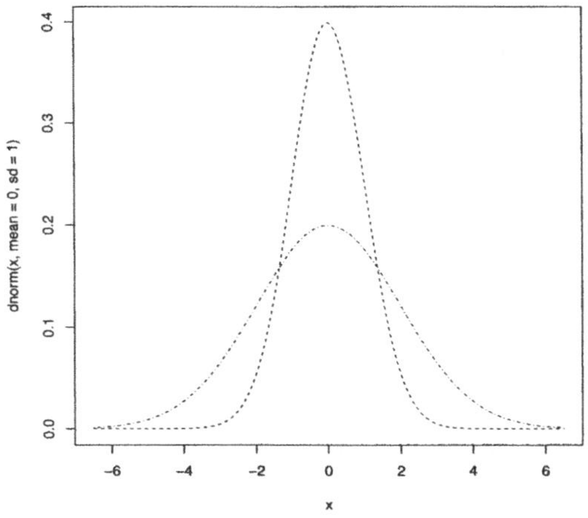

FIGURE 3.8 **PDF of normal random variables with $\mu = 0$ and $\sigma = 1$, $\sigma = 2$.**

This function is calculated by the R function `pnorm`. We are also able to invert the function by `qnorm`, which yields the quantile z_q defined by

$$P\{Z \le z_q\} = q,$$

for a probability level $q \in (0, 1)$. Armed with these functions, we may easily compute the CDF and quantiles for a generic normal. If $X \sim N(\mu, \sigma^2)$ and we need the probability

$$P\{X \le x\},$$

we have just to standardize:

$$P\{X \le x\} = P\left\{\frac{X - \mu}{\sigma} \le \frac{x - \mu}{\sigma}\right\} = \Phi\left(\frac{x - \mu}{\sigma}\right),$$

since $Z = (X - \mu)/\sigma$ is standard normal. If we need a quantile x_q, such that $P\{X \le x_q\} = q$, we find the corresponding quantile z_q and go the other way around by destandardizing:

$$P\{Z \le z_q\} = q \Leftrightarrow P\{\mu + \sigma Z \le \mu + \sigma z_q\} = q \Leftrightarrow P\{X \le \mu + \sigma z_q\} = q.$$

In R, we can avoid all of this if we specify the parameters `mean` and `sd`.

■ **Example 3.2** **Some R functions for the normal distribution**

The following examples show how to use `pnorm` and `qnorm`:

```
> pnorm(10,mean=5,sd=3)
[1] 0.9522096
> pnorm((10-5)/3)
[1] 0.9522096
> qnorm(0.95,mean=5,sd=3)
[1] 9.934561
> z=qnorm(0.95);z
[1] 1.644854
> 5+3*z
[1] 9.934561
```

We may also easily check the following well-known properties:

1. $P\{\mu - \sigma \le X \le \mu + \sigma\} \approx 68.27\%$
2. $P\{\mu - 2\sigma \le X \le \mu + 2\sigma\} \approx 95.45\%$
3. $P\{\mu - 3\sigma \le X \le \mu + 3\sigma\} \approx 99.73\%$

```
> pnorm(1) - pnorm(-1)
[1] 0.6826895
> pnorm(2) - pnorm(-2)
[1] 0.9544997
> pnorm(3) - pnorm(-3)
[1] 0.9973002
```

The last property shows that the normal distribution has rather thin tails, since most realizations are within three standard deviations of the mean, which makes its use sometimes questionable. Indeed, the normal distribution is a sort of benchmark in terms of fat vs. thin tails, a fact that is measured by its kurtosis:

$$\kappa = E\left[\left(\frac{X - \mu}{\sigma}\right)^4\right] = 3.$$

Sometimes, kurtosis is defined as $\kappa - 3$, which should actually be called *excess* kurtosis. Distributions with positive excess kurtosis allow for more frequent extreme events than the normal. Another important feature of the normal distribution is symmetry, which implies a zero skewness:

$$\gamma = E\left[\left(\frac{X - \mu}{\sigma}\right)^3\right] = 0.$$

The assumption that returns on an asset are normally distributed implies their symmetry, which is contradicted by empirical facts, especially when financial derivatives are involved. Hence, a normal model is sometimes too stylized to be credible. Nevertheless, the normal distribution has also a few quite interesting and convenient properties:

- It is a limit distribution according to the central limit theorem: If we sum a large number of independent and identically distributed random variables, the result has a normal distribution in the limit. This makes the normal distribution useful in inferential statistics, for suitably large samples.
- If we add normal random variables, we get another normal random variable. Formally, if we consider n independent normals $X_i \sim \mathsf{N}(\mu_i, \sigma_i^2)$, $i = 1, \ldots, n$, and take an affine combination with coefficients a and b_i, then

$$a + \sum_{i=1}^{n} b_i X_i \sim \mathsf{N}\left(a + \sum_{i=1}^{n} b_i \mu_i, \sum_{i=1}^{n} b_i^2 \sigma_i^2\right).$$

The proof of this property requires more advanced concepts related to moment generating functions.[14]

It is said that the normal distribution is a stable distribution, since a linear combination of normals is still normal. The process can also go the other way around, since stable random variables are infinitely divisible, i.e., can be expressed as sums of similar variables. We just mention that there is a stable distribution that includes the normal as a specific case.[15]

In Section 3.3.4 we will also see some properties of the multivariate generalization of the normal distribution. The normal distribution is also the basis of other distributions obtained by transforming a normal variable or by combining independent standard normals.

3.2.3.1 Lognormal distribution

A random variable X is said to have a lognormal distribution if $\log X$ is normal. In other words, if $Y \sim \mathsf{N}(\mu, \sigma^2)$, then $X = e^Y$ is lognormal, denoted by $X \sim \mathsf{LogN}(\mu, \sigma^2)$. It can be shown that the PDF density of the lognormal can be written as

$$f_X(x) = \frac{1}{x\sigma\sqrt{2\pi}} \exp\left\{-\frac{(\log x - \mu)^2}{2\sigma^2}\right\}, \qquad x > 0. \qquad (3.16)$$

It is important to notice that in this case the two parameters μ and σ^2 cannot be interpreted as in the normal case. In fact, using the moment generating function of the normal, it is easy to show that

$$\mathrm{E}[X] = e^{\mu + \sigma^2/2}, \quad \mathrm{Var}(X) = e^{2\mu + \sigma^2}\left(e^{\sigma^2} - 1\right).$$

---

[14]See Section 8.6.2.

[15]We do not deal with stable distributions in this introductory treatment, since they require more advanced concepts; see, e.g., [15, pp. 129–131].

Note how, due to convexity of the exponential function and Jensen's inequality, if $X \sim \text{LogN}(\mu, \sigma^2)$, then

$$\text{E}[X] = e^{\mu + \sigma^2/2} \geq e^{\mu} = e^{\text{E}[X]}.$$

Another noteworthy fact is that

$$Y \sim \text{N}(-\sigma^2/2, \sigma^2) \quad \Rightarrow \quad \text{E}[X] = 1. \tag{3.17}$$

This suggests using a lognormal variable with unit expected value to model random errors in a multiplicative model, whereas a normal variable with zero expected value would model random errors in an additive model.

Furthermore, since by summing normal variables we still get a normal variable, a similar property holds when we multiply lognormal random variables. Formally, if we consider n independent lognormals $X_i \sim \text{LogN}(\mu_i, \sigma_i^2)$, then

$$\prod_{i=1}^{n} X_i \sim \text{LogN}\left(\sum_{i=1}^{n} \mu_i, \sum_{i=1}^{n} \sigma_i^2\right).$$

This is useful when modeling financial returns, which should cumulate multiplicatively rather than additively. The R family for this distribution is `lnorm`. For instance, we may check Eq. (3.17) empirically:

```
> set.seed(55555)
> sigma <- 2
> X=rlnorm(1000000, meanlog=-(sigma^2)/2, sd=sigma)
> mean(X)
[1] 0.9928838
```

3.2.3.2   Chi-square distribution

Let Z_i, $i = 1, \ldots, n$, be standard and independent normal variables. The random variable X defined as

$$X = Z_1^2 + Z_2^2 + \cdots + Z_n^2$$

is certainly not normal, as it cannot take negative values. This variable is called chi-square with n degrees of freedom and is often denoted by χ_n^2. Recalling that the expected value of a squared standard normal is

$$\text{E}[Z^2] = \text{Var}(Z) + \text{E}^2[Z] = 1,$$

we immediately see that

$$\text{E}[X] = n.$$

It can also be shown that

$$\text{Var}(X) = 2n.$$

Figure 3.9 shows the PDF for a chi-square variable with 4 and 8 degrees of freedom, obtained as follows:

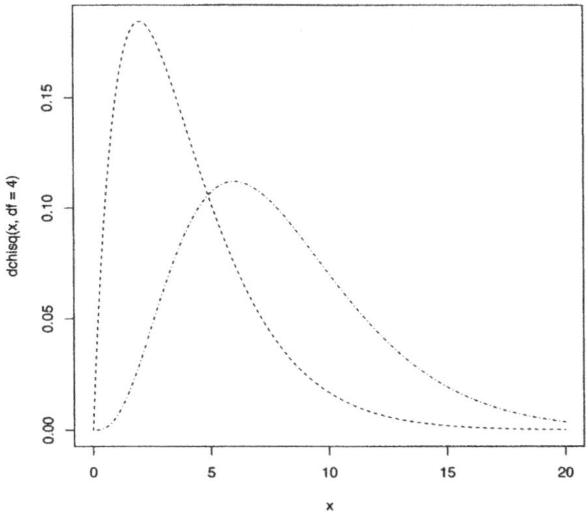

FIGURE 3.9 **PDF of two chi-square random variables with 4 and 8 degrees of freedom, respectively; the variable with 4 degrees of freedom is less uncertain and has a higher mode.**

```
> x <- seq(0,20,by=0.01)
> plot(x,dchisq(x,df=4),type="l",lty="dashed")
> lines(x,dchisq(x,df=8),type="l",lty="dotdash")
```

The chi-square distribution is relevant in inferential statistics, and there is a well-known nonparametric test of goodness of fit owing its name to it. It is worth noting that a chi-square distribution may be regarded as a specific case of a gamma random variable, discussed later. In finance, we also meet the *noncentral* chi-square distribution, which is obtained by the following construction. Let us consider

$$X = \sum_{i=1}^{n} (Z_i + a_i)^2 .$$

The corresponding variable is called noncentral chi-square with n degrees of freedom and noncentrality parameter $\lambda = \sum_{i=1}^{n} a_i^2$. It is possible to generalize this distribution to a noninteger ν, resulting in a $\chi_\nu'^2(\lambda)$ variable. The following formulas for the relevant moments can be shown:

$$\mathrm{E}\left[\chi_\nu'^2(\lambda)\right] = \nu + \lambda, \qquad \mathrm{Var}\left(\chi_\nu'^2(\lambda)\right) = 2(\nu + 2\lambda). \qquad (3.18)$$

The R function family `chisq` can also be used for noncentral case. For instance:

```
> set.seed(55555)
> rchisq(3, df=4.5, ncp=3.6)
```

[1] 17.775371 8.719674 15.453155

from which we see that the degrees of freedom `df` and the noncentrality parameter `ncp` may be noninteger. This distribution is relevant in the simulation of square-root diffusions.[16]

3.2.3.3   Student's t distribution

If Z and χ_n^2 are a standard normal and a chi-square with n degrees of freedom, respectively, and they are also mutually independent, the random variable

$$t_n = \frac{Z}{\sqrt{\chi_n^2/n}}$$

has a Student's t distribution with n degrees of freedom, denoted by t_n. This distribution is widely used in inferential statistics, for computing confidence intervals and testing hypotheses. A notable feature is that it features fatter tails than the normal. Indeed, it can be shown that

$$\mathrm{E}[t_n] = 0\,, \qquad \mathrm{Var}[t_n] = \frac{n}{n-2}.$$

We observe that, for large n, variance goes to 1, but it is not defined for $n \leq 2$. Since this variance is larger than the corresponding variance of the standard normal, the t distribution is more dispersed. In fact, the excess kurtosis is

$$\frac{6}{n-4}$$

for $n > 4$. We may check these fatter tails by plotting its density in Fig. 3.10:

```
> x <- seq(-4.5,4.5,by=0.01)
> plot(x,dnorm(x),type="l",lty="solid")
> lines(x,dt(x,df=1),type="l",lty="dotdash")
> lines(x,dt(x,df=5),type="l",lty="dashed")
```

The plot shows the densities of t_1 and t_5 random variables, along with a standard normal. We observe that the PDF of the t distribution is bell-shaped much like a standard normal; the main differences lie in its heavier tail and in a lower mode. For increasing n, the tails of t_n get thinner and the density tends to a standard normal. In fact, for large n the two distributions are virtually identical. The t distribution can be generalized to the multivariate case, as shown in Section 3.3.4.

3.2.3.4 F distribution

Another relevant distribution can be derived from the normal and plays a fundamental role in analysis of variance and in testing regression models. If we

---

[16]See Section 6.3.2, where we deal with sample path generation for continuous-time stochastic processes.

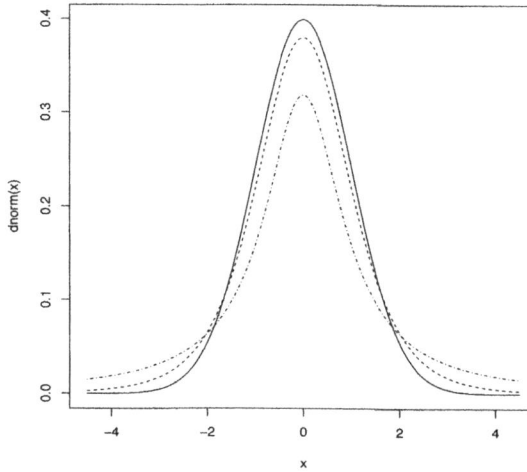

FIGURE 3.10 **PDF of Student's t distribution, with n=1 (dash-dotted line) and n=5 (dashed line), compared with a standard normal (continuous line).**

combine two independent chi-square variables with n and m degrees of freedom, respectively, as follows:

$$F_{n,m} = \frac{\chi_n^2/n}{\chi_m^2/m},$$

we obtain the F distribution with n and m degrees of freedom. R offers functions like `df(x, df1, df2)`; there is a noncentral version of the F distribution as well.

3.2.4 BETA DISTRIBUTION

The beta distribution is continuous and has support on the unit interval $[0, 1]$. It depends on two parameters $\alpha_1, \alpha_2 \geq 0$, defining its shape through the PDF

$$f(x) = \frac{x^{\alpha_1 - 1}(1 - x)^{\alpha_2 - 1}}{\mathrm{B}(\alpha_1, \alpha_2)}, \qquad x \in [0, 1], \tag{3.19}$$

The denominator of the above ratio is just needed in order to normalize the PDF. A straightforward approach is defining the beta function:

$$\mathrm{B}(\alpha_1, \alpha_2) \equiv \int_0^1 x^{\alpha_1 - 1}(1 - x)^{\alpha_2 - 1} \, dx.$$

Alternatively, we may use the following gamma function:

$$\Gamma(\alpha) \equiv \int_0^\infty e^{-x} x^{\alpha-1} dx, \quad \alpha > 0. \tag{3.20}$$

Using integration by parts, it is easy to see that the gamma function satisfies the following recursion:

$$\Gamma(\alpha + 1) = \alpha\Gamma(\alpha). \tag{3.21}$$

This shows that the gamma function is an extension of the factorial function outside the domain of integer numbers. In fact, using Eq. (3.21) for an integer argument and unfolding the recursion, we find

$$\Gamma(n) = (n-1)!$$

The gamma function appears in the definition of different PDFs, including, needless to say, the gamma distribution to be discussed next. It also turns out that

$$B(\alpha_1, \alpha_2) = \frac{\Gamma(\alpha_1)\Gamma(\alpha_1)}{\Gamma(\alpha_1 + \alpha_2)}.$$

The PDFs displayed in Fig. 3.11 are obtained by the following R snapshot; this shows that the uniform and certain triangular distributions are special cases of the beta distribution.

```
> x <- seq(0.001,0.999,by=0.001)
> par(mfrow = c(3,3))
> plot(x,dbeta(x,2,1),type='l')
> plot(x,dbeta(x,1,1),type='l')
> plot(x,dbeta(x,1,2),type='l')
> plot(x,dbeta(x,1,0.8),type='l')
> plot(x,dbeta(x,0.8,0.8),type='l')
> plot(x,dbeta(x,0.8,1),type='l')
> plot(x,dbeta(x,8,3),type='l')
> plot(x,dbeta(x,3,3),type='l')
> plot(x,dbeta(x,3,8),type='l')
```

Possible uses of a beta random variable are related to random percentages, like recovery rates after a default. If needed, the support of the beta distribution can be easily mapped to the interval (a, b). It is also interesting to compare the PDF of the beta distribution in Eq. (3.19) with the PMF of a binomial distribution in Eq. (3.13). Apart from the normalization factors, they look quite similar, at least if we choose integer parameters α_1 and α_2 in the PDF of a beta variable. In fact, these distributions are quite intertwined within the framework of Bayesian statistics, when we consider uncertainty in the parameter p of a Bernoulli random variable; the beta distribution can be used to represent such an uncertainty, both in the prior and the posterior.[17]

---

[17]See Section 4.6 and Chapter 14.

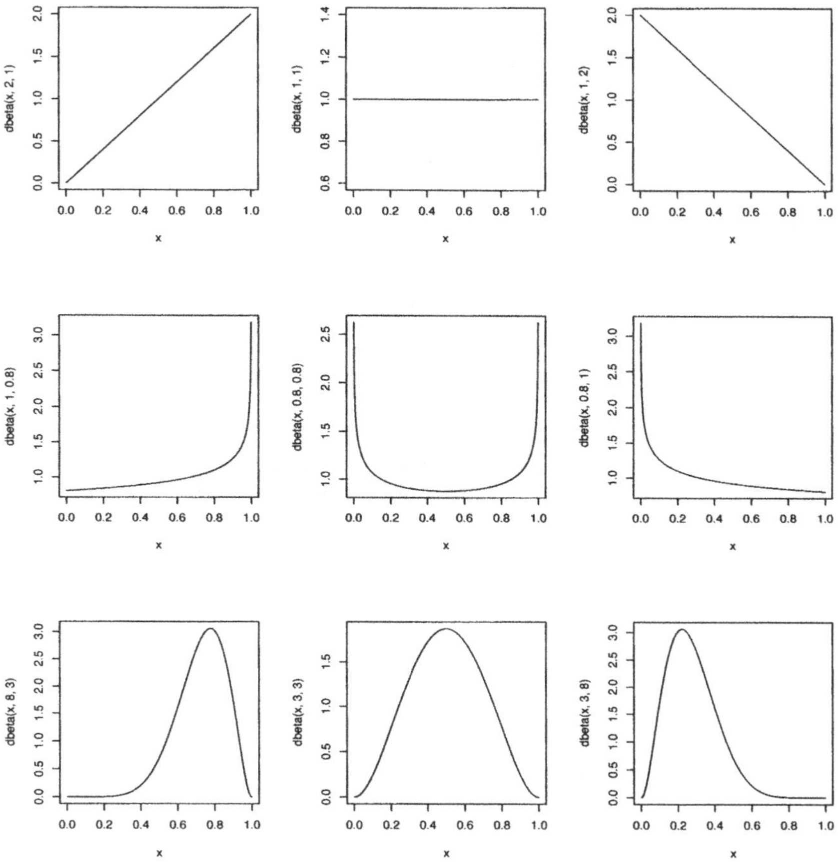

FIGURE 3.11 **PDFs of several beta distributions.**

3.2.5 GAMMA DISTRIBUTION

The gamma distribution is characterized by the PDF

$$f_X(x) = \frac{\lambda^{\alpha} x^{\alpha-1} e^{-\lambda x}}{\Gamma(\alpha)}, \tag{3.22}$$

which depends on parameters α and λ; we use notation $X \sim \text{Gamma}(\alpha, \lambda)$. A notable feature of the gamma distribution is that it includes other distributions as special cases:

1. It is immediate to see that the distribution $\text{Gamma}(1, \lambda)$ boils down to the exponential distribution.

2. A property of the gamma distribution is that if we consider n independent gamma variables $X_i \sim \text{Gamma}(\alpha, \lambda)$, $i = 1, \ldots, n$, then

$$\sum_{i=1}^{n} X_i \sim \text{Gamma}(n\alpha, \lambda).$$

3. By putting the first two facts together, we see that $\text{Gamma}(n, \lambda)$ is just an Erlang distribution.

4. The chi-square distribution with n degrees of freedom is obtained as $\text{Gamma}(n/2, \frac{1}{2})$.

The fundamental moments of a gamma variable are

$$\text{E}[X] = \frac{\alpha}{\lambda}, \quad \text{Var}(X) = \frac{\alpha}{\lambda^2}.$$

This suggests that λ is a scale parameter, whereas α is a shape parameter. This is in fact the case, in the sense that if $X \sim \text{Gamma}(\alpha, 1)$, then $X/\lambda \sim \text{Gamma}(\alpha, \lambda)$. Actually, some authors prefer to use λ as the scale parameter, whereas others prefer using $\beta = 1/\lambda$. R offers both possibilities, as there are two alternatives in the gamma function family:

```
> dgamma(2, shape=3, rate=2)
[1] 0.2930502
> dgamma(2, shape=3, scale=1/2)
[1] 0.2930502
```

The following R snapshot produces the plot in Fig. 3.12, which illustrates the effect of the shape parameter for a normalized scale parameter:

```
> x <- seq(0,7,by=0.01)
> plot(x,dgamma(x,shape=.5,rate=1),type='l',
     ylim=c(0,1.2),ylab='')
> lines(x,dgamma(x,shape=1,rate=1))
> lines(x,dgamma(x,shape=2,rate=1))
> lines(x,dgamma(x,shape=3,rate=1))
```

3.2.6 EMPIRICAL DISTRIBUTIONS

Sometimes, no textbook distribution will fit the available data. In such a case, one could consider resorting to an empirical distribution. In the discrete case, we have just to specify a set of values and the associated probabilities. In the continuous case, we must come up with a sensible way to devise a PDF or a CDF based on a discrete set of observations.

Piecewise linear interpolation is a straightforward way of building a CDF. Given n observations X_i, $i = 1, \ldots, n$, of the random variable X, we first sort them and find the order statistics $X_{(1)} \leq X_{(2)} \leq \cdots \leq X_{(n)}$. Now we build a

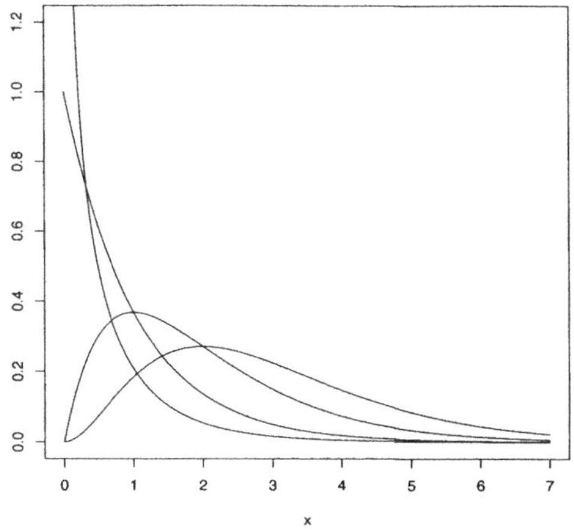

FIGURE 3.12 **PDFs of some gamma distributions.**

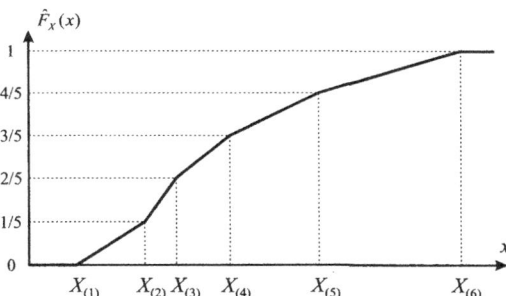

FIGURE 3.13 **Building an empirical CDF by piecewise linear interpolation.**

piecewise linear CDF $\hat{F}_X(\cdot)$ as follows:

$$
\hat{F}_X(x) = \begin{cases} 0, & \text{if } x < X_{(1)}, \\ \dfrac{i-1}{n-1} + \dfrac{x - X_{(i)}}{(n-1)(X_{(i+1)} - X_{(i)})}, & \text{if } X_{(i)} \le x < X_{(i+1)} \\ & \qquad \text{for } i = 1, \ldots, n-1, \\ 1, & \text{if } x \ge X_{(n)}. \end{cases}
$$

(3.23)

In Fig. 3.13 we illustrate an example for $n = 6$. After defining the empirical CDF, it is easy to sample from it using, e.g., the inverse transform method of Section 5.3.

Kernel-based estimation is an alternative, possibly more sophisticated way to build an empirical distribution. Unlike the piecewise-linear interpolation approach, it aims at building a PDF, and it can be interpreted as a way to "smooth" a histogram. Roughly speaking, the idea is to center a suitable elementary density around each observation, to add all of the densities, and to scale the result according to the number of observations in such a way as to obtain a legitimate overall PDF.

R offers a `density` function to this aim, receiving a vector of observations and producing an object of class `density`. The function behavior can be controlled by setting several parameters, among which we emphasize:

- `kernel`, which is used to select the elementary PDF; natural options are `gaussian` (the default), `rectangular`, and `triangular`, but there are others.
- `bw`, which is used to select the bandwidth, i.e., the width of each elementary PDF.

To illustrate the role of bandwidth, let us generate a sample of observations by using a mixture of two normals:

```
> set.seed(55555)
> data <- c(rnorm(40,10,10), rnorm(30,75,20))
> hist(data,nclass=15)
> windows()
> par(mfrow=c(1,2))
> kd1 <- density(data, bw=2)
> plot(kd1,main="")
> kd2 <- density(data, bw=10)
> plot(kd2, main="")
> class(kd1)
[1] "density"
```

The corresponding histogram is displayed in Fig. 3.14, and suggests a bimodal distribution. The kernel `kd1` is indeed of class `density`, and it can easily be plotted, resulting in the two PDFs displayed in Fig. 3.15. We clearly see that a small bandwidth does not smooth data enough and produces a confusing PDF; a larger bandwidth yields a definitely more reasonable result.

An extreme form of empirical distribution, in a sense, is the bootstrap approach sometimes employed in finance. The idea is to take a time series of prices or returns, and to generate future scenarios by sampling past outcomes. There are refinements of this idea, but it should be clear that if we take past observations and we just mix them randomly to come up with sample paths, we are implicitly assuming that longitudinal dependencies and correlations can be ruled out. Hence, we are relying on the efficient market hypothesis. On the other hand, a notable advantage of the bootstrap approach is that, when dealing with multidimensional time series, we automatically account for cross-sectional dependencies. In the next section, indeed, we move on to consider multivariate distributions, and we shall see that modeling such dependencies is far from trivial, and this explains why bootstrap is often advocated.

Finally, two general remarks are in order about empirical distributions:

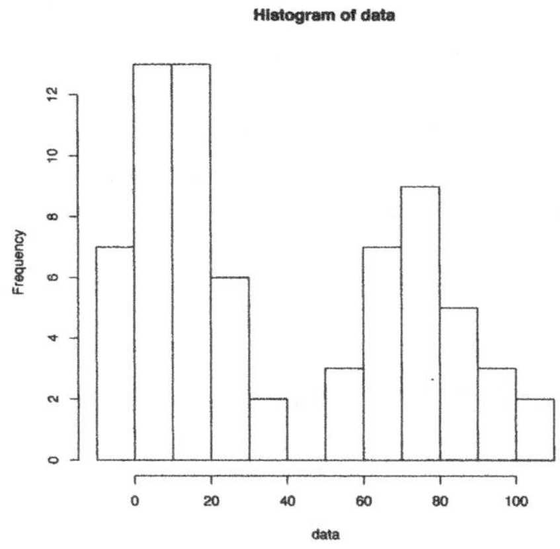

FIGURE 3.14 **Histogram of data generated by a mixture of normals.**

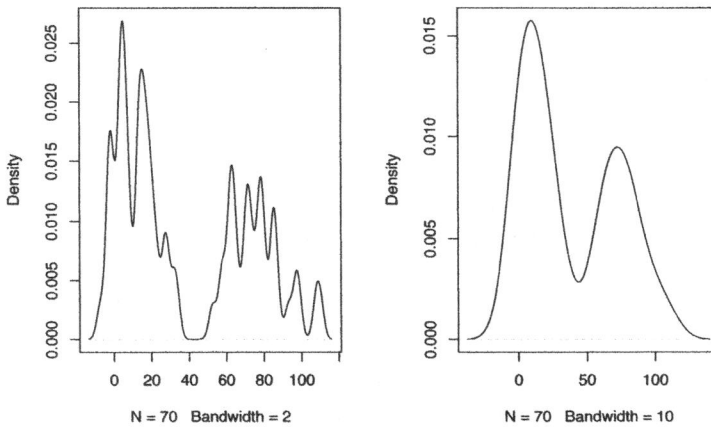

FIGURE 3.15 **Kernel densities for different bandwidths.**

1. Unless suitable countermeasures are adopted, they will never generate anything beyond the range of what was observed in the sample that we use to generate the distribution. This is quite clear from Fig. 3.13 and may rule out extreme scenarios, as the resulting distribution has finite support. One possibility is to append suitable tails in order to correct for this range effect. The PDF produced by kernel estimation approach, in fact, goes a bit beyond what is observed, depending on the selected bandwidth. More specific approaches to select tails have been proposed in the literature.

2. The idea of just fitting the data as they are in order to capture all of their nuances, without the oversimplification of textbook distributions, is quite appealing. However, we should always keep in mind the danger of over-fitting. For instance, if we observe bimodal data, we should wonder if there is some fundamental reason behind that pattern, which should be represented in our uncertainty model, or if it is just the effect of sampling variability.

3.3   Multivariate distributions: Covariance and correlation

In the previous sections we have considered univariate distributions, which provide us with the nuts and bolts for modeling uncertainty. However, the need for modeling dependence is pervasive in finance and economics, and this leads us to consider multivariate distributions. We are interested in two main dimensions of dependence:

- Dependence between two variables observed at the same time instant, i.e., *cross-sectional dependence*. As an example, we are interested in the relationship between return of different assets over the same time horizon. This kind of dependence may be represented explicitly by a possibly complicated multivariate distribution or built implicitly, e.g., by factor models described in Section 3.8.2.
- Dependence between two observations of the same variable at different time instants, i.e., *longitudinal dependence*. For instance, we want to investigate the relationship between returns of the same asset in two consecutive days. Again, this kind of dependence may be represented explicitly by a possibly complicated multivariate distribution or built implicitly, e.g., by time series models described in Section 3.6.

Clearly, the decomposition of dependence in two orthogonal dimensions is an oversimplification. In some cases, we really need to consider both dimensions together, which leads to the analysis of panel data. In other cases, the decomposition may be a sensible modeling simplification. For instance, we might want to check the dependence of two bond defaults that do not really occur at the same time instant, but from a modeling point of view they could be considered as happening at the same time (unless we wish to model causal links between them).

In this section we start by considering the structure of multivariate distributions, emphasizing their complexity and the fact that a characterization of marginal distributions does not tell the whole story, unless we assume independence. Then, we deal with standard concepts like covariance and Pearson's correlation. Showing their potential limitations paves the way for Section 3.4, where we illustrate more advanced ideas revolving around the copula concept.

Finally, we recall a couple of multivariate distributions, namely, the normal and Student's t, which generalize their univariate counterparts.

3.3.1 MULTIVARIATE DISTRIBUTIONS

We know that univariate distributions may be described by the PMF in the discrete case or the PDF in the continuous case. Actually, the CDF is a more general concept, since it is valid in a general setting. So, the starting point to to get the picture about the dependence of two random variables is their joint CDF.

DEFINITION 3.1 (Joint cumulative distribution function) *The joint CDF of random variables X and Y is defined as*

$$F_{X,Y}(x,y) \equiv P\big(\{X \le x\} \cap \{Y \le y\}\big).$$

The streamlined notation $P(X \le x, Y \le y)$ is typically used to denote a joint event.

The joint CDF is a function of two variables, x and y, and it fully characterizes the joint distribution of the random variables X and Y. It is easy to see how the definition can be generalized to n random variables.

The joint CDF is the most general tool to characterize a multivariate distribution. Other concepts may be defined, depending on the character of the involved random variables. In the case of two discrete variables, where X may take values x_i, $i = 1,2,3,\ldots$, and Y may take values y_j, $j = 1,2,3,\ldots$, we define the joint PMF as

$$p_{X,Y}(x_i, y_j) = P(X = x_i, Y = y_j).$$

When dealing with continuous random variables, we may define a joint PDF $f_{X,Y}(x,y)$ such that

$$P\big((X,Y) \in C\big) = \iint\limits_{(x,y)\in C} f_{X,Y}(x,y)\,dx\,dy. \tag{3.24}$$

In general, we cannot be sure that such a function exists. To be precise, we should say that the two random variables are *jointly continuous* when their joint PDF exists. Given the joint PDF, we may find the joint CDF by integration:

$$F_{X,Y}(a,b) = P\big(X \in (-\infty, a], Y \in (-\infty, b]\big) = \int_{-\infty}^{a} \int_{-\infty}^{b} f_{X,Y}(x,y)\,dy\,dx.$$
$$\tag{3.25}$$

Also the joint PMF and PDF can be defined for n random variables. Needless to say, specifying a function of n variables requires a lot of effort unless one imposes some structure:

- One possibility for doing so is to define parametric families of distributions, depending on a limited number of parameters. For instance, multivariate generalizations of normal and t distributions, to be discussed later, essentially depend on first- and second-order moments.

- Another possibility is to characterize each random variable individually, by resorting to the wide array of univariate distributions we are familiar with, and then put the pieces together in order to capture dependence. This kind of decomposition is the idea behind the copula approach.

We can do without the above machinery in one easy case, i.e., when the random variables are independent. In such a case, the events $\{X \leq x\}$ and $\{Y \leq y\}$ are independent, and the CDF factors into the product of two terms:

$$F_{X,Y}(x,y) = \mathrm{P}\{X \leq x\} \cdot \mathrm{P}\{Y \leq y\} = F_X(x)F_Y(y).$$

In general, whatever refers to a single variable, within the context of a multivariate distribution, is called *marginal*. In case of independence, the joint CDF is just the product of marginal CDFs. The same applies to joint PMFs and PDFs. A first question is: Given the joint CDF $F_{X,Y}(x,y)$, how can we find the marginal CDFs $F_X(x)$ and $F_Y(y)$, pertaining to each individual variable, if we do not assume independence? In principle, the task for the CDF is fairly easy. We obtain the marginal CDFs for the two random variables as follows:

$$F_X(x) = \mathrm{P}(X \leq x) = \mathrm{P}(X \leq x, Y \leq +\infty) = F_{X,Y}(x, +\infty).$$

By the same token, $F_Y(y) = F_{X,Y}(+\infty, y)$. In the discrete case, to obtain the marginal PMFs from the joint PMF, we just use the total probability theorem and the fact that events $\{Y = y_j\}$ are disjoint:

$$p_X(x_i) = \mathrm{P}(X = x_i) = \mathrm{P}\left(\bigcup_j \{X = x_i, Y = y_j\}\right)$$
$$= \sum_j \mathrm{P}(X = x_i, Y = y_j) = \sum_j p_{X,Y}(x_i, y_j).$$

If we want to obtain marginal PDFs from the joint PDF, we may work in much the same way as in the discrete case:

$$\mathrm{P}(X \in A) = \mathrm{P}\big(X \in A, Y \in (-\infty, +\infty)\big) = \int_A \int_{-\infty}^{+\infty} f_{X,Y}(x,y)\, dy\, dx.$$

If we introduce the marginal density

$$f_X(x) = \int_{-\infty}^{+\infty} f_{X,Y}(x,y)\, dy,$$

we have

$$\mathrm{P}(X \in A) = \int_A f_X(x)\, dx.$$

The marginal PDF for Y is obtained in the same way, integrating with respect to x.

We see that, given a joint distribution, we may find the marginals. It is tempting to think that, given the two marginals, we may recover the joint distribution in some way. This is *not* true, as the marginal distributions do not say anything about the link among random variables.

▣ **Example 3.3** **Marginals are not enough: Part 1**

Consider the discrete-time stochastic process

$$X_t = t \cdot \epsilon, \qquad t = 0, 1, 2, 3, \dots, \tag{3.26}$$

where $\epsilon \sim N(0, 1)$. We immediately see that each X_t is normal with expected value 0 and variance t^2. This marginal distribution is the same for the process

$$X_t = t \cdot \epsilon_t, \qquad t = 0, 1, 2, 3, \dots, \tag{3.27}$$

where ϵ_t is again standard normal, but we have one such variable for each time instant. Indeed, Eq. (3.26) describes a rather degenerate process, since uncertainty is linked to the realization of a *single* random variable. The following R snapshot produces a set of sample paths for the two processes, illustrated in Figs. 3.16 and 3.17:

```
set.seed(55555)
len <- 50
times <- 0:len
eps <- rnorm(10)
lims <- c(len*min(eps),len*max(eps))
plot(x=times,y=eps[1]*times,type='l',ylim=lims)
points(x=times,y=eps[1]*times)
for (k in 2:10)
  lines(x=times,y=eps[k]*times)
  points(x=times,y=eps[k]*times)

windows()
eps <- rnorm(300)
path <- (0:300)*c(0,eps)
plot(x=0:300,y=path,type='l')
```

Note the use of the variable `lims`, to set the limits of successive plots, and the function `windows()`, to open a new window for plotting without overwriting the previous one.

It is easy to see the two processes have little in common, despite the fact that their marginal PDFs are the same for every time instant.

FIGURE 3.16   **A few sample paths of the stochastic process of Eq. (3.26).**

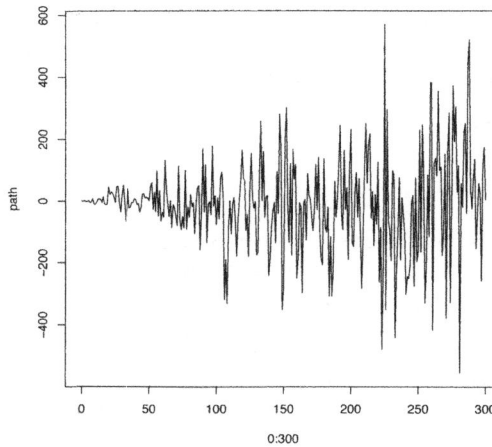

FIGURE 3.17   **A single sample path of the stochastic process of Eq. (3.27).**

The example above deals with dependence over time for a monodimensional stochastic process, i.e., in the longitudinal sense. Specifying marginals for each time instant falls short of capturing the structure of a stochastic process. We will make similar considerations later, when dealing with the Wiener process. The next example illustrates the case of cross-sectional dependence between two random variables: We may have quite different joint distributions, sharing the same marginals.

◾ Example 3.4 **Marginals are not enough: Part 2**

Consider the following joint PDFs (in a moment we will check that they are legitimate densities):

$$f_{X,Y}(x,y) = 1, \qquad\qquad 0 \le x, y \le 1,$$
$$g_{X,Y}(x,y) = 1 + (2x-1)(2y-1), \qquad 0 \le x, y \le 1.$$

They look quite different, but it is not too difficult to see that they yield the same marginals. The first case is easy:

$$f_X(x) = \int_0^1 f_{X,Y}(x,y)\,dy = \int_0^1 1\,dy = 1.$$

By symmetry, we immediately see that $f_Y(y) = 1$ as well. Hence, the two marginals are two uniform distributions on the unit interval $[0,1]$. Now let us tackle the second case. When integrating with respect to y, we should just treat x as a constant:

$$
\begin{aligned}
g_X(x) &= \int_0^1 [1 + (2x-1)(2y-1)]\,dy \\
&= \int_0^1 1\,dy \ + \ (2x-1)\int_0^1 (2y-1)\,dy \\
&= 1 + (2x-1)[y^2 - y]\big|_0^1 = 1.
\end{aligned}
$$

As before, it is easy to see by symmetry that $f_Y(y) = 1$ as well. Again, the two marginals are two uniform distributions, but the links between the two random variables are quite different.

Before closing the example, it is easy to see that $f_{X,Y}(x,y)$ is a legitimate density as it is never negative and

$$\int_0^1 \int_0^1 1\,dx\,dy = 1.$$

Actually, this is just the area of a unit square. Checking the legitimacy of $g_{X,Y}(x,y)$ is a bit more difficult. The easy part is checking that the integral over the unit square is 1. Given the marginals above, we may write

$$\int_0^1 \int_0^1 g_{X,Y}(x,y)\,dy\,dx = \int_0^1 g_X(x)\,dx = \int_0^1 1\,dx = 1.$$

We should also check that the function is never negative; this is left as an exercise.

The two examples point out that there is a gap between two marginals and a joint distribution. The missing link is exactly what characterizes the dependence between the two random variables. This missing link is the subject of a whole branch of probability theory, which is called *copula theory*; a copula is a function capturing the essential nature of the dependence between random variables, separating it from the marginal distributions. Copulas are described in Section 3.4. In the next section we recall a more classical, simpler, but partial characterization of the links between random variables, based on covariance and correlation.

3.3.2 COVARIANCE AND PEARSON'S CORRELATION

If two random variables are not independent, we need a way to measure their degree of dependence. This is implicit in the joint distribution, which may be difficult if not impossible to manage and estimate, in general. The simplest measure in this vein is covariance.

DEFINITION 3.2 (Covariance) *The covariance between random variables X and Y is defined as*

$$\mathrm{Cov}(X, Y) \equiv \mathrm{E}\big[(X - \mathrm{E}[X])(Y - \mathrm{E}[Y])\big].$$

Usually, the covariance between X and Y is denoted by σ_{XY}.

Covariance is the expected value of the product of two deviations from the mean, and its sign depends on the signs of the two factors. We have positive covariance when the events $\{X > \mathrm{E}[X]\}$ and $\{Y > \mathrm{E}[Y]\}$ tend to occur together, as well as the events $\{X < \mathrm{E}[X]\}$ and $\{Y < \mathrm{E}[Y]\}$, because the signs of the two factors in the product tend to be the same. If the signs tend to be different, we have a negative covariance. What covariance tries to capture is whether the two variables tend to be on the same side of their respective means, both above or both below, in which case covariance is positive, or on opposite sides, in which case covariance is negative.

We recall a few properties of covariance, which are easy to check:

$$\mathrm{Cov}(X, Y) = \mathrm{E}[XY] - \mathrm{E}[X] \cdot \mathrm{E}[Y], \qquad (3.28)$$

$$\mathrm{Cov}(X, X) = \mathrm{Var}(X), \qquad (3.29)$$

$$\mathrm{Cov}(X, Y) = \mathrm{Cov}(Y, X), \qquad (3.30)$$

$$\mathrm{Cov}(\alpha + X, Y) = \mathrm{Cov}(X, Y), \quad \text{where } \alpha \in \mathbb{R}, \qquad (3.31)$$

$$\mathrm{Cov}(\alpha X, Y) = \alpha \, \mathrm{Cov}(X, Y), \quad \text{where } \alpha \in \mathbb{R}, \qquad (3.32)$$

$$\mathrm{Cov}(X, Y + Z) = \mathrm{Cov}(X, Y) + \mathrm{Cov}(X, Z). \qquad (3.33)$$

Equation (3.28) yields a convenient way of computing covariance. Equation (3.29) shows that covariance is, as the name suggests, a generalization of variance; this is also reflected in the notation $\sigma_{XX} = \sigma_X^2$. Equation (3.30) points out an important issue: Covariance is a measure of association, but it has nothing to do with cause–effect relationships. Equation (3.31) shows that covariance, just like variance, is not affected by a deterministic shift. Putting Eqs. (3.32) and (3.33) together, we may prove the following theorem.

THEOREM 3.3 (Variance of a sum of random variables) *Given a collection of random variables X_i, $i = 1, \ldots, n$, we obtain*

$$\operatorname{Var}\left(\sum_{i=1}^{n} X_i\right) = \sum_{i=1}^{n} \operatorname{Var}(X_i) + 2 \sum_{i=1}^{n} \sum_{j<i} \operatorname{Cov}(X_i, X_j).$$

Covariance is a generalization of variance. Hence, it is not surprising that it shares a relevant shortcoming: Its value depends on how we measure the underlying quantities. To obtain a measure that does not depend on units of measurement, we may consider the covariance between standardized versions of X and Y:

$$\operatorname{Cov}\left(\frac{X - \mu_X}{\sigma_X}, \frac{Y - \mu_Y}{\sigma_Y}\right) = \frac{\operatorname{Cov}(X, Y)}{\sigma_X \sigma_Y},$$

where μ_X, μ_Y, σ_X, and σ_Y are expected values and standard deviations of X and Y, respectively, and we have used properties of covariance. This motivates the definition of the correlation coefficient

DEFINITION 3.4 (Correlation coefficient) *The correlation coefficient between random variables X and Y is defined as*

$$\rho_{XY} \equiv \frac{\operatorname{Cov}(X, Y)}{\sqrt{\operatorname{Var}(X)}\sqrt{\operatorname{Var}(Y)}} = \frac{\sigma_{XY}}{\sigma_X \sigma_Y}.$$

To be precise, the above definition corresponds to Pearson's correlation; later we will see that there are alternative definitions of similar measures. The coefficient of correlation is adimensional, and it can be easily interpreted, as it can take only values in the interval $[-1, 1]$. While we usually do not prove theorems in this book, we make an exception here as the proof is instructive.

THEOREM 3.5 *The correlation coefficient ρ_{XY} may only take values in the range $[-1, 1]$. If $\rho_{XY} = \pm 1$, then X and Y are related by $Y = a + bX$, where the sign of b is the sign of the correlation coefficient.*

PROOF Consider the following linear combination of X and Y:

$$Z = \frac{X}{\sigma_X} + \frac{Y}{\sigma_Y}.$$

We know that variance cannot be negative; hence

$$\operatorname{Var}\left(\frac{X}{\sigma_X} + \frac{Y}{\sigma_Y}\right) = \operatorname{Var}\left(\frac{X}{\sigma_X}\right) + \operatorname{Var}\left(\frac{Y}{\sigma_Y}\right) + 2\operatorname{Cov}\left(\frac{X}{\sigma_X}, \frac{Y}{\sigma_Y}\right)$$

$$= 1 + 1 + 2\rho_{X,Y} \geq 0.$$

This inequality immediately yields $\rho_{X,Y} \geq -1$. By the same token, consider a slightly different linear combination:

$$\text{Var}\left(\frac{X}{\sigma_X} - \frac{Y}{\sigma_Y}\right) = 1 + 1 - 2\rho_{X,Y} \geq 0 \quad \Rightarrow \quad \rho_{X,Y} \leq 1.$$

We also know that if $\text{Var}(Z) = 0$, then Z must be a constant. In the first case, variance will be zero if $\rho = -1$. Then, we may write

$$Z = \frac{X}{\sigma_X} + \frac{Y}{\sigma_Y} = \alpha$$

for some constant α. Rearranging the equality, we have

$$Y = -X\frac{\sigma_Y}{\sigma_X} + \alpha\sigma_Y.$$

This can be rewritten as $Y = a + bX$, and since standard deviations are non-negative, we see that the slope is negative. Considering the second linear combination yields a similar relationship for the case $\rho = 1$, where the slope b is positive. ∎

Given the theorem above, it is fairly easy to interpret a specific value of correlation:

- A value close to 1 shows a strong degree of positive correlation.
- A value close to -1 shows a strong degree of negative correlation.
- If correlation is zero, we speak of *uncorrelated variables*.

A visual illustration of correlation is given in Fig. 3.18. Each scatterplot shows a sample of 100 joint observations of two random variables for different values of correlation.[18] The effect of correlation is quite evident if we think to "draw a line" going through each cloud of point; the slope of the line corresponds to the sign of the correlation. In the limit case of $\rho = \pm 1$, the observations would exactly lie on a line.

We easily see that if two variables are independent, then their covariance is zero. In fact, independence implies $\text{E}[XY] = \text{E}[X] \cdot \text{E}[Y]$; this can be shown, e.g., by taking advantage of the factorization of the joint PDF into the product of the two marginals. Then, using the property (3.28), we see that independence implies that covariance is zero. However, the converse is *not* true in general, as we may see from the following counterexample.

---

[18] Actually, they are sampled from a joint normal distribution with $\mu_1 = \mu_2 = 10, \sigma_1 = \sigma_2 = 5$. The multivariate normal distribution is introduced in Section 3.3.4.

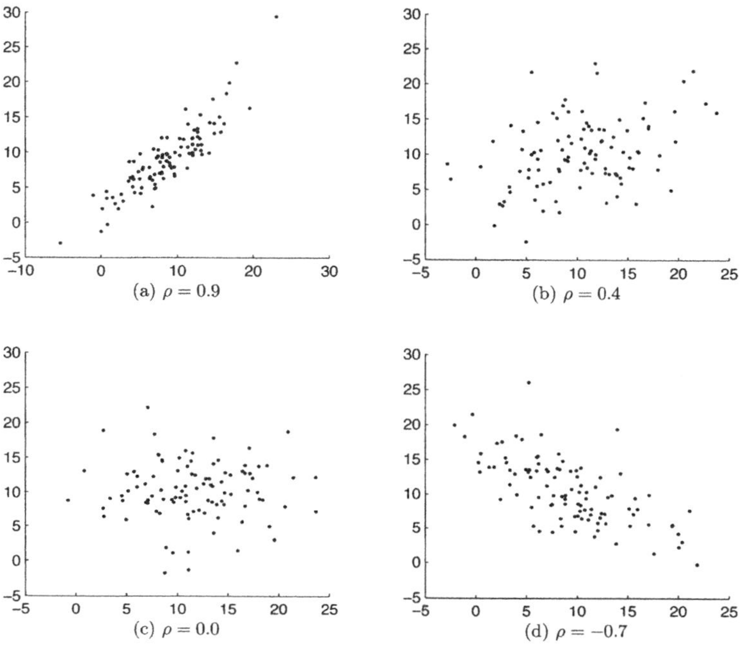

FIGURE 3.18 **Samples of jointly normal variables for different values of the correlation coefficient ρ.**

▣ Example 3.5 **Two dependent but uncorrelated random variables**

Let us consider a uniform random variable on the interval $[-1, 1]$; its expected value is zero, and on its support the density function is constant and given by $f_X(x) = \frac{1}{2}$. Now, define random variable Y as

$$Y = \sqrt{1 - X^2}.$$

Clearly, there is a very strong interdependence between X and Y because, given the realization of X, Y is perfectly predictable. However, their covariance is zero! We know from Eq. (3.28) that

$$\mathrm{Cov}(X, Y) = \mathrm{E}[XY] - \mathrm{E}[X]\mathrm{E}[Y],$$

but the second term is clearly zero, as $\mathrm{E}[X] = 0$. Furthermore,

$$\mathrm{E}[XY] = \int_{-1}^{1} x\sqrt{1 - x^2} \cdot \frac{1}{2}\, dx = 0,$$

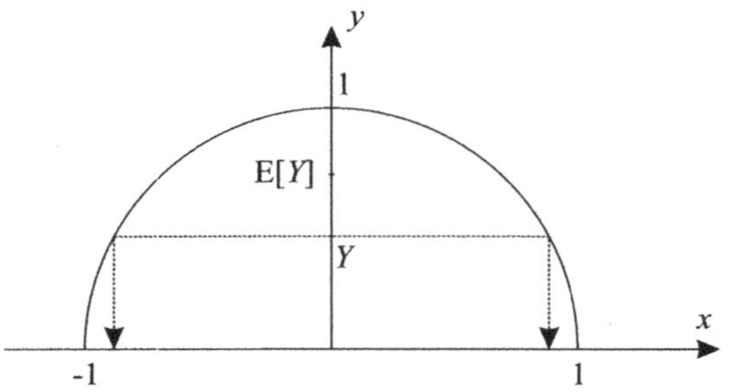

FIGURE 3.19 **A counterexample about covariance.**

because of the symmetry of the integrand function, which is an odd function, in the sense that $f(-x) = -f(x)$. Hence, the first term is zero as well, and the two variables have zero covariance and are uncorrelated. One intuitive way to explain the weird finding of this example is the following. First note that points with coordinates (X, Y) lie on the upper half of the unit circumference $X^2 + Y^2 = 1$. But if $Y < \mathrm{E}[Y]$, we may have either $X > \mathrm{E}[X]$ or $X < \mathrm{E}[X]$. In other words, if Y is above its mean, there is no way to tell whether X will be above or below its mean. This is illustrated in Fig. 3.19. A similar consideration applies when $Y > \mathrm{E}[Y]$.

The example shows that covariance is not really a perfect measure of dependence, as it may be zero in cases in which there is a very strong dependence. In fact, covariance is rather a measure of *concordance* between a pair of random variables, which is related to the signs of the deviations with respect to the mean. Furthermore, covariance measures a *linear* association between random variables. A strong nonlinear link, as the one in Fig. 3.19, may not be detected at all, or only partially.

3.3.3   R FUNCTIONS FOR COVARIANCE AND CORRELATION

In real life, covariances and correlations must be estimated on the basis of empirical data. We will consider estimation issues in the next chapter, but R offers two basic and easy-to-use functions to estimate covariance and correlation. Not quite surprisingly, they are called `cov` and `cor`. To illustrate them, let us generate two independent samples from the standard normal distribution, and combine the observations in order to induce correlation. It is easy to show, and is left as an exercise, that if variables $Z_1, Z_2 \sim \mathsf{N}(0, 1)$ are independent, then the variables X_1 and X_2 defined as

$$X_1 = Z_1, \quad X_2 = \rho Z_1 + \sqrt{1 - \rho^2} Z_2 \qquad (3.34)$$

are still standard normal but have correlation ρ:

```
> set.seed(55555)
> rho <- 0.8
> Z1 <- rnorm(100000)
> Z2 <- rnorm(100000)
> X1 <- Z1
> X2 <- rho*Z1 + sqrt(1-rho^2)*Z2
> cor(Z1,Z2)
[1] -0.00249354
> cor(X1,X2)
[1] 0.8002211
> cov(X1,X2)
[1] 0.7978528
```

Note that covariance and correlation, since the two variables are standard, should be the same, but they are not because of sampling errors. In the above snapshot, we have invoked `cov` and `cor` by passing two vectors as input arguments. By doing so, we obtain covariance and correlation between the corresponding variables. If we are interested in covariance and correlation matrices, we should collect the input data in a matrix:

```
> X <- as.matrix(cbind(2*X1,-3*X2))
> cov(X)
          [,1]      [,2]
[1,]  3.996719 -4.787117
[2,] -4.787117  8.954152
> cor(X)
           [,1]       [,2]
[1,]  1.0000000 -0.8002211
[2,] -0.8002211  1.0000000
> cov2cor(cov(X))
           [,1]       [,2]
[1,]  1.0000000 -0.8002211
[2,] -0.8002211  1.0000000
```

Note the use of the function `cbind` to collect two vectors side by side as columns, and `as.matrix` to transform the result to a matrix. The function `cov2cor` can be used to transform the former into the latter. Once again, we suggest to compare the sample estimates with the theoretical values.

If you try getting online help on `cor`, you will notice that there is an additional parameter, `method = c("pearson", "kendall", "spearman")`. By default, we obtain Pearson's correlations, but there are alternative definitions of correlation, which we discuss later in Section 3.4. To understand the rationale between them, let us consider the following snapshot:

```
> set.seed(55555)
```

```
> X=runif(100000)
> cor(X,X)
[1] 1
> cor(X,X^2)
[1] 0.9683232
> cor(X,X^4)
[1] 0.8660849
```

As expected, the correlation between X and itself is 1, but if we raise X to increasing powers, we decrease correlation. This is due to the nonlinearity in X^2 and X^4, which is lost in a linear measure of correlation. The larger the power, the less linear the resulting relationship, which is reflected in a diminishing value of Pearson's correlation. However, we may well argue that, at a deeper level, there is no substantial difference in the dependence between X and itself and between X and X^4. We will see that Kendall's and Spearman's correlations avoid this issue.[19]

3.3.4 SOME TYPICAL MULTIVARIATE DISTRIBUTIONS

In this section we illustrate two commonly used multivariate distributions, namely, the multivariate counterparts of normal and t distributions. They are similar in the sense that:

1. They are both characterized by the vector of expected values $\boldsymbol{\mu}$ and the covariance matrix $\boldsymbol{\Sigma}$.
2. They belong to the family of elliptical distributions.

The essential difference is that, just like the univariate case, the t distribution has fatter tails, which are related to the number of degrees of freedom.

3.3.4.1 Multivariate normal

An n-dimensional multivariate normal variable $\mathbf{X} \sim \mathsf{N}(\boldsymbol{\mu}, \boldsymbol{\Sigma})$, $\boldsymbol{\mu} \in \mathbb{R}^n$, $\boldsymbol{\Sigma} \in \mathbb{R}^{n,n}$, has PDF

$$f_{\mathbf{X}}(\mathbf{x}) = \frac{1}{\sqrt{(2\pi)^n \det(\boldsymbol{\Sigma})}} \exp\left\{ -\frac{1}{2}(\mathbf{x} - \boldsymbol{\mu})^{\mathsf{T}} \boldsymbol{\Sigma}^{-1} (\mathbf{x} - \boldsymbol{\mu}) \right\}. \qquad (3.35)$$

Strictly speaking, this applies to a nonsingular distribution, i.e., when $\boldsymbol{\Sigma}$ is positive definite. If the covariance matrix is only positive semidefinite, then there is a linear combination of variables, with coefficients $\mathbf{w}$, such that

$$\mathrm{Var}\left(\mathbf{w}^{\mathsf{T}}\mathbf{X}\right) = \mathbf{w}^{\mathsf{T}}\boldsymbol{\Sigma}\mathbf{w} = 0.$$

From a statistical point of view this means that there is some redundancy in the variables, since we may find a linear combination that is constant; from a

---

[19] You are advised *not* to try estimating Kendall's correlation with R on a very large data set, as this may require a lot of computation and is likely to block your computer.

mathematical point of view this implies that the covariance matrix cannot be inverted and we cannot write the PDF as in Eq. (3.35).

We immediately see that, when $\mu = 0$ and $\Sigma = I$, the joint density boils down to

$$f_{\mathbf{Z}}(\mathbf{z}) = \frac{1}{\sqrt{(2\pi)^n}} \exp\left\{-\frac{1}{2}\mathbf{z}^{\mathsf{T}}\mathbf{z}\right\},$$

i.e., the product of n independent standard densities. We may obtain a generic multivariate normal from the standard $\mathbf{Z} \sim N(\mathbf{0}, \mathbf{I})$, and the other way around, by the transformations

$$\mathbf{X} = \mu + \mathbf{A}\mathbf{Z}, \qquad \mathbf{Z} = \mathbf{A}^{-1}(\mathbf{X} - \mu), \tag{3.36}$$

where $\mathbf{A}$ is a matrix such that $\Sigma = \mathbf{A}\mathbf{A}^{\mathsf{T}}$. These formulas are straightforward generalizations of typical standardization (also called *whitening* in this context) and destandardization. Possible choices of $\mathbf{A}$ are the lower triangular Cholesky factor $\mathbf{L}$ of Σ, or its square root $\Sigma^{-1/2}$; unlike the Cholesky factor, the square root is symmetric and is defined by $\Sigma = \Sigma^{-1/2}\Sigma^{-1/2}$.

■ **Example 3.6** **A Cholesky factor**

> Let us consider two standard normal variables with correlation coefficient ρ. The covariance matrix and the corresponding lower triangular Cholesky factor are
>
> $$\Sigma = \begin{bmatrix} 1 & \rho \\ \rho & 1 \end{bmatrix}, \qquad \mathbf{L} = \begin{bmatrix} 1 & 0 \\ \rho & \sqrt{1-\rho^2} \end{bmatrix}.$$
>
> The reader is urged to check that indeed $\Sigma = \mathbf{L}\mathbf{L}^{\mathsf{T}}$, and that using this matrix in Eq. (3.36) generates correlated normals as in Eq. (3.34).

The following snapshot shows two R functions, `chol` and `sqrtm` from package `expm` to carry out the above factorizations:

```
> Sigma <- matrix(c(10,2,2,8),nrow=2,ncol=2)
> A <- chol(Sigma)
> t(A) %*% A
     [,1] [,2]
[1,]   10    2
[2,]    2    8
> library(expm)
> sqrtSigma <- sqrtm(Sigma)
> sqrtSigma
           [,1]       [,2]
[1,] 3.1443790 0.3359774
[2,] 0.3359774 2.8084015
```

```
> sqrtSigma %*% sqrtSigma
     [,1] [,2]
[1,]   10    2
[2,]    2    8
```

Note that `chol` actually returns the upper triangular factor $\mathbf{L}^{\mathsf{T}}$. As we shall see in Chapter 5, these transformations are also used to generate samples of correlated multivariate normals from a set of independent standard variates.

We may generalize the well-known property that a linear combination of (jointly) normals is normal as follows.

PROPERTY 3.6 *Let* $\mathbf{X}_1, \mathbf{X}_2, \ldots, \mathbf{X}_r$ *be* m_i-*dimensional, independent normal variables,* $\mathbf{X}_i \sim \mathsf{N}(\boldsymbol{\mu}_i, \boldsymbol{\Sigma}_i)$, $i = 1, \ldots, r$. *Consider a vector* $\boldsymbol{\alpha} \in \mathbb{R}^n$ *and matrices* $\mathbf{A}_i \in \mathbb{R}^{n, m_i}$, $i = 1, \ldots, r$. *Then*

$$\boldsymbol{\alpha} + \sum_{i=1}^{r} \mathbf{A}_i \mathbf{X} \sim \mathsf{N}\left(\boldsymbol{\alpha} + \sum_{i=1}^{r} \mathbf{A}_i \boldsymbol{\mu}_i, \sum_{i=1}^{r} \mathbf{A}_i \boldsymbol{\Sigma}_i \mathbf{A}_i^{\mathsf{T}} \right).$$

Later, we will appreciate the usefulness of the following theorem.[20]

THEOREM 3.7 (Conditional distribution for multivariate normals) *Let us consider an* n-*dimensional multivariate normal variable* $X \sim \mathsf{N}(\boldsymbol{\mu}, \boldsymbol{\Sigma})$ *and the partition* $\mathbf{X} = [\mathbf{X}_1, \mathbf{X}_2]^{\mathsf{T}}$, *where* $\mathbf{X}_1$ *and* $\mathbf{X}_2$, *have dimensions* n_1 *and* n_2, *respectively, with* $n_1 + n_2 = n$. *Let us define the corresponding partitions*

$$\boldsymbol{\mu} = \begin{bmatrix} \boldsymbol{\mu}_1 \\ \boldsymbol{\mu}_2 \end{bmatrix}, \quad \boldsymbol{\Sigma} = \begin{bmatrix} \boldsymbol{\Sigma}_{11} & \boldsymbol{\Sigma}_{12} \\ \boldsymbol{\Sigma}_{21} & \boldsymbol{\Sigma}_{22} \end{bmatrix},$$

where vectors and matrices have the appropriate dimensions. Then, the conditional distribution of X_1, *given that* $X_2 = x$, *is* $\mathsf{N}(\widetilde{\boldsymbol{\mu}}, \widetilde{\boldsymbol{\Sigma}})$, *where*

$$\widetilde{\boldsymbol{\mu}} = \boldsymbol{\mu}_1 + \boldsymbol{\Sigma}_{12} \boldsymbol{\Sigma}_{22}^{-1} (x - \boldsymbol{\mu}_2), \quad \widetilde{\boldsymbol{\Sigma}} = \boldsymbol{\Sigma}_{11} - \boldsymbol{\Sigma}_{12} \boldsymbol{\Sigma}_{22}^{-1} \boldsymbol{\Sigma}_{21}.$$

Example 3.7 **Conditioning in a bivariate normal**

Consider a bivariate normal distribution consisting of two standard normals Z_1 and Z_2 with correlation ρ. Then

$$\boldsymbol{\mu} = \begin{bmatrix} 0 \\ 0 \end{bmatrix}, \quad \boldsymbol{\Sigma} = \begin{bmatrix} 1 & \rho \\ \rho & 1 \end{bmatrix},$$

---

[20]See, e.g., [7] for a proof.

and by Theorem 3.7 the distribution of Z_1 conditional on $Z_2 = z$ is normal with

$$\widetilde{\mu} = 0 + \rho \cdot 1^{-1} \cdot (z - 0) = \rho z,$$
$$\widetilde{\Sigma} = 1 - \rho \cdot 1^{-1} \cdot \rho = 1 - \rho^2.$$

The last theorem is useful for path generation by the Brownian bridge (Section 6.2.3) and variance reduction by stratification (Section 8.5).

3.3.4.2 Multivariate Student's t distribution

The multivariate t distribution is characterized by the PDF

$$f_{\mathbf{X}}(\mathbf{x}) = \frac{\Gamma\left(\dfrac{\nu + n}{2}\right)}{(\pi\nu)^{n/2} \Gamma\left(\dfrac{\nu}{2}\right) \sqrt{\det(\Sigma)}} \left[1 + \frac{1}{\nu}(\mathbf{x} - \boldsymbol{\mu})^{\mathsf{T}} \Sigma^{-1}(\mathbf{x} - \boldsymbol{\mu})\right]^{-(\nu+n)/2}.$$

(3.37)

With respect to the multivariate normal, there is an additional parameter, ν, corresponding to the degrees of freedom, and the notation $t_\nu(\boldsymbol{\mu}, \Sigma)$ is used. The essential moments are

$$\mathrm{E}[\mathbf{X}] = \boldsymbol{\mu}, \quad \mathrm{Cov}(\mathbf{X}) = \frac{\nu}{\nu - 2}\Sigma.$$

We notice that when $\nu \to \infty$, the covariance matrix tends to Σ, just like in the univariate case. More generally, the PDF of the multidimensional t tends to the PDF of a multivariate normal when $\nu \to \infty$. The matrix Σ, in this case, is referred to as a scale parameter, whereas ν is a shape parameter. The most relevant difference between the multivariate normal and t distributions is seen in the tails, which are influenced by ν. If we set aside the complicated normalization factors involving the gamma function and look inside the big square parentheses in Eq. (3.37), we notice that the level curves of the PDF are essentially determined by the same term as in the normal case:

$$(\mathbf{x} - \boldsymbol{\mu})^{\mathsf{T}} \Sigma^{-1}(\mathbf{x} - \boldsymbol{\mu}).$$

If we fix the value of this quantity and draw the corresponding level curve, we obtain an ellipse, which explains why we speak of elliptical distributions. Both distributions are symmetrical with respect to the expected value.

3.3.4.3 The `mvtnorm` package

The `mvtnorm` package offers functions to calculate the PDF and to generate random samples from multivariate normal and t distributions:

```
library(mvtnorm)
mu <- c(20,10)
sigma <- matrix(c(25,35,35,100), nrow=2)
nu <- 2
pts <- seq(-20,60,0.1)
n <- length(pts)
X <- rep(pts,n)
Y <- rep(pts,rep(n,n))
grid <- cbind(X,Y)
densNorm <- dmvnorm(grid, mu,sigma,log=TRUE)
Z1 <- array(densNorm, dim=c(n,n))
contour(pts,pts,Z1,levels=seq(-10,-4))
windows()
densT <- dmvt(grid,mu,sigma,df=nu,log=TRUE)
Z2 <- array(densT, dim=c(n,n))
contour(pts,pts,Z2,levels=seq(-10,-4))
```

FIGURE 3.20 **Script to illustrate the** `mvtnorm` **package.**

- `dmvnorm(x, mean, sigma, log=FALSE)` gives the PDF for a multivariate normal; the parameter `log` can be used to obtain the log-density, i.e, the logarithm of the PDF, which may be useful to "compress" the range of values.
- The function for the t distribution, `dmvt(x, delta, sigma, df, log = TRUE, type = "shifted")`, has a different default for `log`; the parameter `delta` is interpreted as a shift, i.e., a noncentrality parameter.

The corresponding r functions for random variate generation are `rmvnorm` and `rmvt`; there are also q and p functions, but they are of course more problematic than the corresponding univariate functions. The script in Fig. 3.20 shows how to use R functions to plot the *logarithm* of the density of these two distributions. The resulting contour plots are displayed in Fig. 3.21. We should not be surprised by the negative values, since we are displaying log-densities to make the plots more readable. We immediately observe the elliptic character of both densities and the effect of positive correlation. The most striking difference, as we have remarked, is in the fatter tails of the t distribution.

3.4   Modeling dependence with copulas

The idea behind copulas is to factor an n-dimensional multivariate distribution into two components:

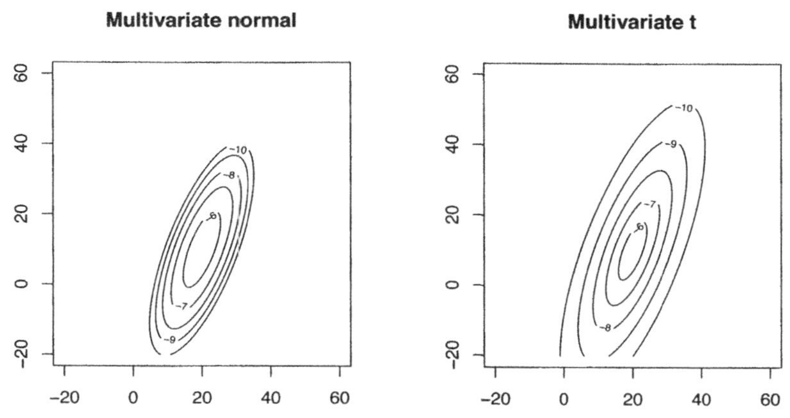

Multivariate normal

Multivariate t

FIGURE 3.21 **Contour plots of the log-density of multivariate normal and t distributions.**

1. A set of n univariate marginal CDFs, $F_1(x_1), \ldots, F_n(x_n)$. Note that CDFs are used, rather than PDFs; incidentally, this allows for more generality, since CDFs have wider applicability than PDFs and PMFs, even though in the following treatment we will only consider continuous distributions.

2. A function $C(u_1, \ldots, u_n)$, called copula, which captures the dependence among the random variables in a "standard" way. The copula maps the n-dimensional hypercube $[0, 1]^n$ into the unit interval $[0, 1]$, and it may be interpreted as the joint CDF of a multivariate distribution with uniform marginals.

The idea is to express the n-dimensional CDF as

$$F_{\mathbf{X}}(x_1, \ldots, x_n) = C\big(F_1(x_1), \ldots, F_n(x_n)\big).$$

Indeed, since $C(\cdots)$ must map a set of marginal probabilities into another probability, it must be a function mapping $[0, 1]^n$ into $[0, 1]$. However, there are additional requirements that we detail below, referring to a two-dimensional copula for the sake of simplicity.

Consider a two-dimensional copula $C(u_1, u_2)$. A reasonable requirement is that, for any $u_1, u_2 \in [0, 1]$,

$$C(0, u_2) = C(u_1, 0) = 0.$$

This amounts to saying that if $u_1 = \mathrm{P}\{X_1 \le x_1\} = 0$, then $F_{\mathbf{X}}(x_1, x_2) = 0$ for whatever value of x_2. In fact, below the support of the distribution of X_1, the CDF must be zero; or, if the support is unbounded below, the limit for

Illustrating a property of multidimensional CDFs.

$x_1 \to -\infty$ must be 0. A similar consideration applies to the second variable x_2, as well as to any variable in a multidimensional case. This leads to the following general definition.

DEFINITION 3.8 (Grounded function) *Consider a function $H(\cdots)$ of n variables, with domain $S_1 \times \cdots \times S_n$, where each set S_k has a smallest element a_k. We say that H is grounded if $H(y_1, \ldots, y_n) = 0$, whenever at least one $y_k = a_k$.*

Thus, we have to require that a copula is a grounded function. In the case of a copula, $S_k = [0, 1]$ and $a_k = 0$. After looking below the support, let us move to the opposite side and have a look above it. There, a CDF must take the value 1; or, if the support is unbounded above, the limit for $x_k \to +\infty$ must be 1. More generally, the copula should have sensible margins.

DEFINITION 3.9 (Margins) *Consider a function $H(\cdots)$ of n variables with domain $S_1 \times \cdots \times S_n$, where each set S_k has a largest element b_k. The one-dimensional margin of H corresponding to the kth variable is the function*

$$H_k(y) \equiv H(b_1, \ldots, b_{k-1}, y, b_{k+1}, \ldots, b_n).$$

It is possible to define higher-dimensional margins, but we will simply refer to one-dimensional margins as margins.

On the basis of this definition, we see that a copula $C(\cdots)$ should have margins $C_k(\cdot)$ such that

$$C_k(u) = u.$$

As a last requirement, consider a two-dimensional CDF and the region illustrated in Fig. 3.22. The darkest rectangle in the figure is the set $[a_1, b_1] \times [a_2, b_2]$. Let us denote this set by S; given the CDF $F_{\mathbf{X}}(x_1, x_2)$, it is easy to see that

$$P\{(X_1, X_2) \in S\} = F_{\mathbf{X}}(b_1, b_2) - F_{\mathbf{X}}(b_1, a_2) - F_{\mathbf{X}}(a_1, b_2) + F_{\mathbf{X}}(a_1, a_2).$$

Since this is a probability, the expression on the right must be positive. This leads to the following definition.

DEFINITION 3.10 (2-increasing function) *Consider a function $H(\cdot, \cdot)$ of two variables, and let $S = [a_1, b_1] \times [a_2, b_2]$ be a rectangle with all vertices in the domain of $H(\cdot, \cdot)$. We define the H-volume of S as*

$$V_H(S) \equiv H(b_1, b_2) - H(b_1, a_2) - H(a_1, b_2) + H(a_1, a_2).$$

We say that the function is 2-increasing if $V_H(S) \geq 0$ for any such rectangle.

The definition may be extended to n-dimensional functions, but we omit the technicalities involved.[21] It should be noted that the 2-increasing property does not imply and is not implied by the fact that the function is nondecreasing in each of its arguments. This is well illustrated by the following counterexamples.[22]

Example 3.8 **Nondecreasing vs. 2-increasing functions**

Consider the function

$$H(x_1, x_2) = \max\{x_1, x_2\}$$

defined on $I^2 = [0, 1]^2$. It is immediate to check that the function is nondecreasing in x_1 and x_2. However, it is not 2-increasing, as the H-volume of I^2 is

$$V_H(I^2) = \max\{1, 1\} - \max\{1, 0\} - \max\{0, 1\} + \max\{0, 0\} = -1.$$

On the contrary, let us consider

$$G(x_1, x_2) = (2x_1 - 1)(2x_2 - 1)$$

defined on $I^2 = [0, 1]^2$. Clearly, the function is decreasing in x_1 for $x_2 < 0.5$, and it is decreasing in x_2 for $x_1 < 0.5$. Nevertheless, it is 2-increasing. To see this, consider the rectangle $S = [a_1, b_1] \times [a_2, b_2]$,

---

[21]See, e.g., [6] or [18].
[22]These counterexamples are borrowed from [18, p. 6].

where $b_1 \geq a_1$ and $b_2 \geq a_2$, and compute its G-volume:

$$
\begin{aligned}
V_G(S) &\equiv (2b_1 - 1)(2b_2 - 1) - (2a_1 - 1)(2b_2 - 1) \\
&\quad - (2b_1 - 1)(2a_2 - 1) + (2a_1 - 1)(2a_2 - 1) \\
&= 4b_1 b_2 - 2b_1 - 2b_2 + 1 \\
&\quad -4a_1 b_2 + 2a_1 + 2b_2 - 1 \\
&\quad -4b_1 a_2 + 2b_1 + 2a_2 - 1 \\
&\quad +4a_1 a_2 - 2a_1 - 2a_2 + 1 \\
&= 4(b_1 - a_1)(b_2 - a_2) \geq 0.
\end{aligned}
$$

All of the above considerations are summarized in the following definition.

DEFINITION 3.11 (Copula) *An n-dimensional copula is a function $C(\cdots)$, defined on the domain $I^n = [0,1]^n$, such that:*

    1. $C(\cdots)$ *is grounded and n-increasing.*

    2. $C(\cdots)$ *has margins C_k, $k = 1, \ldots, n$ satisfying $C_k(u) = u$, for $u \in [0,1]$.*

We have stated requirements such that, given an n-dimensional distribution with marginal CDFs F_k, the function

$$ F_{\mathbf{X}}(x_1, \ldots, x_n) = C\big(F_1(x_1), \ldots, F_n(x_n)\big) \tag{3.38} $$

is a sensible CDF. One might wonder whether we can also go the other way around: Given a joint CDF, is it possible to decompose it in the form of Eq. (3.38)? The answer is positive, thanks to the following fundamental theorem.

THEOREM 3.12 (Sklar's theorem) *Let $F_{\mathbf{X}}$ be an n-dimensional CDF with continuous marginal CDFs $F_1(\cdot), \ldots, F_n(\cdot)$. Then, there exists a unique copula $C(\cdots)$ such that*

$$ F_{\mathbf{X}}(x_1, \ldots, x_n) = C\big(F_1(x_1), \ldots, F_n(x_n)\big), \qquad \forall \mathbf{x}. $$

Note that the above theorem statement asserts uniqueness for *continuous* distributions. In fact, Sklar's theorem also states the existence of a copula for discrete distributions, but there are technical issues concerning uniqueness that we want to avoid here. Since we are just dealing with continuous distributions, we assume invertibility of marginal CDFs and rewrite Eq. (3.38) as

$$ C(u_1, \ldots, u_n) = F_{\mathbf{X}}\big(F_1^{-1}(x_1), \ldots, F_n^{-1}(x_n)\big). \tag{3.39} $$

When looking at this equation, it is important to realize that if we evaluate a marginal CDF $F_k(x)$ for $x = X_k$, where X_k is a random variable characterized by that CDF F_k, then $F_k(X_k)$ is a uniform random variable. To see this, note that, assuming invertibility of the margin and taking $u \in [0, 1]$,

$$P\{F_k(X_k) \le u\} = P\{X_k \le F_k^{-1}(u)\} = F_k\big(F_k^{-1}(u)\big) = u. \qquad (3.40)$$

However, this is just the CDF of a uniform distribution. This fact plays a fundamental role in random variate generation.[23] Indeed, having defined the general concept of a copula, the following tasks must be typically carried out:

1. Choose a copula function, along with marginals, to characterize a multivariate distribution of interest.

2. Estimate the parameters characterizing the copula.

3. Generate random variates based on the selected copula.

We leave the second and the third task to Chapters 4 and 5, respectively. Here we describe a few common copula functions as examples.

The product copula

The product copula is defined as

$$\Pi(u_1, u_2, \ldots, u_n) = u_1 \cdot u_2 \cdots u_n. \qquad (3.41)$$

It is no surprise that, as it turns out, the continuous random variables in the vector $(X_1, \ldots, X_n)^\mathsf{T}$ are independent if and only if their copula is the product copula.

The normal copula

The normal (or Gaussian) copula is obtained by using the CDF of standard normals in Eq. (3.40). Let Φ be the CDF of a standard normal and $\Phi_{\mathbf{R}}^n$ be the joint CDF for an n-dimensional standard normal distribution with correlation matrix $\mathbf{R}$. Then, we may define the normal copula function

$$G_{\mathbf{R}}(\mathbf{u}) = \Phi_{\mathbf{R}}^n\big(\Phi^{-1}(u_1), \ldots, \Phi^{-1}(u_n)\big). \qquad (3.42)$$

Note that this copula cannot be expressed analytically; nevertheless, it is possible to sample from the normal copula.

The t copula

The t copula is built much like the normal copula, on the basis of the CDF of a t distribution with ν degrees of freedom:

$$T_{\nu, \mathbf{R}}(\mathbf{u}) = t_{\nu, \mathbf{R}}^n\big(t_\nu^{-1}(u_1), \ldots, t_\nu^{-1}(u_n)\big), \qquad (3.43)$$

where t_ν is the CDF of a t distribution with ν degrees of freedom and $t_{\nu, \mathbf{R}}^n$ is the CDF of a multivariate t distribution with ν degrees of freedom and correlation matrix $\mathbf{R}$.

---

[23] See the inverse transform method in Chapter 5.

3.4.1   KENDALL'S TAU AND SPEARMAN'S RHO

We have already illustrated a few pitfalls of Pearson's correlation:

- It is only a measure of concordance, rather than dependence; in fact, it essentially captures linear associations.
- It is affected by transformations in the data, even when such transformations do not really affect the underlying association and true dependence.

This has lead to the proposal of different correlation coefficients.

DEFINITION 3.13 (Kendall's tau) *Consider independent and identically distributed random vectors* $\mathbf{X} = [X_1, X_2]^{\mathsf{T}}$ *and* $\mathbf{Y} = [Y_1, Y_2]^{\mathsf{T}}$. *Kendall's tau is defined as*

$$\tau(X_1, X_2) = \mathrm{P}\left\{(X_1 - Y_1)(X_2 - Y_2) > 0\right\} - \mathrm{P}\left\{(X_1 - Y_1)(X_2 - Y_2) < 0\right\}.$$

In plain words, Kendall's tau measures the probability of concordance minus the probability of discordance. Note that in covariance, and as a consequence in defining Pearson's correlation, we take an expected value, which is influenced by the numerical values attained by the random variables. Here we are using a probability in a way that captures some more essential features. This is essentially why tau is not sensitive to monotonic transformations.

DEFINITION 3.14 (Spearman's rho) *Consider a joint CDF* $F_{\mathbf{X}}(x_1, x_2)$, *with marginals* $F_1(\cdot)$ *and* $F_2(\cdot)$. *Spearman's rho is defined as the Pearson's correlation between the transformed random variables* $F_1(X_1)$ *and* $F_2(X_2)$:

$$\rho_S(X_1, X_2) = \rho\big(F_1(X_1), F_2(X_2)\big). \tag{3.44}$$

To get an intuitive grasp of the definition, let us recall that $F_1(X_1)$ and $F_2(X_2)$ are uniformly distributed. Hence, in the definition of ρ_S we pick the two random variables and map them into uniform variables through their CDF, which is a monotonic transformation into the unit interval $[0, 1]$. Hence, we are actually correlating the *ranks* of the realizations of the two variables. Indeed, Spearman's rho is also known as rank correlation. We should mention that sample counterparts have been defined for both Kendall's tau and Spearman's rho, and that there is a deep connection between them and copulas, as well as alternative definitions.[24]

▨ Example 3.9 **Kendall's tau and Spearman's rho**

> The script in Fig. 3.23 illustrates the fact that Spearman's rho and Kendall's tau are not influenced by certain transformations of the un-

---

[24] See, e.g., [6].

```
tryCorrelation <- function(){
  rho <- 0.8
  numObs <- 10000
  X1 <- rnorm(numObs)
  X2 <- rho*X1 + sqrt(1-rho^2)*rnorm(numObs)
  F1 <- exp(X1)
  cat("Correlations",'\n')
  RXX <- cor(X1,X2,method='pearson')
  RFX <- cor(F1,X2,method='pearson')
  cat(" Pearson:  (X1,X2)=", RXX, " (F1,X2)=" , RFX, '\n')
  RXX <- cor(X1,X2,method='spearman')
  RFX <- cor(F1,X2,method='spearman')
  cat(" Spearman: (X1,X2)=", RXX, " (F1,X2)=" , RFX, '\n')
  RXX <- cor(X1,X2,method='kendall')
  RFX <- cor(F1,X2,method='kendall')
  cat(" Kendall:  (X1,X2)=", RXX, " (F1,X2)=" , RFX, '\n')
}
```

FIGURE 3.23 **Illustrating invariance properties of Spearman's rho and Kendall's kappa.**

derlying random variables. For our illustration purposes, we are using their sample counterparts here. We generate a sample of standard normal variables X_1 and X_2, with correlation 0.8. Then we take the exponential of the first sample, $F_1 = e^{X_1}$, and we apply the cor function to estimate different correlations. By the way, in R there is no need to specify the pearson method, as it is the default; hence cor(X1,X2,method='pearson') and cor(X1,X2) are equivalent. Running the script, we obtain

```
> set.seed(55555)
> tryCorrelation()
Correlations
  Pearson:  (X1,X2)= 0.8091203  (F1,X2)= 0.6192021
  Spearman: (X1,X2)= 0.7952385  (F1,X2)= 0.7952385
  Kendall:  (X1,X2)= 0.6000901  (F1,X2)= 0.6000901
```

We observe that Pearson's correlation for the transformed variable is smaller: This is due to the nonlinear nature of the exponential transformation, which in this case weakens the linear relationship between the two random variables. The other two correlation measures are not affected.

3.4.2   TAIL DEPENDENCE

A relevant feature of multivariate distributions, especially for risk management, is tail dependence. The idea is focusing on the joint behavior of random variables when extreme events occur. For the sake of simplicity, we deal here with the upper tail dependence for two random variables X_1 and X_2; the case of lower tail dependence is similar.

DEFINITION 3.15 (Upper tail dependence) *Let us consider two jointly continuous random variables X_1 and X_2 with marginal CDFs $F_1(\cdot)$ and $F_2(\cdot)$. The coefficient of upper tail dependence is defined as*

$$\lambda_U \equiv \lim_{u \to 1^-} P\{X_2 > F_2^{-1}(u) \mid X_1 > F_1^{-1}(u)\}, \qquad (3.45)$$

provided that the limit exists. If $\lambda_U = 0$, then the two random variables are said to be asymptotically independent in the upper tail; otherwise, they are said to be asymptotically dependent in the upper tail.

Since u is reaching its upper limit in the definition, the coefficient λ_U captures the joint behavior of X_1 and X_2 in the limit of the upper-right (North–East) quadrant. It turns out that this property is related to the copula, which means that it is invariant for strictly increasing transformations of the random variables.

Let us recall the following formula from elementary probability:

$$\begin{aligned}
P\{E \cap G\} &= 1 - P\{\overline{E \cap G}\} \\
&= 1 - P\{\overline{E} \cup \overline{G}\} \\
&= 1 - P\{\overline{E}\} - P\{\overline{G}\} + P\{\overline{E} \cap \overline{G}\}, \qquad (3.46)
\end{aligned}$$

where $\overline{E}$ is the complement of event E. It is easy to see that the probability in Eq. (3.45) can be rewritten as

$$\frac{1 - P\{X_1 \le F_1^{-1}(u)\} - P\{X_2 \le F_2^{-1}(u)\} + P\{X_1 \le F_1^{-1}(u), X_2 \le F_2^{-1}(u)\}}{1 - P\{X_1 \le F_1^{-1}(u)\}}.$$
$$(3.47)$$

Therefore, using the copula $C(\cdot, \cdot)$ between X_1 and X_2 and assuming that the limit exists, we may rewrite the above probability as

$$\lambda_U = \lim_{u \to 1^-} \frac{1 - 2u + C(u, u)}{1 - u}. \qquad (3.48)$$

An interesting feature of the normal copula is that, when the correlation coefficient ρ is strictly less than 1, it can be shown that $\lambda_U = 0$, hence, we have tail independence.[25] It is often argued that this fact points out another limitation of the multivariate normal distribution for risk management.

---

[25] See [5] for details.

3.5 Linear regression models: A probabilistic view

Linear regression models are traditionally introduced within an inferential statistics framework. We will see how R can be used to accomplish classical estimation by ordinary least squares in Section 4.4. However, a probabilistic view is also possible, where we consider the problem of approximating a random variable Y by an affine transformation $a + bX$ of another random variable X. This is quite useful in order to better appreciate:

- Certain data reduction methods, which are dealt with later in Section 3.8
- Variance reduction based on control variates, which is discussed in Section 8.3

The key point is how to define a measure of distance between Y and $a + bX$. One possibility is to consider the variance of their difference. If this variance is close to zero, we may argue that the approximation tracks its target pretty well. If we do so and minimize tracking error variance, we formulate the following distance:

$$\text{Var}(Y - a - bX) = \text{Var}(Y - bX) = \sigma_Y^2 + b^2\sigma_X^2 - 2b\sigma_{XY},$$

which is minimized with respect to b by setting

$$b = \frac{\sigma_{XY}}{\sigma_X^2}. \tag{3.49}$$

We immediately see one issue with the above choice: The coefficient a plays no role. A reasonable choice is to select a such that the approximation is unbiased:

$$\text{E}[Y] = \text{E}\,[a + bX] \quad \Rightarrow \quad a = \mu_Y - b\mu_X. \tag{3.50}$$

The reader familiar with linear regression in inferential statistics will find the above formulas quite familiar; in fact, least-squares estimators of the coefficients are just sample counterparts of the expressions in Eqs. (3.49) and (3.50). We may also set the two coefficients together, in a more convincing way, by minimizing the mean square deviation, rather than variance:

$$\min_{a,b} \text{E}\left[(Y - a - bX)^2\right].$$

Using the above calculation and a well-known property of variance, the objective function may be rewritten as follows:

$$\text{E}\left[(Y - a - bX)^2\right] = \text{Var}(Y - a - bX) + \left(\text{E}\,[Y - a - bX]\right)^2$$
$$= \sigma_Y^2 + b^2\sigma_X^2 - 2b\sigma_{XY} + (\mu_Y - a - b\mu_X)^2.$$

Now we enforce the optimality conditions, taking partial derivatives with respect to a and b, respectively. The first condition,

$$-2(\mu_Y - a - b\mu_X) = 0$$

yields the result in Eq. (3.50), whereas the second one is

$$2b\sigma_X^2 - 2\sigma_{XY} - 2(\mu_Y - a - b\mu_X) = 0,$$

which yields the coefficient in Eq. (3.49).

Within this framework, it is also easy to see that we obtain a kind of "orthogonal projection," in the sense that the approximation $a+bX$ and the residual $Y - (a + bX)$ are uncorrelated:

$$
\begin{aligned}
\mathrm{Cov}\,(a + bX, Y - a - bX) &= \mathrm{Cov}\,(bX, Y - bX) \\
&= b\sigma_{XY} - b^2\sigma_X^2 \\
&= b\left(\sigma_{XY} - \frac{\sigma_{XY}}{\sigma_X^2}\sigma_X^2\right) = 0.
\end{aligned}
$$

Thus, the optimal approximation uses the maximum information about Y that can be included in bX; the residual variability is orthogonal.

3.6 Time series models

The previous section was mainly aimed at cross-sectional dependence, i.e., the relationship between different random variables observed at the same time. In this section we move on to longitudinal dependence and consider discrete-time models based on time series. Time series models are an important topic within econometrics, and here we can only afford to scratch the surface. In doing so, we have the following objectives:

- To offer a quick outline of basic models for the sake of unfamiliar readers
- To show some R functions for the analysis and the analysis of time series data
- To illustrate how R can be used to simulate sample paths

In the next chapter we will see some more R functions to estimate time series models. Nevertheless, the tone in this section is a bit different from the rest of the chapter. The theory behind time series is far from trivial, and we will rely on simple simulations to get an intuitive feeling for them. Furthermore, some concepts we use here, such as the analysis of autocorrelation, are a bit halfway between probability and statistics. We include them here as they are part of the toolbox for model building. In fact, applying time series requires the following steps:

1. *Model identification*: We should first select a model structure, i.e., its type and its order.

2. *Parameter estimation*: Given the qualitative model structure, we must fit numerical values for its parameters.

We cover here the first task, and estimation methods are deferred to the next chapter. After a few preliminary considerations, we will move on to consider two fundamental classes of time series models:

- Moving-average models, in Section 3.6.1
- Autoregressive models, in Section 3.6.2

Then, in Section 3.6.3 we will see how these modeling approaches can be integrated in the more general class of ARIMA (autoregressive integrated moving average) processes, also known as Box–Jenkins models. The next step, in Section 3.6.4, is to generalize a bit, moving from scalar time series to multidimensional vector autoregressive models. Finally, in Section 3.6.5 we will present some more specific model classes, like ARCH and GARCH, to represent stochastic volatility. These models are quite relevant for financial applications. As we shall see, time series modeling offers many degrees of freedom, maybe too many; when observing a time series, it may be difficult to figure out which type of model is best-suited to capture the essence of the underlying process. This is why model identification plays a pivotal role.

In its basic form, time series theory deals with *weakly stationary processes*, i.e., time series Y_t with the following properties, related to first- and second-order moments:

1. The expected value of Y_t does not change in time: $E[Y_t] = \mu$.

2. The covariance between Y_t and Y_{t+k} depends only on time lag k, and not on t.

The second condition deserves some elaboration.

DEFINITION 3.16 (Autocovariance and autocorrelation)
Given a weakly stationary stochastic process Y_t, the function

$$\gamma_Y(k) = \text{Cov}(Y_t, Y_{t+k})$$

is called **autocovariance of the process with time lag** k*. The function*

$$\rho_Y(k) = \frac{\gamma_Y(k)}{\sigma^2}$$

is called the **autocorrelation function** *(ACF).*

The definition of autocorrelation relies on the fact that variance is constant as well:

$$\rho_Y(k) = \rho(Y_t, Y_{t+k}) = \frac{\text{Cov}(Y_t, Y_{t+k})}{\sqrt{\text{Var}(Y_t)}\sqrt{\text{Var}(Y_{t+k})}} = \frac{\gamma_Y(k)}{\sigma^2}.$$

In practice, autocorrelation may be estimated by the sample autocorrelation function (SACF), given a sample path Y_t, $t = 1, \ldots, T$:

$$R_k = \frac{\sum_{t=k+1}^{T} \left(Y_t - \overline{Y} \right) \left(Y_{t-k} - \overline{Y} \right)}{\sum_{t=1}^{T} \left(Y_t - \overline{Y} \right)^2}, \tag{3.51}$$

where $\overline{Y}$ is the sample mean of Y_t. The expression in Eq. (3.51) may not look quite convincing, since the numerator and the denominator are sums involving a different number of terms. In particular, the number of terms in the numerator is decreasing in the time lag k. Thus, the estimator looks biased and, for a large value of k, R_k will vanish. However, this is what one expects in real life. Furthermore, although we could account for the true number of terms involved in the numerator, for large k the sum involves very few terms and is not reliable. Indeed, the form of sample autocorrelation in Eq. (3.51) is what is commonly used in statistical software packages, even though alternatives have been proposed.[26] If T is large enough, under the null hypothesis that the true autocorrelations ρ_k are zero, for $k \geq 1$, the statistic $\sqrt{T} R_k$ is approximately normal standard. Since $z_{0.99} = 1.96 \approx 2$, a commonly used approximate rule states that if

$$|R_k| > \frac{2}{\sqrt{T}}, \tag{3.52}$$

then the sample autocorrelation at lag k is statistically significant. For instance, if $T = 100$, autocorrelations outside the interval $[-0.2, 0.2]$ are significant. We should keep in mind that this is an approximate result, holding for a large number T of observations. We may plot the sample autocorrelation function at different lags, obtaining an *autocorrelogram* that can be most useful in pointing out hidden patterns in data.

■ Example 3.10 **Detecting seasonality with autocorrelograms**

> Many time series in economics display seasonality. Consider the time series displayed on the left in Fig. 3.24. A cursory look at the plot does not suggest much structure in the data, but we may try to get a clue by plotting the ACF on the right. However, the ACF does not look quite helpful either. We notice decreasing autocorrelations, many of which are significant; the plot displays two horizontal bands

---

[26]The denominator in Eq. (3.51) is related to the estimate of autocovariance. In the literature, sample autocovariance is typically obtained by dividing the sum by T, even though this results in a biased estimator. If we divide by $T - k$, we might obtain an autocovariance matrix which is not positive semidefinite; see Brockwell and Davis [2], pp. 220–221.

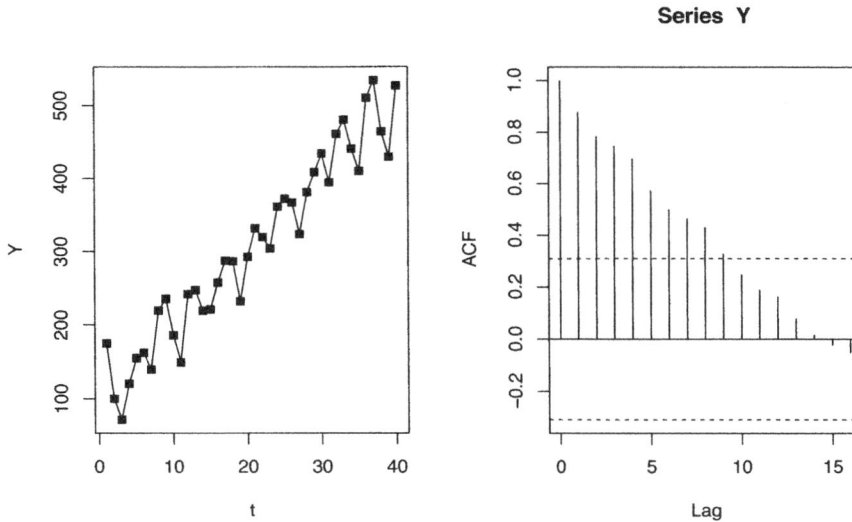

Series Y

FIGURE 3.24 **A seasonal time series and its autocorrelogram.**

outside which autocorrelation is statistically significant. Notice that

$$\frac{2}{\sqrt{40}} = 0.3162278,$$

and the positioning of the two horizontal bands is consistent with Eq. (3.52). In fact, the time series is seasonal and has been generated by simulating the model

$$Y_t = a + bt + S_t + \epsilon_t,$$

which features an additive seasonal component S_t superimposed on a linear trend $a + bt$, plus a noise term ϵ_t. The following R code shows how such a model can be simulated, and how the acf function can be used to plot the SACF:

```
set.seed(55555)
t <- 1:40
factorS <- c(15, 30, -10, -35)
Y <- 100+10*t+factorS[t%%4+1]+rnorm(length(t),0,20)
par(mfrow=c(1,2))
plot(t,Y,type='l')
points(t,Y,pch=15)
acf(Y)
```

The seasonal factors are collected in the vector $(15, 30, -10, -35)$; note that the average factor, for additive models, is zero, and that the indexing `t%%4+1` repeats the four factors in a cycle over the simulation horizon; we recall that the operator `%%` corresponds to the mathematical operator `mod` and yields the remainder of the integer division. The noise terms ϵ_t are normally distributed, which need not be the case in general. The problem actually is in the trend. We should detrend the series in order to see more structure, which can be accomplished by linear regression:

```
mod <- lm(Y~t)
dY <- Y-mod$fitted
plot(t,dY,type='l')
points(t,dY,pch=15)
acf(dY)
par(mfrow=c(1,1))
```

The first statement above uses `lm` to estimate a linear regression model, relating variable Y with time t; the model is assigned to the variable `mod`, which is a rich data structure including `fitted` values, among other things. By subtracting the fitted trendline, we detrend the time series. The new ACF, displayed in Fig. 3.25 shows a quite different pattern: The autocorrelogram is indeed a useful tool to spot seasonality and other hidden patterns in data.

In the above example, we have used an important building block in time series modeling, called *white noise*, which is a sequence of i.i.d. (independent and identically distributed) random variables. If they are normal, we have a Gaussian white noise. In other models, noise may be "colored," i.e., there may be some autocorrelation between the driving random terms ϵ_t. As we have mentioned, the first step in modeling a time series is the identification of the model structure. In the following we outline a few basic ideas that are used to this aim, pointing out the role of autocorrelation in the analysis.

3.6.1 MOVING-AVERAGE PROCESSES

A *finite-order moving-average process of order* q, denoted by $\mathrm{MA}(q)$, can be expressed as

$$Y_t = \mu + \epsilon_t - \theta_1 \epsilon_{t-1} - \theta_2 \epsilon_{t-2} - \cdots - \theta_q \epsilon_{t-q},$$

where random variables ϵ_t are white noise, with $\mathrm{E}[\epsilon_t] = 0$ and $\mathrm{Var}(\epsilon_t) = \sigma^2$. These variables play the role of *random shocks* and drive the process. It is fairly easy to see that the process is weakly stationary. In fact, a first observation is

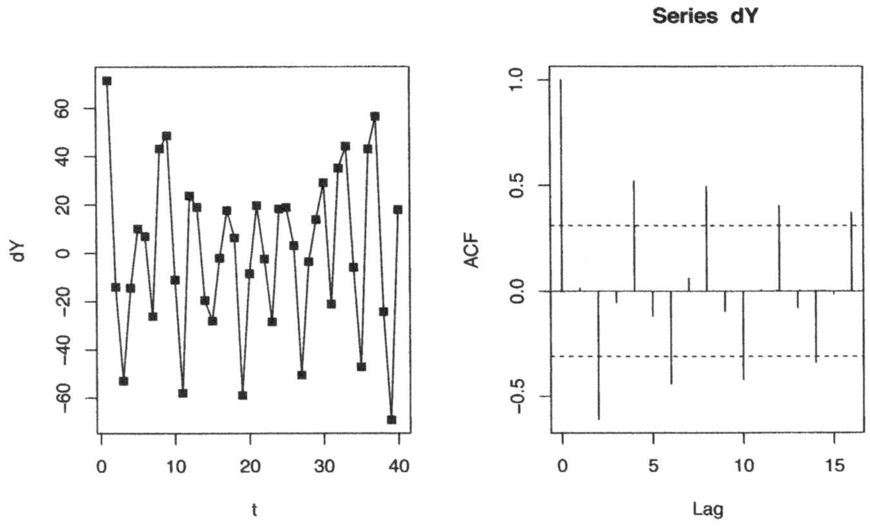

FIGURE 3.25 **Detrended seasonal time series and its autocorrelogram.**

that expected value and variance are constant:

$$\mathrm{E}[Y_t] = \mu + \mathrm{E}[\epsilon_t] - \theta_1 \mathrm{E}[\epsilon_{t-1}] - \cdots - \theta_q \mathrm{E}[\epsilon_{t-q}] = \mu,$$
$$\mathrm{Var}(Y_t) = \mathrm{Var}(\epsilon_t) + \theta_1^2 \mathrm{Var}(\epsilon_{t-1}) + \cdots + \theta_q^2 \mathrm{Var}(\epsilon_{t-q})$$
$$= \sigma^2 \left(1 + \theta_1^2 + \cdots + \theta_q^2\right).$$

The calculation of autocovariance is a bit more involved, but we may take advantage of the uncorrelation of white noise:

$$
\begin{aligned}
\gamma_Y(k) &= \mathrm{Cov}(Y_t, Y_{t+k}) \\
&= \mathrm{E}\left[(\epsilon_t - \theta_1 \epsilon_{t-1} - \cdots - \theta_q \epsilon_{t-q})(\epsilon_{t+k} - \theta_1 \epsilon_{t+k-1} - \cdots - \theta_q \epsilon_{t+k-q})\right] \\
&= \begin{cases} \sigma^2 \left(-\theta_k + \theta_1 \theta_{k+1} + \cdots + \theta_{q-k} \theta_q\right), & k = 1, 2, \ldots, q, \\ 0, & k > q. \end{cases}
\end{aligned}
$$

As a consequence, the autocorrelation function is

$$
\rho_Y(k) = \frac{\gamma_Y(k)}{\gamma_Y(0)} = \begin{cases} \dfrac{-\theta_k + \theta_1 \theta_{k+1} + \cdots + \theta_{q-k} \theta_q}{1 + \theta_1^2 + \cdots + \theta_q^2}, & k = 1, 2, \ldots, q, \\ 0, & k > q. \end{cases}
$$

$$(3.53)$$

Thus, the autocorrelation function depends only on the lag k. We also notice that the autocorrelation function cuts off for lags larger than the order of the process. This makes sense, since the process Y_t is a moving average of the

Series Y

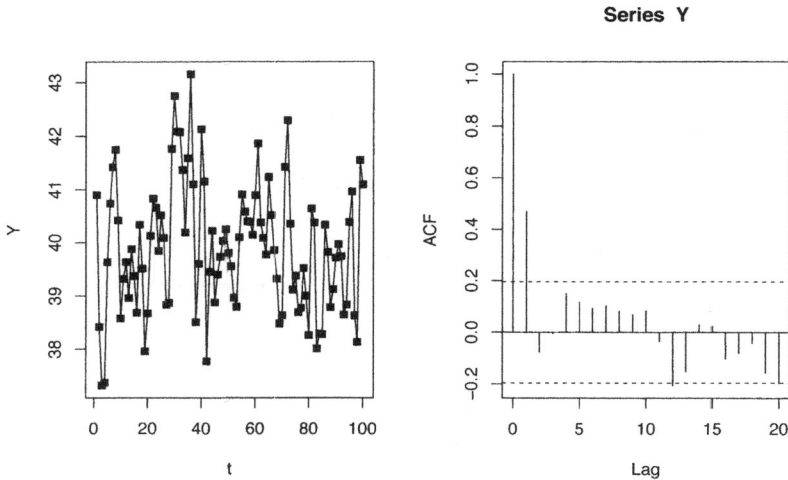

FIGURE 3.26 **Sample path and corresponding sample ACF for the moving-average** process $Y_t = 40 + \epsilon_t + 0.8\epsilon_{t-1}$.

driving white noise ϵ_t. Hence, by checking whether the sample autocorrelation function cuts off after a time lag, we may figure out whether a time series can be modeled as a moving average, as well as its order q. Of course, the sample autocorrelation will not be exactly zero for $k > q$; nevertheless, by using the autocorrelogram and its significance bands, we may get some clue.

■ Example 3.11 Moving-average processes

Let us consider a simple MA(1) process

$$Y_t = 40 + \epsilon_t + 0.8\epsilon_{t-1},$$

where ϵ_t is a sequence of uncorrelated standard normal variables (Gaussian white noise). In Fig. 3.26 we show a sample path and the corresponding sample autocorrelogram obtained by the following R code (note how vectorization avoids a `for` loop):

```
set.seed(55555)
t <- 1:100
epsVet <- rnorm(101)
Y <- 40+epsVet[2:101]+0.8*epsVet[1:100]
par(mfrow=c(1,2))
plot(t,Y,type='l')
points(t,Y,pch=15)
```

```
acf(Y)
```

Incidentally, in this section we are writing our own code to simulate ARMA models for illustrative purposes, but R offers a ready-to-use function `arima.sim`, which will be used later in Section 4.5. The sample autocorrelation looks significant at time lag 1, which is expected, given the nature of the process. Note that, by applying Eq. (3.53), we find that the autocorrelation function, for a $MA(1)$ process $Y_t = \mu + \epsilon_t - \theta_1 \epsilon_{t-1}$, is

$$\rho_Y(1) = \frac{-\theta_1}{1 + \theta_1^2}, \tag{3.54}$$

$$\rho_Y(k) = 0, \qquad k > 1. \tag{3.55}$$

Figure 3.27 shows the sample path and autocorrelogram of a slightly different $MA(1)$ process (the R code is essentially the same as above):

$$Y_t = 40 + \epsilon_t - 0.8\epsilon_{t-1}.$$

The change in sign in θ_1 has a noticeable effect on the sample path; an upswing tends to be followed by a downswing, and vice versa. The autocorrelogram shows a cutoff after time lag 1, and a negative autocorrelation.

If we increase the order of the process, we should expect more significant autocorrelations. In Fig. 3.28, we repeat the exercise for the $MA(2)$ process

$$Y_t = 40 + \epsilon_t + 0.9\epsilon_{t-1} + 0.5\epsilon_{t-2},$$

using the following R code:

```
set.seed(55555)
t <- 1:100
epsVet <- rnorm(102)
Y<-40+epsVet[3:102]+0.8*epsVet[2:101]+0.6*epsVet[1:100]
par(mfrow=c(1,2))
plot(t,Y,type='l')
points(t,Y,pch=15)
acf(Y)
```

We observe that, in this case, the autocorrelation function cuts off after time lag $k = 2$.

We should mention that sample autocorrelograms are a statistical tool. It may well be the case that, for the moving-average processes in the examples,

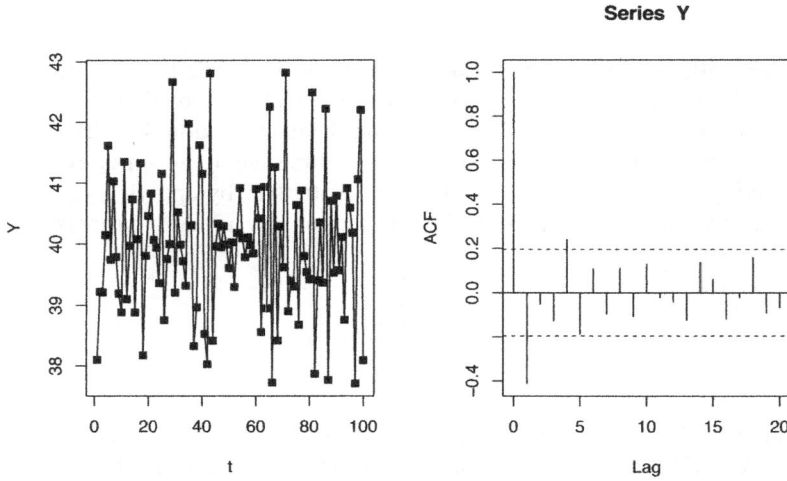

FIGURE 3.27 **Sample path and corresponding ACF for the moving-average process** $Y_t = 40 + \epsilon_t - 0.8\epsilon_{t-1}$.

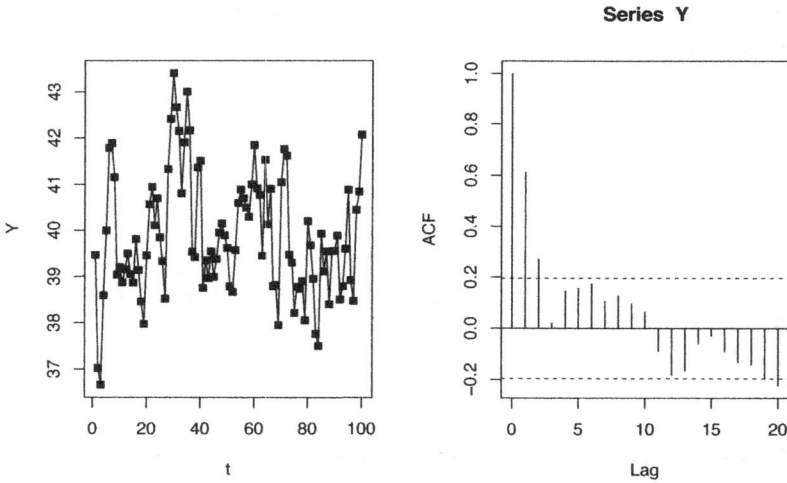

FIGURE 3.28 **Sample path and corresponding ACF for the moving-average process** $Y_t = 40 + \epsilon_t + 0.9\epsilon_{t-1} + 0.5\epsilon_{t-2}$.

we get different pictures when different sample paths are randomly generated. This is a useful experiment to carry out with the help of statistical software.

3.6.2 AUTOREGRESSIVE PROCESSES

In finite-order moving-average processes, only a finite number of past realizations of white noise influence the value of Y_t. This may be a limitation for those processes in which all of the previous realizations have an effect, even though this possibly fades in time. In principle, we could consider an infinite-order moving-average process, but having to do with an infinite sequence of θ_q coefficients does not sound quite practical. Luckily, under some technical conditions, such a process may be rewritten in a compact form involving time-lagged realizations of the Y_t itself. This leads us to the definition of an *autoregressive process* of a given order. The simplest such process is the autoregressive process of order 1, AR(1):

$$Y_t = \delta + \phi Y_{t-1} + \epsilon_t. \tag{3.56}$$

One could wonder under which conditions this process is stationary, since here we cannot use the same arguments as in the moving-average case. A *heuristic* argument to find the expected value $\mu = \mathrm{E}[Y_t]$ is based on taking expectations and dropping the time subscript in Eq. (3.56):

$$\mu = \delta + \phi\mu \quad \Rightarrow \quad \mu = \frac{\delta}{1-\phi}.$$

The argument is not quite correct, as it leads to a sensible result if the process is indeed stationary, which is the case if $|\phi| < 1$. Otherwise, intuition suggests that the process will grow without bounds. The reasoning can be made precise by using the infinite-term representation of Y_t, which is beyond the scope of this book. Using the correct line of reasoning, we may also prove that

$$\gamma_Y(k) = \frac{\sigma^2 \phi^k}{1-\phi^2}, \qquad k = 0, 1, 2, \ldots.$$

In particular, we have

$$\mathrm{Var}(Y_t) = \gamma_Y(0) = \frac{\sigma^2}{1-\phi^2},$$

and we may also observe that, for a stationary AR(1) process,

$$\rho_Y(k) = \frac{\gamma_Y(k)}{\gamma_Y(0)} = \phi^k, \qquad k = 0, 1, 2, \ldots. \tag{3.57}$$

We notice that autocorrelation is decreasing, but it fades away with no sharp cutoff.

▣ Example 3.12 Autoregressive processes

In Figs. 3.29 and 3.30, we show a sample path and the corresponding sample autocorrelogram for the two AR(1) processes

$$Y_t = 8 + 0.8Y_{t-1} + \epsilon_t \quad \text{and} \quad Y_t = 8 - 0.8Y_{t-1} + \epsilon_t,$$

respectively. The R code cannot avoid a `for` loop in this case:

```
set.seed(55555)
t <- 1:100
epsVet <- rnorm(101)
Y <- 40+epsVet[2:101]+0.8*epsVet[1:100]
par(mfrow=c(1,2))
plot(t,Y,type='l')
points(t,Y,pch=15)
acf(Y)
```

Notice that the change in sign in the ϕ coefficient has a significant effect on the sample path, as well as on autocorrelations. In the first case, autocorrelation goes to zero along a relatively smooth path. This need not be the case, as we are working with *sample* autocorrelations. Nevertheless, at least for significant values, we observe a monotonic behavior consistent with Eq. (3.57). The sample path of the second process features evident up- and downswings; we also notice an oscillatory pattern in the autocorrelation.

The autocorrelation behavior of AR processes does not feature the cutoff properties that help us determine the order of an MA process. A tool that has been developed for the identification of AR processes is the *partial autocorrelation function* (PACF). The rationale behind PACF is to measure the degree of association between Y_t and Y_{t-k}, removing the effects of intermediate lags, i.e., $Y_{t-1}, \ldots, Y_{t-k+1}$. We cannot dwell too much on PACF, but we may at least get a better intuitive feeling as follows. Consider three random variables X, Y, and Z, and imagine regressing X and Y on Z:

$$\hat{X} = a_1 + b_1 Z,$$
$$\hat{Y} = a_2 + b_2 Z.$$

Note that we are considering a probabilistic regression, not a sample-based regression. From Section 3.5 we know that

$$b_1 = \frac{\text{Cov}(X, Z)}{\text{Var}(Z)} \quad \text{and} \quad b_2 = \frac{\text{Cov}(Y, Z)}{\text{Var}(Z)}.$$

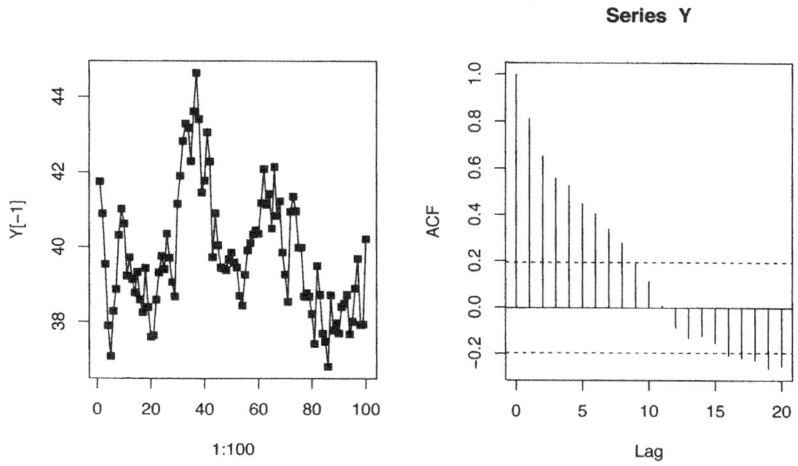

FIGURE 3.29 **Sample path and corresponding ACF for the autoregressive process** $Y_t = 8 + 0.8Y_{t-1} + \epsilon_t.$

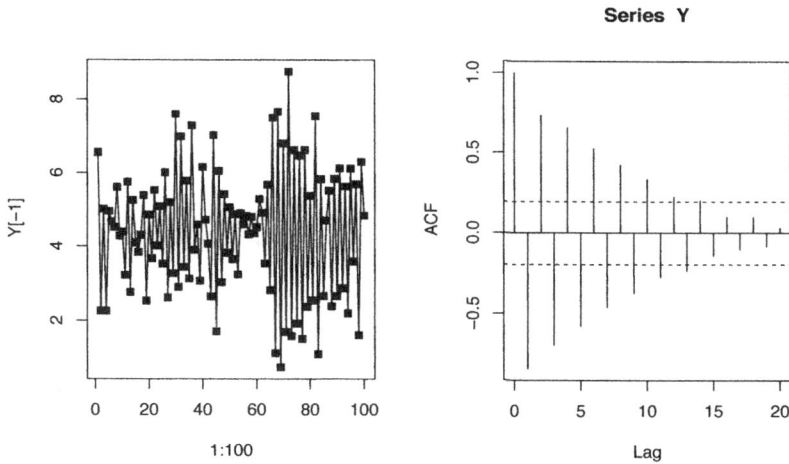

FIGURE 3.30 **Sample path and corresponding ACF for the autoregressive process** $Y_t = 8 - 0.8Y_{t-1} + \epsilon_t.$

Furthermore, we have regression errors

$$X^* = X - \hat{X} = X - (a_1 + b_1 Z),$$
$$Y^* = Y - \hat{Y} = Y - (a_2 + b_2 Z),$$

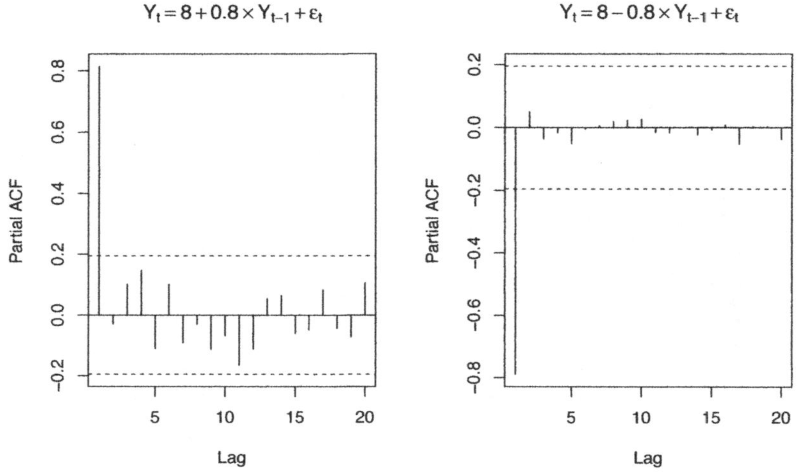

FIGURE 3.31 **Sample partial autocorrelation function for the autoregressive processes of Example 3.12.**

which may be regarded as the random variables X and Y, after the effect of Z is removed. The correlation $\rho(X, Y)$ may be large because of the common factor Z (which plays the role of a "lurking" variable). If we want to get rid of it, we may consider the partial correlation $\rho(X^*, Y^*)$. On the basis of this intuition, we might consider estimating the partial autocorrelation between Y_t and Y_{t-k} by the following linear regression:

$$Y_t = b_0 + b_1 Y_{t-1} + b_2 Y_{t-2} + \cdots + b_{k-1} Y_{t-k+1} + b_k Y_{t-k}.$$

The inclusion of intermediate lagged variables $Y_{t-1}, \ldots, Y_{t-k+1}$ allows to capture their effect explicitly by the regression coefficients $b_1, \ldots, b_{k-1}$. Then, we could use b_k as an estimate of partial autocorrelation. Actually, this need not be the sounder approach, but software packages provide us with ready-to-use functions to estimate the PACF by its sample counterpart. In Fig. 3.31 we show the SPACF for the two AR(1) processes of Example 3.12. The plot was obtained by the `pacf` function. The example also shows how to use `expression` to build a mathematical annotation for the plot, including subscripts, etc.:[27]

```
exprl = expression(Y[t] == 8 + 0.8 %*% Y[t-1] + epsilon[t])
pacf(Y, main = exprl)
```

We see that the SPACF cuts off after lag 1, even though, due to statistical sampling errors, it seems that there is a significant value at larger lags in the

---

[27] Actually, `expression` is much more powerful; please refer to the online help.

first case. This suggests how to use this function to identify the order of an AR model.

3.6.3 ARMA AND ARIMA PROCESSES

Autoregressive and moving-average processes may be merged into ARMA (autoregressive moving-average) processes like

$$Y_t = \delta + \phi_1 Y_{t-1} + \cdots + \phi_p Y_{t-p} + \epsilon_t - \theta_1 \epsilon_{t-1} + \cdots + \theta_q \epsilon_{t-q}. \qquad (3.58)$$

The model above is referred to as ARMA(p, q) process, for self-explanatory reasons. Conditions ensuring stationarity have been developed for ARMA processes, as well as identification and estimation procedures. Clearly, the ARMA modeling framework provides us with plenty of opportunities to fit historical data with a model that can be used to create scenarios by Monte Carlo sampling. However, it applies only to stationary data. It is not too difficult to find real-life examples of data processes that are nonstationary. Just think of stock market indices; most investors really wish that the process is not stationary.

Example 3.13 A nonstationary random walk

A quite common building block in many financial models is the *random walk*. An example of random walk is

$$Y_t = Y_{t-1} + 0.05\eta_t, \qquad (3.59)$$

where η_t is a sequence of independent and standard normal random variables. This is actually an AR process, but from Section 3.6.2 we know that it is nonstationary, as $\phi = 1$. Three sample paths of this process are shown in Fig. 3.32. They look quite different, and a subjective comparison of the sample paths would not suggest that they are realizations of the same stochastic process. However, the ACFs show a common pattern: Autocorrelation fades out very slowly. Indeed, this is a common feature of nonstationary processes. The PACFs display a very strong partial autocorrelation at lag 1, which cuts off immediately in all of the three cases.

Since the theory of stationary MA and AR processes is well developed, it would be nice to find a way to apply it to nonstationary processes as well. A commonly used trick to remove nonstationarity in a time series is *differencing*, by which we consider the time series

$$Y_t' = Y_t - Y_{t-1}. \qquad (3.60)$$

Applying differencing in R is obtained by the `diff` function, which is applied in the following R snapshot to the random walk of Eq. (3.59):

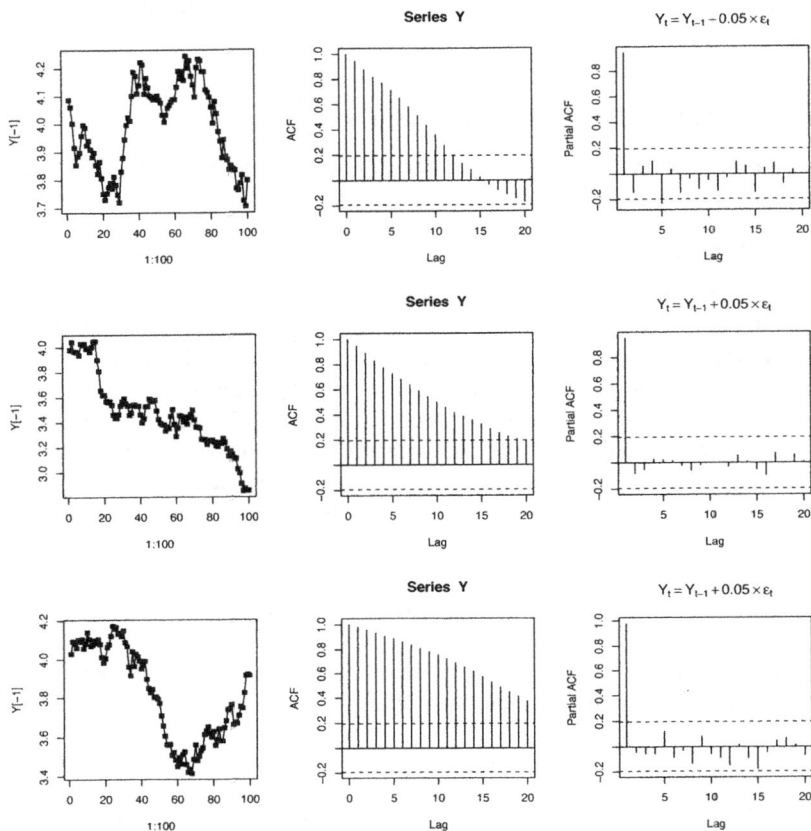

FIGURE 3.32 **Three sample paths and the corresponding ACF and PACF for the random walk** $Y_t = Y_{t-1} + 0.05\eta_t$.

```
set.seed(5)
epsVet <- c(0, rnorm(100)) # first value is dummy
Y <- c(4,numeric(100))
for (t in 2:101)
  Y[t] <- Y[t-1] + 0.05*epsVet[t]
dY <- diff(Y)
par(mfrow=c(2,2))
plot(1:100,Y[-1],type='l')
points(1:100,Y[-1],pch=15)
acf(Y)
plot(1:100,dY,type='l')
points(1:100,dY,pch=15)
acf(dY)
```

Series Y

Series dY

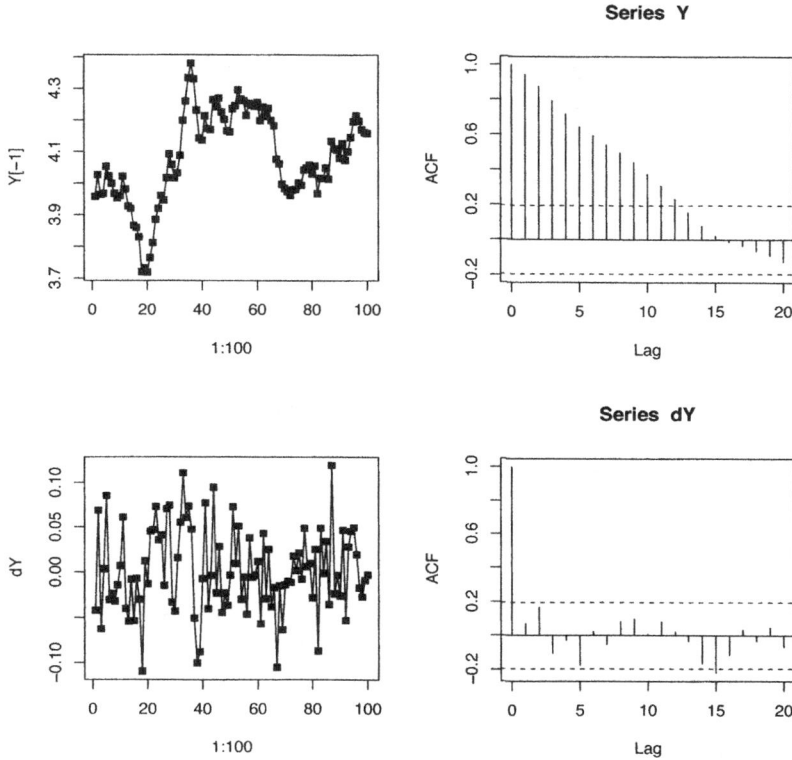

FIGURE 3.33 **The effect of differencing on the simple random walk of Eq. (3.59).**

In Fig. 3.33 we show the sample path and the differenced series, along with their ACFs. The shape of the ACF is not surprising, since the differenced process is just white noise.

Example 3.14 What is nonstationarity, anyway?

A time series with trend

$$Y_t = \alpha + \beta t + \epsilon_t, \qquad (3.61)$$

where ϵ_t is white noise, is clearly nonstationary, as it features a deterministic trend. However, nonstationarity is a subtler concept than that. A little digression is in order to clarify the nature of nonstationarity in a random walk like

$$Y_t = Y_{t-1} + \epsilon_t. \qquad (3.62)$$

The sample paths that we have generated in Example 3.13 show that this random walk does not feature a deterministic trend. The recursive unfolding of Eq. (3.62) yields

$$Y_t = Y_0 + \sum_{k=1}^{t} \epsilon_k.$$

Therefore, we find that

$$E[Y_t \mid Y_0] = Y_0.$$

Hence, we must have a different kind of nonstationarity in the random walk of Eq. (3.62) than in the process described by Eq. (3.61). To investigate the matter, let us consider the expected value of the increment $Y_t' = Y_t - Y_{t-1}$, conditional on Y_{t-1}:

$$E[Y_t' \mid Y_{t-1}] = E[\epsilon_t] = 0.$$

Therefore, given the last observation Y_{t-1}, we cannot predict whether the time series will move up or down. This shows a connection between a stationary random walk and the efficient market hypothesis. Now, let us consider a stationary AR(1) process

$$Y_t = \phi Y_{t-1} + \epsilon_t,$$

where $\phi \in (-1, 1)$. The increment in this case is

$$Y_t' = (\phi - 1)Y_{t-1} + \epsilon_t.$$

Since $(\phi - 1) < 0$, we have

$$\begin{cases} E[Y_t' \mid Y_{t-1}] < 0, & \text{if } Y_{t-1} > 0, \\ E[Y_t' \mid Y_{t-1}] > 0, & \text{if } Y_{t-1} < 0. \end{cases}$$

This suggests that a stationary AR(1) process is *mean reverting*, in the sense that the process tends to return to its expected value; the nonstationary random walk does not enjoy this property.

If we introduce the *backshift operator* B, defined by

$$BY_t = Y_{t-1},$$

we may express the first difference in Eq. (3.60) as

$$Y_t' = Y_t - BY_t = (1 - B)Y_t.$$

Sometimes, differencing must be repeated in order to obtain a stationary time series. We obtain *second-order differencing* by repeated application of (first-order) differencing:

$$
\begin{aligned}
Y_t'' &= Y_t' - Y_{t-1}' \\
&= (Y_t - Y_{t-1}) - (Y_{t-1} - Y_{t-2}) \\
&= Y_t - 2Y_{t-1} + Y_{t-2} \\
&= (1 - 2B + B^2)Y_t \\
&= (1 - B)^2 Y_t.
\end{aligned}
$$

This suggests that we may formally apply the algebra of polynomials to the backshift operator, in order to find differences of arbitrary order. By introducing the operator polynomials

$$
\begin{aligned}
\Phi(B) &= 1 - \phi_1 B - \cdots - \phi_p B^p, \\
\Theta(B) &= 1 - \theta_1 B - \cdots - \theta_q B^q,
\end{aligned}
$$

we can rewrite the ARMA model of Eq. (3.58) in the compact form:

$$
\Phi(B)Y_t = \delta + \Theta(B)\epsilon_t. \tag{3.63}
$$

We may also extend the class of stationary ARMA models, in order to allow for nonstationarity. Doing so, we find the more general class of ARIMA (autoregressive integrated moving average) processes, also known as *Box–Jenkins models*. An $\mathrm{ARIMA}(p, d, q)$ process can be represented as follows:[28]

$$
\Phi(B)(1 - B)^d Y_t = \delta + \Theta(B)\epsilon_t, \tag{3.64}
$$

where $\Phi(B)$ and $\Theta(B)$ are polynomials of order p and q, respectively, and d is a differencing order such that the process Y_t is stationary, whereas, if we take differences of order $(d - 1)$, the process is still nonstationary. The name "integrated" stems from the fact that we obtain the nonstationary process by integrating a stationary one, i.e., by undoing differencing. In most practical applications the order d is 0 or 1. Box–Jenkins models can be extended to cope with seasonality, but a full account of this class of models is beyond the scope of this book; we refer the reader to the references provided at the end of this chapter.

3.6.4 VECTOR AUTOREGRESSIVE MODELS

If you want to generalize the scalar time series models that we have seen so far, in principle, there is no limit to what one can conceive. Simultaneous equations can be built, resulting in quite rich structures. In practice, an overly complicated

---

[28]To be precise, this representation requires that the two polynomials $\Phi(B)$ and $\Theta(B)$ satisfy some additional technical conditions that are beyond the scope of this book.

model need not be the smartest option. On the one hand, there may be issues
with estimation and the assessment of properties such as stationarity. On the
other hand, the predictive power of a seemingly rich model may be disappointing due to overfitting issues. This is why vector autoregressive models, VAR
for short,[29] have been proposed. Let $\mathbf{Y}_t \in \mathbb{R}^k$ be a vector of random variables;
a VAR model can be represented in reduced form as

$$\mathbf{Y}_t = \mathbf{c} + \phi(B)\mathbf{Y}_{t-1} + \epsilon_t, \tag{3.65}$$

where $\phi(B)$ is a polynomial in the backshift operator B and $\mathbf{c}, \epsilon_t \in \mathbb{R}^k$. More
explicitely, a VAR(p) model is

$$\mathbf{Y}_t = \mathbf{c} + \mathbf{\Phi}_1 \mathbf{Y}_{t-1} + \mathbf{\Phi}_2 \mathbf{Y}_{t-2} + \cdots + \mathbf{\Phi}_p \mathbf{Y}_{t-p} + \epsilon_t, \tag{3.66}$$

where $\mathbf{\Phi}_i \in \mathbb{R}^{k,k}$, $i = 1, \ldots, p$. Note that the error term ϵ_t need not be a vector
or independent variables.

Example 3.15 Simulation of a VAR(1) model

Let us consider the bidimensional model

$$Y_{1t} = -0.7 + 0.7Y_{1,t-1} + 0.2Y_{2,t-1} + \epsilon_{1t},$$
$$Y_{2t} = 1.3 + 0.2Y_{1,t-1} + 0.7Y_{2,t-1} + \epsilon_{2t},$$

where the error terms are standard normals with correlation $\rho = 0.5$.
In matrix form, we may write the above equations as

$$\mathbf{Y}_t = \mathbf{c} + \mathbf{\Phi}\mathbf{Y}_{t-1} + \epsilon_t,$$

where

$$\mathbf{c} = \begin{bmatrix} -0.7 \\ 1.3 \end{bmatrix}, \qquad \mathbf{\Phi} = \begin{bmatrix} 0.7 & 0.2 \\ 0.2 & 0.7 \end{bmatrix},$$

and

$$\epsilon \sim \mathsf{N}(\boldsymbol{\mu}, \mathbf{\Sigma}), \qquad \boldsymbol{\mu} = \begin{bmatrix} 0 \\ 0 \end{bmatrix}, \qquad \mathbf{\Sigma} = \begin{bmatrix} 1 & 0.5 \\ 0.5 & 1 \end{bmatrix}.$$

The following R script produces the time series displayed in Fig. 3.34.

```
require(MASS)
tPhi1 <- t(matrix(c(0.7,0.2,0.2,0.7),2,2))
c.vec <- c(-0.7, 1.3)
```

[29]The acronym VAR is a bit unfortunate, as it may be confused with variance or value-at-risk.
To avoid confusion, variance is typically written as Var and value-at-risk as VaR; some authors
use V@R for the latter, and we will follow that practice in this book.

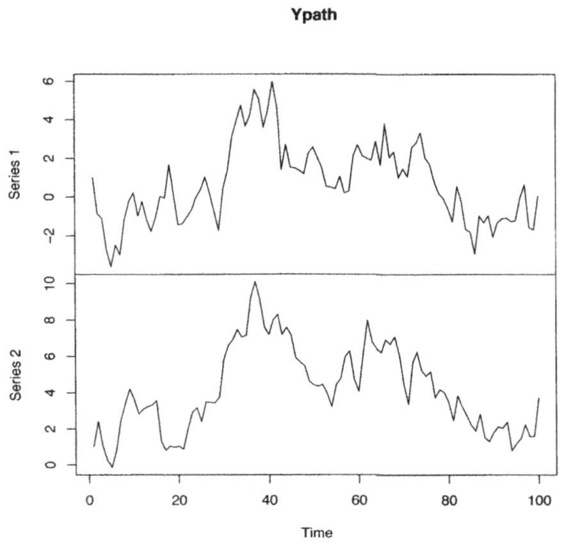

Ypath

FIGURE 3.34 **Bidimensional time series corresponding to a VAR(1) model.**

```
mu.vec <- numeric(2)
cov.mat <- matrix(c(1,0.5,0.5,1),2,2)
set.seed(55555)
T <- 100
Ypath <- matrix(0, T, 2)
Ypath[1,] <- c(1,1) # initial condition
epsPath <- mvrnorm(n=T, mu.vec, cov.mat)
for (t in 2:T)
  Ypath[t,] <- c.vec+Ypath[t-1,]%*%tPhi1+epsPath[t,]
plot.ts(Ypath)
```

Note that, in order to collect the time series in a matrix with rows corresponding to observations, we have to transpose Φ. We also need the MASS package to sample from a multivariate normal.

3.6.5 MODELING STOCHASTIC VOLATILITY

When observing financial time series of returns, the following stylized facts may be noticed:

- Volatility is not constant and there are volatility clusters, i.e., blocks of consecutive periods featuring high or low volatility.

- Returns are not quite symmetric, as there is a negative skew.

The first feature is referred to as heteroskedasticity, in contrast to the homoskedasticity assumption, whereby volatility is constant. To model volatility, it is convenient to build a time series of returns. However, returns over consecutive periods cannot be added, as they should rather be multiplied. This is why a series of log-returns u_t is usually represented:

$$u_t = \log\left(\frac{S_{t+1}}{S_t}\right),$$

where S_t is the price of the relevant financial asset. We want to capture the conditional variance of the time series, which can be expressed as:

$$\sigma_t^2 = \mathrm{Var}\big(u_t \mid u_{t-1}, u_{t-2}, \dots\big) = \mathrm{E}\Big[\big(u_t - \mathrm{E}[u_t]\big)^2 \mid u_{t-1}, u_{t-2}, \dots\Big].$$

It is often the case that $\mathrm{E}[u_t] \approx 0$, if the time bucket is small enough.[30] Hence, the expected return term is typically dropped. To build a model, we might relate volatility to the last shock:

$$\sigma_t^2 = \alpha_0 + \alpha_1 u_{t-1}^2. \tag{3.67}$$

A model like this is called $\mathrm{ARCH}(1)$, where the acronym stands for *autoregressive conditionally heteroskedastic*, and 1 means that only u_t with lag 1 plays a role in Eq. (3.67). This is just a part of an overall model describing u_t. We may build an $\mathrm{ARCH}(q)$ model as

$$\sigma_t^2 = \alpha_0 + \alpha_1 u_{t-1}^2 + \alpha_2 u_{t-2}^2 + \cdots + \alpha_q u_{t-q}^2. \tag{3.68}$$

An ARCH model is, in a sense, an MA-style model for volatility. We may also consider an AR-style model like

$$\sigma_t^2 = \alpha_0 + \alpha_1 u_{t-1}^2 + \beta \sigma_{t-1}^2, \tag{3.69}$$

which is known as $\mathrm{GARCH}(1,1)$, i.e., a generalized ARCH model involving two variables with lag 1.

A GARCH model should be coupled with another equation describing the evolution of a time series of interest. Let us consider a time series of return generated as an expected return plus a random shock u_t:

$$r_t = \mu + u_t,$$

where

$$u_t = \sigma_t \epsilon_t, \qquad \epsilon_t \sim \mathsf{N}(0,1).$$

Note that σ_t is the standard deviation, i.e., the volatility of the return time series, and it follows Eq. (3.69). One may use a fatter-tailed distribution to generate the shocks. The code in Fig. 3.35 generates the two plots displayed in Fig. 3.36.[31]

---

[30]The square-root rule, which we consider in Section 3.7.1, suggests that on very short time spans volatility dominates drift.

[31]The parameters used in the simulation are calibrated on real data in [22, p. 202].

```
set.seed(5555)
mu <- 0.0113
alpha0 <- 0.00092
alpha1 <- 0.086
beta1 <- 0.853
eps <- rnorm(550)
r <- rep(0, 550)
sigma2 <- rep(0, 550)
u <- rep(0, 550)
for (i in 2:550) {
    sigma2[i] <- alpha0 + alpha1*u[i-1]^2 + beta1*sigma2[i-1]
    u[i] <- eps[i] * sqrt(sigma2[i])
    r[i] <- mu + u[i]
}
par(mfrow=c(2,1))
plot(r[51:550], type='l', ylab="return")
plot(sqrt(sigma2[51:550]), type='l', ylab="volatility")
par(mfrow=c(1,1))
```

FIGURE 3.35 **Code to generate a sample path of a GARCH(1,1) model.**

3.7 Stochastic differential equations

Many financial engineering models rely on continuous-time stochastic processes. One reason is that, in fact, time is continuous in the real world, but another one is that this modeling framework allows for the development of a rich theory. The theory is translated into practical models that allow for an analytical solution in some lucky cases, and even if we are not that lucky, we may still resort to numerical methods, including Monte Carlo. In this section we first illustrate a gentle transition from discrete- to continuous-time models, which leads to the introduction of a fundamental building block in continuous-time stochastic modeling, the Wiener process. Continuous-time models are typically expressed as stochastic differential equations, which not only generalize deterministic differential equations but also feature thorny technical issues that require the introduction of a new concept: The stochastic integral in the sense of Itô. This, in turn, leads to an essential result, the celebrated Itô's lemma, which is in some sense an extension of the chain rule for the derivative of composite functions in familiar calculus. In fact, it turns out that the familiar calculus rules do not work in this context and must be suitably modified. This is essential in tackling stochastic differential equations analytically, when possible, and numerically via sample path generation.[32] Then, we introduce the family of Itô stochastic differential equations, which includes geometric Brownian motion, and some

---

[32] Sample path generation for stochastic differential equations is the subject of Chapter 6.

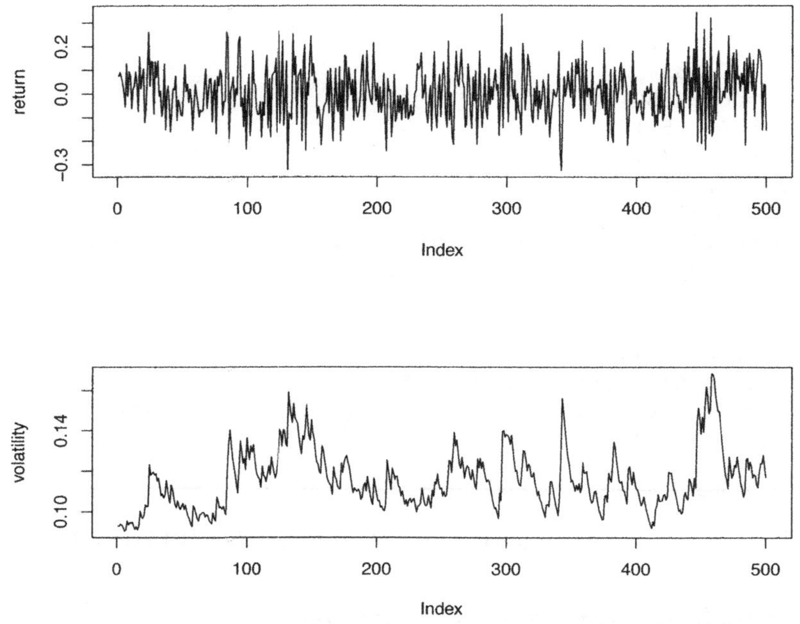

FIGURE 3.36 **Sample path of a GARCH(1,1) model.**

generalizations. In the following, we will mostly consider the price of a stock share as the relevant variable, but we can also model interest rates, exchange rates, macroeconomic variables, etc.

3.7.1 FROM DISCRETE TO CONTINUOUS TIME

It is a good idea to start with a discrete-time model and then derive a continuous-time model heuristically. Consider a time interval $[0, T]$, and imagine that we discretize it by a sequence of time periods of length δt, such that $T = n \cdot \delta t$; discrete-time instants are indexed by $t = 0, 1, 2, \ldots, n$. Let S_t be, for instance, the price of a stock share at time t. One possible and reasonable model for the evolution of this price over time is the multiplicative form:

$$S_{t+1} = u_t S_t, \qquad t = 0, 1, 2, \ldots, \tag{3.70}$$

where u_t is a non-negative random variable and the initial price S_0 is known. If we consider continuous random variables u_t, the model is continuous-state. In simplest model the variables u_t are i.i.d., which may be justified by the efficient market hypothesis, if we believe in it (or by the lack of an alternative model we trust enough). The multiplicative model ensures that prices will stay non-negative, which is an obvious requirement for stock prices. If we used an additive model, such as $S_{t+1} = u_t + S_t$, we should admit negative values for the random variables u_t, in order to represent price drops; therefore, we would not have the guarantee that $S_t \geq 0$. With the multiplicative form, a price drops when $u_t < 1$, but it remains positive. Furthermore, the actual impact of a price change depends on the present stock price; a \$1 increase is a different matter if the current price is \$100 rather than \$5. This is easily accounted for by the multiplicative form, which essentially deals with percent changes in price.

In order to select a sensible probability distribution for the random variables u_t, let us use a typical trick of the trade and consider the natural logarithm of the stock price:

$$\log S_{t+1} = \log S_t + \log u_t = \log S_t + z_t.$$

The random variable z_t is the increment in the logarithm of price, and a common assumption is that it is normally distributed, which implies that u_t is lognormal. Starting from the initial price S_0 and unfolding (3.70) recursively, we find

$$S_t = \prod_{k=0}^{t-1} u_k S_0,$$

which implies that

$$\log S_t = \log S_0 + \sum_{k=0}^{t-1} z_k.$$

Since the sum of normal random variables is still a normal variable, $\log S_t$ is normally distributed, which in turn implies that stock prices are lognormally distributed, according to this model. To put it in another way, the product of lognormals is still lognormal. Using the notation

$$\mathrm{E}[z_t] = \nu, \qquad \mathrm{Var}(z_t) = \sigma^2,$$

we see that

$$\mathrm{E}[\log S_t] = \mathrm{E}\left[\log S_0 + \sum_{k=0}^{t-1} z_k\right]$$

$$= \log S_0 + \sum_{k=0}^{t-1} \mathrm{E}[z_k] = \log S_0 + \nu t, \tag{3.71}$$

$$\mathrm{Var}(\log S_t) = \mathrm{Var}\left(\log S_0 + \sum_{k=0}^{t-1} z_k\right) = \sum_{k=0}^{t-1} \mathrm{Var}(z_k) = t\sigma^2, \tag{3.72}$$

where we take advantage of the intertemporal independence of the variables z_t when calculating variance. The important point to see here is that the expected value and the variance of the increment in the logarithm of the stock price scale linearly with time; this implies that standard deviation scales with the square root of time, whereas expected value scales linearly. This property is known as the square-root rule and implies that on a very short time period volatility, related to a standard deviation, dominates drift, related to an expected value.[33]

Our aim is to build a continuous-time model, so let us consider first what we typically do in a deterministic case to accomplish the task. The familiar approach is to take the limit of a difference equation, for $\delta t \to 0$, and build a differential equation. Informally, if there were no uncertainty, we would recast what we have seen in discrete time as

$$\delta \log S(t) = \log S(t + \delta t) - \log S(t) = \nu \, \delta t.$$

Taking the limit as $\delta t \to 0$, we find a relationship between differentials:

$$d \log S(t) = \nu \, dt.$$

Integrating both differentials over the interval $[0, t]$ yields

$$\int_0^t d \log S(\tau) = \nu \int_0^t d\tau \Rightarrow \log S(t) - \log S(0) = \nu t \Rightarrow S(t) = S(0)e^{\nu t}. \tag{3.73}$$

In the deterministic case, as we remarked earlier in this book, it is customary to write the differential equation as

$$\frac{d \log S(t)}{dt} = \nu$$

or, equivalently, as

$$\frac{dS(t)}{dt} = \nu S(t),$$

where we have used standard calculus, in particular the chain rule of derivatives, to rewrite the differential

$$d \log S(t) = \frac{dS(t)}{S(t)}. \tag{3.74}$$

To make the model stochastic (and more realistic), we have to include a noise term, which entails a few important changes. First, we should write the equation in the form

$$d \log S(t) = \nu \, dt + \sigma \, dW(t), \tag{3.75}$$

where $dW(t)$ can be considered as the increment of a stochastic process over the interval $[t, t + dt]$. Equation (3.75) is an example of a rather tricky object, called a *stochastic differential equation*. It is reasonable to guess that the solution of a stochastic differential equation is a stochastic process, rather than a determin-istic function of time. This topic is quite difficult to deal with rigorously, as

---

[33]This has some implications on the evaluation of risk measures like value-at-risk; see Chapter 13.

it requires some background in measure theory and stochastic calculus, but we may grasp the essentials by relying on intuition and heuristic arguments.

The first thing we need is to investigate which type of continuous-time stochastic process $W(t)$ we can use as a building block. In the next section we introduce such a process, called the Wiener process, which plays more or less the same role as the process z_t above. It turns out that this process is not differentiable, whatever this may mean for a stochastic process; informally, we will see that its sample paths are quite jagged indeed. Hence, we **cannot** write the stochastic differential equation as

$$\frac{d \log S(t)}{dt} = \nu + \sigma \frac{dW(t)}{dt}.$$

Actually, a stochastic differential equation must be interpreted as a shorthand for an *integral* equation much like (3.73), involving the increments of a stochastic process. This calls for the definition of a *stochastic integral* and the related stochastic calculus. A consequence of the definition of the stochastic integral is that working with differentials as in Eq. (3.74) is not possible. We need a way to generalize the chain rule for differentials from the deterministic to the stochastic case. This leads to a fundamental tool of stochastic calculus called *Itô's lemma*.

3.7.2 STANDARD WIENER PROCESS

In the discrete-time model, we have assumed normally distributed increments in logarithmic prices, and we have also seen that the expected value of the increment of the logarithm of price scales linearly with time, whereas standard deviation scales with the square root of time.

In discrete time, we could consider the following process as a building block:

$$W_{t+1} = W_t + \epsilon_t \sqrt{\delta t},$$

where ϵ_t is a sequence of independent standard normal variables. We see that, for $k > j$,

$$W_k - W_j = \sum_{i=j}^{k-1} \epsilon_i \sqrt{\delta t}, \tag{3.76}$$

which implies that

$$E[W_k - W_j] = 0,$$
$$\text{Var}(W_k - W_j) = (k - j)\,\delta t.$$

In order to move to continuous time, we may define the standard Wiener process as a continuous-time stochastic process characterized by the following properties:

1. $W(0) = 0$, which is actually a convention.

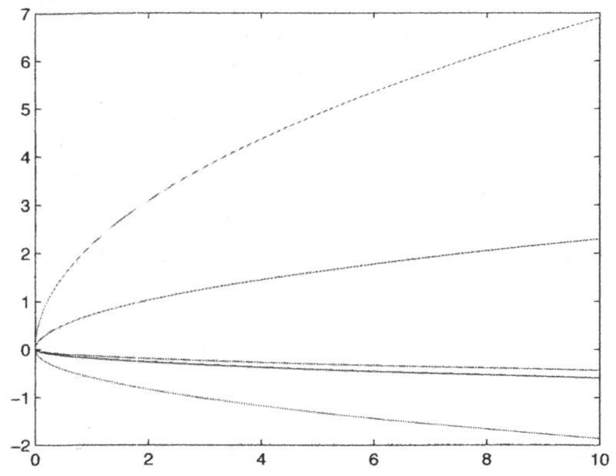

FIGURE 3.37 **Sample paths of a "degenerate" stochastic process.**

2. Given any time interval $[s, t]$, the increment $W(t) - W(s)$ is distributed as $\mathsf{N}(0, t - s)$, a normal random variable with expected value zero and standard deviation $\sqrt{t - s}$. These increments are stationary, in the sense that they do not depend on where the time interval is located, but only on its width.

3. Increments are independent: If we take time instants $t_1 < t_2 \le t_3 < t_4$, defining two nonoverlapping time intervals, then $W(t_2) - W(t_1)$ and $W(t_4) - W(t_3)$ are independent random variables.

Note how the second condition makes sure that standard deviation scales with the square root of time, like in the discrete-time model. To see the importance of the independent increments assumption, let us consider the sample paths of a process defined as $Q(t) = \epsilon\sqrt{t}$, with $\epsilon \sim \mathsf{N}(0, 1)$, which are shown in Fig. 3.37. This is a "degenerate" stochastic process,[34] since knowledge of just one point on a sample path yields the complete knowledge of the whole sample path, which makes the process quite predictable. However, if we only look at the marginal distributions of $Q(t)$, this process just looks like the Wiener process, since

$$\mathrm{E}[Q(t)] = 0 = \mathrm{E}[W(t)],$$
$$\mathrm{Var}[Q(t)] = t = \mathrm{Var}[W(t)].$$

It is lack of independence that makes the difference, as we may also check empirically. We may recast Eq. (3.76) as

$$W_{t+\delta t} = W_t + \epsilon_{t+\delta t}\sqrt{\delta t}, \qquad t = 0, 1, 2, \ldots, \tag{3.77}$$

---

[34]We have already made a quite similar point in Example 3.3.

```
simWiener <- function (T,numSteps,numRepl){
  dt <- T/numSteps
  # generate vector of standard normals
  epsVet <- rnorm(numSteps*numRepl)
  # transform vector to matrix of increments and cumulate
  incrMatrix <- sqrt(dt)*cbind(numeric(numRepl),
                      matrix(epsVet,numRepl,numSteps))
  pathMatrix <- t(apply(incrMatrix,1,cumsum))
  return(pathMatrix)
}

set.seed(5)
T <- 1; numSteps <- 1000; numRepl <- 1
Wpath <- simWiener(T,numSteps,numRepl)
plot(0:numSteps,Wpath,type='l')
```

R programming notes:

1. We avoid `for` loops by generating all of the required standard normals once and using `cumsum` to cumulate the increments.

2. We use `cbind` to add a leading column of zeros, corresponding to the initial value $W(0)$ of the Wiener process, to the matrix of increments. Note that our choice here and in the rest of this book is to store sample paths along matrix rows, rather than along matrix columns.

3. We use `apply` to cumulate the increments in the matrix, because we want a matrix containing sample paths on rows. Since the first column consists of zeros, we use `apply(incrMatrix,1,cumsum)` rather than `apply(incrMatrix,2,cumsum)` to cumulate horizontally. Nevertheless, since we obtain a matrix with a leading *row* of zeros, we have to transpose back the matrix `pathMatrix` before returning it.

FIGURE 3.38 **Simple R function and script to sample the standard Wiener process.**

where the notation $\epsilon_{t+\delta t}$, rather than ϵ_t, may be used to emphasize the fact that this standard normal shock conveys new information. This discretization immediately allows us to write a piece of R code to generate sample paths of the standard Wiener process, which is illustrated in Fig. 3.38 and produces the plot in Fig. 3.39, which is quite different from Fig. 3.37, even though the marginals are the same.

From Fig. 3.39, we observe that sample paths of the Wiener process look continuous, but not differentiable. This may be stated precisely, but not very easily. Introducing continuity and differentiability rigorously calls for specifying a precise concept of stochastic convergence, as we should say that the Wiener process is nowhere differentiable with probability 1. To get an intuitive

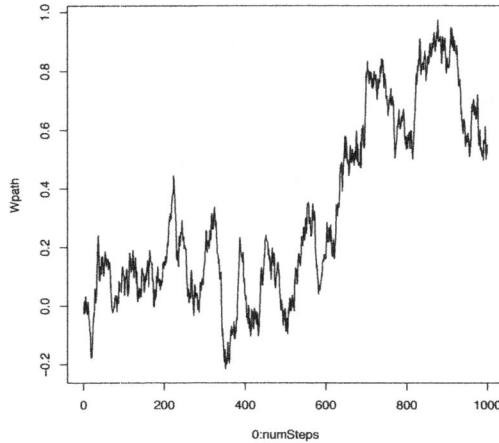

FIGURE 3.39 **A sample path of the standard Wiener process.**

feeling for this fact, let us consider the increment ratio

$$\frac{\delta W(t)}{\delta t} = \frac{W(t + \delta t) - W(t)}{\delta t}.$$

Given the defining properties of the Wiener process, it is easy to see that

$$\text{Var}\left[\frac{\delta W(t)}{\delta t}\right] = \frac{\text{Var}\left[W(t + \delta t) - W(t)\right]}{(\delta t)^2} = \frac{1}{\delta t}.$$

If we take the limit for $\delta t \to 0$, this variance goes to infinity. Strictly speaking, this is no proof of nondifferentiability of $W(t)$, but it does suggest that there is some trouble in using an object like $dW(t)/dt$; indeed, you will never see a notation like this. We only use the *differential* $dW(t)$ of the Wiener process. Informally, we may think of $dW(t)$ as a random variable with distribution $N(0, dt)$. Actually, we should think of this differential as an increment, which may be integrated as follows:

$$\int_s^t dW(\tau) = W(t) - W(s).$$

This looks reasonable, doesn't it? We may even go further and use $W(t)$ as the building block of stochastic differential equations. For instance, given real numbers a and b, we may imagine a stochastic process $X(t)$ satisfying the equation

$$dX(t) = a\,dt + b\,dW(t). \tag{3.78}$$

This equation defines a *generalized Wiener process* and may be solved by straight-forward integration over the time interval $[0, t]$:

$$X(t) - X(0) = \int_0^t dX(\tau) = \int_0^t a \, d\tau + \int_0^t b \, dW(\tau) = at + b \left[W(t) - W(0) \right],$$

which may be rewritten as

$$X(t) = X(0) + at + bW(t).$$

Hence, for the generalized Wiener process we find

$$X(t) \sim \mathsf{N}\big(X(0) + at, b^2 t \big).$$

However, if we consider something more complicated, like

$$dX(t) = a\big(t, X(t)\big) \, dt + b\big(t, X(t)\big) \, dW(t), \qquad (3.79)$$

things are not that intuitive. A process satisfying an equation like (3.79) is called an Itô process. We could argue that the solution should be something like

$$X(t) = X(0) + \int_0^t a\big(s, X(s)\big) \, ds + \int_0^t b\big(\tau, X(\tau)\big) \, dW(\tau). \qquad (3.80)$$

Here the first integral looks like a standard Riemann integral of a function over time, but what about the second one? We need to assign a precise meaning to it, and this leads to the definition of a stochastic integral.

3.7.3 STOCHASTIC INTEGRATION AND ITÔ'S LEMMA

In a stochastic differential equation defining a process $X(t)$, where a Wiener process $W(t)$ is the driving factor, we may assume that the value $X(t)$ depends only on the history of $W(t)$ over the time interval from 0 to t. Technically speaking, we say that process $X(t)$ is *adapted* to process $W(t)$. Now let us consider a stochastic integral like

$$\int_0^T X(t) \, dW(t). \qquad (3.81)$$

How can we assign a meaning to this expression? To begin with, it is reasonable to guess that a stochastic integral is a random variable. If we integrate a deterministic function of time we get a number; so, it is natural to guess that, by integrating a stochastic process over time, we should get a random variable. Furthermore, the stochastic integral above looks related to the sample paths of process $W(t)$, and an approximation could be obtained by partitioning the integration interval into small subintervals defined by points $0 = t_0, t_1, t_2, \ldots, t_n = T$ and considering the sum

$$\sum_{k=0}^{n-1} X(t_k) \left[W(t_{k+1}) - W(t_k) \right]. \qquad (3.82)$$

If you recall how the Riemann integral was introduced in your first calculus course, as the limit of sums over partitions, the above definition does not look too different. Yet, there is a subtle and crucial point. It is very important to notice how we have chosen the time instants in the expression above: $X(t_k)$ is taken at the left extreme of the interval (t_k, t_{k+1}). This is actually *one* possible choice: Why not take the value of X in the midpoint? This choice has an important effect: $X(t_k)$ is a random variable which is independent from the increment $W(t_{k+1}) - W(t_k)$ by which it is multiplied. This choice is what makes the stochastic integral in the sense of Itô a bit peculiar, and it may be motivated as follows.

Example 3.16 Financial motivation of Itô integrals

Consider a set of n assets, whose prices are modeled by stochastic processes $S_i(t)$, $i = 1, \ldots, n$, which are described by stochastic differential equations like (3.79), and assume that we have a portfolio strategy represented by functions $h_i(t)$. These functions represent the number of stock shares of each asset i that we hold at time t. But which functions make sense? An obvious requirement is that functions $h_i(\cdot)$ should not be anticipative: $h_i(t)$ may depend on all the history so far, over the interval $[0, t]$, but clairvoyance should be ruled out. Furthermore, we should think of $h_i(t)$ as the number of shares we hold over a time interval of the form $[t, t + dt)$.

Now, assume that we have an initial wealth that we invest in the portfolio, whose initial value, depending on the portfolio strategy represented by the functions $h_i(\cdot)$, is

$$V_h(0) = \sum_{i=1}^{n} h_i(0) S_i(0) = \mathbf{h}^{\mathsf{T}}(0) \mathbf{S}(0),$$

where we have grouped $h_i(t)$ and $S_i(t)$ in column vectors. What about the dynamics of the portfolio value? If the portfolio is self-financing, i.e., we can trade assets but we do not invest (nor withdraw) any more cash after $t = 0$, it can be shown that the portfolio value will satisfy the equation

$$dV_h(t) = \sum_{i=1}^{n} h_i(t) \, dS_i(t) = \mathbf{h}^{\mathsf{T}}(t) \, d\mathbf{S}(t).$$

This looks fairly intuitive and convincing, but some careful analysis is needed to prove it (see, e.g., [1, Chapter 6]). Then, we may reasonably guess that the wealth at time $t = T$ will be

$$V_h(T) = V_h(0) + \int_0^T \mathbf{h}^{\mathsf{T}}(t) \, d\mathbf{S}(t).$$

However, it is fundamental to interpret the stochastic integral as the limit of an approximation like (3.82), i.e.,

$$\int_0^T \mathbf{h}^\mathsf{T}(t)\, d\mathbf{S}(t) \approx \sum_{k=0}^{n-1} \mathbf{h}^\mathsf{T}(t_k) \left[\mathbf{S}(t_{k+1}) - \mathbf{S}(t_k) \right].$$

The number of stock shares we hold at time t_k does *not* depend on future prices $\mathbf{S}(t_{k+1})$. First we allocate wealth at time t_k, and *then* we observe return over the time interval (t_k, t_{k+1}). This makes financial sense and is why Itô stochastic integrals are defined the way they are.

Now, if we take approximation (3.82) and consider finer and finer partitions of the interval $[0, t]$, letting $n \to \infty$, what do we obtain? The answer is technically involved. We must select a precise concept of stochastic convergence and check that everything makes sense. Using mean square convergence, it can be shown that the definition makes indeed sense and provides a rigorous definition of the stochastic integral in the sense of Itô.

The definition of stochastic integral has some important consequences. To begin with, what is the expected value of the integral in Eq. (3.81)? This is a good starting question, because the stochastic integral is a random variable, and the first and foremost feature of a random variable is its expected value. We may get a clue by considering the approximation (3.82):

$$\mathrm{E}\left[\int_0^T X(t)\, dW(t) \right] \approx \mathrm{E}\left\{ \sum_{k=0}^{n-1} X(t_k) \left[W(t_{k+1}) - W(t_k) \right] \right\}$$

$$= \sum_{k=0}^{n-1} \mathrm{E}\left\{ X(t_k) \left[W(t_{k+1}) - W(t_k) \right] \right\}$$

$$= \sum_{k=0}^{n-1} \mathrm{E}\left[X(t_k) \right] \cdot \mathrm{E}\left[W(t_{k+1}) - W(t_k) \right] = 0,$$

where we have used the independence between $X(t_k)$ and the increments of the Wiener process over the time interval (t_k, t_{k+1}), along with the fact that the expected value of these increments is zero. This shows that the integral of Eq. (3.81) is a random variable with expected value zero, but can we say something more? The definition of stochastic integral does not yield a precise way to compute it practically. We may try, however, to consider a specific case to get some intuition. The following example illustrates one nasty consequence of the way in which we have defined the stochastic integral.

■ Example 3.17 Chain rule and stochastic differentials

Say that we want to "compute" the stochastic integral

$$\int_0^T W(t)\,dW(t).$$

Analogy with ordinary calculus would suggest using the chain rule of differentiation of composite functions to obtain a differential which can be integrated directly. Specifically, we *might guess* that

$$dW^2(t) = 2W(t)\,dW(t).$$

This in turn *would* suggest that

$$\int_0^T W(t)\,dW(t) = \frac{1}{2}\int_0^T dW^2(t) = \frac{1}{2}W^2(T).$$

Unfortunately, this *cannot* be the correct answer, as it contradicts our previous findings. We have just seen that the expected value of an integral of this kind is zero, but:

$$
\begin{aligned}
\mathrm{E}\left[\frac{1}{2}W^2(T)\right] &= \frac{1}{2}\mathrm{E}\left[W^2(T)\right] \\
&= \frac{1}{2}\left\{\mathrm{Var}\left[W(T)\right] + \mathrm{E}^2\left[W(T)\right]\right\} = \frac{T}{2} \neq 0.
\end{aligned}
$$

We see that the two expected values do not match at all, and there must be something wrong somewhere.

Example 3.17 shows that the chain rule does not work in Itô stochastic calculus. To proceed further, we need to find the right rule, and the answer is Itô's lemma. As usual in this section, we cannot follow rigorous arguments, but we may give an informal, yet quite instructive argument (following [12, Chapter 13]). The argument is instructive as it provides us with valuable intuition explaining what went wrong in Example 3.17. Let us recall that an Itô process $X(t)$ satisfies a stochastic differential equation such as

$$dX = a(X,t)\,dt + b(X,t)\,dW, \tag{3.83}$$

which is in some sense the continuous limit of

$$\delta X = a(X,t)\delta t + b(X,t)\epsilon(t)\sqrt{\delta t}, \tag{3.84}$$

where $\epsilon(t) \sim \mathsf{N}(0,1)$. What we actually need is a way to derive a stochastic differential equation for a function $F(X,t)$ of $X(t)$, as this plays the role of the derivative of a composite function, which is what the chain rule is used for.

Let us take a little step back in the realm of deterministic calculus, and consider what is the rationale behind Taylor's expansion of a function $G(x, y)$ of two variables. The key ingredient we need for our reasoning is the formula for the differential of such a function:

$$dG = \frac{\partial G}{\partial x} dx + \frac{\partial G}{\partial y} dy,$$

which indeed may be obtained from Taylor's expansion:

$$\delta G = \frac{\partial G}{\partial x}\delta x + \frac{\partial G}{\partial y}\delta y + \frac{1}{2}\frac{\partial^2 G}{\partial x^2}(\delta x)^2 + \frac{1}{2}\frac{\partial^2 G}{\partial y^2}(\delta y)^2 + \frac{\partial^2 G}{\partial x\,\partial y}\delta x\,\delta y + \cdots$$

when $\delta x, \delta y \to 0$. Now we may apply this Taylor expansion to $F(X, t)$, limiting it to the leading terms. In doing so it is important to notice that the term $\sqrt{\delta t}$ in Eq. (3.84) needs careful treatment when squared. In fact, we have something like

$$(\delta X)^2 = b(X, t)^2 \epsilon^2 \delta t + \cdots,$$

which implies that the term in $(\delta X)^2$ cannot be neglected in the approximation. Since ϵ is a standard normal variable, we have $\mathrm{E}[\epsilon^2] = 1$ and $\mathrm{E}[\epsilon^2\,\delta t] = \delta t$. A delicate point is the following: It can be shown that, as δt tends to zero, the term $\epsilon^2\,\delta t$ can be treated as nonstochastic, and it is equal to its expected value. An informal (far from rigorous) justification relies on the variance of this term:

$$\begin{aligned}
\mathrm{Var}(\epsilon^2\,\delta t) &= (\delta t)^2 \left\{ \mathrm{E}[\epsilon^4] - \mathrm{E}^2[\epsilon^2] \right\} \\
&= (\delta t)^2 \left\{ 3 - \mathrm{Var}^2[\epsilon] \right\} \\
&= (\delta t)^2 \left\{ 3 - 1 \right\} = 2(\delta t)^2,
\end{aligned}$$

which, for $\delta t \to 0$, can be neglected with respect to first-order terms. Here we have used the fact $\mathrm{E}[\epsilon^4] = 3$, which can be checked by using moment generating functions[35] or, cheating a bit, by recalling that the kurtosis of any normal, including a standard one, is 3. A useful way to remember this point is the *formal* rule

$$(dW)^2 = dt. \tag{3.85}$$

Hence, when δt tends to zero, in the Taylor expansion we have

$$(\delta X)^2 \to b(X, t)^2\, dt.$$

Neglecting higher-order terms and taking the limit as both δX and δt tend to zero, we end up with

$$dF = \frac{\partial F}{\partial X} dX + \frac{\partial F}{\partial t} dt + \frac{1}{2}\frac{\partial^2 F}{\partial X^2} b(X, t)^2\, dt,$$

---

[35] See Example 8.9.

which, substituting for dX, becomes the celebrated **Itô's lemma**:

$$dF = \left[a(X,t)\frac{\partial F}{\partial X} + \frac{\partial F}{\partial t} + \frac{1}{2}b(X,t)^2\frac{\partial^2 F}{\partial X^2} \right] dt + b(X,t)\frac{\partial F}{\partial X}dW. \quad (3.86)$$

Although this proof is far from rigorous, we see that all the trouble is due to the term of order $\sqrt{\delta t}$, which is introduced by the increment of the Wiener process.

It is instructive to see that if we set $b(X,t) = 0$ in the Itô differential equation, i.e., if we eliminate randomness, then we step back to familiar ground. In such a case, the stochastic differential equation is actually deterministic and can be rewritten as

$$dx = a(x,t)dt \quad \Rightarrow \quad \frac{dx}{dt} = a(x,t),$$

where we use the notation x rather than X to point out the deterministic nature of function $x(t)$. Then, Itô's lemma boils down to

$$dF = a(x,t)\frac{\partial F}{\partial x}dt + \frac{\partial F}{\partial t}dt, \quad (3.87)$$

which can be rearranged to yield the following chain rule for a function $F(x,t)$:

$$\frac{dF}{dt} = \frac{\partial F}{\partial x}\frac{dx}{dt} + \frac{\partial F}{\partial t}. \quad (3.88)$$

In order to grasp Itô's lemma, let us consider Example 3.17 once more.

Example 3.18 Itô's lemma

In order to compute the stochastic integral of $W^2(t)$, we may apply Itô's lemma to the case $X(t) = W(t)$, by setting $a(X,t) \equiv 0$, $b(X,t) \equiv 1$, and $F(X,t) = X^2(t)$. Hence, we have

$$\frac{\partial F}{\partial t} = 0, \quad (3.89)$$

$$\frac{\partial F}{\partial X} = 2X, \quad (3.90)$$

$$\frac{\partial^2 F}{\partial X^2} = 2. \quad (3.91)$$

It is important to point out that in Eq. (3.89) the partial derivative with respect to time is zero; it is true that $F(X(t),t)$ depends on time through $X(t)$, but here we have no direct dependence on t, thus the *partial* derivative with respect to time vanishes. To put it another way, the dependence of $X(t)$ with respect to time does not play any role when taking the partial derivative with respect to t, because $X(t)$ is held fixed. You may also have a look back at Eq. (3.88) to see the relationship between the partial and a total derivative of $F(X(t), t)$ with respect to t.

The application of Itô's lemma yields

$$dF = d(W^2) = dt + 2W\,dW. \tag{3.92}$$

It is essential to notice that dt is exactly the term which we would *not* expect by applying the usual chain rule. But this is the term that allows us to get the correct expected value of $W^2(T)$. Indeed, by straightforward integration of Eq. (3.92) now we find

$$W^2(T) = W^2(0) + \int_0^T dW^2(t) = 0 + \int_0^T dt + \int_0^T W(t)\,dW(t).$$

Taking the expected values yields

$$\mathrm{E}\big[W^2(T)\big] = T,$$

which is coherent with our findings in Example 3.17.

Let us summarize our findings: With respect to Eq. (3.87), in Itô's lemma we have an extra term in dW, which is expected given the form of the differential equation defining the stochastic process $X(t)$, and an unexpected term

$$\frac{1}{2}b^2\frac{\partial^2 F}{\partial x^2}.$$

In deterministic calculus, second-order derivatives occur in second-order terms linked to $(\delta t)^2$, which can be neglected; but here we have a term of order $\sqrt{dt}$ which must be taken into account even when it is squared.

3.7.4 GEOMETRIC BROWNIAN MOTION

Itô's lemma may be used to find the solution of stochastic differential equations, at least in relatively simple cases. A very important example is geometric Brownian motion. Geometric Brownian motion is defined by a specific choice of the functions in Itô stochastic differential equation:

$$dS(t) = \mu S(t)\,dt + \sigma S(t)\,dW(t),$$

where μ and σ are constant parameters referred to as drift and volatility, respectively. Intuition would suggest to rewrite the equation as

$$\frac{dS(t)}{S(t)} = \mu\,dt + \sigma\,dW(t),$$

and then to consider the differential of $d \log S$, which would be dS/S in deterministic calculus, to find the corresponding integral. However, we have seen that some extra care is needed when dealing with stochastic differential equations. Nevertheless, the above intuition deserves further investigation, so let us apply Itô's lemma and find the stochastic differential equation for $F(S,t) = \log S(t)$. As a first step we compute the partial derivatives:

$$\frac{\partial F}{\partial t} = 0,$$

$$\frac{\partial F}{\partial S} = \frac{1}{S},$$

$$\frac{\partial^2 F}{\partial S^2} = -\frac{1}{S^2}.$$

Then, putting all of them together we find

$$dF = \left(\frac{\partial F}{\partial t} + \mu S \frac{\partial F}{\partial S} + \frac{1}{2} \sigma^2 S^2 \frac{\partial^2 F}{\partial S^2} \right) dt + \sigma S \frac{\partial F}{\partial S} \, dW$$

$$= \left(\mu - \frac{\sigma^2}{2} \right) dt + \sigma \, dW.$$

Now we see that our guess above was not that bad, as this equation may be easily integrated and yields

$$\log S(t) = \log S(0) + \left(\mu - \frac{\sigma^2}{2} \right) t + \sigma W(t).$$

Recalling that $W(t)$ has a normal distribution and can be written as $W(t) = \epsilon \sqrt{t}$, where $\epsilon \sim \mathsf{N}(0,1)$, we conclude that the logarithm of price is normally distributed:

$$\log S(t) \sim \mathsf{N} \left[\log S(0) + \left(\mu - \frac{\sigma^2}{2} \right) t, \quad \sigma^2 t \right].$$

We can rewrite the solution in terms of $S(t)$:

$$S(t) = S(0) \exp \left\{ \left(\mu - \frac{\sigma^2}{2} \right) t + \sigma W(t) \right\},$$

or

$$S(t) = S(0) \exp \left\{ \left(\mu - \frac{\sigma^2}{2} \right) t + \sigma \sqrt{t} \epsilon \right\}. \tag{3.93}$$

On the one hand, this shows that prices, according to the geometric Brownian motion model, are lognormally distributed. On the other hand, now we have a way to generate sample paths, based on a discretization of a prescribed time horizon in time periods of length δt, like we did with the standard Wiener process. A simple R function to do so is given in Fig. 3.40.[36] This function, when given the drift and volatility selected in the script, yields the plot in Fig. 3.41.

```
simGBM <- function (S0,mu,sigma,T,numSteps,numRepl){
  dt <- T/numSteps
  nuT <- (mu-sigma^2/2)*dt
  sigmaT <- sqrt(dt)*sigma
  pathMatrix = matrix(nrow=numRepl,ncol=numSteps+1)
  pathMatrix[,1] <- S0
  for (i in 1:numRepl){
    for (j in 2:(numSteps+1)){
      pathMatrix[i,j] <- pathMatrix[i,j-1]*
                          exp(rnorm(1,nuT,sigmaT))
    }
  }
  return(pathMatrix)
}

S0 <- 50; mu <- 0.1; sigma <- 0.3; T <- 1
numSteps <- 1000; numRepl <- 1
set.seed(5)
path <- simGBM(S0,mu,sigma,T,numSteps,numRepl)
plot(0:numSteps, path, type='l')
```

FIGURE 3.40 **Simple R function to sample geometric Brownian motion.**

By using properties of the lognormal distribution[37] and letting $\nu = \mu - \sigma^2/2$, we also find

$$\mathrm{E}[\log(S(t)/S(0))] = \nu t,$$
$$\mathrm{Var}[\log(S(t)/S(0))] = \sigma^2 t,$$
$$\mathrm{E}[S(t)/S(0)] = e^{\mu t}, \tag{3.94}$$
$$\mathrm{Var}[S(t)/S(0)] = e^{2\mu t}(e^{\sigma^2 t} - 1), \tag{3.95}$$

from which we see that the drift parameter μ is linked to the continuously compounded return. The volatility parameter σ is related to standard deviation of the increment of logarithm of price. The roles of drift and volatility can also be grasped intuitively by considering the following approximation of the differential equation:

$$\frac{\delta S}{S} \approx \mu \, \delta t + \sigma \, \delta W,$$

where $\delta S/S$ is the return of the asset over small time interval δt. According to this approximation, we see that return can be approximated by a normal variable

---

[36]We have already met this function in the first chapter; please refer to Fig. 1.5 for programming notes. Later, we will describe a more efficient version.

[37]See Section 3.2.3.

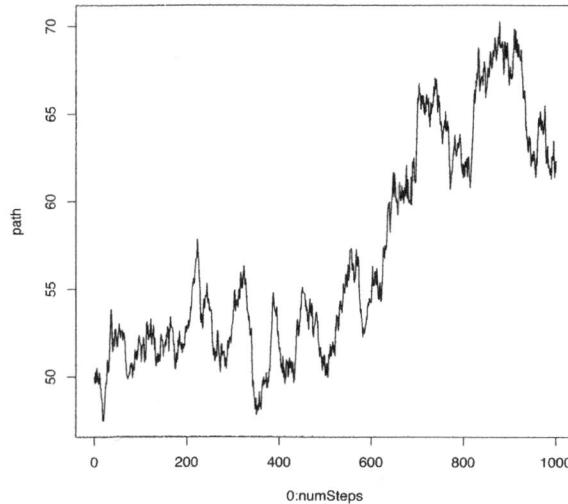

FIGURE 3.41   **A sample path of geometric Brownian motion.**

with expected value $\mu \, \delta t$ and standard deviation $\sigma \sqrt{\delta t}$. Actually, this normal distribution is only a local approximation of the "true" (according to the model) lognormal distribution.

3.7.5   GENERALIZATIONS

Geometric Brownian motion is not the only type of stochastic process relevant in finance, and the Wiener process is not the only relevant building block. One of the main features of these processes is the continuity of sample paths. However, discontinuities do occur sometimes, such as jumps in prices. In this case, different building blocks are used, such as the Poisson process, which counts the events occurring at a certain rate. We should also note that continuous sample paths do not make sense for certain state variables such as credit ratings. Another point is that the lognormal distribution, that we get from geometric Brownian motion, is a consequence of the normality associated with the Wiener process. Distributions with fatter tails are typically observed, questioning the validity of the simple models based on Gaussian diffusions.

Indeed, several alternative processes have been proposed and are commonly used in financial engineering. Here we just mention some possible generalizations, leaving further details to Chapter 6 on path generation and Chapter 11 on pricing derivatives.

Correlated Wiener processes. The need to model the dynamics of a set of asset prices arises when we deal with a portfolio, or when we want to price a rainbow option depending on multiple underlying assets. The simplest model that we may adopt is a direct generalization of geometric Brownian motion. According to this approach, the prices $S_i(t)$ of assets $i = 1, \dots, n$ satisfy the equation

$$dS_i(t) = \mu_i S_i(t)\, dt + \sigma_i S_i(t)\, dW_i(t),$$

where the Wiener processes $W_i(t)$ are not necessarily independent. They are characterized by a set of instantaneous correlation coefficients ρ_{ij}, whose meaning can be grasped by an extension of the formal rule of Eq. (3.85):

$$dW_i \cdot dW_j = \rho_{ij}\, dt.$$

From a Monte Carlo point of view, sample path generation of these processes require the generation of increments

$$\delta W_i = \epsilon_i \sqrt{\delta t}, \qquad \delta W_j = \epsilon_j \sqrt{\delta t},$$

where ϵ_i and ϵ_j are standard normals with correlation ρ_{ij}.

Ornstein–Uhlenbeck processes. In geometric Brownian motion, we have an exponential trend on which a diffusion process is superimposed.[38] This exponential growth makes no sense for several financial variables, which rather feature mean reversion. Then, we may resort to Ornstein–Uhlenbeck processes like

$$dX(t) = \gamma\big[\overline{X} - X(t)\big]\,dt + \sigma\, dW(t).$$

They still are Gaussian diffusions, but we observe that the drift can change sign, pulling the process back to a long-run average $\overline{X}$, with speed γ. The application of this model to a short-term interest rate $r(t)$ yields the Vasicek model:

$$dr(t) = \gamma\big[\overline{r} - r(t)\big]\,dt + \sigma\, dW(t).$$

Square-root diffusions. One drawback of Ornstein–Uhlenbeck processes is that they do not prevent negative values, which make no sense for stock prices and interest rates. A possible adjustment is the following:

$$dr(t) = \gamma\big[\overline{r} - r(t)\big]\,dt + \sigma\sqrt{r(t)}\, dW(t).$$

This is an example of a square-root diffusion, which in the context of short-term interest rates is known as the Cox–Ingersoll–Ross model. For a suitable choice of parameters, it can be shown that a square-root diffusion stays

---

[38] One might well wonder whether an increasing exponential trend makes sense for the prices of stock shares, as we do not observe this in real life. One reason is that dividends are paid, bringing down the share price.

non-negative. This changes the nature of the process, which is not Gaussian anymore, as we will see in Chapter 6. Similar considerations hold when modeling a stochastic and time-varying volatility $\sigma(t)$. Geometric Brownian motion assumes constant volatility, whereas in practice we may observe time periods in which volatility is higher than usual. In a discrete-time setting, ARCH and GARCH models may be used. In continuous-time, one possible model for stochastic volatility is the Heston model, which consists of a pair of stochastic differential equations:

$$dS(t) = \mu S(t)\, dt + \sigma(t) S(t)\, dW_1(t),$$
$$dV(t) = \alpha \big[\bar{V} - V(t)\big]\, dt + \xi \sqrt{V(t)}\, dW_2(t),$$

where $V(t) = \sigma^2(t)$, $\bar{V}$ is a long-term value, and different assumptions can be made on the correlation of the two driving Wiener processes.

Jump–diffusions. In order to account for jumps, we may devise processes with both a diffusion and a jump component, such as

$$X(t) = \alpha t + \sigma W(t) + Y(t),$$

where $Y(t)$ is a compound Poisson process. From a formal point of view, we require the definition of a stochastic integral of a stochastic process $Y(t)$ with respect to a Poisson process $N(t)$:

$$\int_0^t Y(\tau)\, dN(\tau) = \sum_{i=1}^{N(t)} Y(t_i),$$

where $N(t)$ is the number of jumps occurred up to time t in the Poisson process, and t_i, $i = 1, \dots, N(t)$, are the time instants at which jumps occur. We note that the Wiener and the Poisson process, which are the basic building blocks of jump–diffusions, are quite different, but they do have an important common feature: stationary and independent increments. In fact, the class of Lévy processes generalizes both of them by emphasizing this feature.

3.8   Dimensionality reduction

Monte Carlo methods are often the only feasible approach to cope with high-dimensional problems. Nevertheless, they may require an excessive computational effort to produce a reliable answer. Hence, strategies to reduce the dimensionality of a problem are welcome anyway. Moreover, they can also be put to good use when applying alternative approaches, such as low-discrepancy sequences. Apart from computational convenience, there is another driving force behind dimensionality reduction. While in this chapter we are mainly concerned with model building, in the next one we will deal with model estimation. A rich

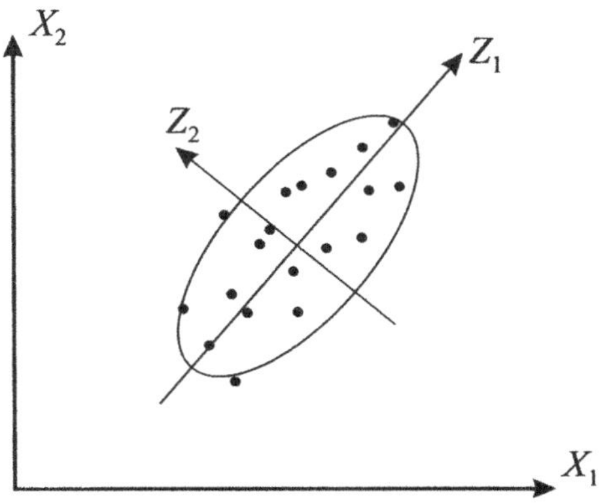

FIGURE 3.42 **Visualization of PCA.**

model may be quite appealing in principle, but it turns into a nightmare if it calls for a huge amount of data for its estimation, and a badly estimated data will make the output of the most sophisticated Monte Carlo simulation utterly useless, if not worse.

Data and dimensionality reduction methods are an important and wide topic in multivariate statistics, and we cannot provide anything but some hints on the most relevant tools for applications in finance and economics, to provide the reader with a feeling for them. In this section we outline two methods:

- Principal component analysis, PCA for short
- Factor models

The second approach should not be confused with factor analysis, which is another multivariate analysis technique for dimensionality reduction. The main difference is that factor analysis techniques deal with latent and nonobservable factors, arguably due to their origin in psychology. The factor models we consider here are quite common in finance and aim at easing some statistical estimation issues when a large number of assets is analyzed.

3.8.1 PRINCIPAL COMPONENT ANALYSIS (PCA)

To introduce principal component analysis (PCA) in the most intuitive way, let us have a look at Fig. 3.42, which shows a scatterplot of observations of two random variables X_1 and X_2. It is clear that the two random variables are positively correlated. Now imagine that, for some reason, we can only afford dealing with a single variable. In order to streamline our model of uncertainty, one possibility would be to get rid of one of the two variables, but which one? Since the variables are related and both display some non-negligible variability, discarding one the two variables looks like a very crude approach to dimensionality reduction. We should try to come up with a single variable trying to capture the most information. Now, consider the two axes in the figure, labeled with variables Z_1 and Z_2. Of course, we may represent the very same set of observations in terms of Z_1 and Z_2. This is just a change of coordinates, accomplished in two steps:

- Data have been centered, which is obtained by subtracting the sample mean from the observations: $\mathbf{X} - \overline{\mathbf{X}}$. This is just a shift in the origin of reference axes, but it is often recommended to improve numerical stability in data analysis algorithms.

- The coordinates Z_1 and Z_2 are obtained by a suitable rotation of the axes.

Vector rotation is accomplished by multiplication by an orthogonal matrix. A square matrix $\mathbf{A}$ is orthogonal if its columns are a set of orthogonal unit vectors,[39] which implies

$$\mathbf{A}^\mathsf{T}\mathbf{A} = \mathbf{A}\mathbf{A}^\mathsf{T} = \mathbf{I}.$$

This also shows that this matrix can be inverted by transposition. To see that multiplication by such a matrix is a rotation, we may observe that it does not change the norm of vectors:

$$\| \mathbf{A}\mathbf{x} \|^2 = (\mathbf{A}\mathbf{x})^\mathsf{T}\mathbf{A}\mathbf{x} = \mathbf{x}^\mathsf{T}\mathbf{A}^\mathsf{T}\mathbf{A}\mathbf{x} = \mathbf{x}^\mathsf{T}\mathbf{x} = \| \mathbf{x} \|^2 \,.$$

Since orthogonal matrices enjoy plenty of nice properties, we might be able to find a linear transformation of variables with some interesting structure. Indeed, in Fig. 3.42 we observe two important facts about the new variables Z_1 and Z_2:

1. They are uncorrelated.
2. The first variable is associated with the largest variance.

Now, if we had to get rid of one variable, we would have no doubt: $\mathbf{Z}_1$ conveys the most information, and the residual one is not only less relevant, but also uncorrelated.

   Now that we have a feeling for what PCA is and why it may be useful, we may investigate it in more depth following two different routes to introduce PCA:

1. As a means of taking linear combinations of variables, in such a way that they are uncorrelated.

2. As a way to combine variables so that the first component has maximal variance, the second one has the next maximal variance, and so forth.

Let us pursue both viewpoints, which actually lead us to same conclusions, as we will see.

---

[39] We should maybe talk about orthonormal vectors and matrices, but we will follow the more common parlance.

3.8.1.1   A geometric view of PCA

Let p be the dimension of the random vector $\mathbf{X}$ and n the number of available joint observations. Such observations are collected into the following matrix:

$$\mathcal{X} = \begin{bmatrix} X_1^{(1)} & X_2^{(1)} & \cdots & X_p^{(1)} \\ X_1^{(2)} & X_2^{(2)} & \cdots & X_p^{(2)} \\ \vdots & \vdots & \ddots & \vdots \\ X_1^{(n)} & X_2^{(n)} & \cdots & X_p^{(n)} \end{bmatrix}.$$

where the element $[\mathcal{X}]_{kj}$ in row k and column j is the component j of observation k, i.e., $X_j^{(k)}$.

A linear data transformation, including centering, can be written as

$$\mathbf{Z} = \mathbf{A}(\mathbf{X} - \overline{\mathbf{X}}),$$

where $\mathbf{A} \in \mathbb{R}^{p,p}$. We assume that data have already been centered, in order to ease notation. Hence,

$$Z_1 = a_{11}X_1 + a_{12}X_2 + \cdots + a_{1p}X_p,$$
$$Z_2 = a_{21}X_1 + a_{22}X_2 + \cdots + a_{2p}X_p,$$
$$\vdots \qquad\qquad \vdots$$
$$Z_p = a_{p1}X_1 + a_{p2}X_2 + \cdots + a_{pp}X_p.$$

The Z_i variables are called *principal components*: Z_1 is the first principal component. Now, let us consider the sample covariance matrix of $\mathbf{X}$, denoted by $\mathbf{S_X}$, with entries

$$S_{ij} = \frac{1}{n-1} \sum_{k=1}^{n} \left(X_i^{(k)} - \overline{X}_i\right) \left(X_j^{(k)} - \overline{X}_j\right).$$

Since we assume centered data, this matrix can be expressed as follows:

$$\mathbf{S_X} = \frac{1}{n-1} \mathcal{X}^\mathsf{T} \mathcal{X}.$$

Now, we may also find the corresponding sample covariance matrix for Z, $\mathbf{S_Z}$. However, we would like to find a matrix $\mathbf{A}$ such that the resulting principal components are uncorrelated; in other words, $\mathbf{S_Z}$ should be diagonal:[40]

$$\mathbf{S_Z} = \mathbf{A} \mathbf{S_X} \mathbf{A}^\mathsf{T} = \begin{bmatrix} S_{Z_1}^2 & & & \\ & S_{Z_2}^2 & & \\ & & \ddots & \\ & & & S_{Z_p}^2 \end{bmatrix},$$

---

[40]See Property 3.6 in Section 3.3.4.1.

where $S^2_{Z_i}$ is the sample variance of each principal component. The matrix $\mathbf{A}$ should diagonalize the sample covariance matrix $\mathbf{S_X}$. To diagonalize $\mathbf{S_X}$, we should consider the product

$$\mathbf{P^T S_X P} = \begin{bmatrix} \lambda_1 & & & \\ & \lambda_2 & & \\ & & \ddots & \\ & & & \lambda_p \end{bmatrix},$$

where matrix $\mathbf{P}$ is orthogonal and its columns consist of the normalized eigenvectors of $\mathbf{S_X}$; since this is symmetric, its eigenvectors are indeed orthogonal. The diagonalized matrix consists of the eigenvalues λ_i, $i = 1, \ldots, p$, of $\mathbf{S_X}$. Putting everything together, we see that the rows $\mathbf{a}_i^\mathsf{T}$, $i = 1, \ldots, p$, of matrix $\mathbf{A}$ should be the normalized eigenvectors:

$$\mathbf{A} = \begin{bmatrix} \mathbf{a}_1^\mathsf{T} \\ \mathbf{a}_2^\mathsf{T} \\ \cdots \\ \mathbf{a}_p^\mathsf{T} \end{bmatrix}.$$

We also observe that the sample variances of the principal components Z_i are the eigenvalues of $\mathbf{S_X}$:

$$S^2_{Z_i} = \lambda_i, \qquad i = 1, \ldots, p.$$

If we sort eigenvalues in decreasing order, $\lambda_1 \geq \lambda_2 \geq \ldots \geq \lambda_p$, we can see that indeed Z_1 is the first principal component, accounting for most variability. Then, the second principal component Z_2 is uncorrelated with Z_1 and is the second in rank. The fraction of variance explained by the first q components is

$$\frac{\lambda_1 + \lambda_2 + \cdots \lambda_q}{\lambda_1 + \lambda_2 + \cdots + \lambda_p} = \frac{\lambda_1 + \lambda_2 + \cdots + \lambda_q}{\sum_{j=1}^p [\mathbf{S_z}]_{jj}}.$$

Taking the first few components, we can account for most variability and reduce the problem dimension by replacing the original variables by the principal components.

3.8.1.2 Another view of PCA

Another view of PCA is obtained by interpreting the first principal component in terms of orthogonal projection. Consider a unit vector $\mathbf{u} \in \mathbb{R}^p$, and imagine projecting the observed vector $\mathbf{X}$ on $\mathbf{u}$. This yields a vector parallel to $\mathbf{u}$, of length $\mathbf{u^T X}$. Since $\mathbf{u}$ has unit length, the projection of observation $\mathbf{X}^{(k)}$ on $\mathbf{u}$ is

$$\mathbf{P}_{\mathbf{X}^{(k)}} = \left(\mathbf{u^T X}^{(k)} \right) \mathbf{u}.$$

We are projecting p-dimensional observations on just one axis, and, of course, we would like to have an approximation that is as good as possible. More precisely, we should find $\mathbf{u}$ in such a way that the distance between the originally observed vector $\mathbf{X}^{(k)}$ and its projection is as small as possible. If we have a sample of n observations, we should minimize the average distance

$$\sum_{k=1}^{n} \| \mathbf{X}^{(k)} - \mathbf{P}_{\mathbf{X}^{(k)}} \|^2,$$

which looks much like a least-squares problem. This amounts to an orthogonal projection of the original vectors on $\mathbf{u}$, where we know that the original and the projected vectors are orthogonal. Hence, we can apply the Pythagorean theorem to rewrite the problem:

$$\| \mathbf{X}^{(k)} - \mathbf{P}_{\mathbf{X}^{(k)}} \|^2 = \| \mathbf{X}^{(k)} \|^2 - \| \mathbf{P}_{\mathbf{X}^{(k)}} \|^2 .$$

In order to minimize the left-hand side of this equality, on the average, we have to maximize

$$\sum_{i=1}^{n} \| \mathbf{P}_{\mathbf{X}_i} \|^2$$

subject to the condition $\| \mathbf{u} \| = 1$. The problem can be restated as

$$\max \quad \mathbf{u}^\mathsf{T} \mathcal{X}^\mathsf{T} \mathcal{X} \mathbf{u}$$
$$\text{s.t.} \quad \mathbf{u}^\mathsf{T} \mathbf{u} = 1.$$

But we know that, assuming that data are centered, the sample covariance matrix is $\mathbf{S}_X = \mathcal{X}^\mathsf{T} \mathcal{X} / (n-1)$; hence, the problem is equivalent to

$$\max \quad \mathbf{u}^\mathsf{T} \mathbf{S_X} \mathbf{u} \tag{3.96}$$
$$\text{s.t.} \quad \mathbf{u}^\mathsf{T} \mathbf{u} = 1. \tag{3.97}$$

In plain English, what we want is finding one dimension on which multidimensional data should be projected, in such a way that the variance of the projected data is maximized. This makes sense from a least-squares perspective, but it also has an intuitive appeal: The dimension along which we maximize variance is the one providing the most information.

To solve the problem above, we may associate the constraint (3.97) with a Lagrange multiplier λ and augment the objective function (3.96) to obtain the Lagrangian function:[41]

$$\mathcal{L}(\mathbf{u}, \lambda) = \mathbf{u}^\mathsf{T} \mathbf{S_X} \mathbf{u} + \lambda(1 - \mathbf{u}^\mathsf{T} \mathbf{u}) = \mathbf{u}^\mathsf{T}(\mathbf{S_X} - \lambda \mathbf{I})\mathbf{u} + \lambda.$$

---

[41] Strictly speaking, we are in trouble here, since we are maximizing a *convex* quadratic form. However, we may replace the equality constraint by an inequality constraint, which results in a concave problem, i.e., a problem in which the optimal solution is on the boundary of the feasible solution. It turns out that the solution we pinpoint using the Lagrange multiplier method is the right one.

The gradient of the Lagrangian function with respect to **u** is

$$2(\mathbf{S_X} - \lambda \mathbf{I})\mathbf{u},$$

and setting it to zero yields the first-order optimality condition

$$\mathbf{S_X u} = \lambda \mathbf{u}.$$

This amounts to saying that λ must be an eigenvalue of the sample covariance matrix, but which one? We can rewrite the objective function (3.96) as follows:

$$\mathbf{u}^\mathsf{T}\mathbf{S_X u} = \lambda \mathbf{u}^\mathsf{T}\mathbf{u} = \lambda.$$

Hence, we see that λ should be the largest eigenvalue of $\mathbf{S_X}$, **u** is the corresponding normalized eigenvector, and we obtain the same result as in the previous section. Furthermore, we should continue on the same route, by asking for another direction in which variance is maximized, subject to the constraint that it is orthogonal to the first direction we found. Since eigenvectors of a symmetric matrix are orthogonal, we see that indeed we will find all of them, in decreasing order of the corresponding eigenvalues.

3.8.1.3   A small numerical example

Principal component analysis in practice is carried out on sampled data, but it may be instructive to consider an example where both the probabilistic and the statistical sides are dealt with.[42] Consider first a random variable with bivariate normal distribution, $\mathbf{X} \sim \mathsf{N}(\mathbf{0}, \boldsymbol{\Sigma})$, where

$$\boldsymbol{\Sigma} = \begin{bmatrix} 1 & \rho \\ \rho & 1 \end{bmatrix}$$

and $\rho > 0$. Thus, X_1 and X_2 are standard normal variables with positive correlation ρ. To find the eigenvalues of $\boldsymbol{\Sigma}$, we must find its characteristic polynomial and solve the corresponding equation

$$\begin{vmatrix} 1 - \lambda & \rho \\ \rho & 1 - \lambda \end{vmatrix} = (1 - \lambda)^2 - \rho^2 = 0.$$

This yields the two eigenvalues $\lambda_1 = 1 + \rho$ and $\lambda_2 = 1 - \rho$. Note that both eigenvalues are non-negative, since $\rho \in [-1, 1]$ is a correlation coefficient. To find the first eigenvector, associated with λ_1, we consider the system of linear equations

$$\begin{bmatrix} 1 & \rho \\ \rho & 1 \end{bmatrix}\begin{bmatrix} u_1 \\ u_2 \end{bmatrix} = (1 + \rho)\begin{bmatrix} u_1 \\ u_2 \end{bmatrix} \quad \Rightarrow \quad \begin{bmatrix} -\rho & \rho \\ \rho & -\rho \end{bmatrix}\begin{bmatrix} u_1 \\ u_2 \end{bmatrix} = \begin{bmatrix} 0 \\ 0 \end{bmatrix}.$$

[42]This example has been adapted from [10, Chapter 9].

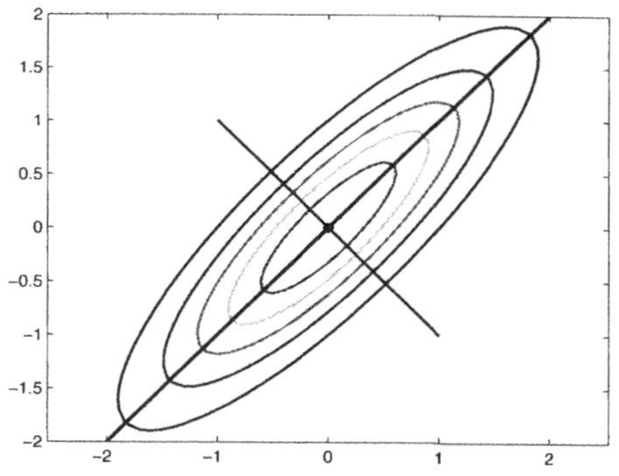

FIGURE 3.43 **Level curves of a multivariate normal with $\rho = 0.85$.**

Clearly, the two equations are linearly dependent and any vector such that $u_1 = u_2$ is an eigenvector. By a similar token, any vector such that $u_1 = -u_2$ is an eigenvector corresponding to λ_2. Two normalized eigenvectors are

$$\gamma_1 = \frac{1}{\sqrt{2}} \begin{bmatrix} 1 \\ 1 \end{bmatrix}, \qquad \gamma_2 = \frac{1}{\sqrt{2}} \begin{bmatrix} 1 \\ -1 \end{bmatrix}.$$

These are the rows of the transformation matrix

$$\mathbf{Z} = \mathbf{A}(\mathbf{X} - \boldsymbol{\mu}) = \frac{1}{\sqrt{2}} \begin{bmatrix} 1 & 1 \\ 1 & -1 \end{bmatrix} \begin{bmatrix} X_1 \\ X_2 \end{bmatrix}.$$

Since we are dealing with standard normals, $\boldsymbol{\mu} = \mathbf{0}$ and the first principal component is

$$Z_1 = \frac{X_1 + X_2}{\sqrt{2}}.$$

The second principal component is

$$Z_2 = \frac{X_1 - X_2}{\sqrt{2}}.$$

As a further check, let us compute the variance of the first principal component:

$$\text{Var}(Z_1) = \tfrac{1}{2}\left[\text{Var}(X_1) + \text{Var}(X_2) + 2\text{Cov}(X_1, X_2)\right] = 1 + \rho = \lambda_1.$$

Figure 3.43 shows the level curves of the joint density of $\mathbf{X}$ when $\rho = 0.85$. Since correlation is positive, the main axis of the ellipses has positive slope. It is easy to see that along that direction we have the largest variability.

Practical PCA is carried out on sampled data, and R provides us with a suitable function called `prcomp`. The R code in Fig. 3.44 samples 2000 points from the above bivariate normal distribution, with $\rho = 0.85$, carries out PCA, and produces the following output:

```
rho <- 0.85
sigma <- matrix(c(1, rho, rho, 1),2,2)
mu <- c(0, 0)
set.seed(55555)
NN <- 2000
library(MASS)
X <- mvrnorm(NN, mu, sigma)
PCA <- prcomp(X, center=TRUE, scale=FALSE, retx=TRUE)
summary(PCA)
print(PCA)
```

FIGURE 3.44 **Code to carry out PCA on sampled data.**

```
> summary(PCA)
Importance of components:
                          PC1      PC2
Standard deviation     1.3574  0.38496
Proportion of Variance 0.9256  0.07444
Cumulative Proportion  0.9256  1.00000
> print(PCA)
Standard deviations:
[1] 1.3573917 0.3849576

Rotation:
            PC1         PC2
[1,] -0.7133313   0.7008270
[2,] -0.7008270  -0.7133313
```

As usual in R, we may get the essential information about an object by using functions `print` or `summary`; here we are applying them on an object, `PCA`, which is a rich data structure including all of the relevant information about the results of the analysis. Let us check these results against the theory, starting with sample statistics, which are fairly close to the theoretical values (we also compute centered data along the way):

```
> m <- colMeans(X);m
[1] -0.013842466 -0.009333299
> Xc <- X - (matrix(1,NN,1) %*% m)
> S <- cov(X);S
           [,1]       [,2]
[1,] 1.0103328 0.8470275
[2,] 0.8470275 0.9803719
```

Then, let us compute eigenvalues and eigenvectors of the sample covariance matrix, and check them against the output of PCA:

```
> vv <- eigen(S)
```

```
> eigval <- vv$values;eigval
[1] 1.8425123 0.1481923
> sqrt(eigval)
[1] 1.3573917 0.3849576
> PCA$sdev
[1] 1.3573917 0.3849576
> eigvet <- vv$vectors;eigvet
           [,1]          [,2]
[1,] -0.7133313  0.7008270
[2,] -0.7008270 -0.7133313
```

Note that if we do not feel comfortable with eigenvalues and eigenvectors, we may also obtain the principal components directly.

```
> A <- PCA$rotation;A
           PC1          PC2
[1,] -0.7133313  0.7008270
[2,] -0.7008270 -0.7133313
```

We clearly see that the first principal component accounts for most variability, precisely:

$$\frac{1.8425123}{1.8425123 + 0.1481923} = 92.56\%.$$

We may also find the transformed variables directly:

```
> z <- PCA$x
> cov(z)
              PC1             PC2
PC1 1.842512e+00  2.636065e-16
PC2 2.636065e-16  1.481923e-01
> head(z)
           PC1          PC2
[1,] -2.3881699 -0.83075134
[2,]  0.6764307 -1.07220758
[3,]  1.5929255 -0.33190052
[4,]  2.3407987  0.06721413
[5,]  1.6718914 -0.10795592
[6,] -0.8683767  0.56002701
> head(Xc %*% A)
           PC1          PC2
[1,] -2.3881699 -0.83075134
[2,]  0.6764307 -1.07220758
[3,]  1.5929255 -0.33190052
[4,]  2.3407987  0.06721413
[5,]  1.6718914 -0.10795592
[6,] -0.8683767  0.56002701
```

Finally, let us draw the scatterplots of the original variables and the principal components, as well as the boxplots of the principal components:

```
> par(mfrow=c(1,2))
```

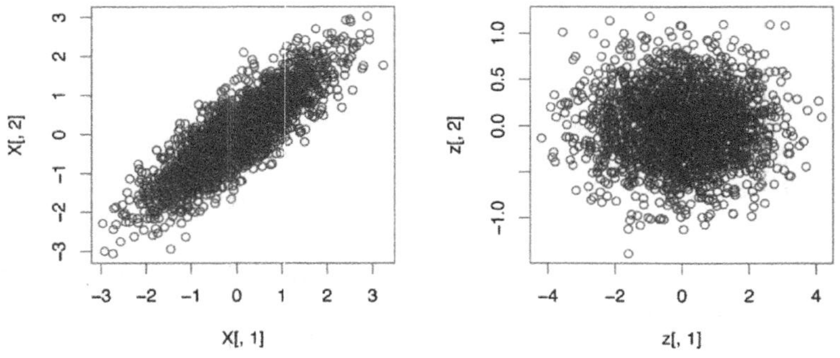

FIGURE 3.45 **Scatterplots of original variables and principal components.**

```
> plot(X[,1], X[,2])
> plot(z[,1], z[,2])
> par(mfrow=c(1,1))
> boxplot(z[,1], z[,2], names=c("Z1","Z2"))
```

The first plot, displayed in Fig. 3.45, shows the positive correlation of the original variables and the lack of correlation of the principal components. Note that the two scales here are different, however, and this hides the relative amounts of variability; the difference in variability is pretty evident in the two boxplots of Fig. 3.46.

3.8.1.4   Applications of PCA

PCA can be applied in a Monte Carlo setting by reducing the number of random variables we have to sample from. A less obvious role is played in the contest of low-discrepancy sequences. As we shall see in Chapter 9, these sequences can replace random sampling in intermediate-sized problems, but may not be as effective in truly high-dimensional settings. Then, one can use low-discrepancy sequences to sample from the principal components, which have the most impact on the output, possibly using random sampling for the remaining ones. We should mention, however, that in order to apply such ideas we have to figure out the relationship between the distribution of the original variables and the principal components, which need not to be an easy task in general.

Another typical application of PCA is in reducing the number of risk factors in financial modeling. Consider, for instance, the *term structure of interest rates*, which specifies a set of (annualized) interest rates applying on different time horizons. To use the most general notation, let $R(0, t, t + \tau)$ be continuously compounded forward interest rate applying on the time period $[t, t + \tau]$, as seen at time 0. This is a forward rate, as it is the rate that can be contracted now for a

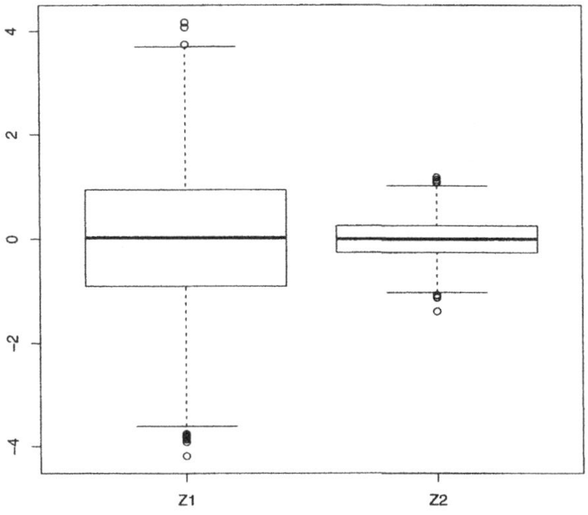

FIGURE 3.46 **Boxplots of principal components.**

time interval starting in the future. If we consider the set of rates $R(t, t, t+\tau)$ as a function of τ, we have the term structure of spot rates, i.e., the rates applying at time t for several maturities τ. Usually, the term structure is increasing in τ, but it moves in time, changing its shape in various ways. Clearly, modeling the term structure requires the definition of a model for each maturity; furthermore, these movements must preserve some basic consistency, in order to prevent arbitrage opportunities. Measuring and managing interest rate risk is complicated by the presence of so many interrelated risk factors, and Monte Carlo simulation can be challenging as well. One possible approach to make the problem more manageable is to replace the term structure by a few key risk factors combining rates at different maturities, possibly obtained by PCA.[43] In typical models, the first few principal components can be interpreted and related to the level, slope, and curvature of the term structure.

3.8.2 FACTOR MODELS

Now let us consider another quite practical question: If we have a large set of jointly distributed random variables, what is the required effort to estimate their covariance structure? Equivalently, how many correlations we need? The

---

[43] See, e.g., Chapters 3 and 6 of [16].

covariance matrix is symmetric; hence, if we have n random variables, we have to estimate n variances and $n(n-1)/2$ covariances. This amounts to

$$n + \frac{n(n-1)}{2} = \frac{n(n+1)}{2}$$

entries. Hence, if $n = 1000$, we should estimate 500,500 entries in the covariance matrix. A daunting task, indeed! If you think that such a case will never occur in practice, please consider a practical portfolio management problem. You might well consider 1000 assets for inclusion in the portfolio. In such a case, can we estimate the covariance matrix? What you know about statistical inference tells that you might need a lot of data to come up with a reliable estimate of a parameter. If you have to estimate a huge number of parameters, you need a huge collection of historical data. Unfortunately, many of them would actually tell us nothing useful: Would you use data from the 1940s to characterize the distribution of returns for IBM stock shares now? We need a completely different approach to reduce our estimation requirements.

The returns of a stock share are influenced by many factors. Some are general economic factors, such as inflation and economic growth. Others are specific factors of a single firm, depending on its management strategy, product portfolio, etc. In between, we may have some factors that are quite relevant for a group of firms within a specific industrial sectors, much less for others; think of the impact of oil prices on energy or telecommunications.

Rather than modeling uncertain returns individually, we might try to take advantage of this structure of general and specific factors. Let us take the idea to an extreme and build a simple model whereby there is one systematic factor common to all of stock shares, and a specific factor for each single firm. Formally, we represent the random return for the stock share of firm i as

$$R_i = \alpha_i + \beta_i R_m + \epsilon_i,$$

where

- α_i and β_i are parameters to be estimated.

- R_m is a random variable representing the systematic risk factor; the subscript m stands for *market*; indeed, financial theory suggests that a suitable systematic factor could be the return of a market portfolio consisting of all stock shares, with a proportion depending on their relative capitalization with respect to the whole market.

- ϵ_i is a random variable representing specific risk, which in financial parlance is also referred to as *idiosyncratic* risk; a natural assumption about these variables is that $E[\epsilon_i] = 0$ (otherwise, we would include the expected value into α_i). Another requirement is that the systematic factor really captures whatever the stock returns have in common, and that the specific factors are independent. Typically, we do not require independence, but only lack of correlation. These requirements can be formalized as

$$\text{Cov}(\epsilon_i, R_m) = 0, \tag{3.98}$$

$$\text{Cov}(\epsilon_i, \epsilon_j) = 0, \qquad i \neq j. \tag{3.99}$$

In the next chapter, where we deal with linear regression, we will see that condition (3.98) is actually ensured by model estimation procedures based on least squares. On the contrary, condition (3.99) is just an assumption, resulting in a so-called *diagonal model*, since the covariance matrix of specific factors is diagonal.

The model above is called single-factor model for obvious reasons, but what are its advantages from a statistical perspective? Let us check how many unknown parameters we should estimate in order to evaluate expected return and variance of return for an arbitrary portfolio. To begin with, observe that for a portfolio with weights w_i, we have

$$R_p = \sum_{i=1}^{n} w_i(\alpha_i + \beta_i R_m + \epsilon_i) = \sum_{i=1}^{n} w_i \alpha_i + R_m \sum_{i=1}^{n} w_i \beta_i + \sum_{i=1}^{n} w_i \epsilon_i.$$

Then,

$$\text{E}[R_p] = \sum_{i=1}^{n} w_i \alpha_i + \text{E}[R_m] \sum_{i=1}^{n} w_i \beta_i + \sum_{i=1}^{n} w_i \text{E}[\epsilon_i] = \sum_{i=1}^{n} w_i \alpha_i + \mu_m \sum_{i=1}^{n} w_i \beta_i,$$

where μ_m is the expected return of the market portfolio (more generally, the expected value of whatever systematic risk factor we choose). From this, we see that we need to estimate:

- n parameters α_i, $i = 1, \ldots, n$
- n parameters β_i, $i = 1, \ldots, n$
- The expected value μ_m

These add up to $2n + 1$ parameters. Variance is a bit trickier, but we may use the diagonality condition (3.99) to eliminate covariances and obtain

$$\text{Var}(R_p) = \text{Var}(R_m) \left(\sum_{i=1}^{n} w_i \beta_i \right)^2 + \sum_{i=1}^{n} w_i^2 \text{Var}(\epsilon_i) + 2 \sum_{i \neq j} w_i w_j \text{Cov}(\epsilon_i, \epsilon_j)$$

$$= \sigma_m^2 \left(\sum_{i=1}^{n} w_i \beta_i \right)^2 + \sum_{i=1}^{n} w_i^2 \sigma_i^2,$$

where σ_m^2 is the variance of the systematic risk factor and σ_i^2 is the variance of each idiosyncratic risk factor, $i = 1, \ldots, n$. They amount to $n + 1$ additional parameters that we should estimate, bringing the total to $3n + 2$ parameters. In the case of $n = 1000$ assets, we have a grand total of 3,002 parameters; this is a large number, anyway, but it pales when compared with the 500,500 entries of the full covariance matrix.

In passing, we note that factor models also play a key role in financial economics. The *capital asset pricing model* (CAPM) is essentially a single-factor model, based on a systematic risk factor related to the return of the market portfolio; if we add some suitable hypotheses to the model, we are led to an important, though controversial, equilibrium model. If more factors are included in the model, plus some hypotheses on the structure of the market and the behavior of investors, we obtain another equilibrium model, *arbitrage pricing theory* (APT). The validity of these models is challenged by alternative views, as they are not just statistical but equilibrium models, relying on questionable assumptions. Nevertheless, the idea of factor models is widely used by practitioners who do not subscribe to CAPM or APT.

From a Monte Carlo point of view, factor models may simplify scenario generation. There might be some computational saving, but this is arguably not the really important point. What is more relevant is that we can provide the simulation model with more reliable parameter estimates and inputs, with a positive impact on output quality.

3.9 Risk-neutral derivative pricing

In this section we outline risk-neutral pricing for financial derivatives, most prominently options, for the sake of the unfamiliar reader. What we cover here is mostly relevant to Chapter 11. If a reader is not directly interested to this topic, but she wants nevertheless to see how Monte Carlo methods are used in this setting as an example, here are the essential results:

- Options (and other derivatives) can be priced by taking the expected value of the discounted payoff.

- The expectation must be taken under a risk-neutral (or risk-adjusted) probability measure.

This justifies the application of Monte Carlo methods to estimate the expected payoff and the option price as a consequence; however, we have to use a modified model for the dynamics of the underlying asset. Typically, this just means changing the drift in a stochastic differential equation.

Options are financial derivatives, i.e., assets whose value depends on some other variable. This can be illustrated by considering a call option, a contract between two counterparts, written on a non-dividend-paying stock: The option holder, i.e., the party buying the option, has the right, but not the obligation, to buy the underlying asset from the option writer, i.e., the party selling the option on the primary market, at a prescribed future time and at a price established now. This is called the *strike price*, denoted by K. Let us denote the current price of the underlying asset by S_0 (known when the option is written) and the future price at a time $t = T$ by S_T; the latter is, at time $t = 0$, a random variable.

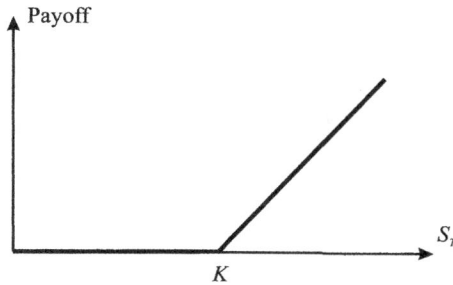

FIGURE 3.47 **Payoff of a call option.**

The time $t = T$ corresponds to the option maturity. A *European-style* option can only be exercised at maturity. On the contrary, *American-style options* can be exercised at any time before their expiration date. The call option will be exercised only if $S_T \geq K$, since there would be no point in using the option to buy at a price K that is larger than the prevailing (spot) market price S_T at time T. Then, the payoff of the option is, from the holder's viewpoint,

$$\max\{S_T - K, 0\}.$$

To interpret this, consider the possibility of exercising the option and buying at price K an asset that can be immediately sold at S_T. Market structure is not really that simple, but some options are settled in cash, so the payoff really has a monetary nature; this occurs for options written on a nontraded market index or on interest rates. This payoff is a piecewise linear function and is illustrated in Fig. 3.47. Put options, unlike calls, grant the holder the right to sell the underlying asset at the strike price. Hence, the payoff of a put option is

$$\max\{K - S_T, 0\}.$$

Put options have a positive payoff when the asset price drops, much like short selling strategies, whereby one borrows and sells an asset, hoping to buy the asset back at a lower price in order to give it back to the lender.

When the future asset price S_T will be realized, at time $t = T$, the payoff will be clearly determined. What is not that clear is the fair price for the contract at time $t = 0$. It is easy to see that a put or call option must have a positive price, as their payoff to the option holder cannot be negative; the option writer, on the contrary, can suffer a huge loss; hence, the latter requires an option premium, paid at time $t = 0$, to enter into this risky contract.

3.9.1 OPTION PRICING IN THE BINOMIAL MODEL

The easiest way to get the essential message about option pricing is by referring to the binomial model of Section 3.2.1, based on multiplicative shocks and a generalization of the Bernoulli random variable. As shown in Fig. 3.48, the

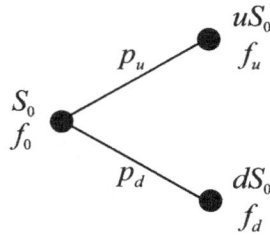

FIGURE 3.48 **One-step binomial model for option pricing.**

future price of the underlying asset can be either uS_0 or dS_0, with probabilities p_u and p_d, respectively. Conditional on the asset price at maturity, the option will provide its holder with a payoff f_T, which can either be f_u or f_d. These values are also illustrated in Fig. 3.48. For instance, in the case of a call option with strike price K, we have

$$f_d = \max\{S_0 d - K, 0\} \quad \text{and} \quad f_u = \max\{S_0 u - K, 0\}.$$

The pricing problem consists of finding a fair price f_0 for the option, i.e., its value at time $t = 0$. Intuitively, one would guess that the two probabilities p_u and p_d play a major role in determining f_0. What is not quite clear is the role that should be played by the overall risk aversion of market participants. However, a simple principle can be exploited in order to simplify our task. The key idea is to build a replicating portfolio, i.e., a portfolio consisting of elementary assets, such that its payoff replicates the option payoff. If we can do so and we know the current prices of the assets in the replicating portfolio, then it is easy to argue that the option value cannot be different from the value of the replicating portfolio. If this were not the case, the intuitive law of one price would be violated. We would have two equivalent portfolios with different prices, and we would just buy the cheaper one and sell the more expensive one to lock in a risk-free profit starting with no money. Such a money making machine is called an arbitrage opportunity, and option pricing relies on the assumption that arbitrage opportunities are ruled out. To be precise, what we need to assume is that they cannot exist in equilibrium, as investors would take immediate advantage of them, in such a way to move prices back to an equilibrium.

The replicating portfolio should certainly include the underlying asset, but we also need an additional one. Financial markets provide us with such an asset, in the form of a risk-free investment. We may consider a bank account yielding a risk-free interest rate r; we assume here that this rate is continuously compounded, so that if we invest \$1 now, we will own \$$e^{rt}$ at a generic time instant t. Equivalently, we may think of a riskless zero-coupon bond, with initial price $B_0 = 1$ and future price $B_1 = e^{rT}$ at time $t = T$, when the option matures. In order to find the number of stock shares and riskless bonds that we should include in the replicating portfolio, we just need to set up a system of two linear equations. Let us denote the number of stock shares by Δ and the number of bonds by Ψ. The initial value of this portfolio is

$$\Pi_0 = \Delta S_0 + \Psi, \qquad (3.100)$$

and its future value, depending on the realized state, will be either

$$\Pi_u = \Delta S_0 u + \Psi e^{rT} \qquad \text{or} \qquad \Pi_d = \Delta S_0 d + \Psi e^{rT}.$$

Note that future value of the riskless bond does not depend on the realized state. If this portfolio has to replicate the option payoff, we should enforce the following two conditions:

$$\begin{aligned} \Delta S_0 u + \Psi e^{rT} &= f_u, \\ \Delta S_0 d + \Psi e^{rT} &= f_d. \end{aligned} \qquad (3.101)$$

Solving the system yields

$$\Delta = \frac{f_u - f_d}{S_0(u - d)},$$

$$\Psi = e^{-rT} \frac{u f_d - d f_u}{u - d}.$$

Hence, by the law of one price, the option value now must be

$$\begin{aligned} f_0 &= \Delta S_0 + \Psi \\ &= \frac{f_u - f_d}{u - d} + e^{-rT} \frac{u f_d - d f_u}{u - d} \\ &= e^{-rT} \left\{ \frac{e^{rT} - d}{u - d} f_u + \frac{u - e^{rT}}{u - d} f_d \right\}. \end{aligned} \qquad (3.102)$$

A striking fact is that this relationship does *not* involve the objective probabilities p_u and p_d. However, let us consider the quantities

$$\pi_u = \frac{e^{rT} - d}{u - d}, \qquad \pi_d = \frac{u - e^{rT}}{u - d},$$

which happen to multiply the option payoffs in Eq. (3.102). We may notice two interesting features of π_u and π_d:

1. They add up to one, i.e., $\pi_u + \pi_d = 1$.
2. They are non-negative, i.e., $\pi_u, \pi_d \geq 0$, provided that $d \leq e^{rT} \leq u$.

The last condition does make economic sense. If we had $e^{rT} < d < u$, the risky stock share would perform better than the risk-free bond in any scenario, and we would just borrow money at the risk-free rate and buy a huge amount of the stock share. On the contrary, if the risk-free rate is always larger than the stock price in the future, one could make easy money by selling the stock share short and investing the proceeds in the risk-free asset. In fact, we can interpret π_u and

π_d as probabilities, and Eq. (3.102) can be interpreted in turn as the discounted expected value of the option payoff,

$$f_0 = e^{-rT}(\pi_u f_u + \pi_d f_d) = e^{-rT}\mathrm{E}^{\mathbb{Q}}\big[f_T\big], \qquad (3.103)$$

provided that we do not use the "objective" probabilities p_u and p_d, but π_u and π_d. Here the notation $\mathrm{E}^{\mathbb{Q}}[\cdot]$ emphasizes this change in the probability measure. These are called *risk-neutral probabilities*, a weird name that we can easily justify. In fact, if we use the probabilities π_u and π_d, the expected value of the underlying asset at maturity is

$$\mathrm{E}^{\mathbb{Q}}\big[S_T\big] = \pi_u S_0 u + \pi_d S_0 d = S_0 e^{rT}.$$

But how is it possible that the expected return of a risky asset is just the risk-free rate? This might happen only in a world where investors do not care about risk, but only about the expected return; in other words, they do not ask for any risk premium. In such a risk-neutral world, the expected return of all assets, at equilibrium, should be the same, and if there is a risk-free asset, its return will be the expected return for all risky assets as well. This observation justifies the name associated with the probabilities that we use in option pricing. Indeed, what we have just illustrated is the tip of a powerful iceberg, the risk-neutral pricing approach, which plays a pivotal role in financial engineering.

3.9.2 A CONTINUOUS-TIME MODEL FOR OPTION PRICING: THE BLACK–SCHOLES–MERTON FORMULA

The binomial model yields an intuitive and enlightening result. However, we cannot certainly represent uncertainty in stock prices by a Bernoulli random variable. If we introduce more states of nature, we have two ways to build a replicating portfolio: Either we introduce more assets, or we introduce a dynamic replicating strategy, i.e., we rebalance the portfolio several times along the option life. The second approach sounds much more promising, as it relies on the same elementary assets as before. In the limit, if we assume rebalancing in continuous time, we can rely on the machinery of stochastic calculus. Let us assume that the underlying asset price follows a geometric Brownian motion,

$$dS_t = \mu S_t\,dt + \sigma S_t\,dW_t,$$

where to streamline notation a bit we use S_t and W_t rather than $S(t)$ and $W(t)$, and let us consider the viewpoint of the option writer, who holds a short position in the call option. Let us denote the fair option price at time t, when the underlying asset price is S_t, by $f(S_t, t)$. Note that we are taking for granted that the option price depends on these two variables only, which makes sense for a path-independent, vanilla option. In the case of Asian options, where the payoff depends on average prices, things are not that simple. Using Itô's lemma, we may write a stochastic differential equation for f:

$$df = \frac{\partial f}{\partial t}\,dt + \frac{\partial f}{\partial S}\,dS + \frac{1}{2}\sigma^2 S^2 \frac{\partial^2 f}{\partial S^2}\,dt. \qquad (3.104)$$

Just as in the binomial case, what we know is the option value at maturity,

$$F(S_T, T) = \max\{S_T - K, 0\},$$

and what we would like to know is $f(S_0, 0)$, the fair option price now. The option holder has a position of value $-f(S_t, t)$ and is subject to considerable risk: If the stock price rockets up, the option holder will exercise, and the writer will have to buy the underlying asset at a high price, unless some countermeasures are taken. One possibility is to buy some stock shares, say Δ, and build a hedged portfolio of value

$$\Pi = -f(S_t, t) + \Delta \cdot S_t.$$

Equation (3.104) does not suggest an immediate way to find the option price, but it would look a little bit nicer without the random term dS. Indeed, it is easy to see that, by choosing

$$\Delta = \frac{\partial f}{\partial S},$$

we can hedge risk by eliminating the random term dS, which includes the increment of the Wiener process. By differentiating Π, we find

$$d\Pi = -df + \Delta \, dS = \left(-\frac{\partial f}{\partial S} + \Delta \right) dS - \left(\frac{\partial f}{\partial t} + \frac{1}{2}\sigma^2 S^2 \frac{\partial^2 f}{\partial S^2} \right) dt. \quad (3.105)$$

With the above choice of Δ, our portfolio is riskless; hence, by no-arbitrage arguments, it must earn the risk-free interest rate r:

$$d\Pi = r\Pi \, dt. \quad (3.106)$$

Eliminating $d\Pi$ between Eqs. (3.105) and (3.106), we obtain

$$\left(\frac{\partial f}{\partial t} + \frac{1}{2}\sigma^2 S^2 \frac{\partial^2 f}{\partial S^2} \right) dt = r \left(f - S\frac{\partial f}{\partial S} \right) dt,$$

and finally

$$\frac{\partial f}{\partial t} + rS\frac{\partial f}{\partial S} + \frac{1}{2}\sigma^2 S^2 \frac{\partial^2 f}{\partial S^2} - rf = 0. \quad (3.107)$$

Now we have a deterministic partial differential equation (PDE for short) describing an option value $f(S, t)$. This is the Black–Scholes–Merton PDE. Such a PDE must be solved with suitable boundary conditions, but before doing so, the following observations are in order about the choice of Δ:

- It is consistent with the value it takes in the binomial case, which is just an increment ratio and becomes a derivative in the continuous case.
- It eliminates the dependence with respect to the true drift μ, which is related to the expected return of the underlying asset; again, this is consistent with the binomial model, where the expected value of the future asset price, based on the objective probabilities, does not plat any role.

What does not quite look consistent with the binomial model is that there we obtain the price as an expected value, whereas here we have a PDE. Actually, they are two sides of the same coin, and the gap can be bridged by one version of the Feynman–Kač formula.

THEOREM 3.17 Feynman–Kač representation theorem. *Consider the partial differential equation*

$$\frac{\partial F}{\partial t} + \mu(x,t)\frac{\partial F}{\partial x} + \frac{1}{2}\sigma^2(x,t)\frac{\partial^2 F}{\partial x^2} = rF,$$

and let $F = F(x,t)$ be a solution satisfying the boundary condition

$$F(T,x) = \Phi(x).$$

Then, under technical conditions, $F(x,t)$ can be represented as

$$F(x,t) = \mathrm{E}_{x,t}\left[\Phi(X_T)\right],$$

where X_t is a stochastic process satisfying the differential equation

$$dX_\tau = \mu(X_\tau,\tau)\,d\tau + \sigma(X_\tau,\tau)\,dW_\tau,$$

with initial condition $X_t = x$.

The notation $\mathrm{E}_{x,t}[\cdot]$ is used to point out that this is a *conditional* expectation, given that at time t the value of the stochastic process is $X_t = x$. From a mathematical viewpoint, the theorem is a consequence of how the stochastic integral in the sense of Itô is defined (see [1] for a clear proof). From a physical point of view, it is a consequence of the connection between Brownian motion (which is a diffusion process) and a certain type of PDEs which can be transformed into the heat equation.

The application of the representation theorem to the Black–Scholes–Merton equation, for an option with payoff function $\Phi(\cdot)$, immediately yields

$$f(S_0,0) = e^{-rT}\mathrm{E}^{\mathbb{Q}}\left[\Phi(S_T)\right],$$

which is consistent with Eq. (3.103). We point out once more that the expectation is taken under a risk-neutral measure, which essentially means that we work as if the stochastic differential equation for S_t were

$$dS_t = rS_t\,dt + \sigma S_t\,dW_t.$$

By recalling the result of Eq. (3.93), we see that under the risk-neutral measure the price S_t is lognormal and may be written as

$$S_t = S_0 \exp\left\{\left(r - \frac{\sigma^2}{2}\right)t + \sigma\sqrt{t}\epsilon\right\},$$

where $\epsilon \sim \mathsf{N}(0,1)$. Then, it is a fairly easy, eve though a bit tedious, exercise in integration to prove the celebrated Black–Scholes–Merton formula (BSM formula for short), which gives the price C of a vanilla European-style call option:

$$C = S_0 \Phi(d_1) - Ke^{-rT}\Phi(d_2), \qquad (3.108)$$

where

$$d_1 = \frac{\log(S_0/K) + (r + \sigma^2/2)T}{\sigma\sqrt{T}},$$

$$d_2 = \frac{\log(S_0/K) + (r - \sigma^2/2)T}{\sigma\sqrt{T}} = d_1 - \sigma\sqrt{T},$$

and $\Phi(x)$ is the cumulative distribution function for the standard normal distribution:

$$\Phi(x) = \frac{1}{\sqrt{2\pi}} \int_{-\infty}^{x} e^{-z^2/2} \, dz.$$

It is easy to use R to implement the above formula and plot the call price for various underlying asset prices and times to maturity. The code in Fig. 3.49 produces the plot in Fig. 3.50. Note that the implementation is a bit naive, in the sense that it does not really work with some input arguments. For instance, when time to maturity is zero and the asset price is the same as the strike price, the function returns NaN. This is the reason why in the last line of the script we plot the option value at maturity by computing the payoff. The reader is invited to improve the code, including some consistency check on the input (for instance, volatility cannot be negative).

The same reasoning can be applied to price a put option, but there is a very useful theorem that solves the problem.

THEOREM 3.18 (Put–call parity) *Consider a call and a put options written on the same non-dividend-paying stock, with the same strike price K and time to maturity T. Then their prices C_0 and P_0 at time $t = 0$ are related as follows:*

$$C_0 + Ke^{-rT} = P_0 + S_0,$$

where S_0 is the current stock price and r is the continuously compounded risk-free interest rate.

PROOF   Consider a portfolio consisting of a call option and an amount of cash given by Ke^{-rT}. At maturity, the value of this portfolio is

$$\max\{S_T - K, 0\} + K = \max\{S_T, K\}. \qquad (3.109)$$

By a similar token, at maturity the value of a portfolio consisting of a put option and one share of the underlying stock is

$$\max\{K - S_T, 0\} + S_T = \max\{K, S_T\}.$$

Since the two portfolios have the same value in any future scenario, they cannot have a different value now, which implies the statement of the theorem. ∎

```
# Black-Scholes-Merton formula for vanilla call
EuVanillaCall <- function(S0,K,T,r,sigma){
  d1 <- (log(S0/K)+(r+sigma^2/2)*T)/(sqrt(T)*sigma)
  d2 <- d1-sqrt(T)*sigma
  value <- S0*pnorm(d1)-K*exp(-r*T)*pnorm(d2)
  return(value)
}

seqS0 <- 30:70; K <- 50; r <- 0.08; sigma <- 0.4
seqT <- seq(from=2, to=0, by=-0.25)
M <- length(seqT)
prices <- matrix(0,nrow=length(seqT),ncol=length(seqS0))
for (t in 1:length(seqT))
  prices[t,] <- EuVanillaCall(seqS0,K,seqT[t],r,sigma)
plot(seqS0, prices[1,], type='l',ylim=c(0,max(prices[1,])))
for (t in (2:(M-1)))
    lines(seqS0, prices[t,])
lines(seqS0, pmax(0,seqS0-K),lwd=2)
```

FIGURE 3.49 **Naive implementation of the Black–Scholes–Merton formula for a vanilla call.**

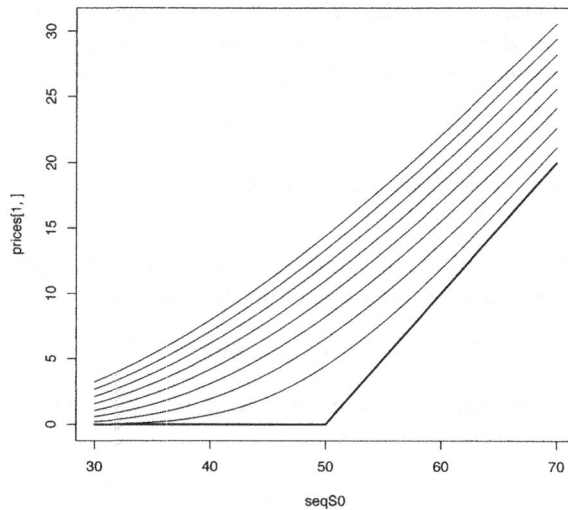

FIGURE 3.50 **Call option prices.**

This proof is instructive and nontechnical, which is why we have included it: It does not rely on any model and is just based on a no-arbitrage consideration, in the form of the law of one price. When we plug the BSM formula for the call option price into Eq. (3.109), we find a similar formula for the put option price:

$$P = Ke^{-rT}\Phi(-d_2) - S_0\Phi N(-d_1), \tag{3.110}$$

where d_1 and d_2 are the same expressions occurring in the call option price. A fundamental advantage of the BSM formula, apart from its computational appeal, is that we may evaluate option sensitivities as well. For instance, the number of shares Δ in the hedging portfolio for a call option is

$$\Delta_C = \frac{\partial C}{\partial S}\bigg|_{S=S_0} = \Phi(d_1). \tag{3.111}$$

This quantity is commonly referred to as "delta," and it may be interpreted as the sensitivity of the option price to changes in the underlying asset price. By differentiating the put–call parity relationship with respect to S_0 we immediately find the corresponding delta for the put option:

$$\Delta_P = \Delta_C - 1. \tag{3.112}$$

Other sensitivities may be obtained, collectively nicknamed as the *greeks*:

$$\Gamma = \frac{\partial^2 C}{\partial S_0^2} = \frac{\phi(d_1)}{S_0\sigma\sqrt{T}}, \tag{3.113}$$

$$\mathcal{V} = \frac{\partial C}{\partial \sigma} = \phi(d_1)S_0\sqrt{T}, \tag{3.114}$$

where $\phi(\cdot)$ is the PDF of a standard normal, not to be confused with the CDF $\Phi(\cdot)$. Equation (3.113) defines and evaluates the option *gamma*, i.e., the second-order derivative of the option price with respect to the underlying asset price. Equation (3.114) defines and evaluates the option *vega*, i.e., the first-order derivative of the option price with respect to volatility. From put–call parity we observe that these two greeks are the same for call and put options.

Another relevant greek is *theta*, which is related to the passage of time:

$$\Theta_C = \frac{\partial C}{\partial t} = -\frac{S_0\phi(d_1)\sigma}{2\sqrt{T}} - rKe^{-rT}\Phi(d2), \tag{3.115}$$

$$\Theta_P = \frac{\partial P}{\partial t} = -\frac{S_0\phi(d_1)\sigma}{2\sqrt{T}} + rKe^{-rT}\Phi(-d2). \tag{3.116}$$

We see that different formulas are required for call and put options, but there is a subtle point here: We have expressed call and option prices at time $t = 0$, as a function of time to maturity T. A more general statement would be recast as a function of time to maturity $T - t$. This is easily accomplished by replacing T with $T - t$ in Eqs. (3.108) and (3.110). It is important to realize that the derivatives defining the option thetas in Eqs. (3.115) and (3.116) are with respect

```
EuVanillaPut <- function(S0,K,T,r,sigma){
  d1 <- (log(S0/K)+(r+sigma^2/2)*T)/(sqrt(T)*sigma)
  d2 <- d1-sqrt(T)*sigma
  value <- K*exp(-r*T)*pnorm(-d2) - S0*pnorm(-d1)
  return(value)
}

BSMdelta <- function(S0,K,T,r,sigma){
  d1 <- (log(S0/K)+(r+sigma^2/2)*T)/(sqrt(T)*sigma)
  return(pnorm(d1))
}

BSMgamma <- function(S0,K,T,r,sigma){
  d1 <- (log(S0/K)+(r+sigma^2/2)*T)/(sqrt(T)*sigma)
  return(dnorm(d1)/(S0*sigma*sqrt(T)))
}

BSMvega <- function(S0,K,T,r,sigma){
  d1 <- (log(S0/K)+(r+sigma^2/2)*T)/(sqrt(T)*sigma)
  return(dnorm(d1)*S0*sqrt(T))
}

BSMthetacall <- function(S0,K,T,r,sigma){
  d1 <- (log(S0/K)+(r+sigma^2/2)*T)/(sqrt(T)*sigma)
  d2 <- d1-sqrt(T)*sigma
  value <- -S0*dnorm(d1)*sigma/(2*sqrt(T))-
           r*K*exp(-r*T)*pnorm(d2)
  return(value)
}

BSMthetaput <- function(S0,K,T,r,sigma){
  d1 <- (log(S0/K)+(r+sigma^2/2)*T)/(sqrt(T)*sigma)
  d2 <- d1-sqrt(T)*sigma
  value <- -S0*dnorm(d1)*sigma/(2*sqrt(T))+
           r*K*exp(-r*T)*pnorm(-d2)
  return(value)
}
```

FIGURE 3.51 **R code to evaluate the put option price and the essential greeks within the BSM framework.**

to t, which is then set to $t = 0$ to find the current theta. Typically, option thetas are negative, which shows that option values are subject to a decay over time.

The option sensitivities, among other things, may be used to evaluate the risk involved in holding a portfolio of options, as we will see in Section 13.3. All of the above formulas are implemented in the R code of Fig. 3.51.

3.9.3 OPTION PRICING IN INCOMPLETE MARKETS

The BSM formula relies on the fact that we can replicate an option. A market in which any option (more generally, any contingent claim) can be replicated, is called a *complete* market. But the obvious question arises: If the BSM formula were correct, why should one trade options? Indeed, in a complete market, one can replicate any derivative, which turns out to be a redundant and useless asset. A fundamental theorem states that if there are no arbitrage opportunities in the market, then there exists a risk-neutral measure that can be used for pricing. But it is market completeness that ensures its uniqueness. However, markets in real life are incomplete, and even if we rule out arbitrage opportunities, we fail to find a *unique* risk-neutral measure for pricing purposes.

A market can be incomplete for many reasons:

- Transaction costs make continuous-time hedging not practical.
- Some risk factors, such as jumps and stochastic volatilities, cannot be hedged.
- The underlying factor of an option is not a tradeable asset.

The last case is quite relevant for interest rate derivatives. We can model uncertainty in interest rates using Itô stochastic differential equations, but we cannot use the interest rate itself in a replicating portfolio. Hence, we cannot build a replicating portfolio using the underlying risk factor. Nevertheless, we still can get rid of risk. In fact, the idea here is not to replicate a risky derivative, but to build a riskless portfolio using two assets depending on the same risk factor.

Assume that we consider the short rate r_t as the underlying risk factor, and let us denote by $Z_1(r_t, t)$ and $Z_2(r_t, t)$ be the prices of two traded assets, whose value depends on the short rate r_t at time t. These assets can be as simple as two zero-coupon bonds, which explains the notation, or more complex interest rate derivatives. Let us assume the following general model for the interest rate:[44]

$$dr_t = m(r_t, t)\, dt + s(r_t, t)\, dW_t, \qquad (3.117)$$

where the drift rate may feature mean reversion. Indeed, the model includes the Vasicek and the Cox–Ingersoll–Ross models of Section 3.7.5. Using Itô's lemma as usual, we may find the stochastic differential equation for the derivative value $Z(r_t, t)$:

$$dZ = \left\{ \frac{\partial Z}{\partial t} + m(r, t)\frac{\partial Z}{\partial r} + \frac{1}{2}s^2(r, t)\frac{\partial^2 Z}{\partial r^2} \right\} dt + s(r, t)\frac{\partial Z}{\partial r}dW_t. \quad (3.118)$$

We may also get rid of risk by setting up a portfolio involving both Z_1 and Z_2 in the right proportions. Let us build a portfolio as follows:

$$\Pi(r, t) = Z_1(r, t) - \Delta Z_2(r, t),$$

---

[44]The treatment below is related to [23].

for a suitable choice of Δ. We note that the above choice implies:

$$\frac{\partial \Pi}{\partial r} = \frac{\partial Z_1}{\partial r} - \Delta \frac{\partial Z_2}{\partial r}, \tag{3.119}$$

$$\frac{\partial^2 \Pi}{\partial r^2} = \frac{\partial^2 Z_1}{\partial r^2} - \Delta \frac{\partial^2 Z_2}{\partial r^2}. \tag{3.120}$$

From the first condition, we see that we obtain $\partial \Pi / \partial r = 0$ if we choose

$$\Delta = \frac{\partial Z_1 / \partial r}{\partial Z_2 / \partial r}.$$

Actually, we are using the same trick as in the BSM case, but now it involves two risky assets depending on the same risk factor. That choice of Δ eliminates the impact of both the underlying risk and the drift of the short rate:

$$d\Pi = \left\{ \frac{\partial \Pi}{\partial t} + \frac{1}{2} s^2(r,t) \frac{\partial^2 \Pi}{\partial r^2} \right\} dt. \tag{3.121}$$

Since this portfolio is risk-free, we can also write

$$d\Pi = r\Pi \, dt. \tag{3.122}$$

By linking Eqs. (3.121) and (3.122), taking advantage of Eqs. (3.119) and (3.120), and rearranging, we find

$$\left\{ \frac{\partial Z_1}{\partial t} + \frac{1}{2} s^2(r,t) \frac{\partial^2 Z_1}{\partial r^2} - rZ_1 \right\} = \Delta \left\{ \frac{\partial Z_2}{\partial t} + \frac{1}{2} s^2(r,t) \frac{\partial^2 Z_2}{\partial r^2} - rZ_2 \right\}. \tag{3.123}$$

Substituting Δ and rearranging again yields the fundamental pricing equation

$$\frac{\frac{\partial Z_1}{\partial t} + \frac{1}{2} s^2(r,t) \frac{\partial^2 Z_1}{\partial r^2} - rZ_1}{\frac{\partial Z_1}{\partial r}} = \frac{\frac{\partial Z_2}{\partial t} + \frac{1}{2} s^2(r,t) \frac{\partial^2 Z_2}{\partial r^2} - rZ_2}{\frac{\partial Z_2}{\partial r}}. \tag{3.124}$$

We note that the drift term $m(r_t, t)$ does not play any role, and this is consistent with the BSM case. However, here we are relating two derivative prices. We can observe that the two ratios above refer to generic derivatives depending on the risk factor r_t, which implies that for any such derivative they are equal to some function of time and the factor itself. Let us define a function $m^*(r_t, t)$ as *minus* the ratios in Eq. (3.124); we change the sign just for the sake convenience, as this allows to write the following pricing PDE:

$$\frac{\partial Z}{\partial t} + m^*(r_t, t) \frac{\partial Z}{\partial r} + \frac{1}{2} s^2(r,t) \frac{\partial^2 Z}{\partial r^2} = rZ. \tag{3.125}$$

We notice that this is quite close to the BSM partial differential equation, but it involves an unknown function $m^*(r_t, t)$. Using the Feynman–Kač theorem, we can think of the derivative price as the expected value of the payoff, if the risk factor follows the risk-adjusted dynamics

$$dr_t = m^*(r_t, t)\, dt + s(r_t, t)\, dW_t. \tag{3.126}$$

Once again, we notice that there is a change in drift, which reflects a change in probability measure. However, due to market incompleteness, we do not know a priori which measure we should use, as uniqueness of the risk-neutral measure requires completeness. The difficulty is that the drift is related to a market price of risk, which does play a role here since we cannot hedge using the underlying asset directly. The way out of the dilemma requires calibrating the model using some traded derivatives. In the case of interest rates, we might use the quoted prices of bonds or other liquid assets to formulate a model calibration problem and solve the corresponding optimization model to estimate the function $m^*(r_t, t)$. Having pinned down the right risk-adjusted measure, we may price other derivatives by using analytical formulas, when available, or by using numerical methods to solve the PDE, like finite differences, or by estimating the price in expectation form by Monte Carlo methods, thanks to the Feynman–Kač representation theorem.

From our viewpoint, we will assume that the model has already been calibrated and that we may proceed to price the derivative asset by estimating the risk-neutral expectation directly. We will see some examples related to interest rates in Section 6.3 and in Chapter 11.

For further reading

- Many of the topics that we have considered in this chapter are reviewed, e.g., in [11].

- There are many excellent textbooks in probability theory; [8] is an example of the top-level ones in terms of quality and readability.

- For some additional information on data dependence and copulas, see [13] or [18].

- A good reference for introductory econometrics in finance, including VAR, ARCH, and GARCH models, is [3]. A deeper treatment is offered in [7].

- As to time series models, you may refer to [2] or the extensive treatment in [9]. A practical introduction focusing on financial time series analysis with R is [22].

- Readers interested in stochastic calculus may refer to [17] for a gentle introduction or to [21] for a deeper treatment.

- Several methods in multivariate statistics, including PCA, are covered in [10] and [20].

- The role of factor models in finance is illustrated, e.g., in [14] and [19].

References

1 T. Björk. *Arbitrage Theory in Continuous Time* (2nd ed.). Oxford University Press, Oxford, 2004.

2 P.J. Brockwell and R.A. Davis. *Time Series: Theory and Methods* (2nd ed.). Springer, New York, 1991.

3 C. Brooks. *Introductory Econometrics for Finance* (2nd ed.). Cambridge University Press, Cambridge, 2008.

4 J.Y. Campbell and L.M. Viceira. *Strategic Asset Allocation*. Oxford University Press, Oxford, 2002.

5 P. Embrechts, C. Kluppelberg, and T. Mikosch. *Modelling Extremal Events for Insurance and Finance*. Springer, Berlin, 2000.

6 P. Embrechts, F. Lindskog, and A. McNeil. Modelling dependence with copulas and applications to risk management. In S.T. Rachev, editor, *Handbook of Heavy Tailed Distributions in Finance*, pp. 329–384. Elsevier, Amsterdam, 2003.

7 W.H. Greene. *Econometric Analysis* (7th ed.). Prentice Hall, Upper Saddle River, NJ, 2011.

8 G. Grimmett and D. Stirzaker. *Probability and Random Processes* (3rd ed.). Oxford University Press, Oxford, 2003.

9 J.D. Hamilton. *Time Series Analysis*. Princeton University Press, Princeton, NJ, 1994.

10 W. Härdle and L. Simar. *Applied Multivariate Statistical Analysis* (2nd ed.). Springer, Berlin, 2007.

11 S. Hubbert. *Essential Mathematics for Market Risk Management*. Wiley, Hoboken, NJ, 2012.

12 J.C. Hull. *Options, Futures, and Other Derivatives* (8th ed.). Prentice Hall, Upper Saddle River, NJ, 2011.

13 H. Joe. *Multivariable Models and Dependence Concepts*. Chapman & Hall/CRC, Boca Raton, FL, 2001.

14 J. Knight and S. Satchell, editors. *Linear Factor Models in Finance*. Elsevier, Amsterdam, 2005.

15 D.P. Kroese, T. Taimre, and Z.I. Botev. *Handbook of Monte Carlo Methods*. Wiley, Hoboken, NJ, 2011.

16 L. Martellini, P. Priaulet, and S. Priaulet. *Fixed-Income Securities: Valuation, Risk Management, and Portfolio Strategies*. Wiley, Chichester, 2003.

17 T. Mikosch. *Elementary Stochastic Calculus with Finance in View*. World Scientific Publishing, Singapore, 1998.

18 R.B. Nelsen. *An Introduction to Copulas*. Springer, New York, 1999.

19 E.E. Qian, R.H. Hua, and E.H. Sorensen, editors. *Quantitative Equity Portfolio Management: Modern Techniques and Applications*. CRC Press, London, 2007.

20 A.C. Rencher. *Methods of Multivariate Analysis* (2nd ed.). Wiley, New York, 2002.

21 S. Shreve. *Stochastic Calculus for Finance (vols. I & II)*. Springer, New York, 2003.

22 R.S. Tsay. *An Introduction to the Analysis of Financial Data with R*. Wiley, Hoboken, NJ, 2013.

23 P. Veronesi. *Fixed Income Securities: Valuation, Risk, and Risk Management*. Wiley, Hoboken, NJ, 2010.

## Chapter Four

Estimation and Fitting

Before embarking in a computationally extensive Monte Carlo study, we should be reasonably confident that we are feeding our simulation tool with sensible input. In the previous chapter we have considered different families of models that can be used as the key ingredient to build a Monte Carlo simulator. Here we consider the quantitative side of the coin: After selecting a model *structure*, how should we estimate its *parameters*? Hence, we step into the domain of inferential statistics, whereby, given a sample of observed data, we engage in increasingly difficult tasks:

- Finding point and interval estimates of basic moments like expected value and variance
- Estimating the parameters of a possibly complicated probability distribution
- Estimating the parameters of a time series model

The first task is typically associated with elementary concepts like confidence intervals, which may be somewhat dangerous and misleading for the newcomer. Computing a confidence interval for the mean of a normal population looks so easy that it is tempting to disregard the pitfalls involved. We will have more to say on this in Chapter 7, where we deal with output analysis. In fact, simulation input and output analysis do share some statistical background.[1] If the output of a Monte Carlo run is used to estimate a probability or a quantile, it may be tempting to apply the same recipes that we adopt to estimate means, which may lead to unreasonable results. Moreover, a probability distribution is not really characterized by expected value and variance. This is the case with a normal distribution $N(\mu, \sigma^2)$, as its parameters have that obvious interpretation, but it is not, e.g., with a gamma distribution. Hence, we need more refined strategies, like moment matching or maximum-likelihood estimation. Maximum likelihood is a powerful principle, which can also be used to estimate the parameters of complicated models, including time series.

We must also be aware that parameter estimation should not be our only concern. Apart from fitting parameters, we should also try to check the fit of the

---

[1]We defer some topics, such as the estimation of probabilities and quantiles to Chapter 7, as from a practical point of view they are more relevant there, even though, from a conceptual point of view, they are also related to the content of this chapter.

overall uncertainty model, which involves nonparametric issues. If we assume that a variable of interest is normally distributed, we should check if this is a defensible, even though debatable, assumption or sheer nonsense. Testing for normality is an example of a nonparametric test. More generally, we may need to compare the fit of alternative distributions, measuring the overall goodness of fit. What is relevant or not depends on the kind of application. For instance, it is possible to build two very different probability distributions having the same expected value, variance, skewness, and kurtosis. Thus, the first few moments do not tell the whole story. This may be not quite relevant if we are just dealing with mean–variance portfolio optimization, but it may be extremely relevant in dealing with extreme risk.

Most books on Monte Carlo methods for finance and economics do not include a chapter on these topics, which is fine if one assumes that readers have a suitable background in econometrics. Other books, such as [10], are aimed at industrial engineers and include an extensive chapter on input analysis. Since this is an introductory book also aimed at engineers and mathematicians, I have deliberately chosen to include such a chapter as well.[2] To make the following treatment as useful as possible to a wide and diversified audience, we will start with really elementary topics, and then we will move on to possibly less standard ones, such as estimation of ARMA and GARCH models, as well as Bayesian parameter estimation. Furthermore, we will use R throughout, so that readers familiar with the underlying concepts, but unfamiliar with R functionalities, can take advantage of the examples. Since it is my firm conviction that understanding the underlying unity of concepts is fundamental for sound and deep learning, it is also important to stress the links of parameter estimation and model fitting with other topics dealt with in this book:

- Some nonstandard optimization methods, which we deal with in Chapter 10, may be needed to estimate parameters or calibrate models, when the resulting optimization model is nonconvex and badly behaved.

- The Kolmogorov–Smirnov test is related to discrepancy concepts, which are at the basis of the low-discrepancy sequences that are discussed in Chapter 9.

- Bayesian estimation is related to Markov chain Monte Carlo methods, illustrated in Chapter 14.

- Linear regression is used in some computational methods for dynamic stochastic optimization, which is in turn fundamental to price American-style options with high dimensionality, as we show in Chapter 11.

- A sound understanding of how confidence intervals arise is important to avoid subtle mistakes in output analysis, as we will stress in Chapter 7.

We start in Section 4.1 with a brief refresher on elementary inferential statistics, i.e., confidence intervals and hypothesis testing for the expected value;

---

[2]Similar considerations apply to the previous chapter, of course.

there, we also outline correlation testing. Then we move on to systematic strategies for parameter estimation in Section 4.2, where we consider moment matching and maximum likelihood. We outline nonparametric tests and checking goodness-of-fit in Section 4.3. We offer a brief refresher on ordinary least-squares methods to estimate linear regression models in Section 4.4, followed by the estimation of time series models in Section 4.5. We close the chapter with Section 4.6, where we compare the orthodox approach to inferential statistics and parameter estimation with the Bayesian view.

The newcomers will certainly realize that model estimation is a huge body of statistical knowledge, which we cannot cover in any satisfactory way in a short chapter. In particular, estimating time series models may require some deep knowledge, as is also the case when estimating a copula. We do not deal with copula estimation here, but later in Section 5.7; there we illustrate how sampling from a copula may be achieved by the R package `copula`, which also includes a function `fitCopula` for fitting purposes. We do not go into details, but we stress that the principle of parameter estimation by maximum likelihood, which we do outline in this chapter, is the theoretical underpinning of many of these estimation approaches.

4.1 Basic inferential statistics in R

In this section we deal with two very basic and related topics: confidence intervals for the mean and elementary hypothesis testing. Our aim is to review the foundations of these statistical tools, in order to get acquainted with the issues involved in more complicated cases, such as testing the significance of correlations, etc. Most readers could probably skip this section, or possibly just skim through it to see which functions R offers to accomplish these tasks. Not surprisingly, given their conceptual links, the relevant R functions boil down to one, `t.test`.

4.1.1 CONFIDENCE INTERVALS

Most of us are introduced to inferential statistics through the calculation of a confidence interval for an expected value μ. This concept is relevant on both ends of a Monte Carlo simulation, as it used to define the input random variables as well as to analyze the output. Given a sample X_i, $i = 1, \ldots, n$ of i.i.d. (independent and identically distributed) random variables, the drill is as follows:

1. Compute sample statistics such as sample mean and sample variance:

$$\bar{X} = \frac{1}{n} \sum_{i=1}^{n} X_i, \qquad S^2 = \frac{1}{n-1} \sum_{i=1}^{n} \left(X_i - \bar{X} \right)^2 .$$

2. Choose a confidence level $(1 - \alpha)$ and pick the corresponding quantile $t_{n-1,1-\alpha/2}$ from a t distribution with $n - 1$ degrees of freedom.

3. Calculate the confidence interval

$$\bar{X} \pm t_{n-1,1-\alpha/2} \frac{S}{\sqrt{n}}. \tag{4.1}$$

This procedure is so easy to perform that one tends to forget that it relies on some important assumptions. The following observations are in order:

- Strictly speaking, the above procedure is correct for normal random variables only. In fact, if variables $X_i \sim N(\mu, \sigma^2)$ *and* they are independent, then it is true that the following standardized statistic is normal:

$$Z = \frac{\bar{X} - \mu}{\sigma/\sqrt{n}} \sim N\left(\mu, \frac{\sigma^2}{n}\right). \tag{4.2}$$

  If we replace σ by its sample counterpart S we find a Student t distribution:

$$T = \frac{\bar{X} - \mu}{S/\sqrt{n}} \sim t_{n-1}, \tag{4.3}$$

  which implies

$$P\left\{-t_{n-1,1-\alpha/2} \leq T \leq t_{n-1,1-\alpha/2}\right\} = 1 - \alpha.$$

  By rearranging this relationship we obtain the confidence interval of Eq. (4.1). A large part of inferential statistics relies on similar distributional results. If we apply the procedure to a different distribution, what we find is at best a good approximation for a suitably large sample; with a small sample and a skewed distribution, we should repeat the drill accounting for the specific features of that distribution.

- It is also very important to stress the role of independence. It is independence in the sample that allows us to write[3]

$$\text{Var}(\bar{X}) = \frac{\sigma^2}{n}.$$

- When estimating sample variance S^2, we recall that division by $n - 1$ is required in order to obtain an unbiased estimator, i.e., to make sure that

$$E[S^2] = \sigma^2.$$

  Unbiasedness, as we insist upon later, is a basic requirement of a sensible estimator.

---

[3] We illustrate the kind of mistakes that we are led to when forgetting the role of independence in Chapter 7.

- When analyzing the output of a Monte Carlo simulation, the sample size is usually rather large. Hence, we usually replace quantiles of the t distribution with quantiles $z_{1-\alpha/2}$ from the standard normal. However, this need not apply to input analysis.

Another danger is a common misinterpretation of confidence intervals. After calculating a confidence interval (l, u), we *cannot* say that

$$P\{\mu \in (l, u)\} = 1 - \alpha. \tag{4.4}$$

This statement does not make any sense, as all of the involved quantities are *numbers*. In the orthodox framework of statistics, μ is an unknown number; the extreme points of the confidence interval, l and u, are numerical realizations of random variables L and U. Hence, the above probabilistic statement makes no sense. We should rephrase the claim of Eq. (4.4) with reference to the random variables L and U, i.e., the lower and upper bound of the interval *before sampling*. Once again: We have to realize that parameters are unknown numbers in orthodox statistics, not random variables.[4] This is illustrated empirically and clarified below by a simple Monte Carlo experiment.

Confidence intervals in R

Building standard confidence intervals in R is accomplished by the function `t.test(x, conf.level = 0.95)`. The function name may look weird at first, but it is justified by the fact that the calculations needed to compute the confidence interval are related to those needed to run a hypothesis test, as we show later. The function receives a vector containing the sample and an optional parameter specifying the confidence level, which is 95% by default:

```
> set.seed(55555)
> X = rnorm(20,mean=10,sd=20)
> t.test(X)
One Sample t-test
data:  X
t = 1.1709, df = 19, p-value = 0.2561
alternative hypothesis: true mean is not equal to 0
95 percent confidence interval:
 -3.960321 14.017022
sample estimates:
mean of x
  5.02835
```

The above output is a bit confusing, as it is cluttered by irrelevant information about a hypothesis test. It might be better to collect the output in a variable and then inspect what we are really interested in:

---

[4]If we want to associate parameters with probabilistic statements, we have to take a Bayesian view; see Section 4.6.1.

```
> ci <- t.test(X,conf.level = 0.99)
> ci$estimate
mean of x
  5.02835
> ci$conf.int
[1] -7.258179 17.314880
attr(,"conf.level")
[1] 0.99
```

As expected, the 99% interval is larger than the 95% interval, and both are rather large due to the small sample size and the large standard deviation with respect to the expected value.

Armed with the above function, we may check the empirical *coverage* of a confidence interval, i.e., the probability that it contains the true expected value. To this aim, it is instructive to compare the result with a normal and an exponential distribution of similar variability.[5] By running the code in Fig. 4.1, we obtain the following results:

```
> coverageExp
[1] 0.9208
> coverageNorm
[1] 0.95
```

As expected, coverage with a skewed distribution is not exactly as specified.

In this section we have only considered confidence intervals for the expected value of a random variable. Different confidence intervals should be used, e.g., when we are faced with the task of estimating variances or probabilities. Another relevant and quite nontrivial case is the estimation of quantiles. Since this kind of application, as far as simulation is concerned, is more related to output analysis, we defer an extended treatement of confidence intervals to Section 7.3.

4.1.2 HYPOTHESIS TESTING

The basic hypothesis test one may wish to run concerns the expected value:

- We test the *null hypothesis* $H_0 : \mu = \mu_0$, for a given μ_0,
- against the *alternative hypothesis* $H_a : \mu \neq \mu_0$.

In the normal case, we rely on the distributional result of Eq. (4.3), where the unknown expected value μ is replaced by the hypothesized value μ_0. This shows that, *if* the null hypothesis is indeed true, then

$$P\left\{ -t_{n-1,1-\alpha/2} \leq \frac{\bar{X} - \mu_0}{S/\sqrt{n}} \leq t_{n-1,1-\alpha/2} \right\} = 1 - \alpha.$$

---

[5]The coefficient of variation is the same in both cases. This is defined as $\sigma/|\mu|$, i.e., the ratio between standard deviation and the absolute value of the expected value.

```
MakeExpSample <- function(n,ev){rexp(n,rate=1/ev)}
MakeNormSample <- function(n,ev){rnorm(n,mean=ev,sd=ev)}
CheckCoverage <- function(ev,nRuns,SampleSize,funSample){
  # note that the last input argument to this function
  # is a function itself, used for sample generation
  count <- 0
  for (i in 1:nRuns){
    X <- funSample(sampleSize, ev)
    CI <- t.test(X, conf.level = 1-alpha)$conf.int
    # count the cases in which we fail to include the
    # true expected value
    if (ev < CI[1] || ev > CI[2]) count <- count+1
  }
  return(1-count/nRuns)
}

set.seed(55555)
nRuns <- 10000
sampleSize <- 20
alpha <- 0.05
ev <- 10
coverageExp <- CheckCoverage(ev, nRuns, SampleSize,
    MakeExpSample)
coverageNorm <- CheckCoverage(ev, nRuns, SampleSize,
    MakeNormSample)
```

FIGURE 4.1 **Checking the coverage of confidence intervals empirically.**

In other words, the standardized test statistic

$$T = \frac{\bar{X} - \mu_0}{S/\sqrt{n}},$$

if the null hypothesis is true, has Student's t distribution with $n - 1$ degrees of freedom, and it should fall within bounds corresponding to quantiles. If T falls outside that interval, there are two possible explanations: It may just be bad luck, or maybe the null hypothesis is wrong. We cannot be sure of either, and we may make two types of error: We may reject a true hypothesis, or we may accept a false one. The elementary approach is conservative and keeps the probability of rejecting a true null hypothesis under control. Therefore, we form a rejection region consisting of two tails,

$$\mathrm{RJ} = \{t : t < -t_{n-1,1-\alpha/2}\} \cup \{t : t > t_{n-1,1-\alpha/2}\},$$

and we reject the null hypothesis if the test statistic $T \in \mathrm{RJ}$. Here α plays the role of a significance level or, better said, the probability of rejecting the null hypothesis if it is true.[6] The rejection region consists of a single tail if the alternative hypothesis is one-sided:

$$H_a: \mu < \mu_0 \quad \Rightarrow \quad RJ = \{t : t < -t_{n-1,1-\alpha}\},$$
$$H_a: \mu > \mu_0 \quad \Rightarrow \quad RJ = \{t : t > t_{n-1,1-\alpha}\}.$$

Note that if we increase α we restrict the rejection region, which implies that, all other factors being equal, it is easier to reject, possibly making a mistake. Statistical folklore dictates that 5% is a sensible choice for the significance level, but this is at best a rule of thumb.

Hypothesis testing is easily accomplished by the t.test function:

```
> set.seed(55555)
> X <- rnorm(10,10,20)
> t.test(X,mu=20)


        One Sample t-test

data:  X
t = -1.4468, df = 9, p-value = 0.1819
alternative hypothesis: true mean is not equal to 20
95 percent confidence interval:
 -6.881385 25.909259
sample estimates:
mean of x
 9.513937
```

In the above snapshot we sample a normal distribution with expected value 10 and test the hypothesis that the expected value is 20. We notice that the confidence interval is large, reflecting considerable variability due to the large standard deviation. By sheer luck, the sample mean is 9.513937, rather close to the true value, but the confidence interval is pretty large. Now, should we accept the null hypothesis or not? One trick is to observe that 20 is within the confidence interval, thus we cannot rule out this value. This way of thinking works here, but it should be discouraged as it does not work well in general, especially with one-tailed testing.

The fundamental output of the procedure is the p-value, which is 0.1819. This is the probability of getting a statistic like that sample mean, or a larger one in absolute value, by pure randomness, assuming that the true expected value is 20. We may compare this p-value against a significance level of 5%: Since the p-value is larger, we do not feel at all confident in rejecting the null hypothesis. In this case, we do know that the null hypothesis is wrong since the true expected value is 10, but from a practical point of view we do not have evidence strong enough. Also note that we cannot really say that we "accept"

---

[6]In fact, what is typically written in textbooks is a hybrid of different viewpoints about the nature of hypothesis testing, and this is why the terminology may sometimes sound weird.

the null hypothesis, as the sample statistic is far from 20; what we can say is that we "fail to reject it."

We may easily change the alternative hypothesis and run one-tailed tests:

```
> t.test(X,mu=20,alternative='less')
        One Sample t-test
data:   X
t = -1.4468, df = 9, p-value = 0.09093
alternative hypothesis: true mean is less than 20
95 percent confidence interval:
    -Inf 22.7997
sample estimates:
mean of x
 9.513937
> t.test(X,mu=20,alternative='greater')
        One Sample t-test
data:   X
t = -1.4468, df = 9, p-value = 0.9091
alternative hypothesis: true mean is greater than 20
95 percent confidence interval:
 -3.771823          Inf
sample estimates:
mean of x
 9.513937
```

The newcomer is strongly invited to compare the two p-values and to interpret their difference.

We may run a host of hypothesis tests in R, which include:

- Testing the difference in the expected values of two large independent samples: `t.test(X,Y)`

- Testing the difference in the expected values of two paired samples: `t.test(X,Y,paired=TRUE)`

- Testing the significance of correlation: `cor.test(X)`, discussed later in Section 4.1.3

The second test is relevant, for instance, in evaluating the performance of two systems or two strategies on a set of simulated scenarios. In such a case, it is fair and makes sense to compare the results on the *same* set of scenarios and to check the differences; hence, the observations in the sample are not independent and should be paired to check if the observed difference is statistically significant.

Unlike trivial tests on the mean of a normal population, other testing procedures may rely on sophisticated distributional results, but they more or less follow the same drill:

- We select some suitable statistic.

- We derive distributional results about that statistic, assuming the null hypothesis is true. These results may rely on some underlying probability distribution and are often approximations valid for large samples.

- Based on these distributional results, the rejection region and the corresponding p-values are determined. Monte Carlo sampling is also used to this purpose, when an approximation is not available, or it is feared that it is a poor one for a limited sample size.

We need not be concerned too much with the theoretical underpinnings of each testing procedure, but we should be aware of the underlying assumptions and the resulting limitations in terms of applicability; then, what we have to look for is the p-value.

4.1.3 CORRELATION TESTING

Testing the significance of correlation is an important task in statistics, and it may also be a useful tool in preparing and analyzing Monte Carlo experiments. A starting point is the following formula for sample covariance:

$$S_{XY} = \frac{1}{n-1} \sum_{i=1}^{n} (X_i - \overline{X})(Y_i - \overline{Y}), \qquad (4.5)$$

where n is the sample size, i.e., the number of observed *pairs* (X_i, Y_i). The above formula is a straightforward sample counterpart of covariance and can also be rewritten as follows:

$$S_{XY} = \frac{1}{n-1} \left(\sum_{i=1}^{n} X_i Y_i - n \overline{X}\,\overline{Y} \right).$$

In passing, we note that when estimating a large covariance matrix, plenty of data are needed to obtain reliable estimates. Such a luxury cannot be afforded quite often, and this motivates the factor models discussed in Section 3.8.2.

To estimate the correlation coefficient ρ_{XY} between X and Y, we may just plug sample covariance S_{XY} and sample standard deviations S_X, S_Y into its definition, resulting in the *sample coefficient of correlation*, or sample correlation for short:

$$R_{XY} = \frac{S_{XY}}{S_X S_Y} = \frac{\displaystyle\sum_{i=1}^{n}(X_i - \overline{X})(Y_i - \overline{Y})}{\sqrt{\displaystyle\sum_{i=1}^{n}\left(X_i - \overline{X}\right)^2} \cdot \sqrt{\displaystyle\sum_{i=1}^{n}\left(Y_i - \overline{Y}\right)^2}}. \qquad (4.6)$$

The factors $n-1$ in S_{XY}, S_X, and S_Y cancel each other, and it can be proved that $-1 \leq r_{XY} \leq +1$, just like its probabilistic counterpart ρ_{XY}.

We should always keep in mind that we are referring to Pearson's correlation, whose limitations have been stressed in Section 3.4.1. Nevertheless, correlation analysis plays an important role in Monte Carlo models:

- In input analysis, we should check if some variables are correlated in order to model them correctly.

- In output analysis, it may be important to check the correlation (or lack thereof) between output estimates, e.g., to find the right batch length in the method of batches. See Section 7.1.

- When applying variance reduction by control variates, it may be important to check the strength of the correlation between the crude Monte Carlo estimator and the control variate that we considering. See Chapter 8.

Whatever the reason why we are estimating a correlation, we should always wonder whether it is statistically significant. To see the point, imagine estimating the correlation when $n = 2$. Since it is possible to fit a straight line going through two points, the resulting correlation is always 1. It is clear that the magnitude of the correlation should be checked against the sample size, and a simple strategy is to test the null hypothesis

$$H_0 : \; \rho_{XY} = 0$$

against the alternative hypothesis

$$H_a : \; \rho_{XY} \neq 0.$$

However, we need a statistic whose distribution under the null hypothesis is fairly manageable. One useful result is that, if the sample is normal, the statistic

$$T = R_{XY} \sqrt{\frac{n-2}{1 - R_{XY}^2}}$$

is approximately distributed as a t variable with $n - 2$ degrees of freedom, for a suitably large sample. This can be exploited to come up with correlation tests.

In R, testing correlations is accomplished by the function `cor.test`. The snapshot in Fig. 4.2 generates two vectors of independent normals, $x1$ and $x2$, to which a common factor is added. This has the effect of producing a correlation coefficient of 0.3262. However, the p-value is 0.1605, which makes the correlation not quite significant, due to the small sample size. The reader is urged to repeat the experiment with $n = 2000$ and check that now the correlation is significant.

4.2 Parameter estimation

Introductory treatments of inferential statistics focus on normal populations. In that case, the two parameters characterizing the distribution, μ and σ^2, coincide with expected value, the first-order moment, and variance, the second-order central moment. Hence, students might believe that parameter estimation is just about calculating sample means and variances. It is easy to see that this is not the case.

```
> set.seed(5)
> Z <- rnorm(20)
> X1 <- rnorm(20,10,5)
> X2 <- rnorm(20,30,8)
> cor.test(X1+5*Z,X2+5*Z)

        Pearson's product-moment correlation

data:  X1 + 5 * Z and X2 + 5 * Z
t = 1.4639, df = 18, p-value = 0.1605
alternative hypothesis: true correlation is not equal to 0
95 percent confidence interval:
 -0.1359711  0.6717381
sample estimates:
      cor
0.3261757
```

FIGURE 4.2 **Testing correlations.**

Example 4.1 **Parameters of a uniform distribution**

Consider a uniform random variable $X \sim \mathsf{U}[a, b]$. We recall that

$$\mathrm{E}[X] = \frac{a+b}{2}, \qquad \mathrm{Var}(X) = \frac{(b-a)^2}{12}.$$

Clearly, the sample mean $\overline{X}$ and the sample variance S^2 do not provide us with direct estimates of the parameters a and b. However, we might consider the following way of transforming the sample statistics into estimates of parameters. If we substitute μ and σ^2 with their estimates, we find

$$a + b = 2\overline{X},$$
$$-a + b = 2\sqrt{3}S.$$

Note that, in taking the square root of variance, we should only consider the positive root, since standard deviation cannot be negative. Solving this system yields the following estimates:

$$\hat{a} = \overline{X} - \sqrt{3}S, \qquad \hat{b} = \overline{X} + \sqrt{3}S.$$

This example suggests a general strategy to estimate parameters:

- Estimate moments on the basis of a random sample.
- Set up a suitable system of equations relating parameters and moments and solve it.

Indeed, this is the starting point of a general parameter estimation approach called *method of moments*. However, a more careful look at the example should raise an issue. Consider the order statistics of a random sample of n observations of a uniform distribution:

$$U_{(1)} \leq U_{(2)} \leq U_{(3)} \leq \cdots \leq U_{(n)}.$$

If we take the above approach, *all* of these observations play a role in estimating a and b. Yet, this is a bit counterintuitive; in order to characterize a uniform distribution, we need a lower and an upper bound on its realizations. So, it seems that only $U_{(1)}$ and $U_{(n)}$, i.e., the smallest and the largest observations, should play a role. We might suspect that there are alternative strategies for finding point estimators and then, possibly, confidence intervals. In this section we outline two approaches: the aforementioned method of moments and the method of maximum likelihood. But before doing so, since there are alternative ways to build estimators, it is just natural to wonder how we can compare them. Therefore, we should first list the desirable properties that make a good estimator.

4.2.1   FEATURES OF POINT ESTIMATORS

We list here a few desirable properties of a point estimator $\hat{\theta}$ for a parameter θ. When comparing alternative estimators, we may have to trade off one property for another. An estimator $\hat{\theta}$ is *unbiased* if

$$\mathrm{E}[\hat{\theta}] = \theta.$$

It is easy to show that the sample mean is an unbiased estimator of the expected value. Biasedness is related to the expected value of an estimator, but what about its variance? Clearly, ceteris paribus, we would like to have an estimator with low variance.

DEFINITION 4.1 (Efficient unbiased estimator) *An unbiased estimator $\hat{\theta}_1$ is more efficient than another unbiased estimator $\hat{\theta}_2$ if*

$$\mathrm{Var}(\hat{\theta}_1) < \mathrm{Var}(\hat{\theta}_2).$$

Note that we must compare unbiased estimators in assessing efficiency. Otherwise, we could obtain a nice estimator with zero variance by just choosing an arbitrary constant. It can be shown that, under suitable hypotheses, the variance of certain estimators has a lower bound. This bound, known as the *Cramér–Rao bound*, is definitely beyond the scope of this book, but the message is clear: We cannot go below a minimal variance. If the variance of an estimator attains that lower bound, then it is efficient. Another desirable property is consistency. An estimator is *consistent* if $\hat{\theta} \to \theta$, where the limit must be understood in some probabilistic sense, when the sample size increases. For instance, the statistic

$$\frac{1}{n}\sum_{i=1}^{n}\left(X_i - \overline{X}\right)^2$$

is not an unbiased estimator of variance σ^2, as we know that we should divide by $n-1$ rather than n. However, it is a consistent estimator, since when $n \to \infty$ there is no difference between dividing by $n-1$ or n. Incidentally, by dividing by n we lose unbiasedness but we retain consistency while reducing variance. Sometimes, we may prefer a biased estimator, if its variance is low, provided that it is consistent.

4.2.2 THE METHOD OF MOMENTS

We have already sketched an application of this method in Example 4.1. To state the approach in more generality, let us introduce the sample moment of order k:

$$M_k \equiv \frac{1}{n}\sum_{i=1}^{n}X_i^k.$$

The sample moment is the sample counterpart of moment $m_k = \mathrm{E}[X^k]$. Let us assume that we need an estimate of k parameters $\theta_1, \theta_2, \ldots, \theta_k$. The method of moments relies on the solution of the following system of equations:

$$\begin{cases} m_1 = g_1(\theta_1, \theta_2, \ldots, \theta_k), \\ m_2 = g_2(\theta_1, \theta_2, \ldots, \theta_k), \\ \quad\vdots \\ m_k = g_k(\theta_1, \theta_2, \ldots, \theta_k). \end{cases}$$

In general, this is a system of nonlinear equations that may be difficult to solve, and we cannot be sure that a unique solution, if any, exists. However, assuming that there is in fact a unique solution, we may just replace moments m_k by sample moments M_k, and solve the system to obtain estimators $\hat{\theta}_1, \hat{\theta}_2, \ldots, \hat{\theta}_k$.

Example 4.2 Application to the normal distribution

Let $(X_1, X_2, \ldots X_n)$ be a random sample from a normal distribution with unknown parameters $(\theta_1, \theta_2) \equiv (\mu, \sigma)$. The relationship between the parameters and the first two moments is $m_1 = \mu$, $m_2 =$

$\sigma^2 + \mu^2$. Plugging sample moments and solving the system yields

$$\hat{\mu} = M_1 = \overline{X},$$

$$\hat{\sigma} = \sqrt{M_2 - \overline{X}^2} = \sqrt{\frac{1}{n}\sum_{i=1}^{n} X_i^2 - \overline{X}^2} = \sqrt{\frac{1}{n}\sum_{i=1}^{n}\left(X_i - \overline{X}\right)^2}.$$

These estimators look familiar enough, but note that we do not obtain an unbiased estimator of variance.

4.2.3 THE METHOD OF MAXIMUM LIKELIHOOD

The method of maximum likelihood is an alternative approach to find estimators in a systematic way. Imagine that a random variable X has a PDF characterized by a single parameter θ, denoted by $f_X(x;\theta)$. If we draw a sample of n i.i.d. variables from this distribution, the joint density is just the product of individual PDFs:

$$f_{X_1,\dots,X_n}(x_1,\dots,x_n;\theta) = f_X(x_1;\theta) \cdot f_X(x_2;\theta) \cdots f_X(x_n;\theta) = \prod_{i=1}^{n} f_X(x_i;\theta).$$

This is a function depending on the parameter θ. Also note that we use lower-case letters x_i to point out that we are referring to a specific set of numerical realizations of the random variable X. If we are interested in estimating θ, *given* a sample $X_i = x_i$, $i = 1,\dots,n$, we may swap the role of variables and parameters and build the *likelihood function*

$$L(\theta) = L(\theta; x_1,\dots,x_n) = f_{X_1,\dots,X_n}(x_1,\dots,x_n;\theta). \tag{4.7}$$

The shorthand notation $L(\theta)$ is used to emphasize that this is a function of the unknown parameter θ, for a given sample of observations. On the basis of this framework, intuition suggests that we should select the parameter yielding the largest value of the likelihood function. For a discrete random variable, the interpretation is more natural: We select the parameter maximizing the probability of what we have indeed observed. For a continuous random variable, we cannot really speak of probabilities, but the rationale behind the method of maximum likelihood should be apparent: We are just trying to find the best explanation of what we observe. The acronym MLE, for maximum-likelihood estimator or estimation, is used to refer to the approach.

■ Example 4.3 **MLE for the exponential distribution**

Let us consider the PDF of an exponential random variable with parameter λ:

$$f_X(x;\lambda) = \lambda e^{-\lambda x}.$$

Given observed values $x_1, \ldots, x_n$, the likelihood function is

$$L(\lambda) = \prod_{i=1}^{n} \lambda e^{-\lambda x} = \lambda^n \exp\left\{-\lambda \sum_{i=1}^{n} x_i\right\}.$$

Quite often, rather than attempting direct maximization of the likelihood function, it is convenient to maximize its logarithm, i.e., the *log-likelihood function*

$$l(\theta) = \ln L(\theta). \tag{4.8}$$

The rationale behind this function is easy to grasp, since by taking the logarithm of the likelihood function, we transform a product of functions into a sum of functions:

$$l(\lambda) = n \ln \lambda - \lambda \sum_{i=1}^{n} x_i.$$

The first-order optimality condition yields

$$\frac{n}{\lambda} - \sum_{i=1}^{n} x_i = 0 \quad \Rightarrow \quad \hat{\lambda} = \frac{1}{\frac{1}{n}\sum_{i=1}^{n} X_i} = \frac{1}{\overline{X}}.$$

The result is rather intuitive, since $\mathrm{E}[X] = 1/\lambda$.

The idea of MLE is immediately extended to the case requiring the estimation of multiple parameters: We just solve a maximization problem involving multiple variables.

■ Example 4.4 **MLE for the normal distribution**

Let us consider maximum-likelihood estimation of the parameters $\theta_1 = \mu$ and $\theta_2 = \sigma^2$ of a normal distribution. Straightforward manipulation of the PDF yields the log-likelihood function:

$$l(\mu, \sigma^2) = -\frac{n}{2}\ln(2\pi) - \frac{n}{2}\ln\sigma^2 - \frac{1}{2\sigma^2}\sum_{i=1}^{n}(x_i - \mu)^2.$$

The first-order optimality condition with respect to μ is

$$\frac{\partial l}{\partial \mu} = \frac{1}{\sigma^2} \sum_{i=1}^{n} (x_i - \mu) = 0 \quad \Rightarrow \quad \hat{\mu} = \frac{1}{n} \sum_{i=1}^{n} x_i.$$

Hence, the maximum-likelihood estimator of μ is just the sample mean. If we apply the first-order condition with respect to σ^2, plugging the estimator $\hat{\mu}$, we obtain

$$\frac{\partial l}{\partial \sigma^2} = -\frac{n}{2} \frac{1}{\sigma^2} + \frac{1}{2\sigma^4} \sum_{i=1}^{n} (x_i - \hat{\mu})^2 \quad \Rightarrow \quad \hat{\sigma^2} = \frac{1}{n} \sum_{i=1}^{n} (x_i - \hat{\mu})^2.$$

In the two examples above, maximum likelihood yields the same estimator as the method of moments. Then, one could well wonder whether there is really any difference between the two approaches. The next example provides us with a partial answer.

▣ Example 4.5 MLE for the uniform distribution

Let us consider a uniform distribution on interval $[0, \theta]$. On the basis of a random sample of n observations, we build the likelihood function

$$L(\theta) = \begin{cases} \dfrac{1}{\theta^n}, & \text{for } 0 \leq x_i \leq \theta,\, i = 1, \ldots, n, \\ 0, & \text{otherwise.} \end{cases}$$

This may look a bit weird at first sight, but it makes perfect sense, as the PDF is zero for $x > \theta$. Indeed, θ cannot be smaller than the largest observation:

$$\theta \geq \max\{x_1, x_2, \ldots, x_n\}.$$

In this case, since there is a constraint on θ, we cannot just take the derivative of $L(\theta)$ and set it to 0 (first-order optimality condition). Nevertheless, it is easy to see that likelihood is maximized by choosing the smallest θ, subject to constraint above. Hence, using the notation of order statistics, we find

$$\hat{\theta} = X_{(n)}.$$

While the application of maximum likelihood to exponential and normal variables looks a bit dull, in the last case we start seeing something worth noting. The estimator is quite different from what we have obtained by using the method of moments. The most striking feature is that this estimator does *not* use the whole sample, but just a single observation. We leave it as an exercise for the reader to check that, in the case of a uniform distribution on the interval $[a, b]$, maximum-likelihood estimation yields

$$\hat{a} = X_{(1)}, \qquad \hat{b} = X_{(n)}.$$

Indeed, the smallest and largest observations are *sufficient statistics* for the parameters a and b of a uniform distribution.[7] We refrain from stating a formal definition of sufficient statistics, but the idea is rather intuitive. Given a random sample $\mathbf{X}$, a sufficient statistic is a function $T(\mathbf{X})$ that captures all of the information we need from the sample in order to estimate a parameter. As a further example, sample mean is a sufficient statistic for the parameter μ of a normal distribution. This concept has far-reaching implications in the theory of inferential statistics.

A last point concerns unbiasedness. In the first two examples, we have obtained biased estimators for variance, even though they are consistent. It can be shown that an unbiased estimator of θ for the uniform distribution on $[0, \theta]$ is

$$\frac{n+1}{n} \cdot X_{(n)}, \tag{4.9}$$

rather than $X_{(n)}$. Again, we see that maximum likelihood yields a less than ideal estimator, even though for $n \to \infty$ there is no real issue. It turns out that maximum-likelihood estimators (MLEs) do have limitations, but a few significant advantages as well. Subject to some technical conditions, the following properties can be shown for MLEs:

- They are consistent.
- They are asymptotically normal.
- They are asymptotically efficient.
- They are invariant, in the sense that, given a function $g(\cdot)$, the MLE of $\gamma = g(\theta)$ is $g(\hat{\theta})$, where $\hat{\theta}$ is the MLE of θ.

As a general rule, finding MLEs requires the solution of an optimization problem by numerical methods, but there is an opportunity here. Whatever constraint we want to enforce on the parameters, depending on domain-specific knowledge, can be easily added, resulting in an optimization problem that can be solved numerically. Plenty of numerical optimization methods are available to this purpose.

---

[7] See [4, p. 375] for a proof.

```
set.seed(55555)
shape <- 2
rate <- 10
X <- rgamma(500, shape=shape, rate=rate)
# ---
library(stats4)
ll <- function(shape,rate) -sum(log(dgamma(X,shape,rate)))
fit1 <- mle(ll, start=list(shape=3,rate=10), nobs=length(X),
            method = "L-BFGS-B", lower=c(0,0))
fit1@coef
confint(fit1)
# ---
library(MASS)
fit2 <- fitdistr(X,"gamma")
fit2$estimate
fit2$sd
```

FIGURE 4.3 **Fitting a gamma distribution.**

4.2.4   DISTRIBUTION FITTING IN R

R offers some tools to fit distributions by maximum likelihood, such as

- mle, contained in the stats4 package,
- fitdistr, contained in the MASS package.

The former function leaves the task of specifying minus the log-likelihood function to the user and produces an object of an S4 class. The second one is a bit more user friendly. The use of these functions is best understood by a simple example related to a gamma distribution. The code in Fig. 4.3 produces the following (abridged) output:

```
> fit1@coef
    shape        rate
 2.233724 11.062195
> confint(fit1)
Profiling...
         2.5 %      97.5 %
shape 1.985328   2.503327
rate  9.684267  12.559251
> fit2$estimate
    shape        rate
 2.233726 11.062211
> fit2$sd
    shape        rate
0.1320868 0.7331218
```

We notice a slight difference in the estimates of shape and rate parameters, due to the optimization algorithms embedded in the two functions. In the case of `mle`, we obtain an S4 object; this is why we have to access its properties by `fit1@coef` rather than by `fit1$coef`. We also select an optimization method, `L-BFGS-B`, so that we may enforce lower bounds on the parameters (otherwise, warnings may be obtained if they get negative in the search process). Also notice the use of a named list to provide initial values for the numerical optimization algorithm, as usual in nonlinear programming. In both cases we have some information on the reliability of the estimate, either via a confidence interval or standard errors.

4.3   Checking the fit of hypothetical distributions

So far, we have been concerned with parameters of probability distributions. We never questioned the fit of the distribution itself against empirical data. For instance, we might assume that a population is normally distributed, and we may estimate and test its expected value and variance. However, normality should not be taken for granted, just like any other claim about the underlying distribution. Sometimes, specific knowledge suggests strong reasons that justify the assumption; otherwise, this should be tested in some way. When we test whether experimental data fit a given probability distribution, we are not really testing a hypothesis about a parameter or two; in fact, we are running a *nonparametric* test.

In this section we illustrate three kinds of approach:

- The chi-square test, which is general purpose and, loosely speaking, checks the fit in terms of histograms and densities.
- The Kolmogorov–Smirnov test, which is is general purpose as well, but checks the fit in terms of the cumulative distribution.
- Two examples ad hoc tests, aimed at checking normality, namely, the Shapiro–Wilks and Jarque–Bera tests.

All of the above procedures will be illustrated using R.

Before describing formal approaches, let us comment on intuitive and graphical checks. Clearly, visual inspection of a histogram and a comparison with the theoretical shape of a PDF may help. However, sometimes the judgement is not so easy to make. To see the point, let us generate a normal sample and look at the resulting histograms in Fig. 4.4:

```
> set.seed(55555)
> par(mfrow=c(1,2))
> X=rnorm(200,10,40)
> hist(X)
> hist(X,nclass=50)
> par(mfrow=c(1,1))
```

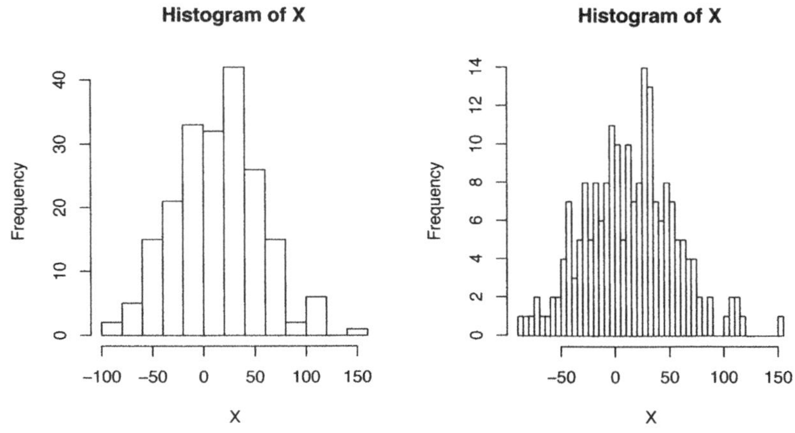

Histogram of X   **Histogram of X**

FIGURE 4.4 **Histograms of a normal sample.**

In the plot on the left of Fig. 4.4, there are very few bins, and the distribution might look compatible with a normal, but other densities as well. If we increase resolution by using more bins, statistical sampling makes the histogram too bumpy to be of any use. By the way, these observations also apply to the chi-square test, which formalizes the intuition of comparing histograms. Sometimes, checking the CDF is a better idea. In fact, another graphical tool relies on plotting the sample quantiles against the theoretical ones, obtaining a Q-Q plot. Using the above sample, we produce the plot in Fig. 4.5 as follows:

```
> qqnorm(X)
> qqline(X)
```

In principle, the points should lie on a straight line, since the sample is in fact normally distributed. This is indeed the case, with the exception of tails, which is to be expected after all as they are always critical.

4.3.1 THE CHI-SQUARE TEST

The idea behind the chi-square test is fairly intuitive and basically relies on a relative frequency histogram, although the technicalities do require some care. The first step is to divide the range of possible observed values in J disjoint intervals, corresponding to bins of a frequency histogram. Given a probability distribution, we can compute the probability p_j, $j = 1, \ldots, J$, that a random variable distributed according to that distribution falls in each bin. If we have n observations, the number of observations that should fall in interval j, if the assumed distribution is indeed the true one, should be $E_j = np_j$. This number should be compared against the number O_j of observations that actually fall

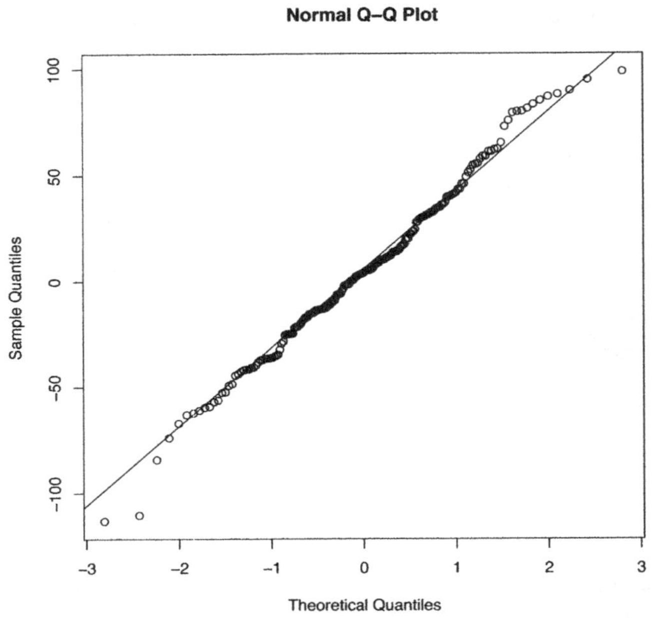

Normal Q–Q Plot

FIGURE 4.5 **Checking the fit of a normal distribution.**

in interval j; a large discrepancy would suggest that the hypothesis about the underlying distribution should be rejected. Just like any statistical test, the chi-square test relies on a distributional property of a statistic. It can be shown that, for a large sample size, the statistic

$$\chi^2 = \sum_{j=1}^{J} \frac{(O_j - E_j)^2}{E_j}$$

has (approximately) a chi-square distribution. We should reject the hypothesis if χ^2 is too large, i.e., if $\chi^2 > \chi^2_{1-\alpha,m}$, where:

- $\chi^2_{1-\alpha,m}$ is a quantile of the chi-square distribution.
- α is the significance level of the test.
- m is the number of degrees of freedom.

What we are missing here is m, which depends on the number of parameters of the distribution that we have estimated using the data. If no parameter has been estimated, i.e., if we have assumed a specific parameterized distribution prior to observing data, the degrees of freedom are $J - 1$; if we have estimated p parameters, we should use $J - p - 1$.

R offers a `chisq.test` function that, among other things, can be used to check the fit of a continuous distribution. To this aim, we have to bin data and compute observed and theoretical frequencies:

```
set.seed(55555)
shape <- 2
rate <- 10
X <- rgamma(500, shape, rate)
breakPts <- c(seq(0,0.8,0.1), 2)
probs <- pgamma(breakPts,shape,rate)
binProbs <- diff(probs)
f.obs <- hist(X, breakPts, plot = FALSE)$counts
chisq.test(x=f.obs,p=binProbs/sum(binProbs))
```

Note that we divide the vector `binProbs` by its sum to make sure that probabilities add up to 1. We also use `hist` as a convenient tool to bin data. The above code results in the following output:

```
        Chi-squared test for given probabilities
data:   f.obs
X-squared = 4.4634, df = 8, p-value = 0.8131
Warning message:
In chisq.test(x = f.obs, p = binProbs/sum(binProbs)) :
  Chi-squared approximation may be incorrect
```

As expected, we cannot reject the null hypothesis, but we get a warning message. Indeed, the idea of the chi-square test is pretty intuitive; however, it relies on approximated distributional results that may be critical. Another tricky point is that the result of the test may depend on the number and placement of bins. Rules of thumb have been proposed and are typically embedded in good statistical software.

4.3.2 THE KOLMOGOROV–SMIRNOV TEST

Continuing the above example, where the variable x stores a sample from a gamma distribution with given `shape` and `rate` parameters, it is very easy to run the Kolmogorov–Smirnov test in R:

```
> ks.test(X,pgamma, shape, rate)
One-sample Kolmogorov-Smirnov test
data:   X
D = 0.0447, p-value = 0.2704
alternative hypothesis: two-sided
```

The Kolmogorov–Smirnov test, unlike the chi-square test, relies on the CDF. Indeed, as you may see above, we have to provide `ks.test` with a function computing cumulative probabilities, `pgamma` in our case, along with all of the relevant parameters.

The idea behind the test relies on a comparison between the assumed CDF $F_X(x)$ and an empirical CDF defined as follows:

$$\hat{F}_n(x) \equiv \frac{\#\{X_i \le x\}}{n},$$

where $\#\{\ldots\}$ denotes the cardinality of a set. The distance is measured in a worst-case sense:[8]

$$D = \sup_x \left| \hat{F}_n(x) - F_X(x) \right|.$$

From a computational point of view, this distance can be calculated as

$$D = \max\{D^+, D^-\},$$

where

$$D^+ = \max_{i=1,\ldots,n} \left\{ \frac{i}{n} - F_X\left(X_{(i)}\right) \right\}, \quad D^- = \max_{i=1,\ldots,n} \left\{ F_X\left(X_{(i)}\right) - \frac{i-1}{n} \right\},$$

and $X_{(i)}$ is the ith order statistic (obtained by sorting values X_i in increasing order). It is clear that a large value D does not support the assumed distribution. What is not so clear is how to devise a rejection region and find the critical values defining its boundary. This depends on the underlying distribution, and Monte Carlo sampling is often used to find such values.

4.3.3 TESTING NORMALITY

The R core includes the Shapiro–Wilk test, implemented in the `shapiro.test` function:

```
> set.seed(55555)
> X=rnorm(20,10,20)
> shapiro.test(X)
        Shapiro-Wilk normality test
data:   X
W = 0.9512, p-value = 0.385
```

In this case, the function returns a rather large p-value, such that we cannot reject the null hypothesis that the distribution is normal. Given this structure, it is clear that we get a conservative test. Strong evidence is needed to reject normality. Let us try a normal and a gamma distribution:

```
> X=rexp(20,10)
> shapiro.test(X)
        Shapiro-Wilk normality test
data:   X
W = 0.9006, p-value = 0.04237
```

---

[8]A similar approach is used in Chapter 9 to assess the star discrepancy of a numerical sequence.

```
set.seed(55555)
X1 <- 1:10
X2 <- rnorm(10,0,10)
eps <- rnorm(10,0,20)
Y <- 30 + 5*X1 + 3*X2 + eps
lmod1 <- lm(Y~X1+X2)
summary(lmod1)
```

FIGURE 4.6 **Fitting a multiple linear regression model.**

```
> X=rgamma(20,2,10)
> shapiro.test(X)
        Shapiro-Wilk normality test
data:  X
W = 0.623, p-value = 5.14e-06
```

Evidence is reasonably strong in the former case, but not as strong as in the latter. The test statistic takes values between 0 and 1, and it compares two estimates of variance; when the statistic is close to 1 we cannot reject normality. In the Jarque–Bera test we take advantage of third- and fourth-order moments, and of the fact that skewness is zero and kurtosis is 3 for a normal. The test statistic is

$$\frac{n}{6}\left[\widehat{\gamma}^2 + \frac{(\hat{\kappa}-3)^2}{4}\right],$$

where $\widehat{\gamma}$ and $\hat{\kappa}$ are estimates of skewness and kurtosis, respectively. Clearly, the test statistic is non-negative and should be close to zero, according to the null hypothesis that the underlying distribution is normal. For a large sample the test statistic is a chi-square variable, which can be used to define a rejection region for the test. However, since this is not quite robust for small and medium size samples, critical values can be obtained by Monte Carlo sampling.

4.4 Estimation of linear regression models by ordinary least squares

Under suitable assumptions, multiple linear regression models may be estimated by ordinary least squares (OLS), which is accomplished in R by the `lm` function. This is illustrated by the code in Fig. 4.6, which produces the following output:

```
Call:
lm(formula = Y ~ X1 + X2)
Residuals:
    Min      1Q  Median      3Q      Max
```

```
-12.396  -9.859  -6.075   8.818  29.831
Coefficients:
            Estimate Std. Error t value Pr(>|t|)
(Intercept)  21.6349    11.0946   1.950 0.092162 .
X1            4.7902     1.7912   2.674 0.031802 *
X2            2.7463     0.4732   5.803 0.000661 ***
---
Signif.codes:  0 '**' 0.001 '**' 0.01 '*' 0.05 '.' 0.1 ' ' 1
Residual standard error: 16 on 7 degrees of freedom
Multiple R-squared: 0.8727,Adjusted R-squared: 0.8363
F-statistic: 23.98 on 2 and 7 DF,  p-value: 0.000737
```

We obtain parameter estimates, along with an assessment of their significance. The p-value of the coefficient for X_1 is significant, despite the high level of noise. The coefficient of determination R^2 is 0.8727, but the adjusted R^2 is 0.8363. We also obtain an overall assessment of the model in the p-value of the F-statistic. The reader is invited to check the effect of reducing the coefficient multiplying X_1 in the data generation process.

We obtain confidence intervals for the estimated coefficients, and we may also test normality and uncorrelation of residuals:

```
> confint(lmod1)
                 2.5 %      97.5 %
(Intercept) -4.5996144 47.869439
X1           0.5546597  9.025779
X2           1.6272928  3.865392
> shapiro.test(lmod1$resid)
Shapiro-Wilk normality test
data:  lmod1$resid
W = 0.8294, p-value = 0.03287
> library(lmtest)
> dwtest(lmod1)
Durbin-Watson test
data:  lmod1
DW = 2.5181, p-value = 0.5605
alternative hypothesis:
true autocorrelation is greater than 0
```

Using `lm` requires the specification of a formula. In the case above, `Y~X1+X2` is self-explanatory, but formulas in R are much more powerful than that. Say that we want to fit a polynomial regression model of order 2:

$$Y = \alpha + \beta_1 x + \beta_2 x^2 + \epsilon.$$

Of course, we can just evaluate each power of the explanatory variable and run a linear regression (the model is nonlinear in variables, but linear in parameters). However, the code in Fig. 4.7 shows a shortcut: by setting `raw=TRUE`, we

```
set.seed(55555)
X <- rnorm(10,0,10)
eps <- rnorm(10,0,20)
Y <- 300 + 5*X + 0.5*X^2 + eps
lmod2 <- lm(Y~poly(X,2,raw=TRUE))
summary(lmod2)
```

FIGURE 4.7 **Fitting a polynomial regression model.**

force R to use familiar polynomials rather than orthogonal ones.[9] We obtain the following output:

```
Call:
lm(formula = Y ~ poly(X, 2, raw = TRUE))
Residuals:
    Min      1Q  Median      3Q     Max
-13.673  -9.653  -5.712   8.714  29.547
Coefficients:
                        Estimate Std. Error t value Pr(>|t|)
(Intercept)            290.82646    7.53560   38.59 2.04e-09 ***
poly(X,2,raw=TRUE)1      4.73459    0.46672   10.14 1.95e-05 ***
poly(X,2,raw=TRUE)2      0.49706    0.04724   10.52 1.53e-05 ***
---
Signif.codes:  0 '**' 0.001 '**' 0.01 '*' 0.05 '.' 0.1 ' ' 1
Residual standard error: 16.02 on 7 degrees of freedom
Multiple R-squared: 0.9664,Adjusted R-squared: 0.9568
F-statistic: 100.7 on 2 and 7 DF,  p-value: 6.951e-06
```

There are cases in which nonlinearity should be used to capture interaction between variables. Consider the model

$$Y = \alpha + \beta_1 X + \beta_2 \delta + \beta_3 X \delta + \epsilon,$$

where $\delta \in \{0,1\}$ is a dummy variable modeling a categorical variable. It is worth noting that when $\delta = 0$, the data generating process is

$$Y = \alpha + \beta_1 X + \epsilon,$$

whereas, when $\delta = 1$, *both* intercept and slope change:

$$Y = (\alpha + \beta_2) + (\beta_1 + \beta_3)X + \epsilon.$$

The code in Fig. 4.8 illustrates such a case. The input data are depicted in Fig. 4.9, along with the fitted lines resulting from setting $\delta = 0$ and $\delta = 1$, respectively. We can see clearly the difference in intercept and slope between the two groups.

```
set.seed(55555)
X1 <- runif(40,min=0,max=50)
X2 <- runif(40,min=0,max=50)
Y1 <- 70 + 2*X1 + rnorm(40,0,10)
Y2 <- 30 + 5*X2 + rnorm(40,0,10)
Y <- c(Y1,Y2)
X <- c(X1,X2)
Z <- c(rep(1,40),rep(0,40))
lmod3 <- lm(Y~X*Z)
summary(lmod3)
plot(X1,Y1,pch=1,col="black",xlim=c(0,max(X)),
     ylim=c(0.8*min(Y),max(Y)),xlab="X",ylab="Y")
points(X2,Y2, pch=2,col="red")
abline(lmod3$coeff[1]+lmod3$coeff[3],
       lmod3$coeff[2]+lmod3$coeff[4],col="red")
abline(lmod3$coeff[1],lmod3$coeff[2],col="black")
```

FIGURE 4.8 **Fitting a regression model with interactions.**

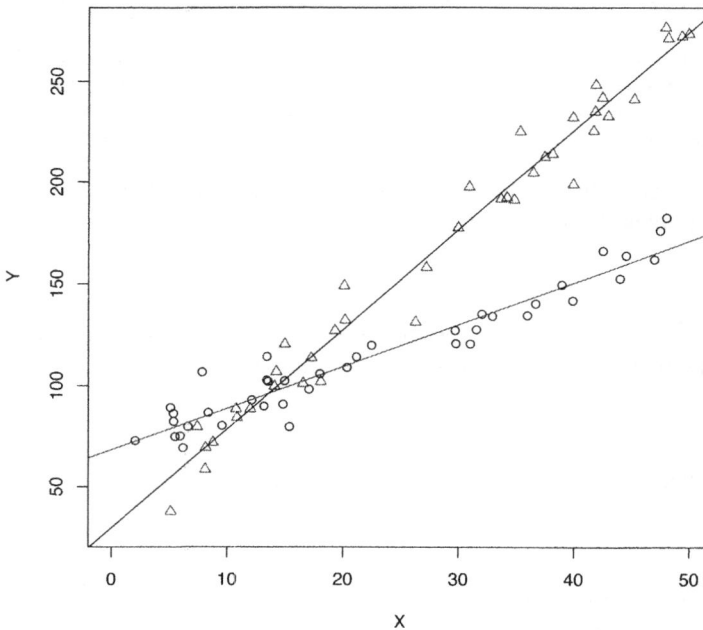

FIGURE 4.9 **A regression model with interactions.**

The following model summary shows that, when using the formula `Y~X*Z`, interactions are automatically accounted for:

```
Call:
lm(formula = Y ~ X * Z)
Residuals:
     Min       1Q    Median        3Q       Max
-27.0281  -5.4642  -0.1196    5.5480   22.4519
Coefficients:
            Estimate Std. Error t value Pr(>|t|)
(Intercept)  29.6160     3.5574   8.325 2.62e-12 ***
X             4.8991     0.1135  43.147  < 2e-16 ***
Z            38.6779     4.6058   8.398 1.90e-12 ***
X:Z          -2.8403     0.1584 -17.928  < 2e-16 ***
---
Signif.codes:  0 '**' 0.001 '**' 0.01 '*' 0.05 '.' 0.1 ' ' 1
Residual standard error: 10.04 on 76 degrees of freedom
Multiple R-squared: 0.9732,    Adjusted R-squared: 0.9722
F-statistic: 921.3 on 3 and 76 DF,  p-value: < 2.2e-16
```

4.5 Fitting time series models

Fitting time series models requires the use of possibly sophisticated procedures, whose first step is selecting the type and order of the model. This model identification step can be accomplished using the concepts that we have outlined in Section 3.6. Given a model structure, estimating its parameters calls for the application, e.g., of maximum-likelihood methods. In this section we just simulate a couple of models and check whether we are able to recover their underlying structure, on the basis of the resulting sample paths, by using R functions.

To simulate an ARIMA model, we may write our own code, as we did in Section 3.6. For the sake of convenience, here we use the function `arima.sim`, which simulates an ARIMA model and accepts, among other things, the following input arguments:

- `model`: a named list with vector-valued components, `ar` and `ma`, specifying the AR and the MA part of the model
- `n`: the time horizon for the simulation
- `n.start`: the number of time buckets that should be simulated and discarded to avoid issues with the transient phase resulting from the initial conditions
- `rand.gen`: a function to generate shocks; by default, `rnorm` is used, but we may use, e.g., a Student's t distribution to allow for slightly fatter tails

---

[9] For an application of polynomial regression to pricing American-style options by Monte Carlo methods, see Section 11.3.

```
set.seed(55555)
library(forecast)
x1 <- arima.sim(n=1500, model=list(ar=c(0.85,-0.45),
        ma=c(-0.23,0.25)), n.start=50,
        rand.gen=function(n) 0.5*rt(n, df=4))
arima(x1,order=c(2,0,2))
auto.arima(x1)
par(mfrow=c(1,2))
ts.plot(x1)
ts.plot(x1[1:100])
par(mfrow=c(1,1))
x2 <- arima.sim(n=1500, list(ar=c(0.85,-0.45),
        ma=c(-0.23,0.25)), n.start=50,
        rand.gen=function(n) 0.5*rt(n,df=4))
arima(x2,order=c(2,0,2))
auto.arima(x2)
```

FIGURE 4.10   **Simulating and fitting an ARMA(2,2) model.**

Then, we estimate the model using two functions:

- arima, which needs a specification of the model order
- auto.arima, provided by the forecast package, which automates the search for the model order

The code in Fig. 4.10 produces the following (edited) output:

```
> arima(x1,order=c(2,0,2))
Series: x1
ARIMA(2,0,2) with non-zero mean
Coefficients:
         ar1      ar2      ma1     ma2   intercept
      0.6477  -0.3853  -0.0880  0.3118   -0.0105
s.e.  0.1364   0.0652   0.1362  0.0517    0.0295
sigma^2 estimated as 0.4749:  log likelihood=-1570.18
AIC=3152.35   AICc=3152.41   BIC=3184.23
> auto.arima(x1)
Series: x1
ARIMA(2,0,2) with zero mean

Coefficients:
         ar1      ar2      ma1     ma2
      0.6474  -0.3849  -0.0875  0.3117
s.e.  0.1363   0.0652   0.1361  0.0517

sigma^2 estimated as 0.475:  log likelihood=-1570.24
AIC=3150.48   AICc=3150.52   BIC=3177.05
```

```
> arima(x2,order=c(2,0,2))
Series: x2
ARIMA(2,0,2) with non-zero mean
Coefficients:
          ar1       ar2       ma1      ma2   intercept
       0.9599   -0.5477   -0.3572   0.2970    -0.0076
s.e.   0.0738    0.0425    0.0772   0.0439     0.0282
sigma^2 estimated as 0.4677:  log likelihood=-1558.8
AIC=3129.6    AICc=3129.66   BIC=3161.48
> auto.arima(x2)
Series: x2
ARIMA(2,0,2) with zero mean
Coefficients:
          ar1       ar2       ma1      ma2
       0.9599   -0.5477   -0.3573   0.2970
s.e.   0.0738    0.0425    0.0772   0.0439
sigma^2 estimated as 0.4678:  log likelihood=-1558.84
AIC=3127.68    AICc=3127.72   BIC=3154.24
```

Note that `arima` needs a specification of the model order; in this case we cheat
and use the actual order used in the simulation, i.e., (2,0,2), where there is no
integration component. The alternative function `auto.arima` is able to find the
correct order in this example, but this is not the case in general. We also note
that, despite the fairly long simulated horizon, different sample paths may lead
to rather different estimates of the coefficients. This illustrates the difficulties in
fitting time series models, but we should consider that the main purpose here is
to capture some essential features of the original time series, in order to produce
realistic scenarios for a Monte Carlo simulation. The code of Fig. 4.10 also uses
the function `ts.plot` to plot the first sample path, along with a zoomed portion
of the initial time interval; see the result in Fig. 4.11.

4.6 Subjective probability: The Bayesian view

So far, we have adopted a rather standard view of parameter estimation, as we
have followed the orthodox approach: Parameters are unknown numbers, which
we try to estimate by squeezing information out of a random sample, in the form
of point estimators and confidence intervals. We also insisted on the fact that,
given a computed confidence interval with some confidence level $(1 - \alpha)$, we
cannot say that the true parameter is contained there with probability $(1 - \alpha)$.
This statement makes no sense, since we are only comparing known numbers
(the realized bounds of the confidence interval) and an unknown number (the
parameter), but no random variable is involved. So, there is no such a thing as
a "probabilistic knowledge" about parameters, and data are the only source of

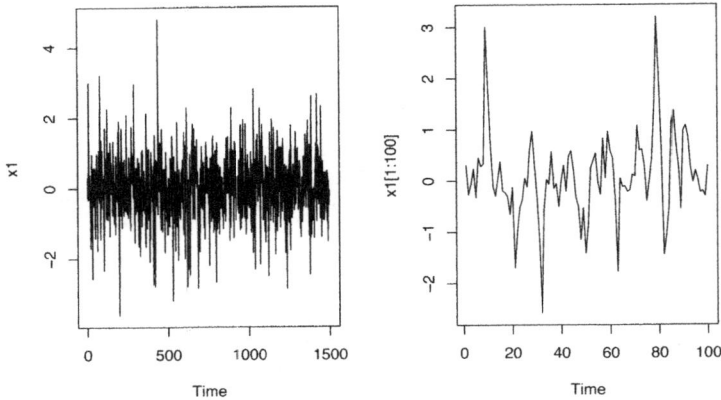

FIGURE 4.11 **Sample path produced by** `arima.sim`; **the plot on the right zooms over the initial portion of the path.**

information; any other knowledge, objective or subjective, is disregarded. The following example illustrates the potential difficulties induced by this view.[10]

▣ Example 4.6 A pathological case

Let X be a uniformly distributed random variable, and let us assume that we do not know where the support of its distribution is located, but we know that its width is 1. Then, $X \sim U(\mu - 0.5, \mu + 0.5)$, where μ is the unknown expected value of X, as well as the midpoint of the support. In order to estimate μ, let us take a sample of only $n = 2$ independent realizations X_1 and X_2 of the random variable. Now, let us consider the order statistics

$$X_{(1)} = \min\{X_1, X_2\}, \qquad X_{(2)} = \max\{X_1, X_2\},$$

and the confidence interval

$$\mathcal{I} = [X_{(1)}, X_{(2)}]. \tag{4.10}$$

Given this way of building a confidence interval, what is the confidence level of $\mathcal{I}$, i.e., the probability $P\{\mu \in \mathcal{I}\}$? Once again, note that we are *not* asking this question by referring to a specific interval

---

[10]This example is taken from [6, p. 45], which in turn refers back to [3].

after sampling, but to a random interval before sampling. Because of symmetry, the two observations have a probability 0.5 of falling to the left or to the right of μ. The confidence interval will not contain μ if both of them fall on the same half of the support. Then, since X_1 and X_2 are independent, we have

$$P\{\mu \notin \mathcal{I}\} = P\{X_1 < \mu, X_2 < \mu\} + P\{X_1 > \mu, X_2 > \mu\}$$
$$= 0.5 \times 0.5 + 0.5 \times 0.5 = 0.5.$$

Therefore, the confidence level for $\mathcal{I}$ is the complement of this probability, i.e., 50%. Now, let us suppose that we observe $X_1 = 0$ and $X_2 = 0.6$. What is the probability that μ is included in the confidence interval $\mathcal{I}$ resulting from Eq. (4.10), i.e., $P\{0 \le \mu \le 0.6\}$? In general, this question does not make any sense, since μ is a number. But in this specific case, we have some additional knowledge leading to the conclusion that the expected value is certainly included in the interval $[0, 0.6]$. In fact, if $X_{(1)} = 0$, we may conclude $\mu \le 0.5$; by the same token, if $X_{(2)} = 0.6$, we may conclude $\mu \ge 0.1$. Since the confidence interval $\mathcal{I}$ includes the interval $[0.1, 0.5]$, we would have good reasons to claim something like $P\{0 \le \mu \le 0.6\} = 1$. But again, this makes no sense in the orthodox framework. By a similar token, if we get $X_1 = 0$ and $X_2 = 0.001$, we would be tempted to say that such a small interval is quite unlikely to include μ. However, within the framework of orthodox statistics, there is no way in which we can express this intuition properly.

On one hand, the example illustrates the need to make our background knowledge explicit: The Bayesian framework is an alternative to the orthodox approach, where it can be argued that unconditional probabilities do not exist, in the sense that probabilities are always conditional on some background knowledge and assumptions. On the other hand, we see the need of a way to express subjective views, which may be revised after collecting empirical data. Bayesian estimation has been proposed to cope with such issues as well.

4.6.1 BAYESIAN ESTIMATION

Consider the problem of estimating a parameter θ, which characterizes the probability distribution of a random avaliable X. We have some prior information about θ that we would like to express in a sensible way. We might assume that the unknown parameter lies anywhere in the unit interval $[0, 1]$, or we might assume that it is close to some number μ, but we are somewhat uncertain about it. Such a knowledge or subjective view may be expressed by a probability density $p(\theta)$, which is called the *prior* distribution of θ. In the first case, we might associate a uniform distribution with θ; in the second case the prior could be a normal distribution with expected value μ and variance σ^2. Note that this is the variance that we associate with the parameter, which is itself a random variable, rather than a number, and not the variance of the random variable X itself.

In Bayesian estimation, the prior is merged with experimental evidence by using Bayes' theorem. Experimental evidence consists of independent observations $X_1, \ldots, X_n$ from the unknown distribution. Here and in the following we mostly assume that random variable X is continuous, and we speak of densities; the case of discrete random variables and probability mass functions is similar. We also assume that the values of the parameter θ are not restricted to a discrete set, so that the prior is a density as well. Hence, let us denote the density of X by $f(x \mid \theta)$, to emphasize its dependence on parameter θ. Since a random sample consists of a sequence of independent random variables, their joint distribution, *conditional* on θ, is

$$f_n(x_1, \ldots, x_n \mid \theta) = f(x_1 \mid \theta) \cdot f(x_2 \mid \theta) \cdots f(x_n \mid \theta).$$

The conditional density $f_n(x_1, \ldots, x_n \mid \theta)$ is also called the *likelihood function*, as it is related to the likelihood of observing the data values $x_1, \ldots, x_n$, given the value of the parameter θ; also notice the similarity with the likelihood function in maximum-likelihood estimation.[11] Note that what really matters here is that the observed random variables $X_1, \ldots, X_n$ are independent *conditionally* on θ. Since we are speaking about $n+1$ random variables, we could also consider the joint density

$$g(x_1, \ldots, x_n, \theta),$$

but this will not be really necessary for what follows. Given the joint conditional distribution $f_n(x_1, \ldots, x_n \mid \theta)$ and the prior $p(\theta)$, we can find the marginal density of $X_1, \ldots, X_n$ by applying the total probability theorem:

$$g_n(x_1, \ldots, x_n) = \int_\Omega f_n(x_1, \ldots, x_n \mid \theta) p(\theta) \, d\theta,$$

where we integrate over the domain Ω on which θ is defined, i.e., the support of the prior distribution. Now what we need is to invert conditioning, i.e., we would like the distribution of θ conditional on the observed values $X_i = x_i$, $i = 1, \ldots, n$, i.e.,

$$p_n(\theta \mid x_1, \ldots, x_n).$$

This *posterior* density should merge the prior and the density of observed data conditional on the parameter. This is obtained by applying Bayes' theorem to densities, which yields

$$p_n\left(\theta \mid x_1, \ldots, x_n\right) = \frac{g(x_1, \ldots, x_n, \theta)}{g_n(x_1, \ldots, x_n)} = \frac{f_n(x_1, \ldots, x_n \mid \theta) p(\theta)}{g_n(x_1, \ldots, x_n)}. \qquad (4.11)$$

---

[11] See Section 4.2.3.

Note that the posterior density involves a denominator term $g_n(x_1, \ldots, x_n)$ that does not depend on θ. Its role is only to normalize the posterior distribution, so that its integral is 1. Sometimes, it might be convenient to rewrite Eq. (4.11) as

$$p_n(\theta \,|\, x_1, \ldots, x_n) \propto f_n(x_1, \ldots, x_n \,|\, \theta) p(\theta), \qquad (4.12)$$

where the symbol $\propto$ means "proportional to." The denominator term does not affect the *shape* of the posterior, which encodes our uncertainty about the parameter θ, and it may be argued that it is not really essential.[12] In plain English, Eq. (4.11) states that the posterior is *proportional* to the product of the likelihood function $f_n(x_1, \ldots, x_n \,|\, \theta)$ and the prior distribution $p(\theta)$:

$$\text{posterior} \propto \text{prior} \times \text{likelihood}.$$

What we are saying is that, given some prior knowledge about the parameter and the distribution of observations conditional on the parameter, we obtain an updated distribution of the parameter, conditional on the actually observed data.

4.6.2 BAYESIAN LEARNING AND COIN FLIPPING

We tend to take for granted that coins are fair, and that the probability of getting heads is $\frac{1}{2}$. Let us consider flipping a possibly unfair coin, with an unknown probability θ of getting heads. In order to learn this unknown value, we flip the coin repeatedly, i.e., we run a sequence of independent Bernoulli trials with unknown parameter θ. If we do not know anything about the coin, we might just assume a uniform prior

$$p(\theta) = 1, \qquad 0 \leq \theta \leq 1.$$

If we flip the coin N times, we know that the probability $f_N(H \,|\, \theta)$ of getting H heads, conditional on θ, is related to the binomial probability distribution:

$$f_N(H \,|\, \theta) \propto \theta^H (1 - \theta)^{N-H}. \qquad (4.13)$$

This is our likelihood function. If we regard this expression as the probability of observing H heads, given θ, this should actually be the probability mass function of a binomial variable with parameters θ and N, but we are disregarding the binomial coefficient, which does not depend on θ and just normalizes the distribution. If we multiply this likelihood function by the prior, which is just 1, we obtain the posterior density for θ, given the number of observed heads:

$$p_N(\theta \,|\, H) \propto \theta^H (1 - \theta)^{N-H}, \qquad 0 \leq \theta \leq 1. \qquad (4.14)$$

Equations (4.13) and (4.14) look like the same thing, because we use a uniform prior, but they are very different in nature. Equation (4.14) gives the posterior

---

[12]This will play a major role in Chapter 14 on Markov chain Monte Carlo.

```
# BAYESIAN COIN FLIPPING
theta <- 0.2
# build a sample path of 1000 unfair coin flips,
# fixing the initial four flips
set.seed(55555)
# the initial portion of the sample path is given
# the rest is random
history <- c(1, 1, 0, 0, ifelse(runif(996)<=theta,1,0))
# cumulate the number of heads at each time step
cumHistory <- c(0, cumsum(history))
# posterior likelihood, when h heads are drawn in n flips
post <- function(x,n,h) (x^h) * ((1-x)^(n-h))
# the nine time instants at which posterior is plotted
times <- c(0:4, 10, 20, 100, 1000)
# x-values to plot posterior densities
x <- seq(from=0, by=0.001, to=1)
# note the use of letters to select nine characters
# to be used later in printing labels for plots
labels <- letters[1:9]
par(mfrow=c(3,3))
for (i in 1:9){
  nFlips <- times[i]
  nHeads <- cumHistory[times[i]+1]
  y <- post(x, nFlips, nHeads)
  # sprintf is used to print a formatted string
  lab <- sprintf('(%s) N=%d; H=%d', labels[i],nFlips,nHeads)
  plot(x,y/max(y),xlim=c(0,1),ylim=c(0,1.1),type='l',ylab='',
       xlab='',main=lab)
}
par(mfrow=c(1,1))
```

FIGURE 4.12 **Sampling a sequence of coin flips and updating the posterior.**

density of θ, conditional on the fact that we observed H heads and $N - H$ tails. If we look at it this way, we recognize the shape of a beta distribution, which is the density of a continuous random variable, the parameter θ, rather than the mass function of a discrete random variable, the number of heads H out of N flips. To normalize the posterior, we should multiply it by the appropriate value of the beta function.[13] Again, this normalization factor does not depend on θ and can be disregarded.

The R code in Fig. 4.12 generates a sequence of 1000 unfair coin flips, with $\theta = 0.2$, and successively updates and plots the posterior density, resulting in the plots of Fig. 4.13. There, we display posterior densities normalized in

---

[13]See Section 3.2.4.

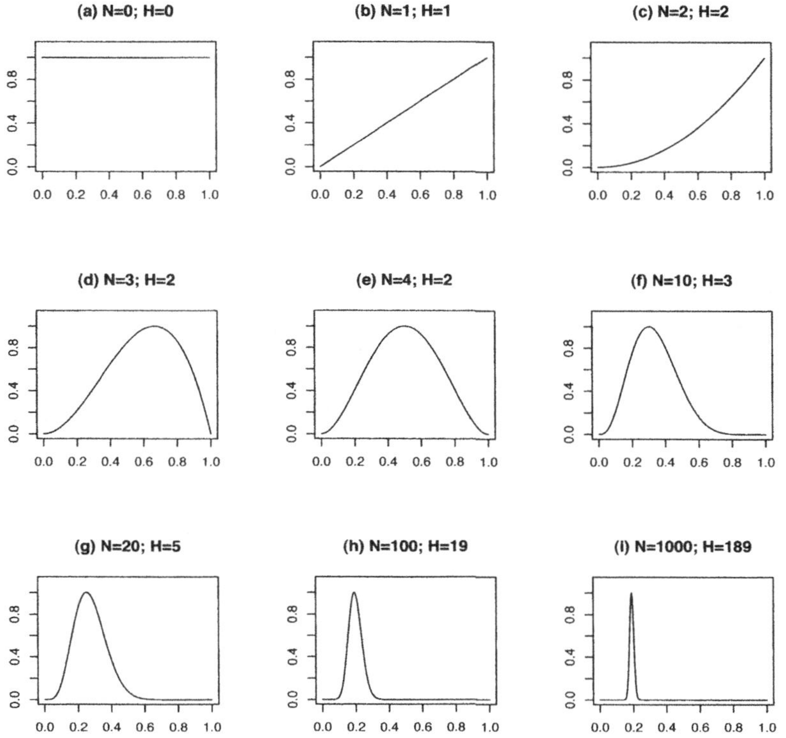

FIGURE 4.13 **Updating the posterior density in coin flipping.**

such a way that their maximum is 1, after flipping the coin N times and having observed H heads. In Fig. 4.13(a) we just see the uniform prior before any flip of the coin. In the simulated history, the first flip lands heads. After observing the first heads, we know for sure that $\theta \neq 0$; indeed, if θ were zero, we could not observe any heads. The posterior is now proportional to a triangle:

$$p_1(\theta \,|\, 1) \propto \theta^1 (1 - \theta)^{1-1} = \theta, \qquad 0 \leq \theta \leq 1.$$

This triangle is shown in Fig. 4.13(b). We observe another heads in the second flip, and so the updated posterior density is a portion of a parabola, as shown in Fig. 4.13(c):

$$p_2(\theta \,|\, 2) \propto \theta^2 (1 - \theta)^{2-2} = \theta^2, \qquad 0 \leq \theta \leq 1.$$

With respect to the triangle, the portion of parabola is more tilted toward $\theta = 1$, because we observed two heads out of two launches. However, when we get tails at the third flip, we rule out $\theta = 1$ as well. Proceeding this way, we get beta distributions concentrating around the true (unknown) value of $\theta = 0.2$.

Armed with the posterior density, how can we build a Bayes' estimator for θ? Figure 4.13 would suggest taking the mode of the posterior, which would spare us the work of normalizing it. However, this need not be the most sensible choice. If we consider the expected value for the posterior distribution, we find

$$E\left[\theta \,|\, X_1 = x_1, \ldots, X_n = x_n\right] = \int_\Omega \theta p_n(\theta \,|\, x_1, \ldots, x_n)\, d\theta. \qquad (4.15)$$

There are different ways of framing the problem, which are a bit beyond the scope of this book,[14] but one thing that we can immediately appreciate is the challenge we face. The estimator above involves what looks like an intimidating integral, but our task is even more difficult in practice, because finding the posterior density may be a challenging computational exercise as well. In fact, given a prior, there is no general way of finding a closed-form posterior; things can really get awkward when *multiple* parameters are involved. Moreover, there is no guarantee that the posterior distribution $p_n(\theta \,|\, x_1, \ldots, x_n)$ will belong to the same family as the prior $p(\theta)$. In general, numerical methods are needed in Bayesian computational statistics and, needless to say, Monte Carlo approaches play a key role here. However, there are some exceptions. A family of distributions is called a *conjugate family of priors* if, whenever the prior is in the family, the posterior is too. For instance, if we choose a beta density as the prior in coin flipping, the posterior is a beta density as well. The following example illustrates the idea further.

Example 4.7 The case of a normal prior

Consider a sample $(X_1, \ldots, X_n)$ from a normal distribution with unknown expected value θ and known variance σ_0^2. Then, given our knowledge about the multivariate normal, and taking advantage of independence among observations, we have the following likelihood function:

$$f_n(x_1, \ldots, x_n \,|\, \theta) = \frac{1}{(2\pi)^{n/2}\sigma_0^n} \exp\left\{-\sum_{i=1}^n \frac{(x_i - \theta)^2}{2\sigma_0^2}\right\}.$$

Let us assume that the prior distribution of θ is normal, too, with expected value μ and σ:

$$p(\theta) = \frac{1}{\sqrt{2\pi}\sigma} \exp\left\{-\frac{(\theta - \mu)^2}{2\sigma^2}\right\}.$$

---

[14]One way to frame the problem is by introducing a *loss function* that accounts for the cost of a wrong estimate. It can be shown that Eq. (4.15) yields the optimal estimator when the loss function has a certain quadratic form.

To find the posterior, we may simplify our work by considering in each function only the part that involves θ, wrapping the rest within a proportionality constant. In more detail, the likelihood function can be written as

$$f_n(x_1, \ldots, x_n \,|\, \theta) \propto \exp \left\{ -\frac{1}{2\sigma_0^2} \sum_{i=1}^{n} (x_i - \theta)^2 \right\}. \qquad (4.16)$$

Then, we may simplify the expression further, by observing that

$$\sum_{i=1}^{n} (x_i - \theta)^2 = \sum_{i=1}^{n} (x_i - \bar{x} + \bar{x} - \theta)^2$$

$$= \sum_{i=1}^{n} (x_i - \bar{x})^2 + \sum_{i=1}^{n} (\bar{x} - \theta)^2$$

$$= \sum_{i=1}^{n} (x_i - \bar{x})^2 + n (\theta - \bar{x})^2,$$

where $\bar{x}$ is the average of x_i, $i = 1, \ldots, n$. Then, we may include terms not depending on θ into the proportionality constant and rewrite Eq. (4.16) as

$$f_n(x_1, \ldots, x_n \,|\, \theta) \propto \exp \left\{ -\frac{n}{2\sigma_0^2} (\theta - \bar{x})^2 \right\}. \qquad (4.17)$$

By a similar token, we may rewrite the prior as

$$p(\theta) \propto \exp \left\{ -\frac{(\theta - \mu)^2}{2\sigma^2} \right\}. \qquad (4.18)$$

By multiplying Eqs. (4.17) and (4.18), we obtain the posterior

$$p_n(\theta \,|\, x_1, \ldots, x_n) \propto \exp \left\{ -\frac{1}{2} \left[\frac{n}{\sigma_0^2} (\theta - \bar{x})^2 + \frac{1}{\sigma^2} (\theta - \mu)^2 \right] \right\}. \qquad (4.19)$$

Again, we should try to include θ within one term; to this aim, we use a bit of tedious algebra and rewrite the argument of the exponential as follows:

$$\frac{n}{\sigma_0^2} (\theta - \bar{x})^2 + \frac{1}{\sigma^2} (\theta - \mu)^2 = \frac{1}{\xi^2} (\theta - \nu)^2 + \frac{n}{\sigma_0^2 + n\sigma^2} (\bar{x} - \mu)^2,$$

where

$$\nu = \frac{n\sigma^2 \bar{x} + \sigma_0^2 \mu}{n\sigma^2 + \sigma_0^2}, \qquad (4.20)$$

$$\xi^2 = \frac{\sigma^2 \sigma_0^2}{n\sigma^2 + \sigma_0^2}. \qquad (4.21)$$

Finally, this leads us to

$$p_n(\theta \,|\, x_1, \ldots, x_n) \propto \exp\left\{-\frac{1}{2\xi^2}(\theta - \nu)^2\right\}. \qquad (4.22)$$

Disregarding the normalization constant, we immediately recognize the familiar shape of a normal density, with expected value ν and variance ξ^2. Then, given an observed sample mean $\overline{X}$ and a prior μ, Eq. (4.20) tells us that the Bayes' estimator of θ can be written as

$$\mathrm{E}\left[\theta \,|\, X_1, \ldots, X_n\right] = \frac{n\sigma^2}{n\sigma^2 + \sigma_0^2}\,\overline{X} + \frac{\sigma_0^2}{n\sigma^2 + \sigma_0^2}\,\mu$$

$$= \frac{n/\sigma_0^2}{n/\sigma_0^2 + 1/\sigma^2}\,\overline{X} + \frac{1/\sigma^2}{n/\sigma_0^2 + 1/\sigma^2}\,\mu. \qquad (4.23)$$

Equation (4.23) has a particularly nice and intuitive interpretation: The posterior estimate is a weighted average of the sample mean $\overline{X}$ (the new evidence) and the prior μ, with weights that are inversely proportional to σ_0^2/n, the variance of sample mean, and σ^2, the variance of the prior. The more reliable a term, the larger its weight in the average.

What may sound a little weird in the example is the assumption that the variance σ_0^2 of X is known, but its expected value is not. Of course, one can extend the framework to cope with estimation of multiple parameters. Furthermore, we have considered two lucky cases, coin flipping and the normal prior for a normal distribution, where the posterior is very easy to compute as it preserves the original shape of the prior. In such cases we speak of conjugate priors. In other cases, we are not so lucky and we need to compute a possibly tough integral to find the normalization constant making the posterior a density. Indeed, Bayesian approaches were not quite practical until computational tools were devised to solve the difficulty by Markov chain Monte Carlo methods. These strategies are described in Chapter 14 and allow to run a simulation even if we do not know a density exactly, but only a likelihood proportional to the density.

For further reading

- From the simulation practitioner's viewpoint, an excellent introduction to input analysis can be found in [10].
- If you are interested in a sound treatment of the theoretical background of inferential statistics and parameter estimation, [2] is recommended.

- We have given an extremely cursory and superficial look at linear regression models and their estimation. If you would like a treatment of linear regression from the statistical point of view, one possibility is [5].

- There are many books on econometrics, covering regression models and more, such as [1] and, at a more advanced level, [7].

- The above references on econometrics also cover time series models. If you want an extensive treatment, including estimation of time series models, you may refer to [8].

- A quite readable introduction to Bayesian estimation can be found in [9], which is replete with good R code as well.

References

1 C. Brooks. *Introductory Econometrics for Finance* (2nd ed.). Cambridge University Press, Cambridge, 2008.

2 G. Casella and R.L. Berger. *Statistical Inference* (2nd ed.). Duxbury, Pacific Grove, CA, 2002.

3 M.H. DeGroot. *Probability and Statistics*. Addison-Wesley, Reading, MA, 1975.

4 M.H. DeGroot and M.J. Schervish. *Probability and Statistics* (3rd ed.). Addison-Wesley, Boston, 2002.

5 N.R. Draper and H. Smith. *Applied Regression Analysis* (3rd ed.). Wiley, New York, 1998.

6 I. Gilboa. *Theory of Decision under Uncertainty*. Cambridge University Press, New York, 2009.

7 W.H. Greene. *Econometric Analysis* (7th ed.). Prentice Hall, Upper Saddle River, NJ, 2011.

8 J.D. Hamilton. *Time Series Analysis*. Princeton University Press, Princeton, NJ, 1994.

9 J.K. Kruschke. *Doing Bayesian Data Analysis: A Tutorial with R and BUGS*. Academic Press, Burlington, MA, 2011.

10 A.M. Law and W.D. Kelton. *Simulation Modeling and Analysis* (3rd ed.). McGraw-Hill, New York, 1999.

## Part Three

Sampling and Path Generation

## Chapter Five

Random Variate Generation

Armed with a suitable model of uncertainty, possibly one of those described in Chapter 3, whose parameters have been estimated as illustrated in Chapter 4, we are ready to run a set of Monte Carlo experiments. To do so, we must feed our simulation program with a stream of random variates mimicking uncertainty. Actually, the job has already been done, at least for R users trusting the rich library of random generators that is available. So, why should we bother with the internal working of these generators and learn how random variates are generated? Indeed, we will not dig too deep in sophisticated algorithms, and we will also refrain from treating generators for a too wide class of probability distributions. Still, there are quite good reasons to get a glimpse of the underpinnings of random variate generation. The aims of this chapter are:

- To gain an understanding of the different options that R offers: for instance, there are alternative generators that can be set and controlled with the RNGkind function.

- To understand how to manage seeds of random generators, which is essential for debugging and controlling experiments.

- To have an idea of what may go wrong with bad generators that have been proposed in the past and may still be around.

- To pave the way for some variance reduction strategies, to be described in Chapter 8.

- To better understand the link between Monte Carlo simulation and integration, which in turn is essential to apply alternative approaches based on the low-discrepancy sequences dealt with in Chapter 9.

- A bit of knowledge about random variate generation by acceptance–rejection is also useful to better appreciate certain methods for computational Bayesian statistics, like Markov chain Monte Carlo; see Chapter 14.

The overall structure of a Monte Carlo simulation is described in Section 5.1. Essentially, a Monte Carlo simulator is a function mapping a stream of (pseudo)-random numbers, i.e., a sequence of i.i.d. variables with uniform distribution on the unit interval (0,1), into a set of estimates of interest. The generation of random numbers is described in Section 5.2. Then, random numbers are transformed in order to create a sample from another, more interesting, probability distribution. We describe a few standard techniques for random variate

253

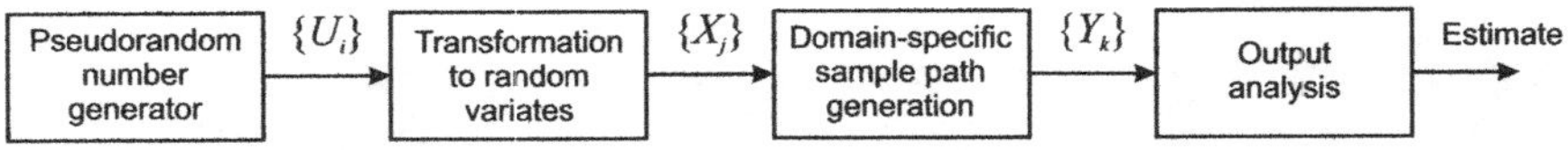

FIGURE 5.1 **Structure of a Monte Carlo simulation.**

generation, like the inverse transform method (Section 5.3) and the acceptance–
rejection method (Section 5.4). For certain specific distributions, it may be
preferable to apply ad hoc methods. The ubiquitous normal distribution is dealt
with in Section 5.5, where we cover both the univariate and the multivariate
case; a few more ad hoc methods are outlined in Section 5.6. Finally, in Section
5.7 we discuss how to generate jointly distributed variables whose dependence
structure is described by a copula. We defer to Chapter 6 the task of generating
sample paths of stochastic processes, typically described by stochastic differential equations, since they involve additional issues about the discretization of
continuous-time models.

A note on terminology. As we have already hinted at, there is nothing random
in a Monte Carlo simulation carried out in R, as it typically involves the generation of pseudorandom numbers, rather than genuinely random ones. We will occasionally stress this point, but, more often than not, we will drop the "pseudo"
for the sake of brevity. Indeed, in both output analysis and variance reduction
methods, we behave as we were truly dealing with random variates. We will use
the term random *variate* in general but random *number* to refer to uniform variates only, as they are actually the primary input of a Monte Carlo simulation.
Later, we will mostly avoid the term *quasi–random* numbers when referring to
deterministic low-discrepancy sequences that are the subject of Chapter 9, as
this name is misleading. Of course, pseudorandom numbers are by necessity
deterministic as well, but low-discrepancy sequences are *conceived* as deterministic.

5.1   The structure of a Monte Carlo simulation

A typical Monte Carlo simulation can be represented as the sequence of building
blocks depicted in Fig. 5.1.

1. The starting point is a stream of random numbers $\{U_i\}$, i.e., a sequence
   of i.i.d. variables uniformly distributed on the unit interval $(0, 1)$. Clearly,
   unless we resort to not quite practical physical random number generators, we must rely on an algorithm that actually generates *pseudo*-random
   numbers. The reasons for doing so are:

   - Efficiency: To run a Monte Carlo simulation we may need a huge set
     of random numbers, and an efficient computer algorithm is the only
     way to achieve that, as physical generators are slower.

- Repeatability: We may need to repeat the experiment, and program debugging would turn into a nightmare in the case of true randomness. In principle, we could generate random numbers by a physical device and then store them, but this would not be practical for very extensive Monte Carlo runs.

2. We do have some good strategies to emulate randomness, which are limited to uniform variates. Hence, even if we need, say, normal random variables, we have to use uniform random numbers as the primary input and then resort to some transformation to obtain more relevant distributions. In the second step, using an array of such methods, we transform the stream of uniform random numbers into a sequence of random variates with the desired distribution, resulting in a stream $\{X_j\}$. We use a different subscript, j rather than i, to stress the fact that more than one random number might be needed to generate one realization of the random variate of interest, as is the case with acceptance–rejection strategies described in the following. Furthermore, the random variate we need may be multidimensional, as in the case of a multivariate normal.

3. The third step is the actual simulation model, where domain-specific dynamics are represented and used to generate a sample path. It may be the case that a sample path consists of the draw of a single random variate, but in general we may need to generate a whole sample path of a possibly complicated stochastic process. Unlike the previous steps, this is highly problem dependent. For instance, we may generate a sample path of stock prices or interest rates and, finally, the payoff of an exotic option.

4. In the fourth and last step, all of the outcomes are collected in order to come up with an estimate, typically in the form of both a point estimate and a confidence interval. It is important, when applying standard inferential statistics methods, that the outcomes form a sample of i.i.d. random variables (see Chapter 7).

Now we may grasp the link between multidimensional integration and Monte Carlo simulation: The structure of Fig. 5.1 points out quite clearly that a typical Monte Carlo simulation is, at least conceptually, a function mapping a vector of n uniform random numbers into a numeric output. More formally, it is a function $g : (0,1)^n \longrightarrow \mathbb{R}$, where the domain is the n-fold Cartesian product of the unit interval,

$$(0,1)^n \equiv \underbrace{(0,1) \times (0,1) \times \cdots \times (0,1)}_{n \text{ times}},$$

i.e., the *unit hypercube*. When estimating an expected value, we are calculating the following integral:

$$\int_{(0,1)^n} g(\mathbf{u}) \, d\mathbf{u}. \tag{5.1}$$

Actually, this view is valid whenever we know exactly the length of the input required to run one Monte Carlo replication, i.e., how many random numbers are needed to get one observation of the output random variable of interest. If so, we know the dimension of the space on which we are integrating. In other cases, as we shall see, we may not really know n. Then, the interpretation as an integral is still valid, but we cannot apply the methods of Chapter 9, which are based on *deterministic* low-discrepancy sequences of numbers replacing the stream of random numbers. This approach requires an integration over a domain of given dimension, and this in turn enforces a few requirements on how the block sequence of Fig. 5.1 is implemented in detail.

5.2 Generating pseudorandom numbers

A pseudorandom number generator is a tool to generate a stream of numbers on the unit interval. To be more precise, some generators produce numbers in the $[0, 1)$ interval, i.e., 0 is a possible output, but 1 is not. For some purposes, the value 0 can be troublesome; this happens, e.g., if we have to transform the stream of random numbers by taking a logarithm. In such cases, numbers in the open interval $(0, 1)$ are preferred. In both cases, we will use the notation $U \sim U(0, 1)$ to refer to the uniform distribution on the unit interval.

There is a rich literature on the generation of random numbers, which is often technically challenging. Fortunately, the average user need not be an expert, but some background knowledge is useful in order to use the available generators properly. In Section 5.2.1 we describe the basic textbook approach, i.e., linear congruential generators, and we point out some of their limitations in Section 5.2.2. Then, we outline a fairly general structure for random number generators in Section 5.2.3, and finally, in Section 5.2.4, we illustrate the principles that are the basis of some generators available in R.

5.2.1 LINEAR CONGRUENTIAL GENERATORS

The standard textbook method to generate $U(0, 1)$ random numbers is the *linear congruential generator* (LCG). This approach was actually implemented in many generators included in commercial software, and it is still possible to choose an LCG. However, LCGs are not considered state-of-the-art anymore and should not be used for high-quality Monte Carlo experiments, especially if these are rather extensive and require a huge stream of input numbers. Nevertheless, LCGs are most useful as an introduction to the structure of more sophisticated generators and illustrate well the pitfalls of pseudorandom number generation.

An LCG actually generates a sequence of non-negative integer numbers Z_i, which are transformed into uniform random numbers. The sequence of integer numbers starts from a user-selected *seed* Z_0; then, given the previous integer number Z_{i-1}, the next number in the sequence is generated as follows:

$$Z_i = (aZ_{i-1} + c) \bmod m, \tag{5.2}$$

```
LCG <- function(a,c,m,seed,N){
  ZSeq <- numeric(N)
  USeq <- numeric(N)
  for (i in 1:N){
    # the operator %% implements mod
    seed <- (a*seed+c) %% m
    ZSeq[i] <- seed
    USeq[i] <- seed/m
  }
  return(list(U=USeq, Z=ZSeq))
}
```

FIGURE 5.2 **Function to generate random numbers by a linear congruential generator.**

```
a <- 5
c <- 3
m <- 16
seed <- 7
N <- 20
seq <- LCG(a,c,m,seed,N)
# unpack the output list into two sequences
USeq <- seq$U
ZSeq <- seq$Z
# sprintf generates an output string by a C-like format
# specification, which is concatenated with a newline
# character and printed by cat
for (k in 1:N)
  cat(sprintf('%3d %3d %7.4f', k, ZSeq[k], USeq[k]), "\n")
```

FIGURE 5.3 **Script to check the linear congruential generator.**

where a (the multiplier), c (the shift), and m (the modulus) are properly chosen integer numbers and "mod" denotes the remainder of integer division (e.g., 15 mod 6 = 3). Finally, the integer number Z_i is transformed into a uniform number

$$U_i = \frac{Z_i}{m}. \tag{5.3}$$

Since Z_i ranges in the set $\{0, 1, 2, 3, \ldots, m - 1\}$, U_i ranges in the unit interval $[0, 1)$. Note that 0 is a possible output, but 1 is not. In Fig. 5.2 we display a function implementing an LCG. The function, given a choice of the parameters a, c, and m, as well as a seed, yields a list consisting of two vectors of length N, where USeq contains the uniform random numbers and ZSeq contains the integer numbers.

In order to check the function LCG and learn a couple of fundamental points, we may run the script of Fig. 5.3 and get the following output:

```
 1    6   0.3750
 2    1   0.0625
 3    8   0.5000
 4   11   0.6875
 5   10   0.6250
 6    5   0.3125
 7   12   0.7500
 8   15   0.9375
 9   14   0.8750
10    9   0.5625
11    0   0.0000
12    3   0.1875
13    2   0.1250
14   13   0.8125
15    4   0.2500
16    7   0.4375
17    6   0.3750
18    1   0.0625
19    8   0.5000
20   11   0.6875
```

On the basis of this simple run, we may observe what follows:

- It is clear that there is nothing random in the sequence generated by an LCG. The sequence is built from an initial number Z_0, the seed of the sequence, which is the initial state of the generator. By starting the sequence from the same seed, we will always get the same sequence. We see quite clearly why we should speak of *pseudo*random numbers. We may also check this point in R, which generates random numbers with the function runif. Actually, runif does not use an LCG (by default), but we may control the seed (better said, the initial state of the generator) as follows:

```
> set.seed(55555)
> runif(5)
[1] 0.9598632 0.6586607 0.3076999 0.9489697 0.1188039
> runif(5)
[1] 0.3415344 0.0415116 0.4067968 0.10739569 0.6403764
> set.seed(55555)
> runif(5)
[1] 0.9598632 0.6586607 0.3076999 0.9489697 0.1188039
```

The function set.seed receives an integer number and sets the initial state of the generator accordingly; we will see later what this state actually is when using more sophisticated generators.

- Since we take a ratio of integer numbers, an LCG actually generates rational numbers in the set

$$\left\{ \frac{0}{m}, \frac{1}{m}, \frac{2}{m}, \frac{3}{m}, \ldots, \frac{m-1}{m} \right\},$$

rather than real numbers. This may look like a significant limitation, but on second thought we should realize that whatever we do on a computer is based on rational numbers, as floating-point numbers are represented, and possibly truncated, by a finite number of bits. In practice, this is not a serious problem, provided that the modulus m is large enough.

- A less obvious point is that the generator is necessarily *periodic*. As we have already observed, we may generate at most m distinct integer numbers Z_i in the range from 0 to $m - 1$, and whenever we repeat a previously generated number, the sequence will be repeated as well (which is not very random at all!). A careful look at the previous output shows that, having started with the initial seed $Z_0 = 7$, we get back there after 16 steps. Of course, an LCG with a period of 16 is not acceptable for any practical purpose; yet, this is not too bad, as 16 is the maximum possible period for a modulus $m = 16$. We do much worse if we select $a = 11$, $c = 5$, and $m = 16$. In this case, starting from $Z_0 = 3$, we get the following sequence of integer numbers Z_i:

$$6, \quad 7, \quad 2, \quad 11, \quad 14, \quad 15, \quad 10, \quad 3, \quad 6, \ldots$$

which has half the maximal period. Thus, since the maximum possible period is m, we should choose a large modulus in order to have a large period. Indeed, a common choice for generators proposed in the literature was

$$2^{31} - 1 = 2,147,483,647.$$

This may look like a very large period, but with faster and faster hardware we run longer and longer simulations, and the above period may not be good enough.

The last observation has been one of the reasons behind the development of alternative generators. Nevertheless, a proper choice of a and c ensures that the period is maximized and that the sequence looks random enough. For instance, a generator proposed in [4] features the following parameters:

$$a = 40692, \qquad c = 0, \qquad m = 2,147,483,399.$$

This LCG has zero shift, which is a rather common choice, as c does not provide too many advantages. Hence, we speak of a *multiplicative* generator; it is also clear that in such a case we must steer away from zero, as the LCG would get stuck there. But what is "random enough" exactly? There are quite sophisticated testing procedures to check random number generators, which are beyond the scope of this book. Anyway, we point out a few important issues in the next section.

5.2.2 DESIRABLE PROPERTIES OF RANDOM NUMBER GENERATORS

The first requirement for a good uniform random number generator is that it is, rather unsurprisingly, *uniform*. A quick check of uniformity can be run by evaluating mean and variance of sample sequences. Since we know that, for a random variable $U \sim \mathsf{U}(0, 1)$, we have

$$\mathrm{E}[U] = \frac{1}{2}, \qquad \mathrm{Var}(U) = \frac{1}{12},$$

a comparison is easy to make. A less crude test would check if, by dividing the unit interval into equally spaced bins, the same number of observations falls into each one of them. In other words, we may draw a histogram or run something like a goodness-of-fit test. However, this is not quite enough to tell a good random number generator. To see the point, consider a sequence like

$$U_i = \frac{i}{m}, \qquad i = 0, 1, \ldots, m - 1,$$

which is obtained by an LCG with $a = c = 1$. This is certainly uniform and has a maximum period. However, we want i.i.d. variables, where now we stress independence. From this point of view, the choice $a = 1$ is a very poor one in general, as we get a sequence like

$$Z_n = (Z_0 + nc) \bmod m,$$

featuring long increasing subsequences.

Given the structure of LCGs, there is no hope of getting a genuinely independent stream, of course; yet, a good LCG should be able to trick statistical testing procedures into "believing" that they produce a sequence of independent observations from the uniform distribution. Without considering sophisticated tests, we can investigate how short sequences of random numbers are distributed in bidimensional or tridimensional spaces. More precisely, imagine that we draw points of the form

$$W_i^{(2)} \equiv (U_i, U_{i+1})$$

on the plane. These points, if the sequence is independent, should be uniformly distributed on the unit square. By the same token, points of the form

$$W_i^{(3)} \equiv (U_i, U_{i+1}, U_{i+2})$$

should be uniformly distributed on the unit cube. As an illustration, let us try the script in Fig. 5.4.[1] The first part of the script uses an LCG with $a = 65$, $c = 1$, and $m = 2048$. We generate the whole sequence of 2048 points, corresponding to the maximum period, and plot points of the form (U_i, U_{i+1}) in the upper half of Fig. 5.5. If we look at the plot, we observe a fairly good filling of the unit

---

[1] These examples are taken from [7, pp. 22–25].

```
m <- 2048
a <- 65
c <- 1
seed <- 0
U <- LCG(a,c,m,seed, 2048)$U
# first pair of plots
par(mfrow = c(2,1))
plot(U[1:(m-1)], U[2:m], pch=20)
plot(U[1:511], U[2:512], pch=20)
# second plot
a <- 1365
c <- 1
U <- LCG(a,c,m,seed, 2048)$U
win.graph()
plot(U[1:(m-1)], U[2:m], pch=20)
```

R programming notes:

1. We instruct R to generate two stacked plots by issuing the command `par(mfrow = c(2,1))`; this tells R that the next plot should be arranged in a 2 × 1 array, plotting row by row.

2. With the next command, `plot(U[1:(m-1)], U[2:m], pch=20)`, we generate two partial sequences shifted by 1, which are displayed by small bullets, selected by `pch=20`. Be sure to understand the difference between `1:(m-1)` and `1:m-1`.

3. `win.graph` is one of the functions that can be used to open new windows for plotting.

FIGURE 5.4 **Script to illustrate the lattice structure of LCGs.**

square, which may be interpreted as a sign of good quality. However, if we only draw the first quarter of the sequence, as we do next, a different pattern emerges. The bottom half of Fig. 5.5 shows that the initial portion of the sequence does not cover the unit square uniformly, pointing out a deficiency of the generator.

The second part of the script, where we set $a = 1365$, yields much worse results, however. In Fig. 5.6 we observe that points are arranged on parallel lines. This kind of pattern is known as *lattice structure* and is typical of LCGs. Actually, this also happens in Fig. 5.5. In general, similar structures can be observed in three dimensions when considering points with coordinates of the form (U_i, U_{i+1}, U_{i+2}). The problem is more evident here, as very few lines with large gaps are observed.

Let us try to understand where the problem comes from, with reference to Fig. 5.6.[2] A quick look at the figure suggests that points (U_i, U_{i+1}) are arranged

---

[2]See, e.g., [6] for a deeper investigation.

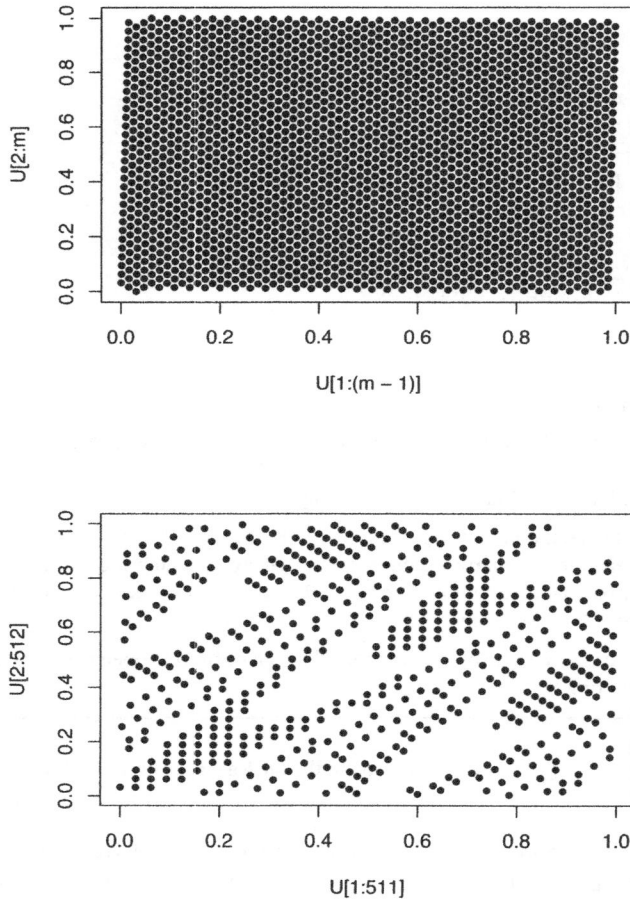

FIGURE 5.5 **Plots obtained by running the script of Fig. 5.4.**

in lines of the form

$$3U_{i+1} + U_i = \alpha_j, \qquad (5.4)$$

for just a few values of the intercept α_j. In order to check this, let us investigate what happens in terms of the corresponding integer numbers Z_i and Z_{i+1}, by noting that

$$Z_{i+1} = (aZ_i + c) \bmod m$$

implies

$$Z_{i+1} = aZ_i + c - km,$$

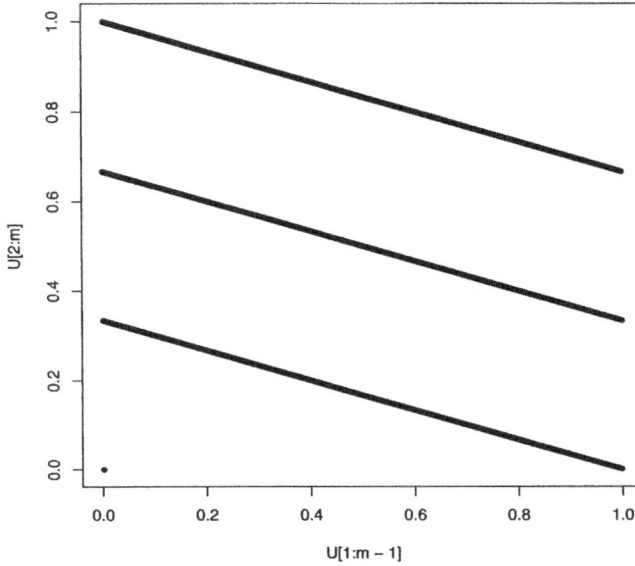

FIGURE 5.6 **Lattice structure in LCGs.**

for some integer k. Hence, with the choice $a = 1365$, $c = 1$, and $m = 2048$, we find

$$3Z_{i+1} + Z_i = 3 \times (1365 \times Z_i + 1 - k \times 2048) + Z_i$$
$$= (3 \times 1365 + 1) \times Z_i - k \times 2048 + 3.$$

Now, if we divide by $m = 2048$ in order to get back to the uniform numbers, we notice that $3 \times 1365 + 1 = 4096$, an integer multiple of the modulus. Indeed, this is what makes this case peculiar: the right-hand side of the previous equation, when divided by 2048, turns out to be the intercept α_j:

$$\frac{4096 \times Z_i - k \times 2048 + 3}{2048} = \beta_j + \frac{3}{2048} = \alpha_j,$$

where β_j is integer. Therefore, we find just a few lines of the form of Eq. (5.4). To see how many, notice that, since $U_i, U_{i+1} \in [0, 1)$, we have

$$3U_{i+1} + U_i \in [0, 4),$$

suggesting that at most four parallel lines may intersect the unit square. Indeed, a look at Fig. 5.6 shows that there are four such lines, but one seems to boil down to an isolated point close to the origin. Using R, it is easy to check that we get back to the initial integer seed $Z_0 = 0$ after $Z_{2047} = 3$, since $(3 \times 1365 + 1) \bmod 2048 = 0$. Therefore, we obtain

$$U_{2047} + U_{2048} = \frac{3}{2048} + 0,$$

which corresponds to a line with $\beta_j = 0$, almost passing through the origin.

Having pointed out some issues with LCGs, we should also mention that they do have some advantages as well:

- They are efficient.
- It is easy to come up with portable generators, i.e., algorithms that can work on different machines.
- They allow for reproducibility by managing a simple seed.
- Sequences may be partitioned into separate streams.

The last point deserves some comment. By "separate streams" we mean different sequences of "independent" random numbers that can either be assigned to model different sources of randomness in a simulation[3] or to different CPUs when running parallel Monte Carlo. We could just assign a different seed to each stream, but with a naive choice there is a danger that they overlap, impairing the independence requirement. With a multiplicative LCG, i.e., when the increment $c = 0$, is easy to skip ahead in the sequence in order to partition the whole sequence in nonoverlapping subsequences:

$$Z_{i+k} = a^k Z_i \bmod m = \big((a^k \bmod m) Z_i\big) \bmod m.$$

So, skipping ahead by k steps just requires computing $a^k \bmod m$ once. All of these requirements should also be met by alternative and more sophisticated generators.

5.2.3 GENERAL STRUCTURE OF RANDOM NUMBER GENERATORS

In the previous section we have shown that LCGs may have several limitations, and that careful testing is needed in order to select a suitable one. Indeed, LCGs were state of the art in the past, but they have been replaced by alternative generators. Nevertheless, all of these generators share the same conceptual structure as LCGs, in which we have:

- A finite set $\mathcal{S}$ of states
- A way to assign the initial state in $\mathcal{S}$
- A state transition function from $\mathcal{S}$ to $\mathcal{S}$
- An output space $\mathcal{U}$

---

[3]This may be useful to apply certain variance reduction strategies.

- An output function from $\mathcal{S}$ to $\mathcal{U}$

In the simple LCG case, the states are just integer numbers and the output space is the unit interval. More recent generators use more complex state spaces. We cannot cover all of them, but we do point out a couple of ideas that have actually been implemented in R.

One possibility is to devise *combined generators*. The idea is to combine multiple generators that, taken individually, would have quite some weaknesses, but improve a lot when combined. One such generator is the Wichman–Hill generator, which combines three LCGs. For instance:

$$
\begin{aligned}
X_i &= (171\,X_{i-1})\,\mathrm{mod}\,m_1, && \text{where } m_1 = 30269,\\
Y_i &= (172\,Y_{i-1})\,\mathrm{mod}\,m_2, && \text{where } m_2 = 30307,\\
Z_i &= (170\,Z_{i-1})\,\mathrm{mod}\,m_3, && \text{where } m_3 = 30232.
\end{aligned}
$$

We see that $\mathcal{S}$ consists of three integer numbers. The output function is

$$
U_i = \left(\frac{X_i}{m_1} + \frac{Y_i}{m_2} + \frac{Z_i}{m_3}\right)\,\mathrm{mod}\,1.
$$

This choice results in much longer periods, and other generators along this line have been proposed. It should be noted that an advantage of a combined generator over an LCG with large modulus m is that we avoid numerical issues with potential overflows.

An alternative approach is to use a richer state space $\mathcal{S}$, as in *multiple recursive generators*:

$$
X_i = (a_1 X_{i-1} + a_2 X_{i-2} + \cdots + a_k X_{i-k})\,\mathrm{mod}\,m.
$$

A class of generators, called *Tausworthe generators* or *linear feedback shift registers*, implements this idea with $m = 2$. In other words, the state consists of bits, a choice that takes advantage of the binary operations of a computer. The output function maps a sequence of bits into a number in the unit interval:

$$
U_i = \sum_{j=1}^{L} X_{is+j-1} 2^{-j},
$$

where s and L are positive integers playing the role of a step size and a word length, respectively. A natural choice for L is 32 or 64, depending on the word length of the machine. The advantage of exploiting the binary arithmetic of hardware, where operations modulo 2 correspond to a XOR (exclusive OR) operation is obvious.[4] Actually, variants of this scheme are used. *Generalized feedback shift registers* (GFSR) use k vectors $\mathbf{X}_i = (x_{i,1}, \ldots, x_{i,L})^{\mathsf{T}}$ of L bits, rather than single bits:

$$
\mathbf{X}_i = (a_1 \mathbf{X}_{i-1} + a_2 \mathbf{X}_{i-2} + \cdots + a_k \mathbf{X}_{i-k})\,\mathrm{mod}\,2. \tag{5.5}
$$

---

[4]The exclusive OR, denoted as $\oplus$, is defined as follows: $(1 \oplus 0) = (0 \oplus 1) = 1$; $(0 \oplus 0) = (1 \oplus 1) = 1$. In the *inclusive* OR case, denoted as $\vee$, we have $(1 \vee 1) = 1$.

Then, we may use an output function like

$$U_i = \sum_{j=1}^{L} x_{i,j} 2^{-j}.$$

One very common generator, the *Mersenne twister*, is obtained by adding a further tweak, i.e., by interpreting a GFSR in matrix terms:

$$\mathbf{Y}_i = \mathbf{A}\mathbf{Y}_{i-1},$$

where $\mathbf{Y}_i$ is a vector of kL bits and $\mathbf{A}$ is a $(kL) \times (kL)$ matrix. In the case of the GFSR of Eq. (5.5), we have

$$\mathbf{A} = \begin{bmatrix} \mathbf{0} & \mathbf{I}_L & \cdots & \mathbf{0} \\ \vdots & \vdots & \ddots & \vdots \\ \mathbf{0} & \mathbf{0} & \cdots & \mathbf{I}_L \\ a_k\mathbf{I}_L & a_{k-1}\mathbf{I}_L & \cdots & a_1\mathbf{I}_L \end{bmatrix},$$

where $\mathbf{I}_L$ is the $L \times L$ identity matrix. Clearly, the upper part of the matrix implements a shift, and the new bit word is generated by the last line. Twisting is obtained by using a different last line in matrix $\mathbf{A}$. A version of Mersenne twister is widely used and, at the time of writing, it is the default generator in R. This version uses $k = 624$ and has the huge period

$$2^{19937} - 1.$$

Incidentally, the name stems from the concept of Mersenne numbers, i.e., numbers of the form $2^p - 1$; for certain values of p, a Mersenne number is also a prime number. Indeed, both $p = 19,937$ and the corresponding Mersenne number are prime.

5.2.4 RANDOM NUMBER GENERATORS IN R

Random numbers can be generated in R by the function `runif`, which generates a vector of given size:

```
> set.seed(55555)
> runif(5)
[1] 0.9598632 0.6586607 0.3076999 0.9489697 0.1188039
```

Notice the use of `set.seed` to set the seed (better said, the state) of the generator. The function `.Random.seed` can be used to inspect the current state of the generator. Its value depends on the kind of generator, which can be selected by the function `RNGkind` by setting the input parameter `kind`; possible values of this parameter are `Wichmann-Hill` and `Mersenne-Twister`, the default.

The following snapshot shows that, with the default setting, the state of the generator is an integer vector consisting of 626 entries, which can be reset by passing a single number to `set.seed`:

```
> mode(.Random.seed)
[1] "numeric"
> class(.Random.seed)
[1] "integer"
> length(.Random.seed)
[1] 626
> set.seed(55555)
> cat(.Random.seed[1:6],"\n")
403 624 -734508337 397372100 1252553461 -911245518
> runif(3)
[1] 0.9598632 0.6586607 0.3076999
> cat(.Random.seed[1:6],"\n")
403 3 -1423130422 -543017428 -426382801 -1354865947
> set.seed(55555)
> cat(.Random.seed[1:6],"\n")
403 624 -734508337 397372100 1252553461 -911245518
```

This corresponds to the state of a Mersenne twister, plus some additional information. The next snapshot shows how to change the generator from Mersenne twister to Wichmann–Hill, as well as the effect on the state:

```
> RNGkind(kind="Mersenne-Twister")
> set.seed(55555)
> cat(.Random.seed[1:6],"\n")
403 624 -734508337 397372100 1252553461 -911245518
> RNGkind(kind="Wichmann-Hill")
> set.seed(55555)
> cat(.Random.seed,"\n")
400 15266 22906 19508
> RNGkind(kind="default")
> set.seed(55555)
> cat(.Random.seed[1:6],"\n")
403 624 -734508337 397372100 1252553461 -911245518
```

Indeed, the state of a combined Wichmann-Hill generator consists of three integer numbers.

5.3 The inverse transform method

Suppose that we are given the CDF $F_X(x) = P\{X \le x\}$, and that we want to generate random variates distributed according to $F_X(\cdot)$. If we are able to invert $F_X(\cdot)$ easily, then we may apply the following inverse transform method:

    1. Draw a random number $U \sim \mathsf{U}(0,1)$

    2. Return $X = F_X^{-1}(U)$

It is easy to see that the random variate X generated by this method is actually characterized by the CDF $F_X(\cdot)$:

$$P\{X \leq x\} = P\{F_X^{-1}(U) \leq x\} = P\{U \leq F_X(x)\} = F_X(x),$$

where we have used the monotonicity of F_X and the fact that U is uniformly distributed.[5]

◼ Example 5.1 Generating exponential variates

A typical distribution which can be easily simulated by the inverse transform method is the exponential distribution. If $X \sim \exp(\mu)$, where $1/\mu$ is the expected value of X, its CDF is

$$F_X(x) = 1 - e^{-\mu x}.$$

A direct application of the inverse transform method yields

$$x = -\frac{1}{\mu} \log(1 - U).$$

Since the distributions of U and $(1 - U)$ are actually the same, it is customary to generate exponential variates by drawing a random number U and by returning $-\log(U)/\mu$. Generating exponential random variates is also one way to simulate a Poisson process, as we shall see in Chapter 6.

The inverse transform method is quite simple conceptually; however, we may not apply it when F_X is not invertible, as it happens with discrete distributions. In this case the cumulative distribution function is piecewise constant, with jumps located where the probability mass is concentrated, i.e., for possible values of the discrete random variable. Nevertheless, we may adapt the method with a little thought. Consider a discrete empirical distribution with a finite support:

$$P\{X = x_j\} = p_j, \qquad j = 1, 2, \ldots, n.$$

---

[5]The reader may appreciate the similarity between this line of reasoning and that of Eq. (3.40), where we showed that if X has CDF $F_X(\cdot)$, then $F_X(X)$ is a uniform random variable.

Then we should generate a uniform random variate U and return X as

$$X = \begin{cases} x_1 & \text{if } U < p_1, \\ x_2 & \text{if } p_1 \leq U < p_1 + p_2, \\ \vdots \\ x_j & \text{if } \sum_{k=1}^{j-1} p_k \leq U < \sum_{k=1}^{j} p_k, \\ \vdots \end{cases}$$

It may be instructive to see how this idea may be implemented in a simple way (not the most efficient one, however). Suppose that we have a distribution defined by probabilities

$$0.1, \quad 0.2, \quad 0.4, \quad 0.2, \quad 0.1,$$

for the values $1, 2, 3, 4, 5$. First, we find the cumulative probabilities,

$$0.1, \quad 0.3, \quad 0.7, \quad 0.9, \quad 1.0,$$

and then we draw a uniform random number, say $U = 0.82$. For each cumulative probability P, we check if $U > P$, which yields the vector

$$[1, \quad 1, \quad 1, \quad 0, \quad 0],$$

where 1 corresponds to "true" and 0 to "false." To select the correct value to return, we must sum the ones in this vector (the total is 3 here) and add 1; in this case we should return the value 4. Using R, this may be accomplished by working directly on vectors, as shown in Fig. 5.7. The code also includes a script to check the results by plotting a histogram, shown in Fig. 5.8, for the specific numerical example that we are considering.

Sometimes, the CDF is invertible in principle, but this is quite costly computationally. In such a case, one possibility is to resort to the acceptance–rejection method we describe next.

5.4 The acceptance–rejection method

Suppose that we must generate random variates according to a PDF $f_X(x)$, and that the difficulty in inverting the corresponding CDF makes the inverse transform method unattractive. To see a concrete example, consider the beta distribution with PDF

$$f_X(x) = \frac{x^{\alpha_1 - 1}(1 - x)^{\alpha_2 - 1}}{\mathrm{B}(\alpha_1, \alpha_2)}, \qquad x \in [0, 1],$$

```
EmpiricalDrnd <- function(values, probs, howmany){
  # make sure probs add up to 1
  probs <- probs/sum(probs)
  # get cumulative probabilities
  cumprobs <- cumsum(probs)
  N <- length(probs)
  sample <- numeric(howmany)
  for (k in 1:howmany){
    loc <- sum(runif(1)>cumprobs)+1
    sample[k] <- values[loc]
  }
  return(sample)
}

set.seed(55555)
values <- 1:5
probs <- c(0.1, 0.3, 0.4, 0.15, 0.05)
sample <- EmpiricalDrnd(values,probs,10000)
freqs <- hist(sample,breaks=0:5,plot=FALSE)
plot(freqs$counts,type='h',lwd=15)
```

R programming notes:

1. The parameter howmany specifies the desired sample size, and the parameters values and probs describe the support and the PMF of the distribution. For efficiency reasons, when a function like this is called repeatedly, a better choice would be to ask for the CDF, forcing the user to compute it once and for all.

2. The variable loc locates the vector index that corresponds to the value to be returned. Note that we have to add 1, since vector indexing in R starts from 1, not 0 like in other languages.

3. Note the use of hist to calculate the frequencies for specified breakpoints, without plotting the histogram automatically. For this purpose we may use plot, taking advantage of its flexibility.

FIGURE 5.7 **Sampling from an empirical discrete distribution.**

for parameters $\alpha_1, \alpha_2 > 1$, where

$$B(\alpha_1, \alpha_2) \equiv \int_0^1 x^{\alpha_1-1}(1-x)^{\alpha_2-1}\, dx$$

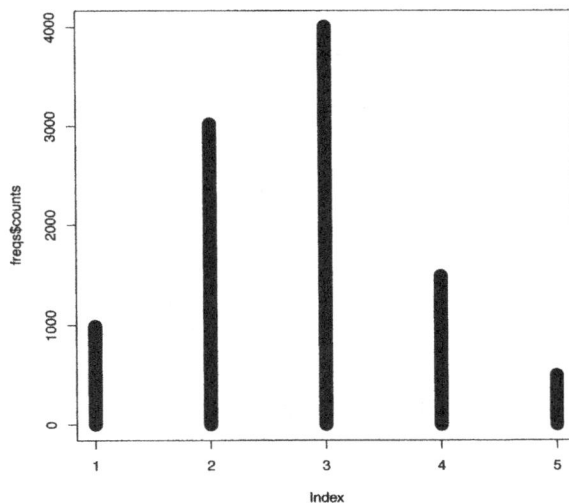

FIGURE 5.8 **Histogram produced by calling** EmpiricalDrnd.

is the beta function needed for normalization.[6] To obtain the CDF we should integrate the PDF, resulting in a function that requires a costly numerical inversion by solving a nonlinear equation.

Assume that we can find a function $t(x)$ such that

$$t(x) \geq f_X(x), \qquad \forall x \in I,$$

where I is the support of $f_X(\cdot)$. The function $t(\cdot)$ is not a probability density, but the related function $g(x) = t(x)/c$ is, provided that we choose

$$c = \int_I t(x) \, dx.$$

If the distribution $g(\cdot)$ is easy to sample from, it can be shown that the following acceptance–rejection method generates a random variate X distributed according to the density $f_X(\cdot)$:

1. Generate $Y \sim g$.
2. Generate $U \sim U(0, 1)$, independent of Y.
3. If $U \leq f_X(Y)/t(Y)$, return $X = Y$; otherwise, repeat the procedure.

---

[6]See Section 3.2.4. The definition of the beta distribution, per se, does not require the above condition on its parameters; however, the PDF would have a vertical asymptote for either $\alpha_1 < 1$ or $\alpha_2 < 1$, which would prevent the application of acceptance–rejection.

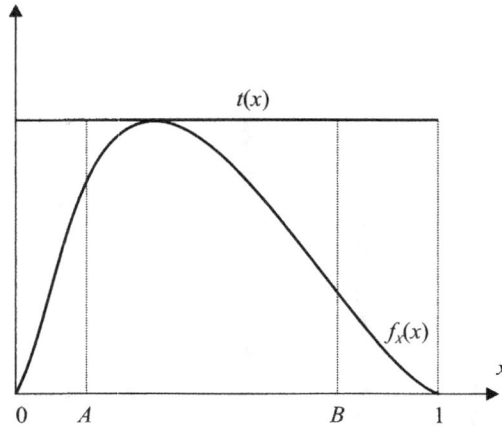

FIGURE 5.9 **Graphical illustration of the acceptance–rejection method.**

The PDF $f_X(\cdot)$ is called the *target* density, and the "easy" density $g(\cdot)$ is called the *instrumental* or *candidate* density. If the support I is bounded, a natural choice of instrumental density is the uniform distribution on I, and we may set

$$c = \max_{x \in I} f_X(x).$$

Before proving the correctness of the method, let us get an intuitive grasp from Fig. 5.9, where we show a target PDF much like a beta distribution, whose support is the unit interval $(0, 1)$. The upper bounding function $t(\cdot)$ is a constant corresponding to the maximum of the target density. Hence, the instrumental density is uniform, and the variables Y spread evenly over the unit interval. Consider point A; since $f_X(A)$ is close to $t(A)$, A is likely to be accepted, as the ratio $f_X(A)/t(A)$ is close to 1. When we consider point B, where the value of the density $f_X(\cdot)$ is small, we see that the ratio $f_X(B)/t(B)$ is small; hence, B is unlikely to be accepted, which is what we would expect. It can also be shown that the average number of iterations to terminate the procedure with an accepted value is c.

Example 5.2 Generating a beta variate

Consider the density

$$f_X(x) = 30(x^2 - 2x^3 + x^4), \qquad x \in [0, 1].$$

The reader is urged to verify that this is indeed a density (actually, it is the beta density with $\alpha_1 = \alpha_2 = 3$). Since the PDF is a fourth-degree polynomial, the CDF is a fifth-degree polynomial. If we apply the inverse transform method, we have to invert such a polynomial at each generation; this inconvenience suggests the application of the acceptance–rejection method. By straightforward calculus we see that

$$\max_{x\in[0,1]} f_X(x) = \frac{30}{16},$$

for $x^* = 0.5$. We may check this result by running the first lines of the script in Fig. 5.10:

```
> f <- function(x) dbeta(x,shape1=3,shape2=3)
> out <- optimize(f,interval=c(0,1),maximum=TRUE)
> out
$maximum
[1] 0.5
$objective
[1] 1.875
```

Using the uniform density as the easy instrumental density $g(\cdot)$, we obtain the following algorithm:

1. Draw two independent and uniformly distributed random variables U_1 and U_2.

2. If $U_2 \leq 16(U_1^2 - 2U_1^3 + U_1^4)$, accept $X = U_1$; otherwise, reject and go back to step 1.

In the script of Fig. 5.10 we also draw the points generated according to this scheme, with coordinates U_1 and U_2; this produces the plot in Fig. 5.11, where rejected points are depicted as empty circles, and accepted points as bullets. We observe that a larger fraction of points is accepted when the PDF is larger, as intuition requires. The average number of iterations to generate one random variate is $\frac{30}{16}$.

Now, let us prove:

1. That the acceptance–rejection algorithm indeed is equivalent to sampling from the PDF $f_X(\cdot)$.

2. That the expected number of iterations to accept Y is c.

The plot of Fig. 5.11 is enlightening, as it shows that what we are doing is based on a common trick in mathematics: embedding a problem into a higher-dimensional space.[7] In fact, our task is to generate a single random variable X

---

[7]The treatment here follows [8, pp. 47–51].

```
f <- function(x) dbeta(x,shape1=3,shape2=3)
out <- optimize(f,interval=c(0,1),maximum=TRUE)
c <- out$objective
x <- seq(from=0,to=1,by=0.01)
plot(x,f(x),type='l')
set.seed(55555)
for (k in 1:1000){
  U1 <- runif(1)
  U2 <- runif(1,min=0,max=c)
  if (U2 <= f(U1))
    points(U1,U2,pch=20)
  else
    points(U1,U2,pch=1)
}
```

FIGURE 5.10 **Script to check acceptance–rejection for Example 5.2.**

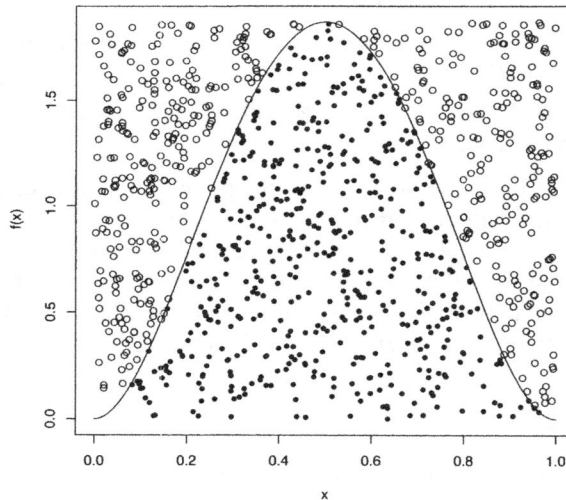

FIGURE 5.11 **The plot produced by the R script of Fig. 5.10.**

according to the PDF $f_X(\cdot)$, but this density may also be interpreted as

$$f_X(x) = \int_0^{f_X(x)} du,$$

i.e., as a marginal density within the joint distribution of two random variables X and U, uniformly distributed on the set $\{(x, u) \mid 0 \le u \le f_X(x)\}$, i.e.,

$$(X, U) \sim \mathsf{U}\{(x, u) \mid 0 \le u \le f_X(x)\}.$$

Thus, sampling in the one-dimensional case is equivalent to sampling the bidimensional distribution. The argument applies to the case in which the upper bounding function $t(\cdot)$ is a constant, as in Example 5.2, so that $g(\cdot)$ boils down to a uniform density and the larger domain is just a box. To generalize the idea to an arbitrary instrumental density, let us calculate the probability $P\{X \in \mathcal{A}\}$, for an arbitrary set[8] $\mathcal{A}$, when X is generated by the acceptance–rejection scheme:

- sample $Y \sim g$;
- sample $U \mid Y = y \sim \mathsf{U}(0, cg(y))$;
- accept y if it satisfies the constraint $u \le f_X(y)$.

Note that X is given by Y conditional on the event "accept", which is equivalent to $\{U \le f_X(Y)\}$. Let $\mathcal{X}$ be the support of $f_X(\cdot)$. We can write:

$$P\{X \in \mathcal{A}\} = P\{Y \in \mathcal{A} \mid U \le f_X(Y)\} = \frac{P\{Y \in \mathcal{A}, U \le f_X(Y)\}}{P\{U \le f_X(Y)\}}$$

$$= \frac{\displaystyle\int_{\mathcal{A}} \int_0^{f_X(y)} \frac{du}{cg(y)} g(y)\, dy}{\displaystyle\int_{\mathcal{X}} \int_0^{f_X(y)} \frac{du}{cg(y)} g(y)\, dy} = \frac{\displaystyle\int_{\mathcal{A}} \frac{f_X(y)}{cg(y)} g(y)\, dy}{\displaystyle\int_{\mathcal{X}} \frac{f_X(y)}{cg(y)} g(y)\, dy}$$

$$= \frac{\displaystyle\int_{\mathcal{A}} f_X(y)\, dy}{\displaystyle\int_{\mathcal{X}} f_X(y)\, dy} = \int_{\mathcal{A}} f_X(y)\, dy.$$

This shows that the density of the accepted Y is indeed $f_X(\cdot)$. Furthermore, we also see that the probability of the event "accept" is

$$P\{U \le f_X(Y)\} = \int_{\mathcal{X}} \int_0^{f_X(y)} \frac{du}{cg(y)} g(y)\, dy = \frac{1}{c}.$$

However, since the iterations in acceptance–rejection are independent and we stop at the first success, the number of iterations follows a geometric distribution with parameter $p = 1/c$. Since a geometric random variable with parameter p has expected value $1/p$,[9] it follows that the expected number of iterations to accept a proposed value is c.

We also observe that:

1. A tight upper bounding function, which is reflected in a small value of c, is needed for efficiency.

---

[8] Strictly speaking, the set should be measurable.
[9] See Section 3.2.1 and Eq. (3.14).

2. A priori, we do not know how many random numbers are needed to generate a single variate by acceptance–rejection.

The second point implies that we do not know a priori the dimension of space on which we are integrating when carrying out Monte Carlo simulation. This does not prevent the application of Monte Carlo sampling, but it is a stumbling block when we use low-discrepancy sequences, as we shall see in Chapter 9. It is also worth mentioning that rejection sampling is not restricted to univariate distributions, and it is can also be applied to sample from a PDF which is not completely known and is given up to a normalization constant. This is quite relevant in computational Bayesian statistics, as we shall see in Chapter 14.

5.5 Generating normal variates

Sampling from a normal distribution is a common need in financial applications. Recall first that if $X \sim \mathsf{N}(0,1)$, then $\mu + \sigma X \sim \mathsf{N}(\mu, \sigma^2)$; hence, we just need a method for generating a stream of independent standard normal variables. If correlation needs to be induced, this can be easily accomplished as we shall explain. Over the years, a number of methods have been proposed to generate standard normal variates. Some are obsolete, like the idea of summing a suitably large number of uniform variates, taking advantage of the central limit theorem. The inverse transform method was not considered efficient in the past, but it is quite practical by now, as efficient approximations have been developed to invert the CDF of the standard normal. In between, there is a set of ad hoc methods, and some are implemented in the R function `rnorm`.

5.5.1 SAMPLING THE STANDARD NORMAL DISTRIBUTION

The general-purpose inverse transform methods may be applied by inverting the standard normal CDF. In R, this is accomplished, for a generic normal, by the function call

```
x <- qnorm(p,mu,sigma)
```

that returns the quantile at probability level p of a variable with expected value mu and standard deviation `sigma`. The following snapshot shows that there is little difference between calling `rnorm` or generating uniform variates and then inverting the normal CDF:

```
> system.time(qnorm(runif(50000000)))
   user  system elapsed
   5.74    0.16    5.89
> system.time(rnorm(50000000))
   user  system elapsed
   5.80    0.05    5.85
```

Indeed, inversion of the CDF is currently the default method in R. Nevertheless, there are faster *ad hoc* methods that can be selected as follows:

```
> set.seed(55555,normal.kind="Ahrens-Dieter")
> system.time(rnorm(50000000))
   user   system elapsed
   3.43     0.05     3.48
```

In the following, we will not cover the generation of normal variates too extensively, but it is nevertheless instructive to get a glimpse of the ingenuity that may be employed to come up with ad hoc methods, and the dangers of bad random variate generators. A well-known approach is the Box–Muller method, which takes advantage of the representation of points on the plane by polar coordinates. Consider two independent variables $X, Y \sim N(0, 1)$, and let (R, θ) be the polar coordinates of the point of Cartesian coordinates (X, Y) in the plane, so that

$$ d = R^2 = X^2 + Y^2, \qquad \theta = \tan^{-1} Y/X. $$

Given independence, the joint density of X and Y is

$$ f(x, y) = \frac{1}{\sqrt{2\pi}} e^{-x^2/2} \frac{1}{\sqrt{2\pi}} e^{-y^2/2} = \frac{1}{2\pi} e^{-(x^2+y^2)/2} = \frac{1}{2\pi} e^{-d/2}. $$

The last expression looks like a product of an exponential density for d and a uniform distribution; the term $1/2\pi$ may be interpreted as the uniform distribution for the angle $\theta \in (0, 2\pi)$. However, we are missing some constant term in order to obtain the exponential density. In fact, to properly express the density in terms of (d, θ), we should take the Jacobian of the transformation from (x, y) to (d, θ) into account.[10] Some calculations yield

$$ J = \begin{vmatrix} \dfrac{\partial d}{\partial x} & \dfrac{\partial d}{\partial y} \\ \dfrac{\partial \theta}{\partial x} & \dfrac{\partial \theta}{\partial y} \end{vmatrix} = 2, $$

and the correct density in the alternative coordinates is

$$ f(d, \theta) = \frac{1}{2} \frac{1}{2\pi} e^{-d/2}. $$

Hence, we may generate R^2 as an exponential variable with mean 2, θ as a uniformly distributed angle, and then transform back into Cartesian coordinates to obtain two independent standard normal variates. The Box–Muller algorithm may be implemented as follows:

1. Generate two independent uniform variates $U_1, U_2 \sim U(0, 1)$.

2. Set $R^2 = -2 \log U_1$ and $\theta = 2\pi U_2$.

---

[10] See, e.g., [10] for details.

3. Return $X = R \cos \theta$, $Y = R \sin \theta$.

In practice, this algorithm may be improved by avoiding the costly evaluation of trigonometric functions and integrating the Box–Muller approach with the rejection approach. The idea results in the following polar rejection method:

1. Generate two independent uniform variates $U_1, U_2 \sim U(0, 1)$.
2. Set $V_1 = 2U_1 - 1$, $V_2 = 2U_2 - 1$, $S = V_1^2 + V_2^2$.
3. If $S > 1$, return to step 1; otherwise, return the independent standard normal variates:

$$X = \sqrt{\frac{-2 \log S}{S}} V_1, \qquad Y = \sqrt{\frac{-2 \log S}{S}} V_2.$$

We refer the reader to [9, Section 5.3] for a justification of the polar rejection method.

Example 5.3 A bad interaction of strategies

We have seen that LCGs may exhibit a lattice structure. Since the Box–Muller transformation is nonlinear, one might wonder if the composition of these two features may yield weird effects. We may check this in a somewhat peculiar case (see [7]), using the R script in Fig. 5.12. The script generates 2046 uniform random numbers for a sequence with modulus $m = 2048$; we discard the last pair, because the generator has maximum period and reverts back to the seed, which is 0 and causes trouble with the logarithm. Vectors U1 and U2 contain odd- and even-numbered random numbers in the sequence. The first part of the resulting plot, displayed in Fig. 5.13, shows poor coverage of the plane. The second part shows that swapping the pairs of random numbers may have a significant effect, whereas with truly random numbers the swap should be irrelevant. Of course, using better LCGs, or better random number generators altogether prevents pathological behavior like this. Nevertheless, it may be preferable to use the inverse transform method to avoid undesirable effects.

5.5.2 SAMPLING A MULTIVARIATE NORMAL DISTRIBUTION

In many financial applications one has to generate variates according to a multivariate normal distribution with (vector) expected value μ and covariance matrix Σ. This task may be accomplished by finding the Cholesky factor for Σ,

```
m <- 2048
a <- 1229
c <- 1
N <- m-2
seed <- 0
U <- LCG(a,c,m,seed,N)$U
U1 <- U[seq(from=1,by=2,to=N-1)]
U2 <- U[seq(from=2,by=2,to=N)]
X <- sqrt(-2*log(U1)) * cos(2*pi*U2)
Y <- sqrt(-2*log(U1)) * sin(2*pi*U2)

par(mfrow=c(2,1))
plot(X,Y,pch=20,cex=0.5)
X <- sqrt(-2*log(U2)) * cos(2*pi*U1)
Y <-sqrt(-2*log(U2)) * sin(2*pi*U1)
plot(X,Y,pch=20,cex=0.5)
par(mfrow=c(1,1))
```

FIGURE 5.12 **Script to check the Box–Muller approach.**

i.e., an upper triangular[11] matrix $\mathbf{U}$ such that $\mathbf{\Sigma} = \mathbf{U}^{\mathsf{T}}\mathbf{U}$. Then we apply the following algorithm:

1. Generate n independent standard normal variates $Z_1, \ldots, Z_n \sim \mathsf{N}(0,1)$.
2. Return $X = \boldsymbol{\mu} + \mathbf{U}^{\mathsf{T}}\mathbf{Z}$, where $\mathbf{Z} = [Z_1, \ldots, Z_n]^{\mathsf{T}}$.

Example 5.4 Sampling using Cholesky factors

A rough code to simulate multivariate normal variables is illustrated in Fig. 5.14. The code builds a matrix, where columns correspond to variables and rows to observations, as usual. Assume that we have the following parameters:

```
> Sigma <- matrix(c(4,1,-2,1,3,1,-2,1,5),nrow=3)
> mu <- c(8,6,10)
> eigen(Sigma)$values
[1] 6.571201 4.143277 1.285521
```

---

[11]In many numerical analysis textbooks, Cholesky decomposition is expressed as $\mathbf{A} = \mathbf{L}\mathbf{L}^{\mathsf{T}}$, for a lower triangular matrix $\mathbf{L}$; R returns an upper triangular matrix $\mathbf{U}$, but the two forms are clearly equivalent.

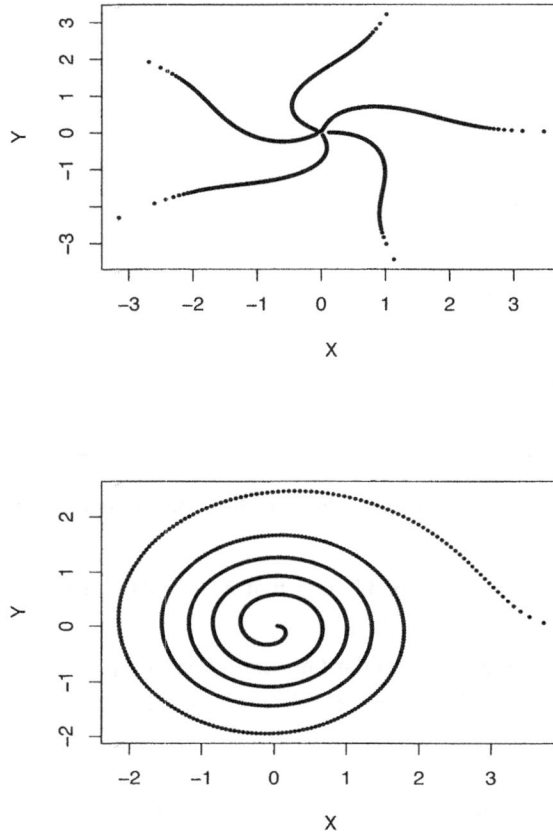

FIGURE 5.13 **Effect of swapping random numbers in the Box–Muller transformation.**

Note that we make sure that the matrix Σ is positive definite, as it should be, by checking its eigenvalues with the function `eigen`. Now we may generate a sample and verify the results:

```
> set.seed(55555)
> Z <- MultiNormrnd(mu,Sigma,10000)
> colMeans(Z)
[1] 7.995686 5.981794 9.987316
> cov(Z)
```

```
MultiNormrnd <- function(mu,sigma,howmany){
  n <- length(mu)
  Z <- matrix(0,nrow=howmany,ncol=n)
  U <- chol(sigma)
  for (i in 1:howmany)
   Z[i,] <- mu + rnorm(n) %*% U
  return(Z)
}
```

FIGURE 5.14 Code to sample from a multivariate normal distribution.

```
                [,1]         [,2]         [,3]
    [1,]   3.9930313  0.9789963  -2.003760
    [2,]   0.9789963  2.9525912   1.034064
    [3,]  -2.0037602  1.0340636   5.021224
```

We should note that Cholesky decomposition is not the only one that we can use.[12] An alternative is the square root matrix $\Sigma^{1/2}$, i.e., a symmetric matrix such that

$$\Sigma = \Sigma^{1/2}\Sigma^{1/2}.$$

One way to find a symmetric square root is the following, taking advantage of matrix diagonalization:

```
> eig <- eigen(Sigma)
> V <- eig$vectors
> V %*% diag(eig$values) %*% t(V)
     [,1] [,2] [,3]
[1,]    4    1   -2
[2,]    1    3    1
[3,]   -2    1    5
> sqrtSigma <- V %*% diag(sqrt(eig$values)) %*% t(V)
> sqrtSigma %*% sqrtSigma
     [,1] [,2] [,3]
[1,]    4    1   -2
[2,]    1    3    1
[3,]   -2    1    5
```

A handier alternative is the function sqrtm from package expm, as we have seen in Section 3.3.4. Anyway, with R there is no need to carry out that amount

---

[12]More on this in Section 13.4.

of work by ourselves, as we can rely on the `mvrnorm` function from the `MASS` package:

```
> library(MASS)
> set.seed(55555)
> Z <- mvrnorm(n=10000,mu=mu,Sigma=Sigma)
> colMeans(Z)
[1]   7.967968   5.993851 10.001322
> cov(Z)
            [,1]         [,2]         [,3]
[1,]   3.9914023 0.9278669 -2.083500
[2,]   0.9278669 2.9171899  1.014537
[3,] -2.0835005 1.0145372  5.089220
```

The `mvtnorm` package offers an alternative function, `rmvnorm`, for this purpose.

5.6   Other ad hoc methods

Many ad hoc methods rely on relationships among distributions.[13] For instance, we know that an Erlang variable $\text{Erl}(n, \lambda)$ is the sum of n independent exponentials X_i with rate λ. Hence, we may use the inverse transform method as follows:

$$\sum_{i=1}^{n} X_i = \sum_{i=1}^{n} \left(-\frac{1}{\lambda} \log U_i \right) = -\frac{1}{\lambda} \log \left(\prod_{i=1}^{n} U_i \right).$$

By a similar token, we could take advantage of the fact that a chi-square variable χ_n^2 with n degrees of freedom is obtained by squaring and summing n independent standard normals. However, we have also seen that these distributions may be regarded as specific subcases of the gamma distribution. We recall the PDF of a $\text{Gamma}(\alpha, \lambda)$ random variable:

$$f_X(x) = \frac{\lambda^\alpha x^{\alpha-1} e^{-\lambda x}}{\Gamma(\alpha)}, \tag{5.6}$$

where α is a shape parameter and λ is a scale parameter. Clearly, there is a notable difference between the cases $\alpha < 1$ and $\alpha \geq 1$. Let us focus on the latter one. Different algorithms have been proposed for such a gamma variable. The following one, which we describe for illustrative purposes only, is based on acceptance–rejection:

1. Set $d = \alpha - \frac{1}{3}$ and $c = 1/\sqrt{9d}$.
2. Generate two independent variates $U \sim \text{U}(0,1)$ and $Z \sim \text{N}(0,1)$, and set $V = (1 + cZ)^3$.
3. If $Z > -1/c$ and $\log U < 0.5Z^2 + d - dV + d \log V$, then return $X = dV$; otherwise go back to step 2.

---

[13] See Sections 3.2.2, 3.2.3, and 3.2.5.

The example also points out the fact that the boundary between ad hoc and general methods is a bit blurred, since we are using acceptance–rejection, but in a very peculiar way. Luckily, we may just rely on the `rgamma` function in R! Another interesting case is the multivariate t distribution $t_\nu(\boldsymbol{\mu}, \boldsymbol{\Sigma})$ (see Section 3.3.4). We may take advantage of the fact that a t variable is related to the ratio of a standard normal and a chi-square, which in turn is a special case of the gamma. To induce correlation, we may use Cholesky factors of the covariance matrix, just like in the case of the normal.

1. Generate a column vector $Z \sim N(\mathbf{0}, \mathbf{I})$ of n independent standard normals.

2. Generate $S \sim \text{Gamma}\left(\frac{\nu}{2}, \frac{1}{2}\right) \equiv \chi^2_\nu$.

3. Compute $\mathbf{Y} = \sqrt{\nu/S}\, \mathbf{Z}$ and the Cholesky factor $\mathbf{U}$ of $\boldsymbol{\Sigma}$.

4. Return $\mathbf{X} = \boldsymbol{\mu} + \mathbf{U}^\mathsf{T}\mathbf{Y}$.

To deal with the multivariate t distribution, we may use R functions from the package `mvtnorm`, such as `rmvt`, which has been introduced in Section 3.3.4.

5.7 Sampling from copulas

Given a copula $C(u_1, \ldots, u_n)$ and marginal CDFs $F_i(\cdot)$, $i = 1, \ldots, n$, we may generate a vector $\mathbf{X}$ of n dependent variables with those marginals by the following scheme:

1. Generate $\mathbf{U} \sim C(u_1, \ldots, u_n)$.

2. Return $\mathbf{X} = \left(F_1^{-1}(U_1), \ldots, F_n^{-1}(U_n)\right)^\mathsf{T}$.

As an instructive example, say that we want to generate a sample of $n = 10{,}000$ points (X_1, X_2), where $X_1 \sim \text{gamma}(2, 1)$ and $X_2 \sim N(0, 1)$, and the dependence is described by the Student's t copula

$$C(u_1, \ldots, u_n) = T_{\nu, \boldsymbol{\Sigma}}\left(T_\nu^{-1}(u_1), \ldots, T_\nu^{-1}(u_n)\right),$$

where we denote by $T_{\nu, \boldsymbol{\Sigma}}(\cdots)$ the CDF of a multivariate t distribution with ν degrees of freedom and by $T_\nu(\cdot)$ the CDF of the corresponding univariate distribution; note that the covariance matrix $\boldsymbol{\Sigma}$ is actually a correlation matrix, in the sense that its diagonal entries are set to 1. To sample from the copula, we need not invert the multidimensional CDF. We rely on the extension of the following property for univariate distribution:[14] If X has CDF $F_X(\cdot)$, then $F_X(X) \sim U(0, 1)$. Thus, we sample $Y \sim t_\nu(\mathbf{0}, \boldsymbol{\Sigma})$ from the multidimensional t distribution, as we have seen in the previous section, and then we compute $\mathbf{U} = (T_\nu(Y_1), \ldots, T_\nu(Y_n))^\mathsf{T}$. The vector $\mathbf{U}$ takes values in the unit hypercube and has uniform marginals, thanks to the aforementioned property;

---

[14]See Eq. 3.40.

```
library(mvtnorm)
set.seed(55555)
mu <- c(0,0)
Sigma <- matrix(c(1,0.8,0.8,1), nrow=2)
nu <- 10
n <- 10000
Y <- rmvt(n,delta=mu,sigma=Sigma,df=nu)
U <- pt(Y,df=nu)
X <- cbind(qgamma(U[,1],shape=2,scale=1), qnorm(U[,2]))
plot(X,pch=20,cex=0.5)
par(mfrow=c(2,1))
hist(X[,1],nclass=100)
hist(X[,2],nclass=100)
par(mfrow=c(1,1))
```

FIGURE 5.15 **Code to sample from a multivariate t copula with gamma and normal marginals.**

however, its components reflect the dependence structure associated with the copula. Finally, we generate **X** by inverting the marginals. The code in Fig. 5.15 accomplishes all of this when $\nu = 10$ and

$$\Sigma = \begin{bmatrix} 1 & 0.8 \\ 0.8 & 1 \end{bmatrix},$$

and it also produces the scatterplot and marginal histograms displayed in Fig. 5.16. Note the nonlinear association between the two variables, which cannot be accounted for by Pearson's correlation. We urge the reader to play with the code in order to see the effect of changing the correlation in the copula.

All of the above can also be performed using the `copula` package. With the code in Fig. 5.17 we accomplish two tasks: First we sample from a given copula; then, given the sample, we fit a copula to see if we can get back the copula. The details are as follows:

1. We use `tCopula` to create a bidimensional t copula, with degrees of freedom 10 and correlation parameter 0.8; the copula is specified as exchangeable by `dispstr="ex"`, i.e., the two dimensions behave in the same way. Note that we could also use `ellipCopula`, which generates an elliptical copula. Such a copula family includes the normal and Student's t, which can be specified by the `family` parameter.

2. We sample from the copula using `rCopula`; the output matrix of observations `U` is transformed into `X` by using the marginals we wish.

3. To go back and fit the copula, we need to transform the observations into pseudo-observations in the unit square; this may be accomplished by the

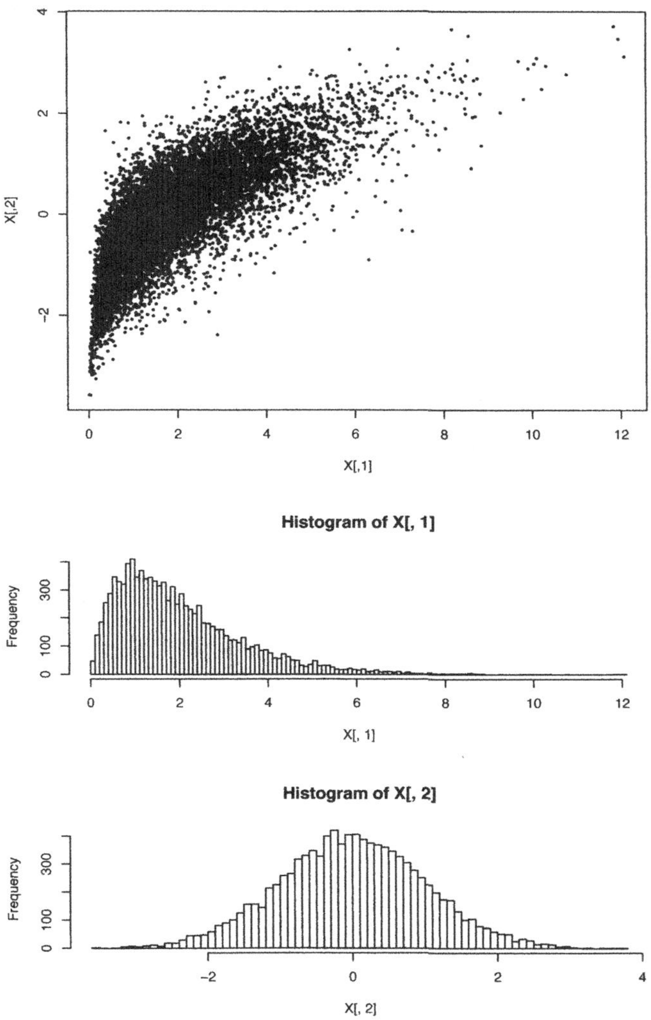

FIGURE 5.16 **Scatterplot and histograms of the marginals for a bidimensional** t **copula.**

function `pobs`. Alternatively, we might fit a marginal and use that. The observations and the pseudo-observations are plotted in Fig. 5.18.

4. Finally, using the pseudo-observations `Uhat`, we fit a bidimensional t copula.

Running the code, we obtain the following fitted copula, which is in good agreement with the one we have used to sample data:

```
library(copula)
tc <- tCopula(param=c(0.8), dim=2, dispstr="ex", df=10)
# the following statement is a possible alternative
# tc <- ellipCopula(param=c(0.8),family='t',dim=2,
#                   dispstr="ex",df=10)
set.seed(55555)
U <- rCopula(10000, tc)
X <- cbind(qgamma(U[,1],shape=2,scale=1), qnorm(U[,2]))
par(mfrow=c(1,2))
plot(X,pch=20,cex=0.5)
Uhat <- pobs(X)
plot(U,pch=20,cex=0.5)
par(mfrow=c(1,1))
fitted.copula <- fitCopula(copula=tCopula(dim = 2),data=Uhat)
```

FIGURE 5.17 **Using the package** copula **to create, sample from, and fit a** *t* **copula.**

```
> fitted.copula
fitCopula() estimation based on 'maximum pseudo-likelihood'
and a sample of size 10000.
      Estimate Std. Error z value Pr(>|z|)
rho.1 0.805305   0.005384   149.6    <2e-16 ***
df    9.161193         NA      NA        NA
---
Signif.codes: 0 '***' 0.001 '**' 0.01 '*' 0.05 '.' 0.1 ' ' 1
The maximized loglikelihood is  5277
Optimization converged
Number of loglikelihood evaluations:
function gradient
      34        11
```

For further reading

- For an extensive treatment of random number generators, see [11].
- Testing procedures for random number generators are described in [5].
- Some books on simulation include extensive chapters on random variate generation. See, e.g., [3] or, at a more advanced level, [2].
- Advanced methods for the generation of random variates are treated at a research level in [1].

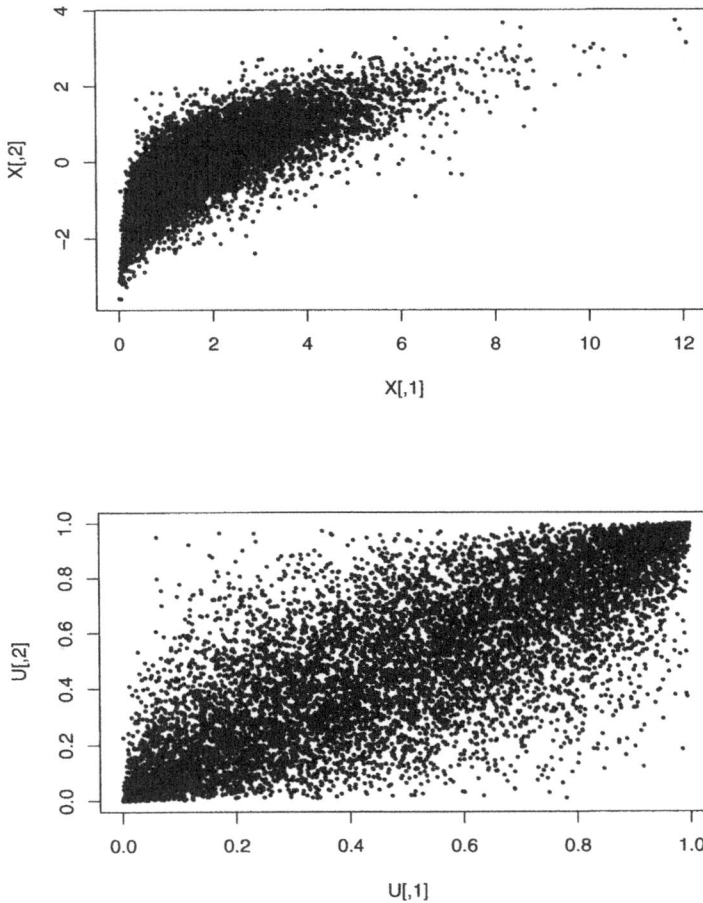

FIGURE 5.18 **Transforming observation to fit a copula.**

References

1 J.H. Ahrens and U. Dieter. Computer methods for sampling from gamma, beta, Poisson and binomial distributions. *Computing*, 12:223–246, 1974.

2 D.P. Kroese, T. Taimre, and Z.I. Botev. *Handbook of Monte Carlo Methods*. Wiley, Hoboken, NJ, 2011.

3 A.M. Law and W.D. Kelton. *Simulation Modeling and Analysis* (3rd ed.). McGraw-Hill, New York, 1999.

4 P. L'Ecuyer. Efficient and portable combined random number generators.

Communications of the ACM, 31:742–749, 1988.

5 C. Lemieux. *Monte Carlo and Quasi–Monte Carlo Sampling*. Springer, New York, 2009.

6 B.D. Ripley. The lattice structure of pseudo-random number generators. *Proceedings of the Royal Society of London; Series A, Mathematical and Physical Sciences*, 389:197–204, 1983.

7 B.D. Ripley. *Stochastic Simulation*. Wiley, New York, 1987.

8 C.P. Robert and G. Casella. *Monte Carlo Statistical Methods*. Springer, New York, 2004.

9 S. Ross. *Simulation* (2nd ed.). Academic Press, San Diego, CA, 1997.

10 S. Ross. *Introduction to Probability Models* (8th ed.). Academic Press, San Diego, CA, 2002.

11 S. Tezuka. *Uniform Random Numbers: Theory and Practice*. Kluwer Academic Publishers, Boston, 1995.

## Chapter Six

Sample Path Generation for Continuous-Time Models

Many models of interest in financial engineering can be represented by continuous-time stochastic differential equations. The application of Monte Carlo methods to this class of models requires sample path generation, which in turn requires a way to discretize them. Numerical analysis offers several ways to discretize deterministic differential equations for simulation purposes, including some very ingenious ones; here we will consider rather simple strategies, but we note that, due to the stochastic character of our applications, we have to tackle both numerical and statistical issues. To motivate the content of the chapter practically, we will mostly consider simple asset pricing problems, leaving more complex examples to Chapter 11 on option pricing, where we also apply variance reduction techniques, and Chapter 13 on risk management and hedging. The risk-neutral pricing approach that we discussed in Section 3.9 allows to price financial instruments by taking suitable expectations, which in turn requires Monte Carlo simulation, when the model is complicated enough to preclude an analytical solution. The following cases, listed in increasing order of complexity, may occur, depending on the nature of the underlying stochastic process and the kind of derivative contract we are evaluating:

1. We just need the value of a random variable at a single point in time and we are able to sample paths exactly. This is the lucky case where we may just need to sample a single random variable, and we do not need to discretize time. A relevant example is vanilla European-style call option in the Black–Scholes–Merton world.

2. We need the value of a random variable at multiple points in time, but we are able to sample paths exactly. Hence, we have to discretize time and the computational effort is increased. In some cases, we need not wonder about the most suitable time discretization, as this is written in the contract. Significant examples are a path-dependent European-style Asian option with discrete-time averaging and a Bermudan-style option, featuring early exercise opportunities but only at prespecified time instants. In both of these cases, the time instants at which we need to sample the value of the underlying asset are stated precisely in the contract. On the contrary, if we are dealing with an American-style derivative, the dis-

cretization step will have an impact on both accuracy and computational effort, as the option may be exercised anywhere before maturity.

3. We are not able to sample paths exactly. This may happen when we step out of the safe domain of geometric Brownian motion (GBM) and deal with mean reversion, stochastic volatility, non-Gaussian processes, etc. In such cases, we not only have to deal with the sampling and estimation errors that are common to all Monte Carlo applications, but with discretization issues as well.

Since here we are dealing with differential equations, the level of mathematical complexity is nontrivial, and we will take an informal view, keeping practical applications in mind. The necessary background on stochastic calculus was given in Section 3.7

We introduce the essential issues of path generation in Section 6.1, where we outline trivial and less trivial discretization strategies, like Euler and Milstein, as well as predictor-corrector methods. In Section 6.2 we investigate the easy case of geometric Brownian motion. There, we also cover the generation of sample paths for correlated processes and the Brownian bridge construction. Simple option pricing examples, concerning derivatives written on non-dividend-paying stocks, will help us along the way. We also apply these basic ideas to the simulation of rather simple interest rate models, like Vasicek and Cox–Ingersoll–Ross short-rate models in Section 6.3; such models involve mean reversion and may result or not in Gaussian processes, but they still feature a constant volatility. We move on and allow for stochastic volatility in Section 6.4. Finally, in Section 6.5 we also introduce jumps. We defer to Chapter 9 the use of low-discrepancy sequences for path generation within the framework of quasi–Monte Carlo simulation.

6.1 Issues in path generation

Let us consider the stochastic process S_t describing, e.g., the price of a non-dividend-paying stock.[1] A wide class of models can be cast into the form of an Itô stochastic differential equation:

$$dS_t = a(S_t, t)\, dt + b(S_t, t)\, dW_t, \tag{6.1}$$

where the function $a(S_t, t)$ defines the drift component, the function $b(S_t, t)$ defines the volatility component, and dW_t is the differential of a standard Wiener process. We know that the standard Wiener process is a Gaussian process with stationary and independent increments such that

$$W_{t+s} - W_t \sim \mathsf{N}(0, s).$$

---

[1] To avoid notational clutter, we will mostly, but not exclusively, use the style S_t rather than $S(t)$ in this chapter. The choice will be purely dictated by convenience.

Therefore, for a small time interval δt, we have

$$\delta W_t = W_{t+\delta t} - W_t \sim \mathsf{N}(0, \delta t),$$

which in practice means that we may simulate small increments as follows:

$$\delta W_t = \epsilon_{t+\delta t}\sqrt{\delta t}, \qquad \epsilon_{t+\delta t} \sim \mathsf{N}(0,1).$$

The notation $\epsilon_{t+\delta t}$ may be preferred to ϵ_t, since it points out clearly that the increment of the Wiener process is generated *after* observing the state at time t; however, this is not quite essential, and in the following we will often just write ϵ, since no confusion should arise. Sampling paths of a standard Wiener process with any time discretization involves no more than the generation of a stream of independent standard normals. When dealing with a simple generalized Wiener process like

$$dS_t = a\, dt + b\, dW_t, \tag{6.2}$$

where a and b are constant, we have just to integrate both sides:

$$\int_t^{t+\delta t} dS_\tau = \int_t^{t+\delta t} a\, d\tau + \int_t^{t+\delta t} b\, dW_\tau. \tag{6.3}$$

This immediately yields the following discretization scheme:

$$S_{t+\delta t} = S_t + a\, \delta t + b\epsilon_{t+\delta t}\sqrt{\delta t}, \tag{6.4}$$

which allows for easy sample path generation. When the discretization step is uniform, it may be better to introduce integer time stamps k, $k = 0, 1, 2, \ldots$, corresponding to time instants $t_k = k\, \delta t$, and rewrite Eq. (6.4) as

$$S_{k+1} = S_k + a\, \delta t + b\epsilon_{k+1}\sqrt{\delta t}, \tag{6.5}$$

where we may insist on writing ϵ_{k+1} to point out that this shock is not known when we observe S_k. We should note that if we generate sample paths of the generalized Wiener process and collect sample statistics, we are prone to sampling errors, in the sense that sample statistics will be subject to statistical fluctuations calling for the evaluation of confidence intervals. Sampling error is due to the random nature of Monte Carlo methods, and it can be mitigated using variance reduction strategies. However, there is no discretization error, in the sense that the marginal distributions of the random variables S_k, both conditional and unconditional, are correct.

By analogy, when faced with the more general Itô process of Eq. (6.1), the simplest discretization approach that comes to mind is

$$\delta S_t = S_{t+\delta t} - S_t = a(S_t, t)\delta t + b(S_t, t)\sqrt{\delta t}\, \epsilon. \tag{6.6}$$

Such a discretization is known as the Euler, or Euler–Maruyama, scheme. Note that we are assuming that functions $a(S_t, t)$ and $b(S_t, t)$ are constant over the time interval $[t, t + \delta t)$. Since this is not really the case, it is clear that we are introducing a discretization error. The issue is faced also when dealing with an ordinary deterministic differential equation, and it can be kept under control by a wide array of numerical analysis techniques. Still, it might be argued that when $\delta t \to 0$, the discretized solution should converge to the "true" solution of the stochastic differential equations. Unfortunately, convergence is a critical concept in stochastic differential equations, and a proper analysis would require a relatively deep mathematical machinery that is beyond the scope of this book.[2] Nevertheless, it is quite easy to see that discretization errors are relevant and may even change the marginal distributions. For instance, consider the geometric Brownian motion described by

$$dS_t = \mu S_t \, dt + \sigma S_t \, dW_t. \tag{6.7}$$

The Euler scheme yields

$$S_{t+\delta t} = (1 + \mu \, \delta t) S_t + \sigma S_t \sqrt{\delta t} \, \epsilon.$$

This is very easy to grasp and to implement, but the marginal distribution of each variable S_t is normal, whereas we know from Section 3.7.3 that it should be lognormal. This is not quite a negligible issue when we model stock prices or interest rates that are supposed not to get negative. In this specific case, we show in Section 6.2 that we may use Itô's lemma and get rid of the discretization error altogether, but this ideal state of the matter cannot always be achieved. When we have to cope with inexact discretization, the Euler scheme is just the simplest choice and an array of more refined discretization schemes have been proposed in the literature. Below we outline a couple of alternatives, namely, the Milstein scheme and predictor-corrector methods.

A final, and possibly less obvious, question is also worth mentioning: Time discretization may induce another form of bias. To see this, imagine pricing an American-style option, which may be exercised at any time instant before and including its expiration date. When pricing it by Monte Carlo, we are actually pricing a Bermudan-style option that can be exercised at a *discrete* set of time instants. Clearly, the value of the Bermudan-style option is a lower bound on the price of the corresponding American-style option; hence, we are introducing a low bias.[3] One possible countermeasure is to adopt Richardson extrapolation, which is a rather common approach in numerical analysis to approximate a limit when $\delta t \to 0$. Given a sequence of prices of Bermudan-style options that can be exercised with higher and higher frequencies, we may approximate a continuous-time exercise opportunity limit.

---

[2] Suitable references are provided at the end of the chapter.

[3] We investigate the sources of bias in American-style option pricing in Section 11.3.1.

6.1.1 EULER VS. MILSTEIN SCHEMES

The Milstein scheme is based on a higher order Taylor expansion of Eq. (6.1):

$$S_{t+\delta t} \approx S_t + a(S_t, t)\,\delta t + b(S_t, t)\,\delta W_t$$
$$+ \frac{1}{2}b(S_t, t)\frac{\partial b}{\partial S}(S_t, t)\left[(\delta W_t)^2 - \delta t\right]. \tag{6.8}$$

We will justify this expansion later; for now, we should note that the formal rule $(dW)^2 = dt$ shows that the last term disappears in the limit, so that we get back to the original differential equation.

Example 6.1 GBM processes by the Milstein scheme

As an immediate illustration, let us apply the scheme of Eq. (6.8) to a GBM process. Since $b(S, t) = \sigma S$ for this kind of process, we have

$$\frac{\partial b}{\partial S}(S_t, t) = \sigma,$$

and Eq. (6.8) becomes

$$S_{t+\delta t} = S_t + \mu S_t\,\delta t + \sigma S_t\,\delta W_t + \frac{1}{2}\sigma^2 S_t\left[(\delta W_t)^2 - \delta t\right].$$

Now, from a computational point of view, we rewrite $\delta W_t = \sqrt{\delta t}\,\epsilon$, which also implies
$$(\delta W_t)^2 = \delta t \cdot \epsilon^2.$$

Finally, we obtain

$$S_{t+\delta t} = S_t\left\{1 + \sigma\sqrt{\delta t}\,\epsilon + \left(\mu + \frac{1}{2}\sigma^2\left[\epsilon^2 - 1\right]\right)\delta t\right\}. \tag{6.9}$$

It is interesting to compare Euler and Milstein approximations against the Taylor expansion of the exact formula based on an exponential:

$$\frac{S_{t+\delta t}}{S_t} = \exp\left\{\left(\mu - \frac{\sigma^2}{2}\right)\delta t + \sigma\,\delta W\right\}$$
$$\approx 1 + \left\{\left(\mu - \frac{\sigma^2}{2}\right)\delta t + \sigma\,\delta W\right\}$$
$$+ \frac{1}{2}\left\{\left(\mu - \frac{\sigma^2}{2}\right)\delta t + \sigma\,\delta W\right\}^2$$
$$\approx 1 + \left\{\left(\mu - \frac{\sigma^2}{2}\right)\delta t + \sigma\,\delta W\right\} + \frac{1}{2}\left(\sigma\,\delta W\right)^2,$$

where in the last line we have neglected terms of order higher than δt. We understand that the Milstein scheme adds some missing term in the Euler approximation

$$\frac{S_{t+\delta t}}{S_t} \approx 1 + \mu\,\delta t + \sigma\,\delta W,$$

so that now the formula is correct and yields an exact expectation and variance up to order δt.

For the sake of simplicity, we provide a justification of the Milstein scheme in the scalar case, assuming that functions $a(\cdot,\cdot)$ and $b(\cdot,\cdot)$ only depend on S, and not on time t. The starting point is recalling that a stochastic differential equation is actually a shorthand for an integral equation

$$S_t = S_0 + \int_0^t a(S_\tau)\,d\tau + \int_0^t b(S_\tau)\,dW_\tau.$$

The Euler scheme approximates integrals over the interval $(t, t+\delta t)$ by "freezing" the integrand functions to their value at time t:

$$\int_t^{t+\delta t} a(S_\tau)\,d\tau \ \approx \ a(S_t)\,\delta t$$

$$\int_t^{t+\delta t} b(S_\tau)\,dW_\tau \ \approx \ b(S_t)\,[W_{t+\delta t} - W_t]. \tag{6.10}$$

We may try to improve on this by finding a better approximation of the diffusion term $b(S_t)$ over the time interval $(t, t+\delta t)$. In order to get a clue, let us consider Itô's lemma for the process $b(S_t)$ (here we do not need to denote derivatives as partial ones, since we have ruled out dependence on time):

$$\begin{aligned} db(X_t) &= b'(S_t)\,dS_t + \frac{1}{2}b''(S_t)b^2(S_t)\,dt \\ &= \left[b'(S_t)a(S_t) + \frac{1}{2}b''(S_t)b^2(S_t) \right] dt + b'(S_t)b(S_t)\,dW_t \\ &= \mu_b(S_t)\,dt + \sigma_b(S_t)\,dW_t, \end{aligned}$$

where we introduce $\mu_b(S_t)$ and $\sigma_b(X_t)$ as shorthands for the drift and volatility functions in the differential equation of $b(S_t)$. If we apply the Euler approximation to $b(S_t)$ over the time interval (t, u), we find

$$\begin{aligned} b(S_u) &\approx b(S_t) + \mu_b(S_t)[u - t] + \sigma_b(S_t)[W_u - W_t] \\ &\approx b(S_t) + b'(S_t)b(S_t)[W_u - W_t], \end{aligned}$$

where we have neglected the drift term, as it is of order $[u - t]$, whereas the increment of the Wiener process is of order $\sqrt{u - t}$. Rather than freezing the integrand in Eq. (6.10), we may substitute the last approximation:

$$\int_t^{t+\delta t} b(S_u)\, dW_u \approx \int_t^{t+\delta t} \{b(S_t) + b'(S_t)b(S_t)[W_u - W_t]\}\, dW_u$$

$$= b(S_t)[W_{t+\delta t} - W_t] + b'(S_t)b(S_t) \int_t^{t+\delta t} [W_u - W_t]\, dW_u. \quad (6.11)$$

The missing piece of the puzzle is the last integral, which we may compute by recalling Example 3.18. Applying that result we immediately find:

$$\int_t^{t+\delta t} [W_u - W_t]\, dW_u = \int_t^{t+\delta t} W_u\, dW_u - W_t \int_t^{t+\delta t} dW_u$$

$$= \frac{1}{2}(W_{t+\delta t}^2 - W_t^2) - \frac{1}{2}\delta t - W_t(W_{t+\delta t} - W_t)$$

$$= \frac{1}{2}[W_{t+\delta t} - W_t]^2 - \frac{1}{2}\delta t.$$

Plugging this into Eq. (6.11) yields

$$\int_t^{t+\delta t} b(S_u)\, dW_u \approx b(S_t)[W_{t+\delta t} - W_t] + \frac{1}{2}b'(S_t)b(S_t)\left([W_{t+\delta t} - W_t]^2 - \delta t\right).$$

Substituting this refined approximation of the diffusion term into the Euler scheme finally yields the Milstein scheme:

$$S_{t+\delta t} \approx S_t + a(S_t)\,\delta t + b(S_t)[W_{t+\delta t} - W_t] + \frac{1}{2}b'(S_t)b(S_t)\left([W_{t+\delta t} - W_t]^2 - \delta t\right).$$

Generalizations of this result are available, for which we refer to the chapter references.

6.1.2 PREDICTOR-CORRECTOR METHODS

When we apply the Euler scheme to the Itô differential equation

$$dS_t = a(S_t, t)\, dt + b(S_t, t)\, dW_t$$

over the time interval $(t, t + \delta t)$, we "freeze" the drift and volatility functions to their initial values and sample

$$S_{t+\delta t} = S_t + a(S_t, t)\,\delta t + b(S_t, t)\sqrt{\delta t}\,\epsilon.$$

The advantage of this scheme is that it is explicit, but we could try to account for the change in the functions $a(\cdot, \cdot)$ and $b(\cdot, \cdot)$ is some way. There is an array of ideas, originally developed for the numerical solution of deterministic (ordinary) differential equations, which can be adapted to the stochastic case, including implicit schemes. An implicit version of the Euler scheme could be

$$S_{t+\delta t} = S_t + a(S_{t+\delta t}, t + \delta t)\,\delta t + b(S_{t+\delta t}, t + \delta t)\sqrt{\delta t}\,\epsilon.$$

The scheme is implicit because now we have to solve an equation to find the new value of $S_{t+\delta t}$. Since this approach could be numerically troublesome, as it may require division by a Wiener increment close to zero, an alternative form is

$$S_{t+\delta t} = S_t + a(S_{t+\delta t}, t + \delta t)\,\delta t + b(S_t, t)\sqrt{\delta t}\,\epsilon, \tag{6.12}$$

where the diffusion term is kept explicit.

▣ Example 6.2 **An implicit Euler scheme**

> If we consider the stochastic differential equation
>
> $$dS_t = aS_t\,dt + b\sqrt{1 - S_t^2}\,dW_t,$$
>
> the scheme of Eq. (6.12) yields
>
> $$S_{t+\delta t} = S_t + aS_{t+\delta t} + b\sqrt{(1 - S_t^2)\delta t}\,\epsilon,$$
>
> which may be rewritten as
>
> $$S_{t+\delta t} = \frac{1}{1 - a\,\delta t}\left[S_t + b\sqrt{(1 - S_t^2)\delta t}\,\epsilon\right].$$

Needless to say, things are not always this easy, not to mention the open question of convergence. A possible alternative is to introduce a correction mechanism comparing the actual and the predicted value of the process in $t + \delta t$. For instance, the value obtained by straightforward application of the Euler scheme may be considered as a predictor

$$\hat{S}_{t+\delta t} = S_t + a(S_t, t)\,\delta t + b(S_t, t)\sqrt{\delta t}\,\epsilon.$$

Since the value of the functions $a(\hat{S}_{t+\delta t}, t + \delta t)$ and $b(\hat{S}_{t+\delta t}, t + \delta t)$ at the end of the timestep are different from what was used in the prediction step, we may introduce a correction as follows:

$$
\begin{aligned}
S_{t+\delta t} = S_t &+ \left[\alpha\,\hat{a}(\hat{S}_{t+\delta t}, t + \delta t) + (1 - \alpha)\,\hat{a}(S_t, t)\right]\delta t \\
&+ \left[\eta\,b(\hat{S}_{t+\delta t}, t + \delta t) + (1 - \eta)\,b(S_t, t)\right]\sqrt{\delta t}\,\epsilon,
\end{aligned} \tag{6.13}
$$

where α and η are weights in the interval $[0, 1]$ and

$$\hat{a}(S,t) = a(S,t) - \eta\, b(S,t) \cdot \frac{\partial b}{\partial S}(S,t). \qquad (6.14)$$

The role played by the weights α and η is rather intuitive, as they blend the initial and terminal values of functions $a(\cdot, \cdot)$ and $b(\cdot, \cdot)$, controlling the amount of correction; often, they are set close to 0.5. The further adjustment of Eq. (6.14) is less intuitive, but we refer the interested reader to [3] for further information, as well as a serious treatment of convergence issues. The aim of this short section was just to get readers acquainted with the richness of discretization approaches, but for the sake of simplicity we will refrain from using advanced path generation methods in what follows. We should also notice that, in the literature, the usefulness of some of these approaches with respect to a simpler Euler scheme with a suitably short time step has been sometimes questioned.

6.2 Simulating geometric Brownian motion

Sometimes, we may use suitable transformations of a differential equation to cast it into a more manageable form. From Section 3.7 we know that, by using Itô's lemma, we may rewrite Eq. (6.7), which defines GBM, as

$$d\log S_t = \left(\mu - \frac{1}{2}\sigma^2\right)dt + \sigma\, dW_t. \qquad (6.15)$$

Now this is a generalized Wiener process that we may easily integrate as we have seen in the previous section. Taking exponentials to get rid of the logarithm yields

$$S_t = S_0 \exp\left(\nu t + \sigma \int_0^t dW(\tau)\right),$$

where $\nu = \mu - \sigma^2/2$. Since the integral of a standard Wiener process is normally distributed, from a computational point of view we may recast the last expression as

$$S_t = S_0 e^{\nu t + \sigma \sqrt{t}\epsilon}, \quad \epsilon \sim \mathsf{N}(0, t),$$

which shows lognormality of prices. From a path generation point of view, the last expression is used as

$$S_{t+\delta t} = S_t \exp\left(\nu\, \delta t + \sigma\sqrt{\delta t}\,\epsilon\right), \qquad (6.16)$$

from which it is easy to generate sample paths of GBMs. A straightforward code is given in Fig. 6.1. Experienced R programmers will certainly notice that the function includes two dreaded nested `for` loops, which does not help efficiency; we will come back to that in a moment. The following snapshot produces the sample paths in Fig. 6.2, showing the impact of volatility:

```
# scalar Geometric Brownian Motion (loop version)
simGBM <- function (S0,mu,sigma,T,numSteps,numRepl){
  dt <- T/numSteps
  # precompute invariant quantities
  nuT <- (mu-sigma^2/2)*dt
  sigmaT <- sqrt(dt)*sigma
  # allocate matrix to store sample paths
  pathMatrix <- matrix(nrow=numRepl,ncol=numSteps+1)
  pathMatrix[,1] <- S0
  for (i in 1:numRepl)
    for (j in 2:(numSteps+1))
      pathMatrix[i,j] <- pathMatrix[i,j-1]*
                              exp(rnorm(1,nuT,sigmaT))
  return(pathMatrix)
}
```

R programming notes:

1. The function creates a matrix of sample paths, where the replications are stored row by row and columns correspond to time instants.

2. The first column contains the initial price for all sample paths.

3. The input arguments are the initial price S0, the drift mu, the volatility sigma, the time horizon T, the number of time steps numSteps, and the number of replications numRepl.

FIGURE 6.1 R code to generate sample paths of a geometric Brownian motion.

```
set.seed(55555)
paths1 <- simGBM(50,0.1,0.1,1,300,5)
paths2 <- simGBM(50,0.1,0.4,1,300,5)
yBounds <- c(min(min(paths1),min(paths2)),
              max(max(paths1),max(paths2)))
par(mfrow=c(1,2))
plot(1:301,paths1[1,],ylim=yBounds,type='l',
     xlab="sigma=10%",ylab='')
for (j in 2:5)
  lines(1:301,paths1[j,])
plot(1:301,paths2[1,],ylim=yBounds,type='l',
     xlab="sigma=40%",ylab='')
for (j in 2:5)
  lines(1:301,paths2[j,])
par(mfrow=c(1,1))
```

The code in Fig. 6.1 is based on two nested for loops, contrary to common recommendations concerning R code efficiency. In order to vectorize the code, it is convenient to rewrite Eq. (6.16) as

$$\log S_{t+\delta t} - \log S_t = \nu\,\delta t + \sigma\sqrt{\delta t}\,\epsilon.$$

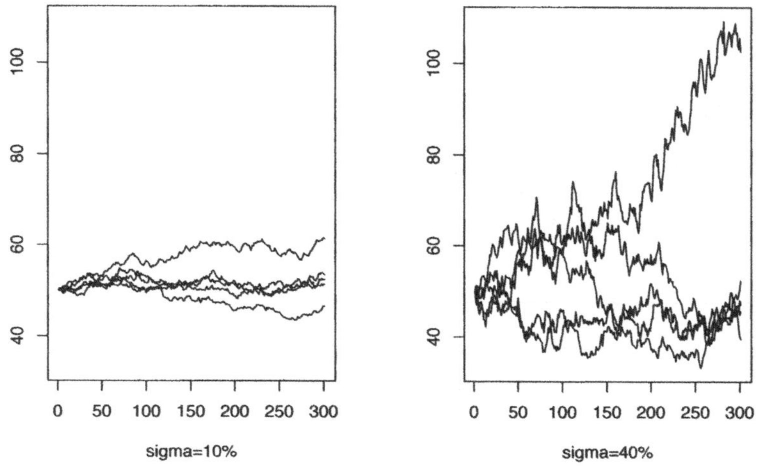

FIGURE 6.2 **Sample paths of GBMs for different values of volatility.**

We generate *differences* in the logarithm of the asset prices and then use the `cumsum` function to obtain cumulative sums of log-increments along each sample path. The resulting function `simvGBM` is illustrated in Fig. 6.3. We may compare the two implementations in terms of speed:

```
> system.time(simGBM(50,0.1,0.1,1,3000,100))
   user   system elapsed
   2.54    0.00    2.54
> system.time(simvGBM(50,0.1,0.1,1,3000,100))
   user   system elapsed
   0.15    0.00    0.16
```

6.2.1 AN APPLICATION: PRICING A VANILLA CALL OPTION

As an immediate application of path generation for GBMs, we may price a vanilla call option on a non-dividend-paying stock. Clearly, there is no point in using Monte Carlo methods in this case, as we may apply the Black–Scholes–Merton formula that we discussed in Section 3.9.2. However, it is quite useful to verify the accuracy of numerical methods in easily solved case to check their viability and possible limitations. In Fig. 6.4 we recall for convenience the function giving the exact option price, and we also propose a very simple function to estimate the price by Monte Carlo. We just need:

 1. To sample lognormal prices using Eq. (6.16).

```
# scalar Geometric Brownian Motions (vectorized version)
simvGBM <- function (S0,mu,sigma,T,numSteps,numRepl){
  dt <- T/numSteps
  nuT <- (mu-sigma^2/2)*dt
  sigmaT <- sqrt(dt)*sigma
  normMatrix <- matrix(rnorm(numSteps*numRepl,nuT,sigmaT),
                       nrow=numRepl)
  logIncrements <- cbind(log(S0),normMatrix)
  logPath <- t(apply(logIncrements,1,cumsum))
  return(exp(logPath))
}
```

FIGURE 6.3 **Vectorized code to generate sample paths of GBM.**

2. To evaluate and discount the corresponding payoff (assuming a constant risk-free rate).

3. To return the estimated price along with a confidence interval, whose confidence level `c.level` is 95% by default. Note that we return a list with fields `value ci` containing the point and the interval estimate, respectively.[4]

The code also includes a little snapshot to verify accuracy. The relevant output is

```
> EuVanillaCall(S0,K,T,r,sigma)
[1] 5.234659
> out$estimate
[1] 5.164366
> out$conf.int
[1] 5.056212 5.272520
> 100*(out$conf.int[2]-out$conf.int[1])/2/out$estimate
[1] 2.09423
```

We observe that the point estimate is not quite precise, even though the true value is included in the 95% confidence interval. Taking the ratio between the half-length of the confidence interval and the point estimate suggests an error above 2%. One crude way to improve the estimate would be to increase the number of replications. We will see in Chapter 8 that variance reduction strategies may be a much better option.

---

[4]We discuss confidence intervals and output analysis in Chapter 7.

```
# exact BSM formula
EuVanillaCall <- function(S0,K,T,r,sigma){
  d1 <- (log(S0/K)+(r+sigma^2/2)*T)/(sqrt(T)*sigma)
  d2 <- d1-sqrt(T)*sigma
  value <- S0*pnorm(d1)-K*exp(-r*T)*pnorm(d2)
  return(value)
}

# Naive MC
BlsMCNaive <- function(S0,K,T,r,sigma,numRepl,c.level=0.95){
  nuT <- (r - 0.5*sigma^2)*T
  sigmaT <- sigma*sqrt(T)
  ST <- S0*exp(rnorm(numRepl, nuT, sigmaT))
  dpayoff <- exp(-r*T)*pmax(0,ST-K)
  aux <- t.test(dpayoff,conf.level=c.level)
  value <- as.numeric(aux$estimate)
  ci <- as.numeric(aux$conf.int)
  return(list(estimate=value,conf.int=ci))
}

S0 <- 50
K <- 60
T <- 1
r <- 0.04
sigma <- 0.4
numRepl <- 50000
EuVanillaCall(S0,K,T,r,sigma)
set.seed(55555)
out <- BlsMCNaive(S0,K,T,r,sigma,numRepl)
out$estimate
out$conf.int
```

FIGURE 6.4 **Exact formula and a simple Monte Carlo pricer for a vanilla call option.**

6.2.2 MULTIDIMENSIONAL GBM

When dealing with options on multiple assets or managing multiple risk factors, we may need to generate a multidimensional geometric Brownian motion. The key issue is how to deal with correlation among risk factors, and there are two modeling choices, which have an impact on how simulation is carried out. For the sake of simplicity we illustrate the two approaches in the bidimensional case, where two Wiener processes drive prices $S_1(t)$ and $S_2(t)$. We may consider two independent standard Wiener processes that enter in both equations, thus inducing a correlation between prices:

$$dS_1(t) = \mu_1 S_1(t)\,dt + \sigma_{11} S_1(t)\,dW_1(t) + \sigma_{12} S_1(t)\,dW_2(t),$$
$$dS_2(t) = \mu_2 S_2(t)\,dt + \sigma_{21} S_2(t)\,dW_1(t) + \sigma_{22} S_2(t)\,dW_2(t).$$

An alternative is to consider instantaneously correlated Wiener processes and write the equations in the somewhat more familiar form:

$$dS_1(t) = \mu_1 S_1(t)\,dt + \sigma_1 S_1(t)\,dW_1(t),$$
$$dS_2(t) = \mu_2 S_2(t)\,dt + \sigma_2 S_2(t)\,dW_2(t).$$

The role of instantaneous correlation ρ can be understood by referring to the following formal rule of stochastic calculus:

$$dW_1 \cdot dW_2 = \rho\,dt.$$

We note that the analogous rule $(dW)^2 = dt$ is actually a specific case of this rule, which may be justified heuristically along the lines of Section 3.7.3. We will take the second approach, which requires the generation of correlated standard normals. We know from Section 5.5.2 that this may be accomplished by taking the Cholesky factorization of the covariance matrix, which in this case is actually a correlation matrix, since we are dealing with standard normals. We illustrate the idea first in the bidimensional case, to reinforce intuition. In fact, when only two processes are involved, we may generate increments based on the correlated standard normal variables ϵ_1 and ϵ_2 obtained as a transformation of independent variables $Z_1, Z_2 \sim \mathsf{N}(0,1)$:

$$\epsilon_1 = Z_1,$$
$$\epsilon_2 = \rho Z_1 + \sqrt{1 - \rho^2} Z_2.$$

The simple code in Fig. 6.5 generates a bidimensional sample path using the above transformation. The following R snapshot produces the plots in Fig. 6.6, which illustrate the impact of correlation:

```
> set.seed(55555)
> S0 <- c(30,50)
> mus <- c(0.1,0.1)
> sigmas <- c(0.3,0.3)
> paths1 <- sim2GBM(S0,mus,sigmas,0.3,1,300)
> paths2 <- sim2GBM(S0,mus,sigmas,0.9,1,300)
> yBounds <- c(min(min(paths1),min(paths2)),
+              max(max(paths1),max(paths2)))
> par(mfrow=c(1,2))
> plot(1:301,paths1[1,],ylim=yBounds,type='l',
       xlab="rho=0.4",ylab='')
> lines(1:301,paths1[2,])
> plot(1:301,paths2[1,],ylim=yBounds,type='l',
       xlab="rho=0.9",ylab='')
> lines(1:301,paths2[2,])
> par(mfrow=c(1,1))
```

```
# bidimensional GBM: receives drifts, volatilities,
# instantaneous correlation and number of steps
# naive implementation
sim2GBM <- function(S0,mus,sigmas,rho,T,numSteps){
  dt <- T/numSteps
  nuTs <- (mus-sigmas^2/2)*dt
  sigmaTs <- sqrt(dt)*sigmas
  pathMatrix <- matrix(nrow=2,ncol=numSteps+1)
  pathMatrix[,1] <- S0
  sqrtRho <- sqrt(1-rho^2)
  for (j in 2:(numSteps+1)){
    eps1 <- rnorm(1)
    pathMatrix[1,j] <- pathMatrix[1,j-1]*
                       exp(nuTs[1]+sigmaTs[1]*eps1)
    eps2 <- rho*eps1+sqrtRho*rnorm(1)
    pathMatrix[2,j] <- pathMatrix[2,j-1]*
                       exp(nuTs[2]+sigmaTs[2]*eps2)
  }
  return(pathMatrix)
}
```

FIGURE 6.5 **Simple R code to generate sample paths of bidimensional GBM.**

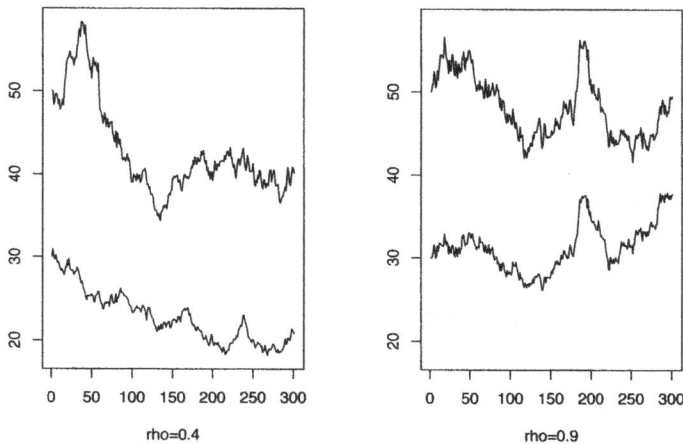

rho=0.4   rho=0.9

FIGURE 6.6 **Sample paths of a bidimensional GBM for different values of instantaneous correlation.**

```
# multidimensional GBM: input vector of drifts, vector of
# volatilities, and matrix of instantaneous correlations
simMVGBM <- function(S0Vet,muVet,sigmaVet,corrMatrix,T,
                     numSteps,numRepl){
  # precompute invariant quantities
  dt <- T/numSteps
  nuT <- (muVet-sigmaVet^2/2)*dt
  sigmaT <- sqrt(dt)*sigmaVet
  # find Cholesky factor: transposition is necessary
  # to obtain the lower triangular factor
  lowerCholFactor <- t(chol(corrMatrix))
  numFactors <- length(muVet)
  # prepare three-dimensional array
  pathArray <- array(dim=c(numRepl,numSteps+1,numFactors))
  for (k in 1:numFactors) pathArray[,1,k] <- S0Vet[k]
  for (i in 1:numRepl){
    for (j in 2:(numSteps+1)){
      # generate independent standard normals
      # and then correlate them
      zVet <- rnorm(numFactors)
      epsVet <- lowerCholFactor %*% zVet
      # new sample vector
      pathArray[i,j,] <- pathArray[i,j-1,]*
                              exp(nuT+sigmaT*epsVet)
    }
  }
  return(pathArray)
}
```

FIGURE 6.7 **Generating sample paths of multidimensional GBMs.**

The previous implementation is rather simple as it involves a single replication. The code in Fig. 6.7 is a bit more involved since, to store multiple replications, we need a tridimensional array; this is created by the function `array`, which requires a vector specifying the array size along each dimension.

6.2.3 THE BROWNIAN BRIDGE

In the previous sections we have generated asset paths according to a natural scheme, which proceeds forward in time. Actually, the Wiener process enjoys some peculiar properties which allow us to generate the sample paths in a different way. Consider a time interval with left and right end points t_l and t_r, respectively, and an intermediate time instant s, such that $t_l < s < t_r$. In standard path generation, we would generate the Wiener process in the natural order: $W(t_l)$, $W(s)$, and finally $W(t_r)$. Using the so-called Brownian bridge, we may generate $W(s)$ as the last value, conditional on the values $W(t_l) = w_l$ and $W(t_r) = w_r$. The conditional value of $W(s)$ is a normal variable with expected value

$$\frac{(t_r - s)w_l + (s - t_l)w_r}{t_r - t_l} \tag{6.17}$$

and variance

$$\frac{(t_r - s)(s - t_l)}{t_r - t_l}. \tag{6.18}$$

This is a consequence of the properties of the conditional distribution of a multivariate normal distribution, which we have stated in Theorem 3.7. Before proving Eqs. (6.17) and (6.18), it is important to realize that they are quite intuitive: The conditional expected value of $W(s)$ is obtained by linear interpolation through w_l and w_r; the variance is low near the two end points t_l and t_r and is at its maximum in the middle of the interval.

A preliminary step to prove Eqs. (6.17) and (6.18) is to find the covariance between $W(t)$ and $W(u)$, when $t < u$:

$$\begin{aligned}
\mathrm{Cov}\big(W(t), W(u)\big) &= \mathrm{Cov}\big(W(t), W(t) + W(u) - W(t)\big) \\
&= \mathrm{Cov}\big(W(t), W(t)\big) + \mathrm{Cov}\big(W(t), W(u) - W(t)\big) \\
&= t + 0 = t.
\end{aligned}$$

The statement can be generalized as $\mathrm{Cov}\big(W(t), W(u)\big) = \min\{t, u\}$. Then, the joint distribution of the three values that we are considering is

$$\begin{bmatrix} W(s) \\ W(t_l) \\ W(t_r) \end{bmatrix} \sim \mathsf{N}\left(\mathbf{0}, \begin{bmatrix} s & t_l & s \\ t_l & t_l & t_l \\ s & t_l & t_r \end{bmatrix}\right).$$

Note that we have listed the time instant in view of the application of Theorem 3.7, since we need the value in s conditional on the values in t_l and t_r. Partitioning vector of expected values and the covariance matrix accordingly yields

$$\mathrm{E}\Big[W(s) \mid W(t_r) = w_r, W(t_l) = w_l\Big] = 0 + \begin{bmatrix} t_l & s \end{bmatrix} \begin{bmatrix} t_l & t_l \\ t_l & t_r \end{bmatrix}^{-1} \begin{bmatrix} w_l \\ w_r \end{bmatrix}$$

and

$$\mathrm{Var}\Big[W(s) \mid W(t_r) = w_r, W(t_l) = w_l\Big] = s - \begin{bmatrix} t_l & s \end{bmatrix} \begin{bmatrix} t_l & t_l \\ t_l & t_r \end{bmatrix}^{-1} \begin{bmatrix} t_l \\ s \end{bmatrix}.$$

By plugging the inverse matrix

$$\begin{bmatrix} t_l & t_l \\ t_l & t_r \end{bmatrix}^{-1} = \frac{1}{t_r - t_l} \begin{bmatrix} t_r/t_l & -1 \\ -1 & 1 \end{bmatrix}$$

into the above formulas we obtain Eqs. (6.17) and (6.18).

By using the Brownian bridge, we may generate sample paths by a sort of bisection strategy. Given $W(0) = 0$, we sample $W(T)$; then we sample $W(T/2)$. Given $W(0)$ and $W(T/2)$ we sample $W(T/4)$; given $W(T/2)$ and $W(T)$ we sample $W(3T/4)$, etc. Actually, we may generate sample paths in any order we wish, with nonhomogeneous time steps. One could wonder why this complicated construction could be useful. There are at least two reasons:

1. It may help in using variance reduction by stratification, to be discussed in Chapter 8. It is difficult to use stratification in multiple dimensions, but we may use stratification just on the terminal value of the asset price, and maybe an intermediate point. Then we generate the remaining values along the sample path by using the bridge.

2. The Brownian bridge construction is also useful when used in conjunction with low-discrepancy sequences, as we shall see in Chapter 9. Such sequences may not work well in very high dimensional domains, but using Brownian bridge we may use high-quality sequences to outline the paths of the Wiener process, by sampling points acting as milestones; then we can fill the trajectory by Monte Carlo sampling.

In Fig. 6.8 we illustrate an R function to generate sample paths of the standard Wiener process using the Brownian bridge technique, but only in the specific case in which the time interval $[0, T]$ is successively bisected (i.e., the number of intervals is a power of 2). In this case, we may simplify the formulas above to sample $W(s)$: If we let $\delta t = t_l - t_r$, we obtain

$$W(S) = \frac{w_r + w_l}{2} + \frac{1}{2}\sqrt{\delta t}\,\epsilon,$$

where ϵ is a standard normal variable. The function receives the length T of the time interval and the number numSteps of subintervals in which it must be partitioned, and it returns a vector containing one sample path. Assume that the number of intervals is 8 (a power of 2). Then we must carry out 3 bisections.

- Given the initial condition $W(t_0) = 0$, we must first sample $W(t_8)$, which means "jumping" over an interval of length T, which is the initial value TJump in the program. Since we store elements in a vector of nine elements (starting with index 1, and including $W(t_0)$), we must jump eight places in this vector to store the new value. The number of places to jump is stored in IJump.

- Then we start the outermost for loop. In the first pass we must only sample $W(t_4)$, given $W(t_0)$ and $W(t_8)$. Given positions left<-1 and right<-IJump+1, we must generate a new value and store it in position i<-IJump/2+1, which is 4+1 = 5 in this case. Here we generate only one value, and we divide both time and index jumps by 2.

- In the second iteration we must sample $W(t_2)$, given $W(t_0)$ and $W(t_4)$, and $W(t_6)$, given $W(t_4)$ and $W(t_8)$. The nested loop will be executed twice, and indexes left, right, and i are incremented by 4.

```
WienerBridge <- function(T, numSteps){
  numBisections <- log2(numSteps)
  if (round(numBisections) != numBisections){
    cat('ERROR: numSteps must be a power of 2\n')
    return(NaN)
  }
  path <- numeric(numSteps+1)
  path[1] <- 0
  path[numSteps+1] <- sqrt(T)*rnorm(1)
  TJump <- T
  IJump <- numSteps
  for (k in 1:numBisections) {
    left <- 1
    i <- IJump/2 + 1
    right <- IJump + 1
    for (j in 1:2^(k-1)){
      a <- 0.5*(path[left] + path[right])
      b <- 0.5*sqrt(TJump)
      path[i] <- a + b*rnorm(1)
      right <- right + IJump
      left <- left + IJump
      i <- i + IJump
    }
    IJump <- IJump/2
    TJump <- TJump/2
  }
  return(path)
}
```

FIGURE 6.8 **Implementing path generation by a Brownian bridge for the standard Wiener process.**

- In the third and final iteration we generate the remaining four values.

The construction above is illustrated in Fig. 6.9. Figure 6.10 includes a script to check that the marginal distributions of the stochastic process that we generate are the correct ones. Expected values should be zero, and standard deviation should be the square root of time:

```
> m <- colMeans(Wpaths[,-1]);m
[1] 5.234951e-05 1.624087e-03 3.127389e-03 1.542460e-03
> sdev <- apply(Wpaths[,-1],2,sd);sdev
[1] 0.4993618 0.7080760 0.8653553 0.9971483
> sqrt((1:numSteps)*T/numSteps)
[1] 0.5000000 0.7071068 0.8660254 1.0000000
```

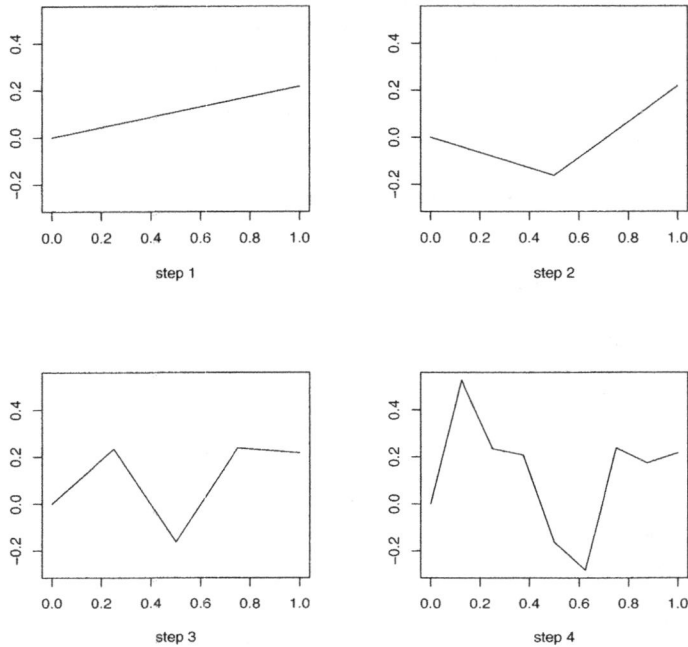

FIGURE 6.9 **Illustrating the Brownian bridge construction by successive bisection.**

We see that, within sampling errors, the result looks correct. Given a way to generate the standard Wiener process, it is easy to simulate geometric Brownian motion. The function is given in Fig. 6.11 and uses a similar approach as the vectorized version `simvGBM` of Fig. 6.3. One thing we should note is the use of the function `diff` to generate the vector of `Increments` in the logarithmic asset price. In fact, in standard Monte Carlo we generate the underlying Wiener process by successive increments; with the Brownian bridge construction we build directly the values of the process at different time instants, and we must use the function `diff` to obtain the relative differences. In some sense, `diff` works in the opposite way to `cumsum`.

6.3   Sample paths of short-term interest rates

Modeling interest rates is a quite difficult endeavor, as it involves different rates applying to different maturities. The simplest class of models deals with only one interest rate applying to a very short time horizon. Let r_t be the risk-free interest rate applying to the time interval $(t, t + dt)$. This may be called the

```
set.seed(55555)
numRepl <- 100000
T <- 1
numSteps <- 4
Wpaths <- matrix(0, nrow=numRepl, ncol=1+numSteps)
for (i in 1:numRepl)
  Wpaths[i,] <- WienerBridge(T, numSteps)
m <- colMeans(Wpaths[,-1]);m
sdev <- apply(Wpaths[,-1],2,sd);sdev
sqrt((1:numSteps)*T/numSteps)
```

FIGURE 6.10 Checking the path generation for the standard Wiener process by a Brownian bridge.

```
GBMBridge <- function(S0, mu, sigma, T, numSteps, numRepl){
  numBisections <- log2(numSteps)
  if (round(numBisections) != numBisections){
    cat('ERROR: numSteps must be a power of 2\n')
    return(NaN)
  }
  dt <- T/numSteps
  nudt <- (mu-0.5*sigma^2)*dt
  Paths <- matrix(0, nrow=numRepl, ncol=numSteps+1)
  for (k in 1:numRepl){
    W <- WienerBridge(T,numSteps)
    Increments <- nudt + sigma*diff(W)
    LogPath <- cumsum(c(log(S0) , Increments))
    Paths[k,] <- exp(LogPath)
  }
  Paths[,1] <- S0
  return(Paths)
}
```

FIGURE 6.11 Simulating geometric Brownian motion by the Brownian bridge.

instantaneous interest rate, although it is often referred to as the *short rate*. There are different models for the short rate, but in this section we consider only the two that we have already mentioned in previous chapters:

1. The Vasicek model, characterized by a stochastic differential equation featuring mean reversion:

$$dr_t = \gamma(\bar{r} - r_t)\,dt + \sigma\,dW_t. \qquad (6.19)$$

2. The Cox–Ingersoll–Ross (CIR) model, which is quite similar to the Vasicek model, but involves a slight change in the volatility term:

$$dr_t = \gamma(\bar{r} - r_t)\, dt + \sqrt{\alpha r_t}\, dW_t. \tag{6.20}$$

In the following, we show that this slight difference has a significant consequence:

- Short rates are normally distributed in the Vasicek model; the CIR model involves a more complicated noncentral chi-square distribution (see Section 3.2.3.2).
- The easier distribution of the Vasicek model results in better analytical tractability; we are able to price bonds and also some options analytically, whereas the CIR model involves more complicated formulas, when available.
- However, the Vasicek model can be criticized as the normal distribution allows for negative interest rates, whereas the volatility term in the CIR models avoids this difficulty, provided that

$$\gamma \cdot \bar{r} > \frac{1}{2}\alpha.$$

The intuition is that when the short rate r_t gets close to zero, the volatility term in Eq. (6.20) vanishes; if the pull toward the long-term average rate is strong enough, the short rate will stay positive.

In both cases, generating sample paths is not too difficult, as we shall see, since we know the exact distribution of interest rates. More involved models call for more sophisticated path generation strategies, possibly involving discretization errors.

6.3.1 THE VASICEK SHORT-RATE MODEL

The process described in Eq. (6.19) is not a GBM; actually, it is an example of an Ornstein–Uhlenbeck process. To try and solve the equation, we may resort to a typical trick of the trade, related to the method of integrating factors in deterministic ordinary differential equations. The trick is to apply Itò's lemma to the process

$$f(r_t, t) = r_t e^{\gamma t}.$$

The usual drill requires the calculation of partial derivatives:

$$\frac{\partial f}{\partial t} = \gamma r_t e^{\gamma t}, \quad \frac{\partial f}{\partial r_t} = e^{\gamma t}, \quad \frac{\partial^2 f}{\partial r_t^2} = 0.$$

Hence, the differential equation for $f(r_t, t)$ is

$$df = \left[\gamma r_t e^{\gamma t} + \gamma(\bar{r} - r_t)e^{\gamma t}\right] dt + \sigma e^{\gamma t}\, dW_t = \gamma \bar{r} e^{\gamma t}\, dt + \sigma e^{\gamma t}\, dW_t.$$

This equation does not look too bad and can be integrated over the interval $(0, t)$:

$$r_t e^{\gamma t} - r_0 = \gamma \bar{r} \int_0^t e^{\gamma \tau} \, d\tau + \sigma \int_0^t e^{\gamma \tau} \, dW_\tau,$$

which can be rewritten as

$$r_t = r_0 e^{-\gamma t} + \bar{r}(1 - e^{-\gamma t}) + \sigma \int_0^t e^{\gamma(\tau - t)} \, dW_\tau.$$

This expression includes a stochastic integral involving the Wiener process that, as we know, is normally distributed and has zero expected value. Thus, we may immediately conclude that the short rate r_t under the Vasicek model is normally distributed with expected value:[5]

$$\mathrm{E}[r_t] = r_0 e^{-\gamma t} + \bar{r}(1 - e^{-\gamma t}). \tag{6.21}$$

Finding variance is a bit trickier and requires the following result.

THEOREM 6.1 (Itô's isometry) *Let X_t be a stochastic process adapted to the standard Wiener process W_t. Then,*

$$\mathrm{E}\left[\left(\int_0^t X_\tau \, dW_\tau\right)^2\right] = \mathrm{E}\left[\int_0^t X_\tau^2 \, d\tau\right].$$

Actually, the result can be proven in a more general form, but it allows us to find the variance of the short rate:

$$\mathrm{Var}(r_t) \equiv \mathrm{E}\left[(r_t - \mathrm{E}[r_t])^2\right] = \mathrm{E}\left[\left(\sigma \int_0^t e^{\gamma(\tau - t)} \, dW_\tau\right)^2\right]$$

$$= \mathrm{E}\left[\sigma^2 \int_0^t e^{2\gamma(\tau - t)} \, d\tau\right] = \frac{\sigma^2}{2\gamma}\left(1 - e^{-2\gamma t}\right). \tag{6.22}$$

It is important to get an intuitive feeling for Eqs. (6.21) and (6.22):

- At time $t = 0$ the short rate is r_0 and variance is zero.
- When $t \to +\infty$, the short rate tends to the long-term average value $\bar{r}$; the larger the speed of mean reversion γ, the faster the convergence.
- Long-term variance is given by $\sigma^2/(2\gamma)$; as expected, it is influenced by the volatility σ; however, also γ plays a role, and a strong mean reversion tends to kill volatility.

In order to use the above knowledge to generate sample paths with time step δt, we should express $r_{t+\delta t}$ conditional on r_t:

$$r_{t+\delta t} = r_t e^{-\gamma \delta t} + \bar{r}(1 - e^{-\gamma \delta t}) + \epsilon \cdot \sigma \sqrt{\frac{1 - e^{-2\gamma t}}{2\gamma}},$$

---

[5]By the way, we can find this expectation by setting $\sigma = 0$ and solving the resulting deterministic equation.

where $\epsilon \sim \mathsf{N}(0, 1)$.

Now we may generate sample paths by the code illustrated in Fig. 6.12. The code produces the plots in Fig. 6.13. We clearly see the impact of the mean reversion coefficient γ. In the plots for $\gamma = 0.1$ we also notice how short rates can indeed get negative in the Vasicek model.

6.3.2   THE COX–INGERSOLL–ROSS SHORT-RATE MODEL

The CIR model relies on a square-root diffusion process, which is not as easy to deal with as the Ornstein–Uhlenbeck process underlying the Vasicek model. However, by studying the transition density of the square-root diffusion process, a link with the noncentral chi-square distribution[6] can be established. We state the following result without proof.

THEOREM 6.2 (CIR short-rate model) *The transition law from r_0 to r_t for the CIR model can be expressed as*

$$r_t = \frac{\alpha(1 - e^{-\gamma t})}{4\gamma} \chi_\nu'^2(\lambda), \qquad (6.23)$$

where the degrees of freedom are $\nu = 4\bar{r}\gamma/\alpha$ and the noncentrality parameter is

$$\lambda = \frac{4\gamma e^{-\gamma t}}{\alpha(1 - e^{-\gamma t})} r_0.$$

By applying the theorem to the transition from r_t to $r_{t+\delta t}$, we immediately obtain a path generation mechanism. It is also useful to find the conditional expectation and variance of r_t given r_0, which is easily obtained by recalling Eq. (3.18). For the expected value we have

$$
\begin{aligned}
\mathrm{E}\big[r_t \mid r_0\big] &= \frac{\alpha(1 - e^{-\gamma t})}{4\gamma}(\nu + \lambda) \\
&= \frac{\alpha(1 - e^{-\gamma t})}{4\gamma}\left[\frac{4\bar{r}\gamma}{\alpha} + \frac{4\gamma e^{-\gamma t}}{\alpha(1 - e^{-\gamma t})} r_0\right] \\
&= r_0 e^{-\gamma t} + \bar{r}(1 - e^{-\gamma t}).
\end{aligned}
\qquad (6.24)
$$

We notice that the expected value has the same intuitive form as in the Vasicek model. Variance is a bit trickier:

$$
\begin{aligned}
\mathrm{Var}\big(r_t \mid r_0\big) &= \left[\frac{\alpha(1 - e^{-\gamma t})}{4\gamma}\right]^2 \cdot 2(\nu + 2\lambda) \\
&= \left[\frac{\alpha(1 - e^{-\gamma t})}{4\gamma}\right]^2 \cdot 2\left[\frac{4\bar{r}\gamma}{\alpha} + 2\frac{4\gamma e^{-\gamma t}}{\alpha(1 - e^{-\gamma t})} r_0\right] \\
&= \frac{r_0 \alpha}{\gamma}\left(e^{-\gamma t} - e^{-2\gamma t}\right) + \frac{\bar{r}\alpha}{2\gamma}\left(1 - e^{-\gamma t}\right)^2.
\end{aligned}
\qquad (6.25)
$$

---

[6]See Section 3.2.3.2.

```r
simVasicek<-function(r0,rbar,gamma,sigma,T,numSteps,numRepl){
  dt <- T/numSteps
  # precompute time invariant quantities
  expg <- exp(-gamma*dt)
  auxc <- rbar*(1-expg)
  vol <- sigma*sqrt((1-expg^2)/(2*gamma))
  paths <- matrix(0,nrow=numRepl,ncol=numSteps+1)
  paths[,1] <- r0
  epsMat <- matrix(rnorm(numRepl*numSteps),nrow=numRepl)
  for (i in 1:numRepl){
    for (t in 1:numSteps)
      paths[i,t+1] <- paths[i,t]*expg+auxc+vol*epsMat[i,t]
  }
  return(paths)
}

# sample and plot
r0 <- 0.02; rbar <- 0.05; gamma <- 0.1
sigma <- 0.015; T <- 1
numSteps <- 250; numRepl <- 40
set.seed(55555)
par(mfrow=c(2,1))

paths <- simVasicek(r0,rbar,gamma,sigma,T,numSteps,numRepl)
plot(paths[1,], ylim=range(paths,rbar), type='l',
     col='mediumorchid2',ylab='',xlab=paste('gamma=',gamma))
for(i in 2:numRepl)
  lines(paths[i,], col=colors()[floor(runif(1,1,657))] )
abline( h=rbar, col='black',lwd=2)
text(0, rbar+0.005, paste("long-term rate =",rbar),adj=0)

gamma <- 5
set.seed(55555)
paths <- simVasicek(r0,rbar,gamma,sigma,T,numSteps,numRepl)
plot(paths[1,], ylim=range(paths,rbar), type='l',
     col='mediumorchid2',ylab='',xlab=paste('gamma=',gamma))
for(i in 2:numRepl)
  lines(paths[i,], col=colors()[floor(runif(1,1,657))] )
abline( h=rbar, col='black',lwd=2)
text(0, rbar+0.005, paste("long-term rate =",rbar),adj=0)

par(mfrow=c(1,1))
```

FIGURE 6.12 **Simulating short rates by the Vasicek model.**

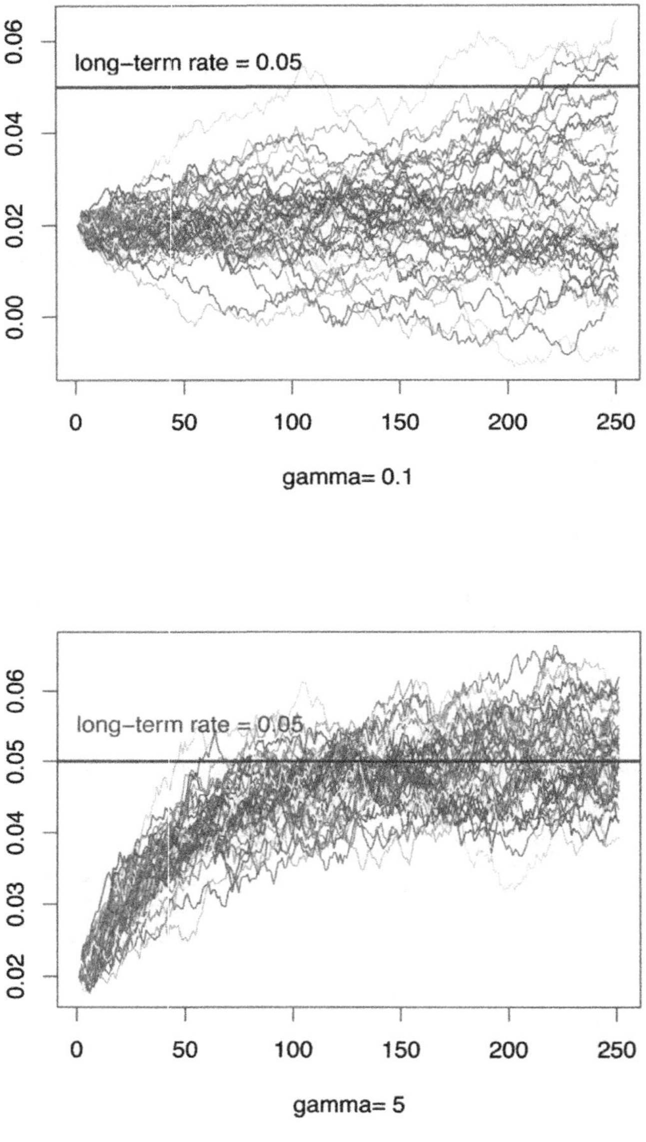

FIGURE 6.13 **Illustrating the effect of mean reversion on short rates in the Vasicek model.**

On the basis of Theorem 6.2 it is easy to build an R function to generate sample paths for the CIR short-rate model. It is also interesting to compare the exact sample paths against those generated by the straightforward Euler discretization

$$r_{t+\delta t} = \gamma \bar{r} \, \delta t + (1 - \gamma \, \delta t) \, r_t + \sqrt{\alpha r_t \, \delta t} \, \epsilon, \qquad (6.26)$$

where $\epsilon \sim \mathsf{N}(0,1)$ as usual. The code displayed in Fig. 6.14 includes two such functions. Figure 6.15 shows a script to compare the exact distribution against a one-step Euler approximation. For the cases $T = 0.25$ and $T = 1$ we sample the rates obtained in one step, with the following parameters:[7]

$$\bar{r} = 0.05, \quad r_0 = 0.04, \quad \gamma = 0.2, \quad \alpha = 0.01.$$

Note that this setting of α corresponds to a volatility coefficient $\sigma = 0.1$ (see Eq. (6.20)). In Fig. 6.16 we compare the kernel densities estimated by sampling 100,000 replications; the densities estimated on the basis of the Euler scheme are drawn with dashed lines. We observe that for $T = 0.25$ there is a close match, but for $T = 1$ the Euler rates, which are in fact normally distributed, may get negative.

6.4 Dealing with stochastic volatility

In geometric Brownian motion models, the volatility σ is constant. On the contrary, an empirical finding is that volatility is stochastic, and clusters of high vs. low volatility can be observed over time. In discrete time, this may be modeled along the lines of GARCH time series models that we have seen in Chapter 3. We may also consider continuous-time models, and one possible approach relies on the square-root diffusion process that we have introduced in Section 6.3.2 for short-term interest rates. One such model is the Heston model:

$$\frac{dS_t}{S_t} = \mu \, dt + \sqrt{V_t} \, dW_t^a, \qquad (6.27)$$

$$dV_t = \alpha(b - V_t) \, dt + \sigma \sqrt{V_t} \, dW_t^b. \qquad (6.28)$$

Equation (6.27) is basically a generalization of GBM, and Eq. (6.28) models variance V_t as a mean-reverting square-root diffusion. An easy way to generate sample paths for this process relies on a simple Euler discretization; we have already seen that this may result in negative values for quantities that are non-negative by definition. The more advanced path generation strategies that we have discussed earlier in the chapter may be exploited to remedy the situation. We defer details to Chapter 11, where we will see how to price a path-dependent, lookback option under stochastic volatility.

[7]We are replicating the results discussed in [1, pp. 124–125].

```
# exact approach
simCIR <- function(r0,rbar,gamma,alpha,T,
                   numSteps,numRepl){
  dt <- T/numSteps
  # precompute time invariant quantities
  expg <- alpha*(1-exp(-gamma*dt))/(4*gamma)
  nu <- 4*rbar*gamma/alpha
  lambdaC <- exp(-gamma*dt)/expg
  paths <- matrix(0,nrow=numRepl,ncol=numSteps+1)
  paths[,1] <- r0
  for (i in 1:numRepl){
    for (t in 1:numSteps){
      chi <- rchisq(1,df=nu,ncp=lambdaC*paths[i,t])
      paths[i,t+1] <- expg*chi}}
  return(paths)
}

# Euler approximation
simEulerCIR <- function(r0,rbar,gamma,alpha,T,
                        numSteps,numRepl){
  dt <- T/numSteps
  # precompute time invariant quantities
  aux1 <- gamma*rbar*dt
  aux2 <- 1-gamma*dt
  aux3 <- sqrt(alpha*dt)
  paths <- matrix(0,nrow=numRepl,ncol=numSteps+1)
  paths[,1] <- r0
  epsMat <- matrix(rnorm(numRepl*numSteps),nrow=numRepl)
  for (i in 1:numRepl){
    for (t in 1:numSteps){
      paths[i,t+1] <- aux1+aux2*paths[i,t]+
        sqrt(paths[i,t])*aux3*epsMat[i,t]}}
  return(paths)
}
```

FIGURE 6.14 **Simulating short rates by the CIR model: exact approach and Euler approximation.**

6.5 Dealing with jumps

Diffusion processes are continuous (with probability 1), and they are often criticized for this very reason, as sudden jumps are indeed observed in financial markets. A simple approach to incorporate jumps is to add the increments of a compound Poisson process, which is characterized by a rate λ and a distribution of the jump size. In practice, this means that jump epochs are separated by a stream of independent exponential variables with rate λ, and that at each such

```
# get exact density and plot against estimate
# based on Euler scheme (using kernel estimates)
rbar <- 0.05
r0 <- 0.04
gamma <- 0.2
alpha <- 0.1^2
par(mfrow=c(2,1))
# case T=0.25
set.seed(55555)
exact1 <- simCIR(r0,rbar,gamma,alpha,0.25,1,100000)
euler1 <- simEulerCIR(r0,rbar,gamma,alpha,0.25,1,100000)
plot(density(exact1[,2]),ylab='',xlab='rates',
     main='T=0.25',xlim=range(exact1[,2],euler1[,2]))
lines(density(euler1[,2]),lty='dashed')
# case T=1
set.seed(55555)
exact2 <- simCIR(r0,rbar,gamma,alpha,1,1,100000)
euler2 <- simEulerCIR(r0,rbar,gamma,alpha,1,1,100000)
plot(density(exact2[,2]),ylab='',xlab='rates',
     main='T=1',xlim=range(exact2[,2],euler2[,2]))
lines(density(euler2[,2]),lty='dashed')
par(mfrow=c(1,1))
```

FIGURE 6.15 **Comparing kernel estimates of CIR interest rates by an exact approach and the Euler approximation.**

epoch we have to add a random jump size to the increment of the diffusion component. For instance, we may consider a generalized Wiener process, described by the differential equation

$$dX_t = \beta\,dt + \sigma\,dW_t,$$

which can be easily solved:

$$X_t = X_0 + \beta t + \sigma W_t.$$

How can we add normally distributed jumps $N(\nu, \xi)$? The theoretical framework we need is provided by the theory of Lévy processes, which generalize both diffusions and Poisson (pure jump) processes. The essential features of Lévy processes are:

   1. Increments are independent.
   2. Increments are stationary.
   3. The process is stochastically continuous.
   4. The initial value is zero (with probability 1).

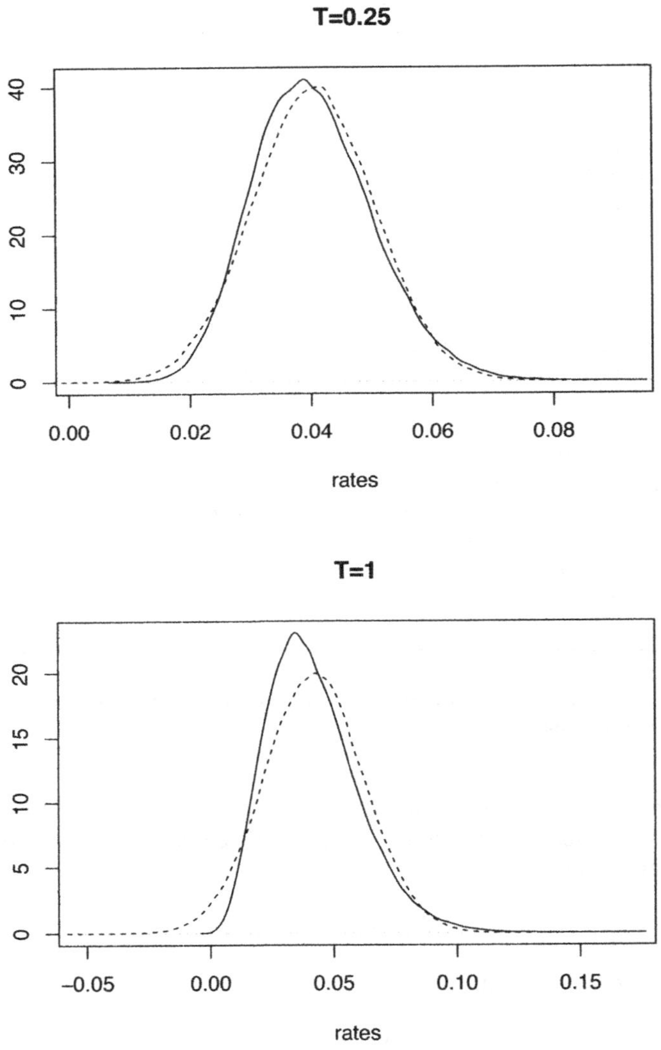

T=0.25

T=1

FIGURE 6.16 **Kernel densities for exact CIR rates (continuous lines) and rates sampled by the Euler scheme (dashed lines).**

These are the same properties of the standard Wiener process, and the only point worth mentioning is property 3, which may seem at odds with discontinuous sample paths of Poisson processes. A correct statement of the property, in a one-dimensional case, is

$$\lim_{h \to 0} P\left\{ |X_{t+h} - X_t| \geq \epsilon \right\} = 0, \qquad \text{for all } \epsilon > 0.$$

Hence, jumps are not actually ruled out, but they cannot be "dense." For instance, in a Poisson process the probability that there is more than one jump in an infinitesimal time interval dt is negligible.

A key result is the Lévy–Itô decomposition, whose proper statement is beyond the scope of this book.[8] Loosely speaking, the theorem states that we may decompose a Lévy process into separate components, accounting for diffusion and jump subprocesses. Let us illustrate how we may take advantage of this theorem by adding a jump component to the above generalized Wiener process. To this aim, we shall take advantage of the following properties of the Poisson process:

- The number N of jumps in the time interval $[0, T]$ is a Poisson random variable with parameter λT.
- Conditional on N, the jump epochs are a sequence of N i.i.d. uniform variables on $[0, T]$.

Therefore, in order to generate a sample path of this jump–diffusion process, we may:

1. Generate the increments of the diffusion component as usual.
2. Sample the number of jumps as a Poisson variable.
3. Generate the time epochs as uniform random variables.
4. Sample the jump sizes and add them to the corresponding increments of the diffusion process.

The code in Fig. 6.17 applies the above ideas and produces the sample path displayed in Fig. 6.18. Following the traditional convention to plot discontinuous functions, we have added a bullet–circle pair corresponding to each point of discontinuity. Note that the process is continuous from the right.

For further reading

- For a sound theoretical treatment of numerical methods to solve stochastic differential equations, see [3].
- Several examples, at a more advanced level with respect to this chapter are described in [1] and [5]. See also [4], which requires some advanced measure theoretic background.
- An interesting discussion of transformation methods and the Milstein scheme is given in [2].

---

[8] See, e.g., [5, p. 209].

```
set.seed(55555)
# generalized Wiener process plus jumps
T <- 1; numSteps <- 1000; h <- T/numSteps
# parameters of the jump component
lambda <- 7; nu <- 0; xi <- 2
# parameters of the diffusion component
beta <- 10;  sigma <- 2
# sample the diffusion increments
dX <- beta*h + sigma*sqrt(h)*rnorm(numSteps)
# sample and add the jump component
N <- rpois(1, lambda*T)
jumpPos <- ceiling(runif(N)*T*numSteps)
jumpSize <- rnorm(N,nu,xi)
dX[jumpPos] <- dX[jumpPos] + jumpSize
# add components and cumulate sample path
X <- c(0, cumsum(dX))
if (N==0){
  plot(0:numSteps,X,type='l')
}else{
  plot(0:numSteps,X,type='l')
  points(jumpPos,X[jumpPos])
  points(jumpPos,X[jumpPos+1],pch=19)
}
```

FIGURE 6.17 **Simulating a jump–diffusion process.**

References

1 P. Glasserman. *Monte Carlo Methods in Financial Engineering*. Springer, New York, 2004.

2 P. Jaeckel. *Monte Carlo Methods in Finance*. Wiley, Chichester, 2002.

3 P.E. Kloeden and E. Platen. *Numerical Solution of Stochastic Differential Equations*. Springer, Berlin, 1992.

4 R. Korn, E. Korn, and G. Kroisandt. *Monte Carlo Methods and Models in Finance and Insurance*. CRC Press, Boca Raton, FL, 2010.

5 D.P. Kroese, T. Taimre, and Z.I. Botev. *Handbook of Monte Carlo Methods*. Wiley, Hoboken, NJ, 2011.

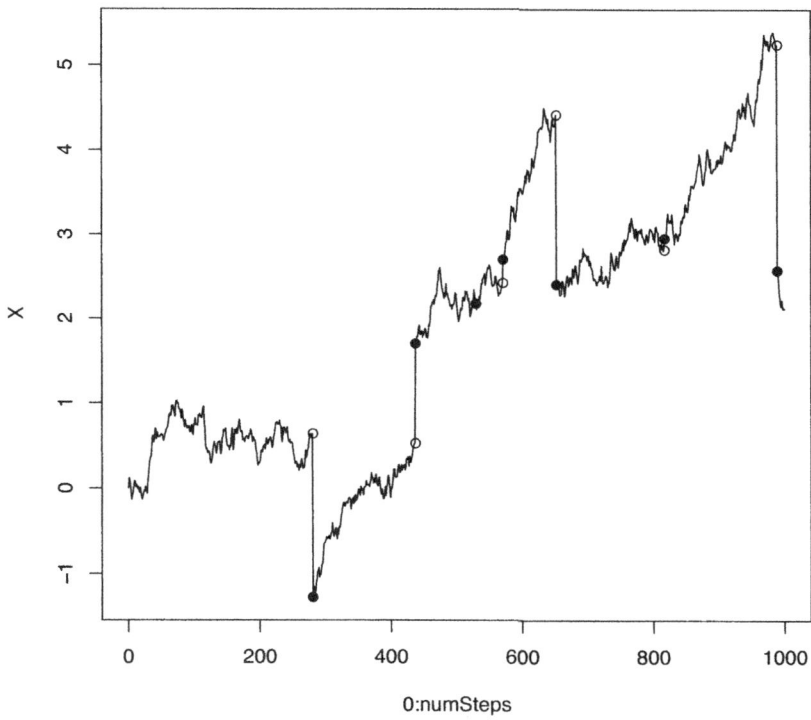

FIGURE 6.18 **Sample path of a jump–diffusion process.**

## Part Four

Output Analysis and Efficiency Improvement

## Chapter Seven

Output Analysis

There are several reasons why we should carefully check the output of a Monte Carlo simulation:

1. Model validation, i.e., check that the model makes economic and/or financial sense, possibly with respect to empirical evidence.

2. Model verification, i.e., check that the computer code we have written implements the model, right or wrong, that we had in mind.

3. Statistical inference, i.e., check the accuracy of the estimates that we obtain by running the model implemented in the code.

In this chapter we deal with the last point. To motivate the development below, let us refer back to the queueing example that we introduced in Section 1.3.3.1, where Lindley's recursion is used to estimate the average waiting time in an $M/M/1$ queue, i.e., a queue where the distributions of interarrival and service times are both memoryless, i.e., exponential.[1] For the sake of convenience, we report the R code in Fig. 7.1; the function has been slightly modified, in order to return the whole vector of recorded waiting times, rather than just its mean.

   With the default input arguments, the average server utilization is

$$\frac{\lambda}{\mu} = \frac{1}{1.1} = 90.91\%,$$

and 10,000 customers are simulated. Is this sample size enough for practical purposes? The following runs suggest some criticality, as the sample means show a considerable variability:

```
> set.seed(55555)
> mean(MM1_Queue_V())
[1] 8.04929
> mean(MM1_Queue_V())
[1] 7.015979
> mean(MM1_Queue_V())
[1] 8.922044
> mean(MM1_Queue_V())
[1] 8.043148
```

---

[1] The memoryless property of the exponential distribution is discussed in Section 3.2.2.

```
MM1_Queue_V <- function(lambda=1, mu=1.1, howmany=10000){
    W = numeric(howmany)
    for (j in 1:howmany) {
        intTime = rexp(1,rate=lambda)
        servTime = rexp(1,rate=mu)
        W[j] = max(0, W[j-1] + servTime - intTime)
    }
    return(W)
}
```

FIGURE 7.1 **Function to simulate a simple $M/M/1$ queue.**

```
> mean(MM1_Queue_V())
[1] 7.250367
> mean(MM1_Queue_V())
[1] 10.7721
> mean(MM1_Queue_V())
[1] 6.290644
> mean(MM1_Queue_V())
[1] 12.91079
```

Now we have to face two related questions:

- How can we quantify the accuracy of estimates for a given sample size?
- How can we find the sample size to obtain a given degree of accuracy?

Statistical inference provides us with a useful tool to answer the first question, namely, confidence intervals; it is fairly easy to reverse-engineer a confidence interval in order to estimate the sample size, i.e., the number of replications, that we need to achieve a selected degree of accuracy. Confidence intervals are clearly related to parameter estimation, which has already been dealt with in Chapter 4. However, we want to show a few potential pitfalls and mistakes that are incurred if one applies those deceptively simple formulas without due care; we provide a couple of enlightening examples in Section 7.1. Then, in Section 7.2, we illustrate how sample sizes may be related with measures of estimate precision. Following common practice, we have introduced confidence intervals with reference to a mean (better said, expected value). However, what we have to estimate is not always an expected value. For instance, we often need estimates of probabilities or quantiles. It is true that a probability is the expected value of an indicator function, but since its value is constrained in the interval $[0, 1]$, some care may be needed. As to quantiles, when we estimate quantiles for small or large probability levels, as is common in risk management, we may face a difficult task. We address the related issues in Section 7.3. The essential message of this chapter, which we summarize in Section 7.4, is both good and bad news: Monte Carlo methods enjoy considerable flexibility and generality, but they are inherently inefficient. By pointing this out, we lay down the motivation for variance reduction strategies and low-discrepancy sequences, which are dealt with in Chapters 8 and 9, respectively.

7.1 Pitfalls in output analysis

We know from Chapter 4 that, given a sample $\{X_1, X_2, \ldots, X_n\}$ of n i.i.d. variables, we may estimate sample mean and sample variance:

$$\bar{X}(n) = \frac{1}{n} \sum_{i=1}^{n} X_i, \qquad S^2(n) = \frac{1}{n-1} \sum_{i=1}^{n} \left[X_i - \bar{X}(n) \right]^2.$$

Given the purpose of this chapter, here we point out the dependence of the above statistics on the sample size n. Then, elementary inferential statistics suggests the confidence interval

$$\bar{X}(n) \pm z_{1-\alpha/2} \frac{S(n)}{\sqrt{n}}, \tag{7.1}$$

at confidence level $1 - \alpha/2$, where $z_{1-\alpha/2}$ is the quantile of the standard normal distribution corresponding to the selected confidence level. As we have already noted, we use the quantile of the normal distribution, since in Monte Carlo simulation the sample size is typically large enough to warrant its use rather than the quantile of the t distribution.

Then, it should be an easy matter to use R to find a confidence interval for the average waiting time in the $M/M/1$ queue. Given the variability that we have observed, we expect quite large confidence intervals:

```
> set.seed(55555)
> as.numeric(t.test(MM1_Queue_V())$conf.int)
[1] 7.886045 8.212535
> as.numeric(t.test(MM1_Queue_V())$conf.int)
[1] 6.881556 7.150402
> as.numeric(t.test(MM1_Queue_V())$conf.int)
[1] 8.739323 9.104764
> as.numeric(t.test(MM1_Queue_V())$conf.int)
[1] 7.876257 8.210038
> as.numeric(t.test(MM1_Queue_V())$conf.int)
[1] 7.077575 7.423160
> as.numeric(t.test(MM1_Queue_V())$conf.int)
[1] 10.5591 10.9851
> as.numeric(t.test(MM1_Queue_V())$conf.int)
[1] 6.167307 6.413980
```

We clearly notice a disturbing pattern: The extreme points of the confidence intervals do jump around, as we expected, but the confidence intervals look overly "confident," as they are rather narrow and do not overlap with each other. In other words, it seems that we are overstating the precision of the estimate.

In fact, what we are doing is *dead wrong*, and we are actually making three mistakes:

- We are using a formula assuming independence of observations, whereas Lindley's recursion clearly shows that the waiting time W_j depends on waiting time W_{j-1}, which makes intuitive sense: If we wait in a queue for a long time, the customer behind us will wait a for a long time, too.
- The random variables are not independent, but they are not even identically distributed: The first waiting time is clearly zero, as the system is initially empty. There is an initial transient phase, and we should wait for the system to reach a steady state before starting to collect statistics. This is not too relevant for a single queue, but it is for a complex system of interacting queues.
- We are using a confidence interval that is meant for normal samples, but there is no reason to believe that waiting times are normally distributed.

By far, the most serious mistake is the first one. Indeed, writing the confidence interval as in Eq. (7.1) is somewhat misleading, as it obscures the fact that it depends on the standard error of estimate, i.e., the standard deviation of the sample mean. The standard deviation of the sample mean can be written as $S/\sqrt{n}$ only under the assumption of independence (actually, lack of correlation is enough). Unfortunately, there is a strong positive correlation between consecutive waiting times, especially when the utilization rate is high; as a consequence, the familiar formula severely understates variance, and this leads to small, but wrong confidence intervals. To clearly see the point, let us write the variance of the average waiting time explicitly:

$$\mathrm{Var}\left(\frac{1}{n}\sum_{i=1}^{n} W_i\right) = \frac{1}{n^2}\left[\sum_{i=1}^{n}\mathrm{Var}(W_i) + 2\sum_{j>i}\mathrm{Cov}(W_i, W_j)\right]$$

$$= \frac{1}{n^2}\left[n\sigma^2 + 2\sum_{j>i}\rho_{ij}\sigma^2\right]$$

$$= \frac{\sigma^2}{n}\left[1 + \frac{2}{n}\sum_{j>i}\rho_{ij}\right],$$

where σ^2 is the variance of each waiting time W_i, and ρ_{ij} is the correlation between the waiting times of customers i and j. We see that, when there is a significant positive correlation, the familiar formula σ^2/n may grossly understate the variance of the sample mean. The strength of the correlation may be checked by the following snapshot:

```
> set.seed(55555)
> W1 <- MM1_Queue_V()
> W2 <- MM1_Queue_V(lambda=1,mu=2)
> par(mfrow=c(1,2))
> acf(W1, main="utilization = 90.91%")
```

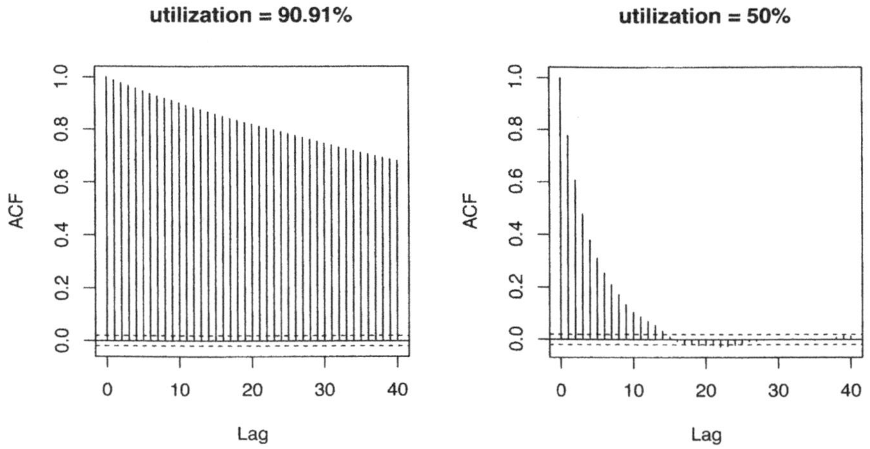

FIGURE 7.2 Impact of server utilization on the autocorrelation of waiting times.

```
> acf(W2, main="utilization = 50%")
> par(mfrow=c(1,1))
```

Running this code produces the two autocorrelation functions[2] for waiting times, displayed in Fig. 7.2. The autorrelation is quite significant for a high utilization rate, whereas it fades away quicker when utilization rate is 50%. This explains why we cannot compute confidence intervals in a standard way. But then, what can we do? In queueing simulations the batch method is often used. The idea is to simulate batches of consecutive customers, collecting sample means. For instance, we may build batches of 10,000 customers and gather the batch means

$$\overline{W}_i = \sum_{k=(i-1)\times 10,000+1}^{i\times 10,000} W_k.$$

The batch mean $\overline{W}_1$ includes customers from 1 to 10,000, the batch mean $\overline{W}_2$ includes customers from 10,001 to 20,000, etc. If batches are long enough, the batch means are practically uncorrelated; this can be checked by plotting the ACF. Furthermore, since they are sums of many waiting times, we can hope that their distribution will be closer to a normal, thus providing further justification for the use of standard confidence intervals. We should also mention that batching observations and calculating confidence intervals on the basis of batch means may also be useful in financial applications, when there are concerns about lack of normality in the observations.

---

[2]We have introduced the autocorrelation function in Section 3.6.

7.1.1 BIAS AND DEPENDENCE ISSUES: A FINANCIAL EXAMPLE

The queueing example of the previous section is a nice and simple example to illustrate pitfalls in using standard confidence intervals without due care but, admittedly, is not quite interesting from a financial point of view. So, let us consider the pricing of an *as-you-like-it* option (also known as *chooser* option)[3]. The option is European-style and has maturity T_2. At time $T_1 < T_2$ you may choose if the option is actually a call or a put; the strike price K is fixed at time $t = 0$. Clearly, at time T_1 we should compare the values of the two options, put and call, and choose the more valuable one. This can be done by using the Black–Scholes–Merton (BSM) formula to evaluate the price of call and put options with initial underlying price $S(T_1)$ and time to maturity $T_2 - T_1$. This means that, conditional on $S_1 = S(T_1)$, we may find the exact expected value of the payoff at time T_2, under the risk-neutral probability. What's more, it turns out that it is actually possible to price a chooser option exactly, at least in the BSM world. To begin with, let us find the critical price S_1^* for which we are indifferent between a call and a put option; to this aim, we just need to consider put–call parity at time $t = T_1$:[4]

$$C_1 + Ke^{-r(T_2-T_1)} = P_1 + S_1 \quad \Rightarrow \quad S_1^* = Ke^{-r(T_2-T_1)}.$$

Now let us consider the strike price

$$K^* = Ke^{-r(T_2-T_1)}$$

and a portfolio consisting of:

1. A vanilla call option with maturity T_2 and strike K
2. A vanilla put option with maturity T_1 and strike K^*

Now, at time T_1 there are two kinds of scenario:

- If $S_1 > K^*$, then the put expires worthless and we keep the call. Therefore, at maturity, we will receive the payoff associated with the call, which indeed corresponds to the optimal choice if $S_1 > K^*$.
- If $S_1 < K^*$, then the put is exercised and at time T_1 we have a position

$$C_1 + Ke^{-r(T_2-T_1)} - S_1.$$

At time T_2 the position is worth

$$C_2 + K - S_2 = \max\{S_2 - K, 0\} + K - S_2 = \max\{K - S_2, 0\}.$$

Therefore, at maturity, we receive the payoff associated with the put, which indeed corresponds to the optimal choice if $S_1 < K^*$.

---

[3] The background on risk-neutral option pricing is given in Section 3.9.
[4] See Theorem 3.18.

```
AYLI <- function(S0,K,r,T1,T2,sigma){
  Kc <- K*exp(-r*(T2-T1))
  value <- EuVanillaCall(S0,K,T2,r,sigma) +
           EuVanillaPut(S0,Kc,T1,r,sigma)
  return(value)
}
```

FIGURE 7.3 **Code to price an as-you-like-it (chooser) option exactly in the BSM world.**

The code to price the chooser option exactly is given in Fig. 7.3 and relies on functions implementing the BSM formula for vanilla calls and puts. It is instructive to compare the price of the chooser option against the two vanilla options:

```
> S0 <- 50
> K <- 50
> r <- 0.05
> T1 <- 2/12
> T2 <- 7/12
> sigma <- 0.4
> EuVanillaCall(S0,K,T2,r,sigma)
[1] 6.728749
> EuVanillaPut(S0,K,T2,r,sigma)
[1] 5.291478
> AYLI(S0,K,r,T1,T2,sigma)
[1] 9.267982
```

As expected, the possibility of choosing may be quite valuable. Even though it is easy to find the price of a simple chooser option, it is extremely instructive to write a pure sampling-based Monte Carlo code. In the present case, this is not quite trivial, as we are not only estimating things: We must make a decision at time T_1. This decision is similar to the early exercise decision we must make with American-style options. To get a feeling for the issues involved, let us consider the scenario tree in Fig. 7.4, where we should associate a price of the underlying asset with each node. Starting from the initial node, with price S_0, we generate four observations of price $S(T_1)$, and for each of these, we sample three prices $S(T_2)$. We have $4 \times 3 = 12$ scenarios, but they are tree-structured. We need this structure, because the decision at time T_1 (either we like the put or we like the call) must be the same for all scenarios departing from each node at time T_1. Without this structure, our decisions would be based on perfect foresight about the future price at time T_2. This nonanticipativity concept is fundamental in dynamic stochastic optimization and in pricing American-style options, as we shall see in Sections 10.5.2 and 11.3.

    A crude Monte Carlo code to price the option is displayed in Fig. 7.5. Here NRepl1 is the number of scenarios that we branch at time T_1 and NRepl2 is

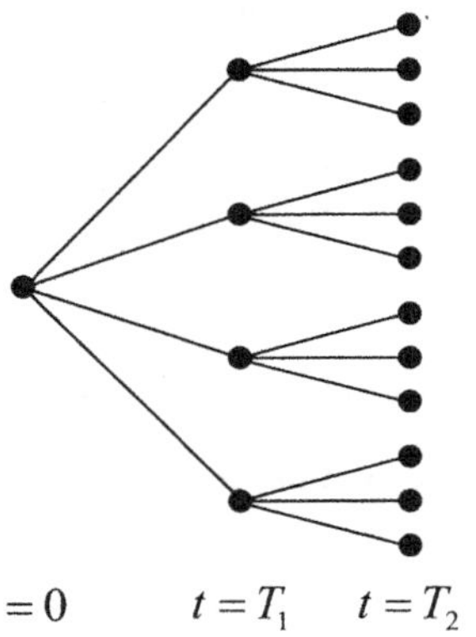

FIGURE 7.4 **Scenario tree for the as-you-like-it option.**

the number of branches at time T_2, for each node at time T_1; hence, the overall number of scenarios is the product of NRepl1 and NRepl2. The vector DiscountedPayoffs has a size corresponding to the overall number of scenarios. For each node at T_1, which is generated as usual with geometric Brownian motion, we generate nodes at time T_2, and we compare the two estimates of expected payoff that we are going to earn if we choose the call or the put, respectively. Then, we select one of the two alternatives and we fill a corresponding block (of size NRepl2) in the vector of discounted payoffs. Finally, we compute sample mean and confidence intervals as usual.

Now, let us run the simulator a few times, comparing the confidence intervals that we obtain against the exact price (9.267982):

```
> NRepl1 <- 1000
> NRepl2 <- 1000
> set.seed(55555)
> AYLIMC(S0,K,r,T1,T2,sigma,NRepl1,NRepl2)$conf.int
[1] 9.246557 9.289328
> AYLIMC(S0,K,r,T1,T2,sigma,NRepl1,NRepl2)$conf.int
[1] 9.025521 9.067313
> AYLIMC(S0,K,r,T1,T2,sigma,NRepl1,NRepl2)$conf.int
[1] 9.408449 9.452174
> AYLIMC(S0,K,r,T1,T2,sigma,NRepl1,NRepl2)$conf.int
[1] 9.260760 9.303662
> AYLIMC(S0,K,r,T1,T2,sigma,NRepl1,NRepl2)$conf.int
[1] 9.434346 9.478341
> AYLIMC(S0,K,r,T1,T2,sigma,NRepl1,NRepl2)$conf.int
[1] 9.144339 9.187482
```

Since we are computing a 95% confidence interval, we would expect to miss the correct value 1 time out of 20 on the average. However, it seems that this happens a tad too often and that we are overstating the reliability of Monte Carlo

```
AYLIMC <- function(S0,K,r,T1,T2,sigma,NRepl1,NRepl2){
  # compute auxiliary quantities outside the loop
  DeltaT <- T2-T1
  muT1 <- (r-sigma^2/2)*T1
  muT2 <- (r-sigma^2/2)*(T2-T1)
  siT1 <- sigma*sqrt(T1)
  siT2 <- sigma*sqrt(T2-T1)
  # vector to contain payoffs
  DiscountedPayoffs <- numeric(NRepl1*NRepl2)
  # sample at time T1
  Samples1 <- rnorm(NRepl1)
  PriceT1 <- S0*exp(muT1 + siT1*Samples1)
  for (k in 1:NRepl1){
    Samples2 <- rnorm(NRepl2)
    PriceT2 <- PriceT1[k]*exp(muT2 + siT2*Samples2)
    ValueCall <- exp(-r*DeltaT)*mean(pmax(PriceT2-K, 0))
    ValuePut <- exp(-r*DeltaT)*mean(pmax(K-PriceT2, 0))
     if (ValueCall > ValuePut)
       DiscountedPayoffs[(1+(k-1)*NRepl2):(k*NRepl2)] <-
          exp(-r*T2)*pmax(PriceT2-K, 0)
     else
       DiscountedPayoffs[(1+(k-1)*NRepl2):(k*NRepl2)] <-
          exp(-r*T2)*pmax(K-PriceT2, 0)
  }
  aux <- t.test(DiscountedPayoffs)
  price <- as.numeric(aux$estimate)
  conf.int <- as.numeric(aux$conf.int)
  return(list(price=price, conf.int=conf.int))
}
```

FIGURE 7.5 **Crude Monte Carlo code to price an as-you-like-it option.**

estimates. Indeed, it is not just a matter of sampling errors, as there are a few things that are dead wrong (again!) in the code above:

- The replications are not really independent: we use *all* of the replications branching from a node at time T_1 to decide for a call or a put.
- There is an unclear source of bias in the estimates. On the one hand, there is a low bias since we are not really deciding optimally between the call and the put; on the other hand we are relying on a limited set of scenarios, rather than the true continuous distribution; thus, we are possibly facing less uncertainty in the sample than we would do in the true world.

Once again, we see that independence among replications is not to be taken for granted. Furthermore, we also observe that when a Monte Carlo simulation involves some decision making, bias issues may arise. This is quite relevant, e.g., when pricing American-style options, as we will see in Section 11.3.1.

7.2 Setting the number of replications

As we have mentioned, we may try to quantify the quality of our estimator by considering the expected value of the squared error of estimate:

$$E[(\bar{X}(n) - \mu)^2] = \text{Var}[\bar{X}(n)] = \frac{\sigma^2}{n},$$

where σ^2 may be estimated by the sample variance $S^2(n)$. Note that the above formula holds if we assume that the sample mean is an unbiased estimator of the parameter μ we are interested in. Clearly, increasing the number n of replications improves the estimate, but how can we reasonably set the value of n?

Suppose that we are interested in controlling the *absolute* error in such a way that, with probability $(1 - \alpha)$,

$$|\bar{X}(n) - \mu| \leq \beta,$$

where β is the maximum acceptable tolerance. In fact, the confidence interval (7.1) is just built in such a way that

$$P\{\bar{X}(n) - H \leq \mu \leq \bar{X}(n) + H\} \approx 1 - \alpha,$$

where we denote the half-length of the confidence interval by

$$H = z_{1-\alpha/2}\sqrt{\frac{S^2(n)}{n}}.$$

This implies that, with probability $1 - \alpha$, we have

$$\bar{X}(n) - \mu \leq H \quad \text{and} \quad \mu - \bar{X}(n) \leq H \quad \Rightarrow \quad |\bar{X}(n) - \mu| \leq H.$$

Hence, linking H to β, we should simply run replications until H is less than or equal to the tolerance β, so that the number n must satisfy

$$z_{1-\alpha/2}\sqrt{\frac{S^2(n)}{n}} \leq \beta. \tag{7.2}$$

Actually, we are chasing our tail a bit here, since we cannot estimate the sample variance $S^2(n)$ until the number n has been set. One way out is to run a suitable number k of pilot replications, in order to come up with an estimate $S^2(k)$. Then we may apply Eq. (7.2) using $S^2(k)$ to determine n. After running the n replications, it is advisable to check that Eq. (7.2) holds with the new estimate $S^2(n)$. Alternatively, we may simply add replications, updating the sample variance, until the criterion is met; however, with this approach we do not control the amount of computation we are willing to spend.

If you are interested in controlling the *relative* error, so that

$$\frac{\mid \bar{X}(n) - \mu \mid}{\mid \mu \mid} \leq \gamma$$

holds with probability $(1 - \alpha)$, things are a little more involved. The difficulty is that we may run replications until the half-length H satisfies

$$\frac{H}{\mid \bar{X}(n) \mid} \leq \gamma,$$

but in this inequality we are using the estimate $\bar{X}(n)$ rather than the unknown parameter μ. Nevertheless, if the inequality above holds, we may write

$$
\begin{aligned}
1 - \alpha &\approx \mathrm{P}\left\{\frac{\mid \bar{X}(n) - \mu \mid}{\mid \bar{X}(n) \mid} \leq \frac{H}{\mid \bar{X}(n) \mid}\right\} \\
&\leq \mathrm{P}\left\{\mid \bar{X}(n) - \mu \mid \leq \gamma \mid \bar{X}(n) \mid\right\} \\
&= \mathrm{P}\left\{\mid \bar{X}(n) - \mu \mid \leq \gamma \mid \bar{X}(n) - \mu + \mu \mid\right\} \\
&\leq \mathrm{P}\left\{\mid \bar{X}(n) - \mu \mid \leq \gamma \mid \bar{X}(n) - \mu \mid + \gamma \mid \mu \mid\right\} \qquad (7.3) \\
&= \mathrm{P}\left\{\frac{\mid \bar{X}(n) - \mu \mid}{\mid \mu \mid} \leq \frac{\gamma}{1 - \gamma}\right\}, \qquad (7.4)
\end{aligned}
$$

where inequality (7.3) follows from the triangle inequality and Eq. (7.4) is obtained by a slight rearrangement. Therefore, we conclude that if we proceed without care, the actual relative error that we obtain is bounded by $\gamma/(1 - \gamma)$, which is larger than the desired bound γ. Therefore, we should choose n such that the following tighter criterion is met:

$$\frac{z_{1-\alpha/2}\sqrt{S^2(n)/n}}{\mid \bar{X}(n) \mid} \leq \gamma', \qquad (7.5)$$

where

$$\gamma' = \frac{\gamma}{1 + \gamma} < \gamma.$$

Again, we should run a few pilot replications in order to get a preliminary estimate of the sample variance $S^2(n)$.

7.3   A world beyond averages

The standard form of a confidence interval for the mean of a normal population is a cornerstone of inferential statistics. The unfortunate side effect is the tendency to use it in settings for which the recipe is not quite appropriate. Consider the problem of estimating the probability

$$\pi = \mathrm{P}\{\mathbf{X} \in A\},$$

where $\mathbf{X}$ is a possibly high-dimensional vector of random variables. It is natural to sample i.i.d. observations $\mathbf{X}_i$, $i = 1, \ldots, n$, and collect the random variables

$$\theta_i = \begin{cases} 1, & \text{if } \mathbf{X}_i \in A, \\ 0, & \text{if } \mathbf{X}_i \notin A. \end{cases}$$

These are i.i.d. observations of a Bernoulli random variable with parameter π, and it is natural to build an estimator

$$p = \frac{1}{n} \sum_{i=1}^{n} \theta_i. \tag{7.6}$$

Then, np is just a binomial variable with parameters n and π, and

$$\mathrm{E}[p] = \pi, \qquad \mathrm{Var}(p) = \frac{\pi(1-\pi)}{n}.$$

We see that there is an obvious link between expected value and variance, and it would make little sense to use the usual sample variance S^2. In fact, building a confidence interval for π calls for same care. However, if the sample size n is large enough, we may take advantage of the fact that the binomial distribution is a sum of i.i.d. variables and tends to a normal distribution, thanks to the central limit theorem. In other words, the statistic

$$\frac{p - \pi}{\sqrt{p(1-p)/n}}$$

is approximately $\mathrm{N}(0, 1)$, which justifies the approximate confidence interval

$$p \pm z_{1-\alpha_2} \sqrt{\frac{\pi(1-\pi)}{n}}. \tag{7.7}$$

Hence, at the very least, we should use the standard confidence interval by using a sensible estimate of variance. More refined methods are available in the literature, but R solves the problem by offering an appropriate function, `binom.test`.

Another delicate estimation problem concerns quantiles. This is relevant, for instance, in the estimation of risk measures like value-at-risk, as we shall see in Chapter 13. In principle, the task is not too hard and can be carried out as follows:

1. Use Monte Carlo to generate the output sample $\{X_1, \ldots, X_n\}$.

2. Build the empirical CDF

$$\hat{F}_{X,n}(x) = \frac{1}{n} \sum_{i=1}^{n} \mathbf{1}_{\{X_1 \leq x\}}, \tag{7.8}$$

where $\mathbf{1}_{\{\cdot\}}$ is the indicator function for an event.

3. Find the generalized inverse (see below) of the empirical CDF and return the estimated quantile at probability level α

$$\hat{q}_{\alpha,n} = \hat{F}_{X,n}^{-1}(x).$$

Since the empirical CDF is piecewise constant, with jumps of size $1/n$, it cannot be inverted as usual, and we resort to the generalized inverse. In this case this is just accomplished by sorting the observations X_i, which yields the order statistics $X_{(i)}$, and pick the observation in position

$$k = \min\left\{i \in \{1,\dots,n\} \;\middle|\; \frac{i}{n} \geq \alpha\right\}. \tag{7.9}$$

Alternative procedures can be devised, possibly based on linear interpolation, but the deep trouble is that if we are estimating extreme quantiles, for small or large values of α, we may get rather unreliable estimates. Hence, we need a way to build a confidence interval. The theory here is far from trivial (see the references at the end of the chapter), but there is a central limit theorem for quantiles, stating that the statistic

$$\sqrt{n}(\hat{q}_{\alpha,n} - q_\alpha)$$

converges in distribution to a normal random variable $N(0, \sigma^2)$, where

$$\sigma^2 = \frac{\alpha(1-\alpha)}{f_X(q_\alpha)}$$

and $f_X(\cdot)$ is the PDF corresponding to the true CDF $F_{X,n}(\cdot)$. This suggests the confidence interval

$$\hat{q}_{\alpha,n} \pm z_{1-\alpha/2}\frac{\alpha(1-\alpha)}{f_X(q_\alpha)\sqrt{n}}.$$

The problem is that, clearly, $f_X(q_\alpha)$ itself must be estimated, and it can be very small, resulting in large confidence intervals. In Section 13.4 we shall see an example of how the difficulty can be circumvented by recasting the problem a bit and using variance reduction methods.

7.4 Good and bad news

Equation (7.1) shows that the half-length of the confidence interval is determined by the ratio

$$\frac{\sigma}{\sqrt{n}},$$

i.e., by the variability in the output performance measure and the sample size. This is both good and bad news:

- On the one hand, we notice that no role is played by the problem dimensionality, i.e., by how many random variables are involved in the simulation or, if you prefer, by the dimension of the space in which we are computing an integral. Actually, the dimension does play a role in terms of computational effort, and it typically creeps into the variability σ itself. Nevertheless, this is certainly good news and it contributes to explain why Monte Carlo methods are widely used.

- On the other hand, the rate of improvement in the quality of our estimate, i.e., the rate of decrease of the error, is something like $O(1/\sqrt{n})$. Unfortunately, the square root is a concave function, and an economist would say that it features diminishing marginal returns. In practice, this means that the larger the sample size, the better, but the rate of improvement is slower and slower as we keep adding observations.

To appreciate the latter point, let us assume that with a sample size n we have a certain precision, and that we want to improve that by one order of magnitude. For instance, we might find a confidence interval for an option price like

$$(1.238, \; 1.285),$$

in which it seems that we got the first decimal digit right, but we also want to get the cents right. This means that we should divide the interval half-length by 10, which implies that we need $100 \times n$ replications. Thus, a brute-force Monte Carlo simulation may take quite some amount of computation to yield an acceptable estimate. One way to overcome this issue is to adopt a clever sampling strategy in order to reduce the variance σ^2 of the observations themselves; the other one is to adopt a quasi–Monte Carlo approach based on low-discrepancy sequences. The next two chapters pursue these two strategies, respectively.

For further reading

- An excellent, yet readable chapter on simulation output analysis can be found in [5].

- For a higher level and more general treatment, you may refer to [1, Chapter 3].

- For the estimation of probabilities, a standard reference is [2]; this topic and other things are covered in [3].

- See [4] for issues related to quantile estimation.

References

1 S. Asmussen and P.W. Glynn. *Stochastic Simulation: Algorithms and Analysis*. Springer, New York, 2010.

2 C.J. Clopper and E.S. Pearson. The use of confidence or fiducial limits illustrated in the case of the binomial. *Biometrika*, 26:404–413, 1934.

3 W.J. Conover. *Practical Nonparametric Statistics* (3rd ed.). Wiley, Hoboken, NJ, 1999.

4 P.W. Glynn. Importance sampling for Monte Carlo estimation of quantiles. In *Proceedings of the 2nd International Workshop on Mathematical Methods in Stochastic Simulation and Experimental Design*, pp. 180–185, 1996. See `http://www.stanford.edu/~glynn/papers/1996/G96.html`.

5 A.M. Law and W.D. Kelton. *Simulation Modeling and Analysis* (3rd ed.). McGraw-Hill, New York, 1999.

## Chapter Eight

Variance Reduction Methods

In Chapter 7 we have discussed output analysis, and we have seen that an obvious way to improve the accuracy of an estimate $\bar{X}(n)$ based on a sample of n replications X_i, $i = 1, \ldots, n$, is to increase the sample size n, since $\text{Var}(\bar{X}(n)) = \text{Var}(X_i)/n$. However, we have also seen that the width of a confidence interval, assuming that replications are genuinely i.i.d., decreases according to a square-root law involving $\sqrt{n}$, which is rather bad news. Increasing the number of replications is less and less effective, and this brute force strategy may result in a remarkable computational burden. A less obvious strategy is to reduce $\text{Var}(X_i)$. At first sight, this seems like cheating, as we have to change the estimator in some way, possibly introducing bias. The variance reduction strategies that we explore in this chapter aim at improving the efficiency of Monte Carlo methods, sometimes quite dramatically, without introducing any bias. This means that, given a required accuracy, we may reduce the computational burden needed to attain it; or, going the other way around, we may improve accuracy for a given computational budget. In Chapter 9 we also consider another strategy, which consists of using deterministic low-discrepancy sequences to drive sampling, rather than pseudorandom numbers.

Some variance reduction strategies are just technical tricks to improve sampling, and they can even be automated in software tools. Other strategies are based on general principles, but require a high degree of customization to work. In some cases, they replace part of the numerical integration process with a partially analytical integration. As the reader may imagine, the more sophisticated strategies require a lot of domain-specific knowledge, may be rather difficult to conceive, but are much more effective. We begin in Section 8.1 with antithetic sampling, which is arguably the simplest strategy and serves well to introduce the topic of this chapter. Antithetic sampling is related to common random numbers, the subject of Section 8.2, which is a bit different in nature, as it deals with differences of random variables; we will actually appreciate common random numbers in Chapter 12 on estimating sensitivities, but they are best introduced just after antithetic sampling. Antithetic sampling and common random numbers, just like low-discrepancy sequences, work on the input stream of numbers feeding the simulation. Control variates, on the contrary, are based on an augmented estimator, which incorporates some more analytical knowledge that we may have at our disposal. As we shall see in Section 8.3, control variates are quite akin to linear regression models, as they try to take advantage of correlation between random variables. Another helpful tool is the conditional variance formula; in Sections 8.4 and 8.5 we will show how it can be exploited in conditional Monte Carlo and stratified sampling, respectively. Finally, a change of probability measure is the principle behind the importance sampling approach introduced in Section 8.6. Intuitively, the idea is to make rare but significant events more likely.

It is our aim to introduce the above strategies by simple, yet concrete examples. Therefore, in this chapter we will use toy integration problems, estimating π and pricing a vanilla call option, for illustration purposes. In practice, we certainly need no Monte Carlo method for these problems, but it is useful to get a basic understanding of variance reduction strategies by trying them on elementary examples, where we may also afford the luxury of comparing numerical results against the exact ones. Later, we will tackle more interesting applications, mostly in Chapter 11 on option pricing; we will also see an application to risk measurement in Chapter 13. Extremely sophisticated customization may be needed to cope with these more realistic problems.

8.1 Antithetic sampling

The antithetic sampling approach is very easy to apply and does not rely on any deep knowledge about the specific problem we are tackling. In crude Monte Carlo, we generate a sample consisting of independent observations. However, inducing some correlation in a clever way may be helpful. Consider a sequence of paired replications $\left(X_1^{(i)}, X_2^{(i)} \right)$, $i = 1, \ldots, n$:

$$\begin{pmatrix} X_1^{(1)} \\ X_2^{(1)} \end{pmatrix}, \begin{pmatrix} X_1^{(2)} \\ X_2^{(2)} \end{pmatrix}, \ldots, \begin{pmatrix} X_1^{(n)} \\ X_2^{(n)} \end{pmatrix}.$$

We allow for correlation within each pair, but observations are otherwise "horizontally" independent. This means that if we pick any element of a pair, this is independent from any element of other pairs. Formally, $X_j^{(i_1)}$ and $X_k^{(i_2)}$ are independent however we choose $j, k = 1, 2$, provided $i_1 \neq i_2$. Thus, the pair-averaged observations

$$X^{(i)} = \frac{X_1^{(i)} + X_2^{(i)}}{2}$$

are independent and can be used to build a confidence interval as usual. Note that we cannot build a confidence interval without averaging within pairs, as we do not require "vertical" independence. Let us consider the sample mean $\bar{X}(n)$

based on the pair-averaged observations $X^{(i)}$; its variance is given by

$$
\begin{aligned}
\operatorname{Var}[\bar{X}(n)] &= \frac{\operatorname{Var}\left(X^{(i)}\right)}{n} \\
&= \frac{\operatorname{Var}\left(X_1^{(i)}\right) + \operatorname{Var}\left(X_2^{(i)}\right) + 2\operatorname{Cov}\left(X_1^{(i)}, X_2^{(i)}\right)}{4n} \\
&= \frac{\operatorname{Var}(X)}{2n}\left(1 + \rho(X_1, X_2)\right).
\end{aligned}
\tag{8.1}
$$

In the first line we use independence among pairs, in the second line we account for covariance within a pair, and finally we arrive at Eq. (8.1), which shows the potential role of correlation. In order to reduce the variance of the sample mean, we should take negatively correlated replications within each pair. Needless to say, it is not fair to claim that variance is reduced because of the factor $2n$, since we have affectively doubled the sample size; we must compare Eq. (8.1) against the variance $\operatorname{Var}(X_i)/(2n)$ of an independent sample of size $2n$. Now the issue is how we can induce negative correlation. Each observation $X_{1,2}^{(i)}$ is obtained by generating possibly many random variates, and all of them ultimately depend on a stream of uniform pseudorandom numbers. Hence, we may consider the idea of inducing a strong negative correlation between input streams of random numbers. This is easily achieved by using a random number sequence $\{U_k\}$ for the first replication in each pair, and then $\{1 - U_k\}$ in the second one. Since the input streams are negatively correlated, we hope that the output observations will be, too.

Example 8.1 Using antithetic sampling in integration

In Example 2.1 we used plain Monte Carlo integration to estimate

$$
I = \int_0^1 e^x \, dx = e - 1 \approx 1.7183.
$$

A small sample consisting of only 100 observations, as expected, does not provide us with a reliable estimate:

```
> set.seed(55555)
> X <- exp(runif(100))
> T <- t.test(X)
> T$estimate
mean of x
 1.717044
> CI <- as.numeric(T$conf.int)
> CI
[1] 1.621329 1.812759
> CI[2]-CI[1]
[1] 0.1914295
```

```
> (CI[2]-CI[1])/2/(exp(1)-1)
[1] 0.05570375
```

Antithetic sampling is easily accomplished here: We must store random numbers and take their complements to 1. In order to have a fair comparison, we consider 50 antithetic pairs, which entails sampling 100 function values as before:

```
> set.seed(55555)
> U1 <- runif(50)
> U2 <- 1-U1
> X <- 0.5*(exp(U1)+exp(U2))
> T <- t.test(X)
> T$estimate
mean of x
  1.71449
> CI <- as.numeric(T$conf.int)
> CI
[1] 1.699722 1.729258
> CI[2]-CI[1]
[1] 0.02953576
> (CI[2]-CI[1])/2/(exp(1)-1)
[1] 0.008594562
```

Now the confidence interval is much smaller and, despite the limited sample size, the estimate is fairly reliable.

The antithetic sampling method looks quite easy to apply and, in the example above, it works pretty well. May we always expect a similar pattern? Unfortunately, the answer is no. The reason why the approach works pretty well in the example is twofold:

1. There is a strong (positive) correlation between U and e^U over the interval $[0, 1]$, because the function is almost linear there. This means that a strong correlation in the simulation input is preserved and turns into a strong correlation in the simulation output. We should not expect impressive results with more complicated and nonlinear functions.

2. Another reason is that the exponential function is monotonic increasing. As we shall see shortly, monotonicity is an important condition for antithetic sampling.

It is instructive to consider a simple example, showing that antithetic sampling may actually backfire and increase variance.

■ Example 8.2 **A counterexample to antithetic sampling**

Consider the function $h(x)$, defined as

$$h(x) = \begin{cases} 0, & x < 0, \\ 2x, & 0 \leq x \leq 0.5, \\ 2 - 2x, & 0.5 \leq x \leq 1, \\ 0, & x > 1, \end{cases}$$

and suppose that we want to take a Monte Carlo approach to estimate

$$\int_0^1 h(x)\,dx.$$

The function we want to integrate is obviously a triangle with both basis and height equal to 1; note that, unlike the exponential function of Example 8.1, this is not a monotone function with respect to x. It is easy to compute the integral as the area of a triangle:

$$\int_0^1 h(x)\,dx \Rightarrow \mathrm{E}[h(U)] = \int_0^1 h(u) \cdot 1\,du = \frac{1}{2}.$$

Now let

$$X_I = \frac{h(U_1) + h(U_2)}{2}$$

be an estimator based on two independent uniform random variables U_1 and U_2, and let

$$X_A = \frac{h(U) + h(1 - U)}{2}$$

be the pair-averaged sample built by antithetic sampling. We may compare the variances of the two estimators:

$$\mathrm{Var}(X_I) = \frac{\mathrm{Var}\big[h(U)\big]}{2},$$

$$\mathrm{Var}(X_A) = \frac{\mathrm{Var}\big[h(U)\big]}{2} + \frac{\mathrm{Cov}\big[h(U), h(1 - U)\big]}{2}.$$

The difference between the two variances is

$$\Delta = \mathrm{Var}(X_A) - \mathrm{Var}(X_I) = \frac{\mathrm{Cov}\big[h(U), h(1 - U)\big]}{2}$$

$$= \frac{1}{2}\Big\{ \mathrm{E}\big[h(U)h(1 - U)\big] - \mathrm{E}\big[h(U)\big]\mathrm{E}\big[h(1 - U)\big] \Big\}.$$

But in this case, due to the shape of h, we have

$$\mathrm{E}\big[h(U)\big] = \mathrm{E}\big[h(1-U)\big] = \tfrac{1}{2}$$

and

$$\mathrm{E}\big[h(U)h(1-U)\big]$$
$$= \int_0^{1/2} 2u \cdot (2 - 2(1-u))\, du + \int_{1/2}^1 2(1-u) \cdot (2 - 2u)\, du$$
$$= \int_0^{1/2} 4u^2\, du + \int_{1/2}^1 (2 - 2u)^2\, du = \frac{1}{3}.$$

Therefore, $\mathrm{Cov}[h(U), h(1-U)] = \tfrac{1}{3} - \tfrac{1}{4} = \tfrac{1}{12}$ and $\Delta = \tfrac{1}{24} > 0$. In fact, antithetic sampling actually increases variance in this case, and there is a trivial explanation. The two antithetic observations within each pair have the same value $h(U) = h(1-U)$, and so $\mathrm{Cov}[h(U), h(1-U)] = \mathrm{Cov}[h(U), h(U)] = \mathrm{Var}[h(U)]$. In this (pathological) case, the variance of the single observation is doubled by applying antithetic sampling.

What is wrong with Example 8.2? The variance of the antithetic pair is actually increased due to the nonmonotonicity of $h(x)$. In fact, while it is true that the random numbers $\{U_i\}$ and $\{1-U_i\}$ are negatively correlated, there is no guarantee that the same holds for $X_i^{(1)}$ and $X_i^{(2)}$ in general. To be sure that the negative correlation in the input random numbers yields a negative correlation in the observed output, we must require a monotonic relationship between them. The exponential function is a monotonic function, but the triangle function of the second example is not.

Another side of the coin is related to the mechanism we use to generate random variates. The inverse transform method is based on the CDF, which is a monotonic function; hence, there is a monotonic relationship between the input random numbers and the random variates generated. This is not necessarily the case with the acceptance–rejection method or the Box–Muller method, which were discussed in Chapter 5. Luckily, when we need normal variates, we may simply generate a sequence $\epsilon_i \sim \mathsf{N}(0,1)$ and use the sequence $-\epsilon_i$ for the antithetic sample.

In order to illustrate antithetic sampling with normal variates, and to see a more concrete example, let us consider the use of antithetic sampling to price a simple vanilla call option. We have seen how such a simple option can be priced using Monte Carlo sampling. It is easy to modify the code in Fig. 6.4 to introduce antithetic sampling; we just need to store a vector of normal variates to use them twice. The resulting code is shown in Fig. 8.1. The following

```
BlsMCAV <- function(S0,K,T,r,sigma,numPairs,c.level=0.95){
  nuT <- (r - 0.5*sigma^2)*T
  sigmaT <- sigma*sqrt(T)
  veteps <- rnorm(numPairs)
  payoff1 <- pmax(0,S0*exp(nuT+sigmaT*veteps)-K)
  payoff2 <- pmax(0,S0*exp(nuT-sigmaT*veteps)-K)
  dpayoff <- exp(-r*T)*(payoff1+payoff2)/2
  aux <- t.test(dpayoff,conf.level=c.level)
  value <- as.numeric(aux$estimate)
  ci <- as.numeric(aux$conf.int)
  return(list(estimate=value,conf.int=ci))
}
```

FIGURE 8.1 **Pricing a vanilla call option with antithetic sampling.**

snapshot checks variance reduction in the case of an at-the-money option:

```
> S0 <- 50
> K <- 50
> sigma <- 0.4
> r <- 0.05
> T <- 5/12
> numRepl <- 50000
> exact <- EuVanillaCall(S0,K,T,r,sigma); exact
[1] 5.614995
> set.seed(55555)
> out1 <- BlsMCNaive(S0,K,T,r,sigma,numRepl); out1
$estimate
[1] 5.568167
$conf.int
[1] 5.488962 5.647371
> set.seed(55555)
> out2 <- BlsMCAV(S0,K,T,r,sigma,numRepl/2); out2
$estimate
[1] 5.596806
$conf.int
[1] 5.533952 5.659661
```

Note that, for the sake of a fair comparison, we divide the number of replications by two when using antithetic sampling. Indeed, there is a slight reduction in variance, but it is far from impressive (and, anyway, one run does not mean too much). The situation looks a bit better for a deeply in-the-money option:

```
> K <- 25
> exact <- EuVanillaCall(S0,K,T,r,sigma); exact
[1] 25.52311
> set.seed(55555)
```

```
> out1 <- BlsMCNaive(S0,K,T,r,sigma,numRepl); out1
$estimate
[1] 25.4772
$conf.int
[1] 25.36299 25.59142
> set.seed(55555)
> out2 <- BlsMCAV(S0,K,T,r,sigma,numRepl/2); out2
$estimate
[1] 25.51441
$conf.int
[1] 25.48482 25.54401
```

A possible explanation is that in this second case we are sampling the payoff
function where the nonlinearity plays a lesser role. We leave a serious computational study to the reader, but we should not be surprised by the fact that
antithetic sampling, which is almost a simple programming trick, does not necessarily produce impressive improvements. Nevertheless, the idea does work
well in some cases, and sometimes it can also be fruitfully integrated with the
approaches that we describe in the following.

8.2   Common random numbers

The common random numbers technique is very similar to antithetic sampling,
but it is applied in a different situation. Suppose that we use Monte Carlo simulation to estimate a value depending on a parameter α. In formulas, we are
trying to estimate something like

$$h(\alpha) = E_\omega \big[f(\alpha; \omega) \big],$$

where we have emphasized randomness by ω, which refers to random events
underlying random variables. We could also be interested in evaluating the
sensitivity of this value with respect to the parameter α:

$$\frac{dh(\alpha)}{d\alpha}.$$

This would be of interest when dealing with option sensitivities beyond the
Black–Scholes–Merton model. Clearly, we cannot compute the derivative analytically; otherwise, we would not use simulation to estimate the function $h(\cdot)$
in the first place. So, the simplest idea would be using simulation to estimate
the value of the finite difference,

$$\frac{h(\alpha + \delta\alpha) - h(\alpha)}{\delta\alpha},$$

for a small value of the increment $\delta\alpha$. However, what we can really do is to
generate observations of the difference

$$\frac{f(\alpha + \delta\alpha; \omega) - f(\alpha; \omega)}{\delta\alpha},$$

and to estimate its expected value. Unfortunately, when the increment $\delta\alpha$ is small, it is difficult to tell if the difference we observe is due to random noise or to the variation in the parameter α. A similar problem arises when we want to compare two portfolio management policies on a set of scenarios; in this case, too, what we need is an estimate of the expected value of the difference between two random variables. In this setting, is it sensible to compare the two performances on the basis of different sets of scenarios?

Let us abstract a little and consider the difference of two random variables

$$Z = X_1 - X_2,$$

where, in general, $E[X_1] \neq E[X_2]$, since they come from simulating two different systems, possibly differing only in the value of a single parameter. By Monte Carlo simulation we get a sequence of independent random variables,

$$Z_j = X_{1,j} - X_{2,j},$$

and use statistical techniques to build a confidence interval for $E[X_1 - X_2]$. To improve our estimate, it would be useful to reduce the variance of the sampled Z_j. If we do not assume independence between $X_{1,j}$ and $X_{2,j}$ (for the same j), we find something quite similar to antithetic sampling, with a change in sign:

$$\text{Var}(X_{1j} - X_{2j}) = \text{Var}(X_{1j}) + \text{Var}(X_{2j}) - 2\,\text{Cov}(X_{1j}, X_{2j}).$$

Then, in order to reduce variance, we may try inducing a positive correlation between X_{1j} and X_{2j}. This can be obtained by using the same stream of random numbers in simulating both X_1 and X_2. The technique works much like antithetic sampling, and the same monotonicity assumption is required to ensure that the technique does not backfire. We will see an application of these concepts in Section 12.1, where we apply Monte Carlo sampling to estimate option price sensitivities.

8.3   Control variates

Antithetic sampling and common random numbers are two very simple techniques that, provided the monotonicity assumption is valid, do not require much knowledge about the systems we are simulating. Better results might be obtained by taking advantage of deeper, domain-specific knowledge. Suppose that we want to estimate $\theta = E[X]$, and that there is another random variable Y, with a *known* expected value ν, which is somehow correlated with X. Such a case occurs when we use Monte Carlo simulation to price an option for which an analytical formula is not known: θ is the unknown price of the option, and ν could be the price of a corresponding vanilla option. The variable Y is called the *control variate* and represents additional, domain-specific knowledge. The correlation Y may be exploited by adopting the controlled estimator

$$X_C = X + c(Y - \nu),$$

where c is a parameter that we must choose. Intuitively, when we run a simulation and observe that our estimate of $E[Y]$ is too large:

$$\hat{\nu} > \nu.$$

Then, we may argue that the estimate $\hat{\theta}$ should be increased or reduced accordingly, depending on the sign of the correlation between X and Y. For instance, if correlation is positive and $\hat{\nu} > \nu$, we should reduce $\hat{\theta}$, because if the sampled values for Y are larger than their average, the sampled values for X are probably too. The contrary applies when correlation is negative. Indeed, it is easy to see that

$$E[X_C] = \theta, \tag{8.2}$$
$$\text{Var}(X_C) = \text{Var}(X) + c^2\text{Var}(Y) + 2c\,\text{Cov}(X,Y). \tag{8.3}$$

The first formula says that the controlled estimator is, for any choice of the control parameter c, an unbiased estimator of θ. The second formula suggests that by a suitable choice of c, we could reduce the variance of the estimator. We could even minimize the variance by choosing the optimal value for c:

$$c^* = -\frac{\text{Cov}(X,Y)}{\text{Var}(Y)}, \tag{8.4}$$

in which case we find

$$\frac{\text{Var}(X_C^*)}{\text{Var}(X)} = 1 - \rho_{XY}^2,$$

where ρ_{XY} is the correlation between X and Y. Note that the sign of c depends on the sign of this correlation, in a way that conforms to our intuition.

The form of the optimal control parameter in Eq. (8.4) looks suspiciously like the slope coefficient in a linear regression model.[1] To see the connection, let us consider the estimation error

$$X - E[X] = \eta.$$

The variance of the random variable η may be large for difficult estimation problems. In linear regression we aim at explain variability of X by introducing an explanatory variable Y:

$$X - E[X] = b(Y - E[Y]) + \epsilon.$$

With respect to the usual notation, here we are swapping the roles of X and Y. Furthermore, we use only the slope coefficient b in the regression model, since we are centering random variables with respect to their expected values. Apart from these unsubstantial changes, we clearly see that the least-squares estimate of b is what we need to maximize explained variability, which turns into Eq. (8.4) by a slight rearrangement.

---

[1] See Section 3.5 for a probabilistic view of linear regression.

```
set.seed(55555)
numPilot <- 50
U1 <- runif(numPilot)
exp1 <- exp(U1)
c <- -cov(U1,exp1)/var(U1)
numReplication <- 150
U2 <- runif(numReplication)
exp2 <- exp(U2)
Xc <- exp2 + c*(U2 - 0.5)
outCV <- t.test(Xc)
out <- t.test(exp(runif(numPilot+numReplication)))
as.numeric(out$estimate)
as.numeric(out$conf.int)
as.numeric(outCV$estimate)
as.numeric(outCV$conf.int)
cor(U1,exp1)
```

FIGURE 8.2 **Using a control variate in a toy integration problem.**

In practice, the optimal value of c must be estimated, since $\mathrm{Cov}(X,Y)$ and possibly $\mathrm{Var}(Y)$ are not known. This may be accomplished by running a set of pilot replications to estimate them. It would be tempting to use these replications both to select c^* and to estimate θ, in order to save precious CPU time. However, doing so would induce some bias in the estimate of θ, since in this case c^* cannot be considered as a given number, but a random variable depending on X itself. This invalidates the arguments leading to Eqs. (8.2) and (8.3). Therefore, unless suitable statistical techniques (e.g., jackknifing) are used, which are beyond the scope of this book, the pilot replications should be discarded.

Example 8.3 **Using control variates in integration**

We can illustrate control variates by referring again to the problem of Example 8.1, i.e., the integration of the exponential function on the unit interval,

$$\mathrm{E}\!\left[e^U\right] = \int_0^1 e^u\, du.$$

A possible control variate in this case is the driving uniform variable U itself, with an expected value of 0.5. We might even afford the luxury of computing c^* exactly, since the variance of the control variate is $\frac{1}{12}$, and with a little effort we may actually calculate the correlation between U and e^U. Since this is not true in general, let us stick to statistical estimates only. The code in Fig. 8.2 produces the following output:

```
> as.numeric(out$estimate)
[1] 1.713364
> as.numeric(out$conf.int)
[1] 1.646735 1.779992
> as.numeric(outCV$estimate)
[1] 1.713169
> as.numeric(outCV$conf.int)
[1] 1.703453 1.722885
> cor(U1,exp1)
[1] 0.9938869
```

A comparison against the naive estimator shows that the reduction in variance is quite remarkable, and it is easily explained by the strength in the correlation. We may not expect such a result in general, unless a strong control variate is available.

As a reality check, let us try the vanilla call example again. A natural control variate is the price of the underlying asset at maturity, whose expected value under the risk-neutral measure is just $S_0 e^{rT}$. The code is shown in Fig. 8.3. Let us try this approach and compare it against naive Monte Carlo:

```
> S0 <- 50
> K <- 50
> sigma <- 0.4
> r <- 0.05
> T <- 5/12
> numRepl <- 50000
> numPilot <- 10000
> exact <- EuVanillaCall(S0,K,T,r,sigma); exact
[1] 5.614995
> set.seed(55555)
> out1 <- BlsMCNaive(S0,K,T,r,sigma,numRepl+numPilot); out1
$estimate
[1] 5.570385
$conf.int
[1] 5.498177 5.642593
> set.seed(55555)
> out2 <- BlsMCCV(S0,K,T,r,sigma,numRepl,numPilot); out2
$estimate
[1] 5.585363
$conf.int
[1] 5.552163 5.618563
```

```
BlsMCCV <- function(S0,K,T,r,sigma,numRepl,numPilot,
                    c.level=0.95){
  # invariant quantities
  nuT <- (r - 0.5*sigma^2)*T
  sigmaT <- sigma*sqrt(T)
  # pilot replication first
  ST <- S0*exp(rnorm(numPilot,nuT,sigmaT))
  optVals <- exp(-r*T) * pmax( 0 , ST-K)
  expOptVal <- S0 * exp(r*T)
  varOptVal <- S0^2 * exp(2*r*T) * (exp(T * sigma^2) - 1)
  cstar <- -cov(ST,optVals)/varOptVal
  # now run MC with controlled estimator
  newST <- S0*exp(rnorm(numRepl,nuT,sigmaT))
  newOptVals <- exp(-r*T) * pmax( 0 , newST-K)
  controlledEst <- newOptVals+cstar*(newST-expOptVal)
  aux <- t.test(controlledEst,conf.level=c.level)
  value <- as.numeric(aux$estimate)
  ci <- as.numeric(aux$conf.int)
  return(list(estimate=value,conf.int=ci))
}
```

FIGURE 8.3 **Pricing a vanilla call option by control variates.**

With this problem instance, it seems that control variates is more effective than antithetic sampling. The control variates approach may be generalized to as many control variates as we wish, with a possible improvement in the quality of the estimates. Of course, this requires more domain-specific knowledge and more effort in setting the control parameters. For an impressive result obtained by a clever control variate, see the case of arithmetic and geometric Asian options in Section 11.2.

8.4   Conditional Monte Carlo

When faced with the task of calculating an expectation, a common technique in probability theory is the application of conditioning with respect to another random variable. Formally $E[X]$ can be expressed as

$$E[X] = E\Big[E[X \mid Y]\Big]. \qquad (8.5)$$

Equation (8.5) is known as the *law of iterated expectations*, among other names, and it suggests the use of $E[X \mid Y]$ as an estimator rather than X. It is important to realize that this conditional expectation is a random variable, and that this makes sense if, given Y, we have a straightforward recipe to calculate the conditional expectation. What we should do is to sample Y and apply this recipe, and Eq. (8.5) makes sure that we have an unbiased estimate. However, what makes sure that by doing so we can reduce variance? The answer lies in the related conditional variance formula:

$$\text{Var}(X) = \text{E}\Big[\text{Var}(X \mid Y)\Big] + \text{Var}\Big(\text{E}[X \mid Y]\Big). \tag{8.6}$$

We observe that all of the terms in Eq. (8.6) involve a variance and are nonnegative. Therefore the conditional variance formula implies the following:

$$\text{Var}(X) \geq \text{Var}\Big(\text{E}[X \mid Y]\Big), \tag{8.7}$$

$$\text{Var}(X) \geq \text{E}\Big[\text{Var}(X \mid Y)\Big]. \tag{8.8}$$

The first inequality states that $\text{E}[X \mid Y]$ cannot have a larger variance than the crude estimator X, and it justifies the idea of conditional Monte Carlo, The second inequality is the rationale behind variance reduction by stratification, which is discussed in the next section. Unlike antithetic sampling, variance reduction by conditioning requires some careful thinking and is strongly problem dependent.

To illustrate the approach, let us consider the expected value $\text{E}\big[e^{-X^2}\big]$, where X is an affine transformation of a Student's t variable with n degrees of freedom:[2]

$$X = \mu + \sigma T_n, \qquad T_n \sim \mathsf{t}_n.$$

In practice, the above transformation is similar to the destandardization of a standard normal, which is replaced by a heavier-tailed variable. If X were normal, things would be much easier. Nevertheless, we may recall that a Student's t variable is related to a normal:

$$T_n = \frac{Z}{\sqrt{\chi_n^2/n}}.$$

Hence, we may rewrite X in terms of a standard normal Z and a chi-square variable:

$$X = \mu + \sigma\frac{Z}{\sqrt{Y}}, \qquad Y = \frac{\chi_n^2}{n}.$$

Thus, conditional on $Y = y$, X is normal, with a variance depending on y: $X \mid Y = y \sim \mathsf{N}(\mu, \sigma^2/y)$. Now, let us consider the conditional expected value

$$\text{E}\Big[e^{-X^2}\Big|Y = y\Big].$$

If we introduce $\xi^2 = \sigma^2/y$, the conditional expectation can be written as

$$\frac{1}{\sqrt{2\pi}\,\xi} \int_{-\infty}^{+\infty} \exp\left\{ -x^2 - \frac{1}{2}\frac{(x-\mu)^2}{\xi^2} \right\} dx. \tag{8.9}$$

---

[2]This example is borrowed from [5, pp. 107–108].

A typical trick of the trade to deal with this kind of integrand functions is completing the square in the argument of the exponential, in order to come up with something related to the PDF of a normal variable:

$$
x^2 + \frac{1}{2}\frac{(x-\mu)^2}{\xi^2} = \frac{x^2\left(1+2\xi^2\right) - 2\mu x + \mu^2}{2\xi^2}
$$

$$
= \left\{ x^2 - 2\frac{\mu}{1+2\xi^2}x + \frac{\mu^2}{1+2\xi^2} \right\}\frac{1+2\xi^2}{2\xi^2}
$$

$$
= \left\{ \left(x - \frac{\mu}{1+2\xi^2} \right)^2 + \frac{\mu^2}{1+2\xi^2} - \frac{\mu^2}{(1+2\xi^2)^2} \right\}\frac{1+2\xi^2}{2\xi^2}
$$

$$
= \frac{\left(x - \dfrac{\mu}{1+2\xi^2} \right)^2}{2\dfrac{\xi^2}{1+2\xi^2}} + \frac{\mu^2}{2\xi^2}\left\{ 1 - \frac{1}{1+2\xi^2} \right\}
$$

$$
= \frac{\left(x - \dfrac{\mu}{1+2\xi^2} \right)^2}{2\dfrac{\xi^2}{1+2\xi^2}} + \frac{\mu^2}{1+2\xi^2}.
$$

When we take the exponential of the last expression, we see that the second term does not depend on x and can be taken outside the integral, leading to a multiplicative factor

$$
\exp\left\{ -\frac{\mu^2}{1+2\xi^2} \right\}.
$$

The first term leads to an integral of a normal density with expected value

$$
\frac{\mu}{1+2\xi^2}
$$

and variance

$$
\frac{\xi^2}{1+2\xi^2}.
$$

The integral of the density is 1, of course, but this change of variance requires some adjustment in the integral of Eq. (8.9), i.e., division by

$$
\sqrt{1+2\xi^2}.
$$

Putting it all together, and substituting back $\xi^2 = \sigma^2/y$, the conditional expectation is

$$
\mathrm{E}\!\left[e^{-X^2} \,\middle|\, Y = y \right] = \exp\left\{ -\frac{\mu^2}{1+2\sigma^2/y} \right\}\frac{1}{\sqrt{1+2\sigma^2/y}}.
$$

Now, what we have to do in order to apply conditional Monte Carlo is to sample a chi-square variable Y with n degrees of freedom and plug the value in the

```
numRepl <- 100000
df <- 5
mu <- 3
sigma <- 0.5
set.seed(55555)
# crude MC
X <- mu+sigma*rt(numRepl,df)
expX <- exp(-X^2)
out <- t.test(expX)
# Conditional MC
Y <- sqrt(rchisq(numRepl,df)/df)
aux <- 1+2*(sigma/Y)^2
expXc <- exp(-mu^2/aux)/sqrt(aux)
outC <- t.test(expXc)
as.numeric(out$estimate)
as.numeric(out$conf.int)
as.numeric(outC$estimate)
as.numeric(outC$conf.int)

plot1 <- cumsum(expX)/(1:numRepl)
plot2 <- cumsum(expXc)/(1:numRepl)
plot(plot1, type='l', ylim=c(0.006,0.009))
lines(plot2, col='red')
```

FIGURE 8.4 **Dealing with a function of a Student's *t* variable by conditional Monte Carlo.**

above conditional expectation. This is carried out by the R code of Fig. 8.4, for $\mu = 3$, $\sigma = 0.5$, and $n = 5$ degrees of freedom. We use `rchisq` to sample the chi-square distribution. In the code we also apply crude Monte Carlo, using `rt` to sample the Student's variable, to compare the results. Indeed, the comparison between the two confidence intervals shows the improvement obtained by conditioning:

```
> as.numeric(out$estimate)
[1] 0.007806775
> as.numeric(out$conf.int)
[1] 0.007472332 0.008141218
> as.numeric(outC$estimate)
[1] 0.007683399
> as.numeric(outC$conf.int)
[1] 0.007590393 0.007776404
```

It is also interesting to see how the two sample means evolve when adding replications successively. This may be obtained by cumulating and plotting sums, as is done in the last lines of the R script, which produces the plot in Fig. 8.5. The difference in convergence speed to the correct value is quite evident.

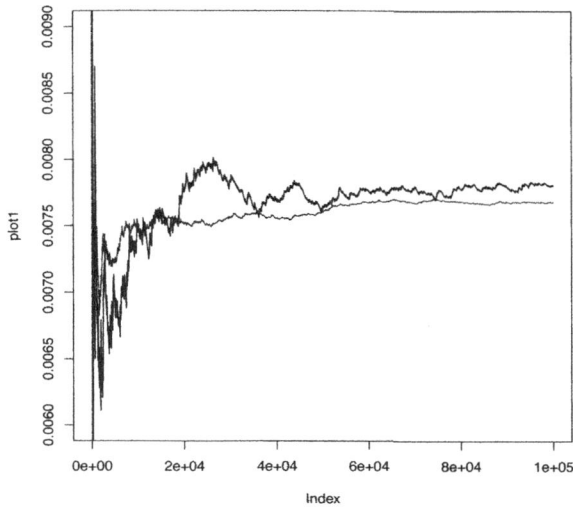

FIGURE 8.5 **Showing the effect of conditioning on the convergence of estimates.**

8.5   Stratified sampling

Stratified sampling, like conditional Monte Carlo, relies on conditioning arguments, but it has a different perspective. We are interested in estimating $\theta = \mathrm{E}[X]$, and let us suppose that X depends on another random variable Y, which may take a finite set of values y_j, $j = 1, \ldots, m$, and has the known PMF:

$$P\{Y = y_j\} = p_j, \qquad j = 1, \ldots, m.$$

Using conditioning, we see that

$$\mathrm{E}[X] = \sum_{j=1}^{m} \mathrm{E}[X \,|\, Y = y_j]p_j.$$

This suggests the idea of sampling X conditional on Y, and then to assemble a stratified estimator:

$$\hat{\theta}^s = \sum_{j=1}^{m} p_j \left(\frac{1}{N_j} \sum_{k=1}^{N_j} X_{jk} \right), \qquad (8.10)$$

where N_j is the number of independent observations allocated to stratum j, and X_{jk} is the kth observation of X in stratum j, conditional on $Y = y_j$. We speak of strata since the values y_j induce a partitioning the sample space into a mutually exclusive and collectively exhaustive collection of subsets. It is a good idea to see a simple application of the idea immediately.

▦ Example 8.4 **A stratified integral**

As a simple example of stratification, consider using simulation to compute

$$\theta = \int_0^1 h(x)\,dx = \mathrm{E}[h(U)].$$

In crude Monte Carlo simulation we draw N uniform random numbers $U_k \sim \mathsf{U}(0,1)$ and compute the sample mean

$$\frac{1}{N}\sum_{k=1}^{N} h(U_k).$$

An improved estimator with respect to crude sampling may be obtained by partitioning the integration interval $[0,1]$ into a set of m disjoint subintervals

$$\left[\frac{j-1}{m},\frac{j}{m}\right], \qquad j = 1,\ldots,m. \tag{8.11}$$

Strictly speaking, these are not disjoint intervals, but since the probability of singletons is zero, this is not quite relevant. Each event $Y = y_j$ corresponds to a random number falling in the jth subinterval; in this case we might choose $p_j = 1/m$. Note that a uniform variable on the unit interval $(0,1)$ can be mapped into a uniform variable on the intervals of Eq. (8.11) by the affine transformation $(U+j-1)/m$. Hence, for each stratum $j = 1,\ldots,m$ we generate N_j random numbers $U_k \sim \mathsf{U}(0,1)$ and estimate

$$\hat{\theta}_j = \frac{1}{N_j}\sum_{k=1}^{N_j} h\left(\frac{U_k+j-1}{m}\right).$$

Then we build the overall estimator:

$$\hat{\theta}^s = \sum_{j=1}^{m} \hat{\theta}_j p_j.$$

Clearly, we may apply the approach only if we know the probabilities p_j for each stratum and are able to sample conditionally on Y. Assuming this, the following questions arise:

1. How can we allocate a budget of N observations into N_j observations for each stratum, such that $\sum_{j=1}^{m} N_j = N$?

2. How can we have the guarantee that, by doing so, we do reduce the overall variance?

We will give an answer to the first question later. For now, let us assume that we follow a simple proportional allocation policy:

$$N_j = p_j N, \qquad j = 1, \ldots, m,$$

subject to rounding, which we neglect. Let us denote by σ_j^2 the conditional variance

$$\sigma_j^2 = \text{Var}(X \mid Y = y_j).$$

Given the independence among replication, we see that

$$\text{Var}(\hat{\theta}^s) = \sum_{j=1}^m p_j^2 \left(\frac{1}{N_j^2} \sum_{k=1}^{N_j} \sigma_j^2 \right) = \sum_{j=1}^m \frac{p_j^2 \sigma_j^2}{N_j}. \qquad (8.12)$$

If we assume proportional allocation of observation to strata, we find

$$\text{Var}(\hat{\theta}_s) = \sum_{j=1}^m \frac{p_j^2 \sigma_j^2}{p_j N} = \frac{1}{N} \sum_{j=1}^m p_j \sigma_j^2 = \frac{1}{N} \text{E}\Big[\text{Var}(X \mid Y) \Big] \leq \frac{1}{N} \text{Var}(X),$$

where the last inequality follows from the inequality (8.8), which in turn is a consequence of the conditional variance formula (8.6). Hence, by using proportional allocation we may indeed reduce variance. A further improvement can be obtained, at least in principle, by an optimal allocation of observations to strata. This may be obtained by solving the nonlinear programming problem:

$$\min \quad \sum_{j=1}^m \frac{p_j^2 \sigma_j^2}{N_j}$$

$$\text{s.t.} \quad \sum_{j=1}^m N_j = N,$$

$$N_j \geq 0.$$

To be precise, we should require that the decision variables N_j take only integer values. In practice, when N is large enough, we may relax integer variables to continuous values, and then round the optimal solution. If we assume an interior optimal solution, i.e., $N_j^* > 0$, we may relax the non-negativity bounds and apply the Lagrange multiplier approach. First, we associate a multiplier λ with the budget constraint and build the Lagrangian function

$$\mathcal{L}(N_1, \ldots, N_m, \lambda) = \sum_{j=1}^m \frac{p_j^2 \sigma_j^2}{N_j} + \lambda \left(\sum_{j=1}^m N_j - N \right),$$

and then write the stationarity conditions with respect to each decision variable:

$$-\frac{p_j^2 \sigma_j^2}{N_j^2} + \lambda = 0, \qquad j = 1, \ldots, m.$$

This condition implies that N_j should be proportional to $p_j\sigma_j$:

$$\frac{N_j}{N_l} = \frac{p_j\sigma_j}{p_l\sigma_l}, \qquad j,l = 1,\dots,m.$$

Taking the budget constraint into account, we find the optimal budget allocation

$$N_j^* = N\frac{p_j\sigma_j}{\sum_{l=1}^{m} p_l\sigma_l},$$

and the corresponding minimal variance is

$$\frac{1}{N}\left(\sum_{j=1}^{m} p_j\sigma_j\right)^2.$$

An obvious difficulty in this reasoning is that it relies on the knowledge of the conditional variances σ_j^2. Since it is quite unlikely that we know them, in order to apply optimal budget allocation we need some pilot runs to estimate the sample conditional variances:

$$S_j^2 = \frac{1}{N_j - 1}\sum_{k=1}^{N_j}\left(X_{jk} - \bar{X}_j\right)^2,$$

where

$$\bar{X}_j = \frac{1}{N_j}\sum_{k=1}^{N_j} X_{jk}$$

is the sample mean for stratum j. Furthermore, we may also build an approximate confidence interval at confidence level $(1 - \alpha)$ as follows:

$$\hat{\theta}^s \pm z_{1-\alpha/2}\sqrt{\sum_{j=1}^{m}\frac{p_j^2 S_j^2}{N_j}}. \tag{8.13}$$

Example 8.5 Estimating π by stratification

In Section 1.1 we considered the use of Monte Carlo methods to estimate π, without much of a success. To improve things, let us consider the integral

$$I = \int_0^1 \sqrt{1 - x^2}\,dx.$$

Since $x^2 + y^2 = 1$ is the well-known equation describing a unit circle, we see that I is the area of a quarter of the unit circle. Therefore, we have

$$\pi = 4\int_0^1 \sqrt{1 - x^2}\,dx.$$

In Fig. 8.6 we show three R functions to estimate π using the above integral. The first function uses crude Monte Carlo, the second one uses straightforward stratification, and the third one integrates stratification and antithetic sampling. By running the three functions with a sample size 5000 we notice that stratification improves the quality of the estimate:

```
> set.seed(55555)
> estPiNaiveMC(5000)
$estimate
[1] 3.138237
$conf.int
[1] 3.113241 3.163233
> estPiStratified(5000)
$estimate
[1] 3.141587
$conf.int
[1] 3.116832 3.166343
> estPiStratifiedAnti(5000)
$estimate
[1] 3.141593
$conf.int
[1] 3.116838 3.166348
> pi
[1] 3.141593
```

In the functions using stratification, the number of strata is just the number of replications. To be fair, we are cheating a bit when integrating stratification with antithetic sampling, as we are doubling the number of observations. Nevertheless, we see that we get an estimate which is correct within at least five decimals. Actually, the natural objection is that with this approach we are reverting back to integration quadrature on a grid of points. Indeed, as we explain later, we cannot afford the luxury of stratifying with respect to several random variables; nevertheless, using some tricks we may still apply the approach to practically relevant cases.

Example 8.6 Option pricing by stratification

Let us apply stratification to option pricing in the simplest case, the vanilla call option. It is natural to stratify with respect to the asset price at maturity, which in turn requires stratification of a standard

```
estPiNaiveMC <- function(numRepl, c.level=0.95){
  U <- runif(numRepl)
  X <- 4*sqrt(1-U^2)
  aux <- t.test(X,conf.level=c.level)
  value <- as.numeric(aux$estimate)
  ci <- as.numeric(aux$conf.int)
  return(list(estimate=value,conf.int=ci))
}

estPiStratified <- function(numRepl, c.level=0.95){
  U <- runif(numRepl)
  jVet <- (1:numRepl)
  X <- 4*sqrt(1-((U+jVet-1)/numRepl)^2)
  aux <- t.test(X,conf.level=c.level)
  value <- as.numeric(aux$estimate)
  ci <- as.numeric(aux$conf.int)
  return(list(estimate=value,conf.int=ci))
}

estPiStratifiedAnti <- function(numRepl, c.level=0.95){
  U <- runif(numRepl)
  jVet <- (1:numRepl)
  X <- 2*(sqrt(1-((U+jVet-1)/numRepl)^2) +
          sqrt(1-((jVet-U)/numRepl)^2))
  aux <- t.test(X,conf.level=c.level)
  value <- as.numeric(aux$estimate)
  ci <- as.numeric(aux$conf.int)
  return(list(estimate=value,conf.int=ci))
}
```

FIGURE 8.6 Using stratification to estimate π.

normal. To accomplish this task, we may stratify a uniform random number as we did in Example 8.4, and then obtain the standard normals by the inverse transform method. This is accomplished by the R code of Fig. 8.7, which produces the following (edited) output:

```
> EuVanillaCall(S0,K,T,r,sigma)
[1] 5.614995
> BlsMCNaive(S0,K,T,r,sigma,numRepl)
$estimate
[1] 5.74592
$conf.int
```

```r
BsmStrat <- function(S0,K,T,r,sigma,numRepl,numStrats,
                     c.level=0.95){
  nuT <- (r - 0.5*sigma^2)*T
  sigmaT <- sigma*sqrt(T)
  numPerStrat <- ceiling(numRepl/numStrats)
  meanStrat <- numeric(numStrats)
  varStrat <- numeric(numStrats)
  for (k in 1:numStrats) {
    U <- (k-1)/numStrats+runif(numPerStrat)/numStrats
    ST <- S0*exp(nuT+sigmaT*qnorm(U))
    dPayoff <- exp(-r*T)*pmax(0,ST-K)
    meanStrat[k] <- mean(dPayoff)
    varStrat[k] <- var(dPayoff)
  }
  meanTot <- mean(meanStrat)
  varTot <- mean(varStrat)
  alpha <- (100-c.level)/100
  half <- qnorm(1-alpha/2)*sqrt(varTot)
  ci <- c(meanTot-half,meanTot+half)
  return(list(estimate=meanTot,conf.int=ci))
}

S0 <- 50; K <- 50; sigma <- 0.4;
r <- 0.05; T <- 5/12
numRepl <- 10000
EuVanillaCall(S0,K,T,r,sigma)
set.seed(55555)
BlsMCNaive(S0,K,T,r,sigma,numRepl)
numStrats <- 25
set.seed(55555)
BsmStrat(S0,K,T,r,sigma,numRepl,numStrats)
```

FIGURE 8.7 **Pricing a vanilla call option by stratified sampling.**

```
[1] 5.562113 5.929728
> BsmStrat(S0,K,T,r,sigma,numRepl,numStrats)
$estimate
[1] 5.636611
$conf.int
[1] 5.614289 5.658932
```

In all of the above examples, the beneficial effect of stratification is rather evident. One may ask, however, how can we take advantage of stratification in a more relevant situation. After all, by straightforward stratification we are, in a sense, pushing back toward numerical integration on a grid, and it seems rather difficult to stratify with respect to several variables. Indeed, in pricing an option depending on several factors or the prices along a whole sample path, we cannot stratify with respect to all of them. However, we may use some tricks, such as the Brownian bridge of Section 6.2.3, to stratify with respect to some milestone prices, and then fill the gaps in each sample path using naive Monte Carlo.

8.6 Importance sampling

Unlike other variance reduction methods, importance sampling is based on the idea of "distorting" the underlying probability measure. Consider the problem of estimating

$$\theta = \mathrm{E}[h(\mathbf{X})] = \int h(\mathbf{x}) f(\mathbf{x})\, d\mathbf{x},$$

where $\mathbf{X}$ is a random vector with joint density $f(\cdot)$. If we know another density $g(\cdot)$ such that $f(\mathbf{x}) = 0$ whenever $g(\mathbf{x}) = 0$, we may write

$$\theta = \int \frac{h(\mathbf{x}) f(\mathbf{x})}{g(\mathbf{x})} g(\mathbf{x})\, d\mathbf{x} = \mathrm{E}_g\!\left[\frac{h(\mathbf{X}) f(\mathbf{X})}{g(\mathbf{X})}\right], \qquad (8.14)$$

where the notation $\mathrm{E}_g[\cdot]$ is used to stress the fact that the last expected value is taken with respect to another measure. The ratio $f(\mathbf{x})/g(\mathbf{x})$ is used to correct the change in probability measure, and it is typically called a *likelihood ratio*: When sampling $\mathbf{X}$ randomly, this ratio will be a random variable.[3] The requirement that $f(\mathbf{x}) = 0$ whenever $g(\mathbf{x}) = 0$ is needed to avoid trouble with division by zero, but it is just a technicality. The real issue is how to select a density $g(\cdot)$ that results in a significant reduction in variance. Actually, there is no guarantee that variance will be reduced, as a poor choice may well increase variance.

To get a clue, let us introduce the notation

$$\theta = \mathrm{E}_f[h(\mathbf{X})]$$

and assume for simplicity that $h(\mathbf{x}) \geq 0$. As we have pointed out, there are two possible ways of estimating θ:

$$\mathrm{E}_f[h(\mathbf{X})] = \int h(\mathbf{x}) f(\mathbf{x})\, d\mathbf{x} = \int \frac{h(\mathbf{x}) f(\mathbf{x})}{g(\mathbf{x})} g(\mathbf{x})\, d\mathbf{x}$$

$$= \int h^*(\mathbf{x}) g(\mathbf{x})\, d\mathbf{x} = \mathrm{E}_g[h^*(\mathbf{X})],$$

---

[3]Readers with a background in stochastic calculus would probably use the term "Radon–Nikodym derivative."

where $h^*(\mathbf{X}) = h(\mathbf{x})f(\mathbf{x})/g(\mathbf{x})$. The two estimators have the same expectation, but what about the variance? Using the well-known properties of the variance, we find

$$\mathrm{Var}_f[h(\mathbf{X})] = \int h^2(\mathbf{x})f(\mathbf{x})\,d\mathbf{x} - \theta^2,$$

$$\mathrm{Var}_g[h^*(\mathbf{X})] = \int h^2(\mathbf{x})\frac{f(\mathbf{x})}{g(\mathbf{x})}f(\mathbf{x})\,d\mathbf{x} - \theta^2.$$

From the second equation, it is easy to see that the choice

$$g(\mathbf{x}) = \frac{h(\mathbf{x})f(\mathbf{x})}{\theta} \qquad (8.15)$$

leads to the ideal condition $\mathrm{Var}_g[h^*(\mathbf{X})] \equiv 0$. We may find an exact estimate with *one* observation! Clearly, there must be a fly somewhere in the ointment. Indeed, this density is *truly* ideal, as it requires the knowledge of θ itself. Nevertheless, Eq. (8.15) does provide us with some guidance, as it suggests that we should use a density matching the product $h(\mathbf{x})f(\mathbf{x})$ as far as possible. By the way, you may see why we have required the condition $h(\mathbf{x}) \geq 0$, since a density cannot be negative.[4] The problem is that $f(\cdot)$ is a density, but its product with $h(\cdot)$ is not; we miss a normalization constant, which is precisely what we want to compute. Not all hope is lost, however:

1. In Chapter 14 on Markov chain Monte Carlo we will see that there are methods to sample from a density which is known up to a constant.

2. We may try to come up with an approximation of the ideal density.

As a further hint, let us consider the difference between the two variances:

$$\Delta\mathrm{Var} = \mathrm{Var}_f[h(\mathbf{X})] - \mathrm{Var}_g[h^*(\mathbf{X})] = \int h^2(\mathbf{x})\left[1 - \frac{f(\mathbf{x})}{g(\mathbf{x})}\right]f(\mathbf{x})\,d\mathbf{x}.$$

We would like to obtain $\Delta\mathrm{Var} > 0$. Unfortunately, while the factor $h^2(\mathbf{x})f(\mathbf{x})$ is non-negative, the term within brackets has to be negative somewhere, since both $f(\cdot)$ and $g(\cdot)$ are densities. We should select a new density g such that

$$\begin{cases} g(\mathbf{x}) > f(\mathbf{x}), & \text{when the term } h^2(\mathbf{x})f(\mathbf{x}) \text{ is large,} \\ g(\mathbf{x}) < f(\mathbf{x}), & \text{when the term } h^2(\mathbf{x})f(\mathbf{x}) \text{ is small.} \end{cases}$$

In fact, the name "importance sampling" derives from this observation.

---

[4] See, e.g., [6, p. 122] to see how to deal with a generic function h.

Example 8.7 Using importance sampling to estimate π

Let us consider again the estimation problem of Example 8.5, where we have noted that

$$\pi = 4 \int_0^1 \sqrt{1 - x^2}\, dx.$$

To improve our estimate, we may apply importance sampling, following an approach described in [1]. A possible idea to approximate the ideal probability distribution is to divide the integration interval $[0, 1]$ into L equally spaced subintervals of width $1/L$. The extreme points of the kth subinterval ($k = 1, \dots, L$) are $(k-1)/L$ and k/L, and the midpoint of this subinterval is $s_k = (k-1)/L + 1/(2L)$. A rough estimate of the integral is obtained by computing

$$\frac{\sum_{k=1}^L h(s_k)}{L} = \tilde{\theta} \approx \theta.$$

Then, to find an approximation of the ideal density $g(x)$, we could use something like

$$\tilde{g}(x) \equiv \frac{h(x)f(x)}{\tilde{\theta}} = \frac{h(x)L}{\sum_{k=1}^L h(s_k)},$$

where we have used the fact $f(x) = 1$ (uniform distribution). Unfortunately, this need not be a density integrating to 1 over the unit interval. In order to avoid this difficulty and to simplify sampling, we may define a probability of sampling from a subinterval and use a uniform density within each subinterval. To this end, let us consider the quantities

$$q_k = \frac{h(s_k)}{\sum_{j=1}^L h(s_j)}, \qquad k = 1, \dots, L.$$

Clearly, $\sum_k q_k = 1$ and $q_k \geq 0$, since our function h is non-negative; hence, the numbers q_k may be interpreted as probabilities. In our case, they may be used as the probabilities of selecting a sample point from the kth subinterval. To summarize, and to cast the problem within the general framework, we have

$$h(x) = \sqrt{1 - x^2},$$
$$f(x) = 1,$$
$$g(x) = Lq_k, \qquad (k-1)/L \leq x < k/L.$$

Here, $g(x)$ is a piecewise constant density; the L factor multiplying the q_k in $g(x)$ is just needed to obtain the uniform density over an interval of length $1/L$.

The resulting code is illustrated in Fig. 8.8, where m is the sample size and L is the number of subintervals. The code is fairly simple, and subintervals are selected as described in the last part of Section 5.3, where we have seen how to sample discrete empirical distributions by the function EmpiricalDrnd of Fig. 5.7:

```
> pi
[1] 3.141593
> set.seed(55555)
> estpiIS(1000,10)
[1] 3.135357
> estpiIS(1000,10)
[1] 3.142377
> estpiIS(1000,10)
[1] 3.14807
> estpiIS(1000,10)
[1] 3.148633
> estpiIS(1000,10)
[1] 3.140776
```

We see that the improved code, although not a very sensible way to compute π, yields a remarkable reduction in variance with respect to crude Monte Carlo.

The approach that we have just pursued looks suspiciously like stratified sampling. Actually, there is a subtle difference. In stratified sampling we define a set of strata, which correspond to events of known probability; here we have not used strata with known probability, as we have used sampling to estimate the probabilities q_k.

Example 8.8 Dancing on the tail of a normal

Let us consider the expected value $E[h(X)]$, where $X \sim N(0,1)$ and

$$h(x) = \begin{cases} 0, & x < 3, \\ 1000 \times (x-3), & 3 \le x < 4, \\ -1000 \times (x-5), & 4 \le x \le 5, \\ 0, & x > 5. \end{cases}$$

We observe that the function $h(\cdot)$ is a triangle with support on the interval $(3,5)$ and a fairly large maximum at $x = 4$. However, since a standard normal has little density outside the interval $(-3,3)$, we have large function values where the probability is low. Crude Monte

```
estpiIS <- function(m,L){
  # define left end points of subintervals
  s <- seq(from=0, by=1/L, to=1-1/L) + 1/(2*L)
  hvals <- sqrt(1 - s^2)
  # get cumulative probabilities
  cs <- cumsum(hvals)
  est <- numeric(m)
  for (j in 1:m){
    # locate subinterval
    loc <- sum(runif(1)*cs[L] > cs) + 1
    # sample uniformly within subinterval
    x <- (loc-1)/L + runif(1)/L
    p <- hvals[loc]/cs[L]
    est[j] <- sqrt(1 - x^2)/(p*L)
  }
  z <- 4*sum(est)/m
  return(z)
}
```

FIGURE 8.8 **Using importance sampling to estimate π.**

Carlo will generate almost all of the observations where the function value is zero, with a corresponding waste of effort. A huge sample size is needed to get an accurate estimate. Therefore, we may try to shift the mode of the density in order to match the maximum of the function. Hence, let us consider a distribution $N(\theta, 1)$, with $\theta = 4$. Finding the likelihood ratio is fairly easy:

$$\frac{f(x)}{g(x)} = \frac{\frac{1}{\sqrt{2\pi}} \exp\left(-x^2/2\right)}{\frac{1}{\sqrt{2\pi}} \exp\left(-(x-\theta)^2/2\right)} = \exp\left\{-\frac{x^2 - (x-\theta)^2}{2}\right\}$$

$$= \exp\left\{-\theta x + \frac{\theta^2}{2}\right\}. \tag{8.16}$$

In Fig. 8.9 we show R code to evaluate the triangle function (please note how vectorization is achieved) and to compare the outcome of crude Monte Carlo against importance sampling, producing the following output:

```
> as.numeric(out$estimate)
[1] 0.4058854
```

```
triangle <- function(x){
  out <- numeric(length(x))
  idx1 <- which((x>3)&(x<=4))
  idx2 <- which((x>4)&(x<5))
  out[idx1] <- 1000*(x[idx1]-3)
  out[idx2] <- -1000*(x[idx2]-5)
  return(out)
}

set.seed(55555)
z <- rnorm(10000)
out <- t.test(triangle(z))
likRatio <- function(x,theta){exp(-theta*x+theta^2/2)}
theta <- 4
x <- z+theta
outC <- t.test(triangle(x)*likRatio(x,theta))
as.numeric(out$estimate)
as.numeric(out$conf.int)
as.numeric(outC$estimate)
as.numeric(outC$conf.int)
```

FIGURE 8.9 **Shifting the expected value of a normal distribution.**

```
> as.numeric(out$conf.int)
[1] 0.1120629 0.6997080
> as.numeric(outC$estimate)
[1] 0.3631287
> as.numeric(outC$conf.int)
[1] 0.3527247 0.3735328
```

We see how a very large confidence interval results from crude Monte Carlo, whereas shifting the mean of the normal produces a much more accurate estimate.

On the one hand, Example 8.8 shows a possible heuristic to find a distorted density, by matching it with the maximum of the integrand function. On the other hand, we see how rare events may be dealt with by importance sampling. We elaborate on the theme in the next section, using a more practically relevant example.

8.6.1   IMPORTANCE SAMPLING AND RARE EVENTS

Importance sampling is often used when small probabilities are involved. Consider, for instance, a random vector $\mathbf{X}$ with joint density f, and suppose that we want to estimate

$$\theta = \mathrm{E}[h(\mathbf{X}) \mid \mathbf{X} \in \mathcal{A}],$$

where $\{\mathbf{X} \in \mathcal{A}\}$ is a rare event with a small but unknown probability $\mathrm{P}\{\mathbf{X} \in \mathcal{A}\}$. Such an event could be the occurrence of a large loss, which is relevant in risk management. The conditional density is

$$f(\mathbf{x}|\mathbf{X} \in \mathcal{A}) = \frac{f(\mathbf{x})}{\mathrm{P}\{\mathbf{X} \in \mathcal{A}\}}$$

for $\mathbf{x} \in \mathcal{A}$. By defining the indicator function

$$\mathbf{1}_{\mathcal{A}}(\mathbf{X}) = \begin{cases} 1, & \text{if } \mathbf{X} \in \mathcal{A}, \\ 0, & \text{if } \mathbf{X} \notin \mathcal{A}, \end{cases}$$

we may rewrite θ as

$$\theta = \frac{\displaystyle\int_{\mathbf{x}\in\mathcal{A}} h(\mathbf{x})f(\mathbf{x}) \, d\mathbf{x}}{\mathrm{P}\{\mathbf{X} \in \mathcal{A}\}} = \frac{\mathrm{E}\big[h(\mathbf{X})\mathbf{1}_{\mathcal{A}}(\mathbf{X})\big]}{\mathrm{E}\big[\mathbf{1}_{\mathcal{A}}(\mathbf{X})\big]}.$$

If we use crude Monte Carlo simulation, many samples will be wasted, as the event $\{\mathbf{X} \in \mathcal{A}\}$ will rarely occur. Now, assume that there is a density $g(\cdot)$ such that this event is more likely under the corresponding probability measure. Then, we generate the observations $\mathbf{X}_i$ according to $g(\cdot)$ and estimate

$$\hat{\theta} = \frac{\displaystyle\sum_{i=1}^{k} \frac{h(\mathbf{X}_i)\mathbf{1}_{\mathcal{A}}(\mathbf{X}_i)f(\mathbf{X}_i)}{g(\mathbf{X}_i)}}{\displaystyle\sum_{i=1}^{k} \frac{\mathbf{1}_{\mathcal{A}}(\mathbf{X}_i)f(\mathbf{X}_i)}{g(\mathbf{X}_i)}}.$$

Let us illustrate the idea with a simple but quite instructive example: Pricing a deeply out-of-the-money vanilla call option. If S_0 is the initial price of the underlying asset, we know that its expected value at maturity is, according to geometric Brownian motion model under the risk-neutral measure, $S_0 e^{rT}$. If this expected value is small with respect to the strike price K, it is unlikely that the option will be in-the-money at maturity, and with crude Monte Carlo many replications are wasted, because the payoff is zero in most of them. We should change the drift in order to increase the probability that the payoff is positive. One possible choice is a drift μ such that the expected value of S_T is the strike price:

$$S_0 e^{\mu T} = K \quad \Rightarrow \quad \mu = \frac{1}{T} \log\left(\frac{K}{S_0}\right).$$

While under the risk-neutral measure we sample $S_T = S_0 e^Z$ by generating normal variates

$$Z \sim \mathsf{N}\left[\left(r - \frac{\sigma^2}{2}\right)T,\ \sigma\sqrt{T}\right],$$

we should sample $S_T = S_0 e^Y$ by generating

$$Y \sim \mathsf{N}\left[\log\left(\frac{K}{S_0}\right) - \frac{\sigma^2 T}{2},\ \sigma\sqrt{T}\right].$$

This in turn requires sampling standard normal variates ϵ and then using

$$Y = \log\left(\frac{K}{S_0}\right) - \frac{\sigma^2 T}{2} + \sigma\sqrt{T}\epsilon.$$

Now the tricky part is to compute the likelihood ratio. For the sake of clarity, assume that we sample Y from a normal distribution $\mathsf{N}(\beta, \xi)$ whereas the original distribution is $\mathsf{N}(\alpha, \xi)$. Then, the ratio of the two densities can be found much like we did in Example 8.8:

$$\frac{\frac{1}{\sqrt{2\pi}\xi}\exp\left\{-\frac{(Y-\alpha)^2}{2\xi^2}\right\}}{\frac{1}{\sqrt{2\pi}\xi}\exp\left\{-\frac{(Y-\beta)^2}{2\xi^2}\right\}} = \exp\left\{-\frac{(Y-\alpha)^2 - (Y-\beta)^2}{2\xi^2}\right\}$$

$$= \exp\left\{\frac{2(\alpha-\beta)Y - \alpha^2 + \beta^2}{2\xi^2}\right\}. \tag{8.17}$$

Now it is easy to extend the naive Monte Carlo code of function `BlsMCNaive`, see Fig. 6.4, to the function `BlsMCIS` displayed in Fig. 8.10. We may check the efficiency gain of importance sampling by running the script in Fig. 8.11. In the script we estimate the price with both crude Monte Carlo and importance sampling, in order to compare the percentage errors with respect to the exact price. We price the same option 100 times, in order to collect the average percentage error and the worst-case percentage error. We reset the random variate generator twice in order to use exactly the same stream of standard normal variates. Running the script, we obtain

```
Average Percentage Error:
  MC    =   2.248%
  MC+IS =   0.305%
Worst Case Percentage Error:
  MC    =   6.178%
  MC+IS =   1.328%
```

We notice that, as we hoped, importance sampling is able to reduce the average error. What is probably even more striking is that the worst-case error may be quite large in naive Monte Carlo, whereas it is more reasonable with importance sampling, which indeed seems more robust. We should note, however, that this improvement is not to be expected for at-the-money options.

```
BlsMCIS <- function(S0,K,T,r,sigma,numRepl,c.level=0.95){
  nuT <- (r - 0.5*sigma^2)*T
  sigmaT <- sigma*sqrt(T)
  ISnuT <- log(K/S0) - 0.5*sigma^2*T
  norms <- rnorm(numRepl,ISnuT,sigmaT)
  ISRatios <- exp( (2*(nuT - ISnuT)*norms - nuT^2 -
              ISnuT^2)/2/sigmaT^2)
  dpayoff <- exp(-r*T)*pmax(0, (S0*exp(norms)-K))
  aux <- t.test(dpayoff*ISRatios,conf.level=c.level)
  value <- as.numeric(aux$estimate)
  ci <- as.numeric(aux$conf.int)
  return(list(estimate=value,conf.int=ci))
}
```

FIGURE 8.10 **Using importance sampling to price a deeply out-of-the-money vanilla call option.**

```
S0 <- 50; K <- 80; r <- 0.05;
sigma <- 0.4; T <- 5/12
numRepl <- 100000
numTimes <- 100
MCError <- numeric(numTimes)
MCISError <- numeric(numTimes)
truePrice <- EuVanillaCall(S0,K,T,r,sigma)
set.seed(55555)
for (k in 1:numTimes){
  MCPrice <- BlsMCNaive(S0,K,T,r,sigma,numRepl)$estimate
  MCError[k] <- abs(MCPrice - truePrice)/truePrice
}
set.seed(55555)
for (k in 1:numTimes){
  MCISPrice <- BlsMCIS(S0,K,T,r,sigma,numRepl)$estimate
  MCISError[k] <- abs(MCISPrice - truePrice)/truePrice
}
cat('Average Percentage Error:\n')
cat(sprintf(' MC   = %6.3f%%', 100*mean(MCError)), '\n')
cat(sprintf(' MC+IS = %6.3f%%', 100*mean(MCISError)), '\n')
cat('Worst Case Percentage Error:\n')
cat(sprintf(' MC   = %6.3f%%', 100*max(MCError)), '\n')
cat(sprintf(' MC+IS = %6.3f%%', 100*max(MCISError)),'\n')
```

FIGURE 8.11 **Comparing the estimation errors obtained by crude Monte Carlo and importance sampling for a deeply out-of-the-money vanilla call option.**

8.6.2 A DIGRESSION: MOMENT AND CUMULANT GENERATING FUNCTIONS

In the previous section we have used a heuristic argument to come up with an importance sampling measure to improve pricing accuracy for a deeply out-of-the-money option. In the next section we will consider a systematic way to find distorted densities for use within importance sampling. Here we pause a little to introduce a couple of relevant concepts for the sake of the unfamiliar reader. As usual, we will not strive for full generality and mathematical rigor, as our aim is just to point out the practical relevance of some concepts from probability theory.

DEFINITION 8.1 (Moment generating function) *Given a continuous random variable X with density $f_X(\cdot)$, its moment generating function is defined as*

$$M_X(\theta) \equiv \mathrm{E}\left[e^{\theta X}\right] = \int_{-\infty}^{+\infty} e^{\theta x} f_X(x)\, dx. \qquad (8.18)$$

The interest of the moment generating function lies in the fact that, as the name suggests, it can be used to compute moments of a random variable, since the following fact can be proved:

$$m_n \equiv \mathrm{E}\left[X^n\right] = M_X^{(n)}(\theta)\Big|_{\theta=0},$$

which means that to compute the moment of order n of the random variable X, denoted by m_n, we can take the derivative of order n of its moment generating function and evaluate it for $\theta = 0$. To see this, let us consider the series expansion of an exponential:

$$e^{\theta x} = 1 + \theta x + \frac{\theta^2 x^2}{2!} + \frac{\theta^3 x^3}{3!} + \cdots + \frac{\theta^n x^n}{n!} + \cdots.$$

If we plug a random variable X into this expression and take an expectation, we find

$$M_X(\theta) \equiv \mathrm{E}\left[e^{\theta X}\right] = 1 + \theta m_1 + \frac{\theta^2 m_2}{2!} + \frac{\theta^3 m_3}{3!} + \cdots + \frac{\theta^n m_n}{n!} + \cdots.$$

By taking the first order derivative, it is easy to see that

$$M_X^{(1)}(\theta) = m_1 + \theta m_2 + \frac{\theta^2 m_3}{2!} + \cdots.$$

By setting $\theta = 0$ we obtain $m_1 = M_X^{(1)}(0)$. By the same token, if we take the second order derivative we obtain

$$M_X^{(2)}(\theta) = m_2 + \theta m_3 + \cdots,$$

which implies $m_2 = M_X^{(2)}(0)$, and so on. Now the problem is to find the moment generating function for the most common distributions. The next example outlines the case of a normal random variable.

◼ Example 8.9 **Moment generating function for normal variables**

It is not too difficult to prove by an integration exercise that the moment generating function of a normal random variable $X \sim \mathsf{N}(\mu, \sigma^2)$ is

$$M_X(\theta) = \exp\left\{\theta\mu + \tfrac{1}{2}\theta^2\sigma^2\right\}. \tag{8.19}$$

The proof relies on typical tricks of the trade, like completing the square within an exponential, much like we have seen in Section 8.4. Then, it is a simple, even though a bit tedious, exercise to prove that for a normal variable X we have

$$\mathrm{E}\left[X^3\right] = 0, \qquad \mathrm{E}\left[X^4\right] = 3.$$

The latter result is a key step to prove that the kurtosis of a normal variable is 3.

DEFINITION 8.2 (Cumulant generating function) *The cumulant generating function of a random variable X with moment generating function $M_X(\theta)$ is defined as*

$$\psi_X(\theta) = \log M_X(\theta) = \log \mathrm{E}\left[e^{\theta X}\right]. \tag{8.20}$$

For instance, given Eq. (8.19), we immediately see that the cumulant generating function of a normal variable $X \sim \mathsf{N}(\mu, \sigma^2)$ is

$$\psi_X(\theta) = \theta\mu + \tfrac{1}{2}\theta^2\sigma^2. \tag{8.21}$$

We should also mention that the cumulant generating function, if it exists, is convex. The cumulant generating function plays a relevant role in exponential tilting for importance sampling.

8.6.3 EXPONENTIAL TILTING

Exponential tilting, also known as exponential twisting, is a common strategy to define distributions for use within importance sampling. The idea is to multiply the original density[5] $f(x)$ by an exponential function depending on a parameter θ. The result, $e^{\theta x}f(x)$, is not a density and needs normalization to define a new density depending on the tilting parameter θ:

$$f_\theta(x) = \frac{e^{\theta x}f(x)}{\mathrm{E}[e^{\theta X}]}. \tag{8.22}$$

We immediately recognize the denominator of this ratio as the moment generating function $M_X(\theta)$. By recalling the definition of the cumulant generating

---

[5]To avoid notational clutter, here we denote the density of a random variable X by $f(x)$ rather than $f_X(x)$.

function, we may rewrite Eq. (8.22) as

$$f_\theta(x) = e^{\theta x - \psi(\theta)} f(x). \tag{8.23}$$

This relationship provides us with the likelihood ratio that we need to apply importance sampling. In fact, from Eq. (8.14) we know that the expected value $E[h(X)]$ under density $f(\cdot)$ can be rewritten as

$$E[h(X)] = E_\theta \left[\frac{h(X)f(X)}{f_\theta(X)} \right] = E_\theta \left[e^{-\theta X + \psi(\theta)} h(X) \right]. \tag{8.24}$$

Thus, in order to apply importance sampling with exponential tilting, we have to:

1. Sample X from the tilted density $f_\theta(\cdot)$.
2. Use the estimator provided by Eq. (8.24), which involves the tilting parameter θ and the cumulant generating function $\psi(\theta)$.

In order to accomplish the first step above, we should wonder which kind of random variable we find, if we tilt its distribution exponentially. The following example illustrates yet another nice property of the normal distribution.

Example 8.10 **Tilting a normal distribution**

Let us find the density of a tilted normal variable $X \sim N(\mu, \sigma^2)$:

$$f_\theta(x) = \frac{e^{\theta x} f(x)}{M_X(\theta)} = \frac{e^{\theta x} \dfrac{1}{\sqrt{2\pi}\,\sigma} \exp\left\{ -\dfrac{1}{2} \left(\dfrac{x - \mu}{\sigma} \right)^2 \right\}}{\exp\left(\theta\mu + \dfrac{1}{2}\theta^2\sigma^2 \right)}.$$

By putting the exponentials together, completing squares, and rearranging, we end up with

$$f_\theta(x) = \frac{1}{\sqrt{2\pi}\,\sigma} \exp\left\{ -\frac{1}{2} \left[\frac{x - (\mu + \theta\sigma^2)}{\sigma} \right]^2 \right\},$$

which shows that the tilted variable is again a normal variable, but with a shifted expected value: $X_\theta \sim N(\mu + \theta\sigma^2, \sigma^2)$.

The next example deals with the case in which, in order to run a Monte Carlo simulation, we have to generate a sequence of normal variables.

⬛ Example 8.11 **Likelihood ratios of tilted distributions**

Consider a simulation which is driven buy a sequence of n i.i.d. variables $X_1, \ldots, X_n$. Thanks to independence, the likelihood ratio is related to a product of densities, which can be expressed in terms of the cumulant generating function:

$$\prod_{i=1}^{n} \frac{f(X_i)}{f_\theta(X_i)} = \prod_{i=1}^{n} \frac{f(X_i)}{\exp(\theta X_i - \psi(\theta)) f(X_i)}$$
$$= \exp\left\{ -\theta \sum_{i=1}^{n} X_i + n\psi(\theta) \right\}. \qquad (8.25)$$

Now the open question is how to select a suitable tilting parameter θ. Typically, problem-dependent arguments are used. We have seen an example when dealing with a deeply out-of-the-money call option, where we argued that, under the importance sampling measure, the expected value of the underlying asset price at maturity should be the option strike price. Hence, let us say that, using such intuitive arguments, we conclude that the expected value of X under the tilted distribution should be $E_\theta[X] = \beta$. To accomplish this objective, we rely on a useful result. If we take the derivative of the cumulant generating function with respect to θ, we find

$$\psi'(\theta) = \frac{E[Xe^{\theta X}]}{E[e^{\theta X}]} = E[Xe^{\theta X - \psi(\theta)}] = E_\theta[X], \qquad (8.26)$$

where we have used Eq. (8.24) the other way around. Therefore, we may solve the equation

$$\psi'(\theta) = \beta$$

to find the corresponding tilting parameter. The material of this section will play a fundamental role in Section 13.4, where we use importance sampling to estimate a risk measure, value-at-risk, by Monte Carlo simulation.

For further reading

- Every intermediate to advanced book on Monte Carlo methods has a chapter on variance reduction strategies; see, e.g., [4].
- More specific references, offering interesting examples, are [3] and [7].
- We did not cover all of the variance reduction strategies that have been proposed in the literature:
  - For Latin hypercube sampling see, e.g., [4].

- Specific control variates for option pricing, based on hedging arguments, are discussed in [2].
- The cross-entropy method is discussed in [4] and [7].
- Moment matching is applied to option pricing in [3].

References

1 I. Beichl and F. Sullivan. The importance of importance sampling. *Computing in Science and Engineering*, 1:71–73, March–April 1999.

2 L. Clewlow and C. Strickland. *Implementing Derivatives Models*. Wiley, Chichester, West Sussex, England, 1998.

3 P. Glasserman. *Monte Carlo Methods in Financial Engineering*. Springer, New York, 2004.

4 D.P. Kroese, T. Taimre, and Z.I. Botev. *Handbook of Monte Carlo Methods*. Wiley, Hoboken, NJ, 2011.

5 C.P. Robert and G. Casella. *Introducing Monte Carlo Methods with R*. Springer, New York, 2011.

6 R.Y. Rubinstein. *Simulation and the Monte Carlo Method*. Wiley, Chichester, 1981.

7 H. Wang. *Monte Carlo Simulation with Applications in Finance*. CRC Press, Boca Raton, FL, 2012.

## Chapter Nine

Low-Discrepancy Sequences

In the previous chapter we have reviewed the main variance reduction strategies that are used in financial and economic applications. Since those approaches are based on probabilistic concepts, they implicitly assume that random sampling in Monte Carlo methods is *really* random. However, the pseudorandom numbers produced by an LCG or by more sophisticated algorithms are not random at all. Hence, one could take a philosophical view and wonder about the very validity of variance reduction methods, and even the Monte Carlo approach itself. Taking a more pragmatic view, and considering the fact that Monte Carlo methods have proven their value over the years, we should conclude that this shows that there are some deterministic number sequences that work well in generating samples. It is also useful to remember that the aim of a Monte Carlo simulation is actually to estimate a multidimensional integral on the unit hypercube:

$$\int_{(0,1)^m} h(\mathbf{u}) \, d\mathbf{u}.$$

The function $h(\cdot)$ may be so complicated that we cannot express it analytically, but this is of no concern conceptually. We need a stream of i.i.d. random numbers to fill the integration domain in a satisfactory manner. When a regular grid derived from classical product rules for numerical integration (see Chapter 2) is not feasible, we may fill it by random numbers, but we could also resort to alternative deterministic sequences that are, in some sense, evenly distributed. This idea may be made more precise by defining the *discrepancy* of a sequence of numbers. In a sense, low-discrepancy sequences are half-way between deterministic grid-based methods and Monte Carlo sampling, as they are *meant* to be deterministic, whereas pseudorandom numbers are deterministic for the lack of a better alternative, yet they do not look regular at all, unlike grids. Low-discrepancy sequences are the foundation of quasi–Monte Carlo methods; sometimes, the term "quasi-random" sequence is used, but we will mostly avoid it as it is a bit misleading.

We first introduce the concept of discrepancy in Section 9.1. Then we cover two well-known low-discrepancy sequences: Halton sequences in Section 9.2 and Sobol sequences in Section 9.3. These two sequences do not exhaust the list of possible approaches, yet they are adequate to grasp the basic concepts and understand the pitfalls of quasi–Monte Carlo simulation. The deterministic nature of low-discrepancy sequences has one major drawback with respect to

379

random sampling: There is no obvious way to assess the quality of estimates, since the usual confidence intervals do not make any sense. There are theoretical bounds on the integration error, but they may be very weak and of little practical use. A possible remedy is the randomization of low-discrepancy sequences. Furthermore, there are scrambling strategies to improve the quality of standard low-discrepancy sequences. We outline these matters in Section 9.4, without entering into theoretical details; some scrambling strategies are offered in R functions, and we just aim at providing the reader with the minimal background for their use. Low-discrepancy sequences play in quasi–Monte Carlo methods the same role as pseudorandom sequences do in Monte Carlo methods. When we need observations from another distribution, most notably the normal, we resort to some techniques that we have introduced in Chapter 5 for random variate generation. We illustrate such transformations and sample path generation by low-discrepancy sequences in Section 9.5. Finally, we should also mention that:

- Sampling by low-discrepancy sequences does not only play a role in simulation, but also in global optimization, i.e., the solution of hard optimization problems.

- Other deterministic sampling approaches are available, relying on good lattice rules; see Section 2.4.3.

9.1 Low-discrepancy sequences

In the first chapters of the book we have insisted on the fact that a Monte Carlo simulator is essentially a complicated function mapping a stream of independent uniform random variables into an estimator of a quantity of interest. Whenever the dimension m of the integration domain is fixed, we are actually computing an integral over the unit hypercube $I^m = (0, 1)^m \subset \mathbb{R}^m$. This is the case when we use, for instance, the inverse transform method to obtain random variates from uniform pseudorandom numbers. However, this does not hold when we resort to acceptance–rejection, because with this kind of method we do not know a priori how many uniform random variables are needed to generate a single observation of the distribution of interest. Indeed, this is the reason why acceptance–rejection strategies are not compatible with the methods of this chapter. A good integration strategy should "cover" the unit hypercube in the most even and uniform way; note that, in this framework, we need not be concerned with independence of random variates, since this is actually guaranteed by a good coverage of I^m. However, we need to clarify what a "good coverage" is, which in turn requires a well-defined way to measure its quality.

Assume that we want to generate a sequence $\mathcal{L}_N$ of N points in I^m,

$$\mathcal{L}_N = \left(\mathbf{x}^{(1)}, \mathbf{x}^{(2)}, \ldots, \mathbf{x}^{(N)} \right),$$

in order to cover that domain in a satisfactory way. If the points are well distributed, the number of them included in any subset G of I^m should be roughly proportional to its volume vol(G). For instance, if $N = 100$, $m = 2$, and we consider the square $G = [0, 0.5] \times [0, 0.5] \subset I^2$, we would expect to find 25 points in set G, since its area is $0.5 \times 0.5 = 0.25$. Whatever N, about 25% of the total number of points should fall there. However, this should apply to any subset of I^2. The concept of discrepancy aims at formalizing this intuition in a mathematically tractable way. Since dealing with arbitrary subsets of the unit hypercube is not quite feasible, we should restrict their form. Given a vector $\mathbf{x} = [x_1, x_2, \ldots, x_m]^\mathsf{T}$, consider the rectangular subset $G_\mathbf{x}$ defined as

$$G_\mathbf{x} = [0, x_1) \times [0, x_2) \times \cdots \times [0, x_m), \qquad (9.1)$$

which has volume $x_1 x_2 \cdots x_m$. If we denote by $S(\mathcal{L}, G)$ a function counting the number of points in the sequence $\mathcal{L}$ that are contained in a subset $G \subset I^m$, a possible definition of discrepancy is

$$D^*(\mathcal{L}_N) = D^*\big(\mathbf{x}^{(1)}, \ldots, \mathbf{x}^{(N)}\big) = \sup_{\mathbf{x} \in I^m} \Big| S(\mathcal{L}_N, G_\mathbf{x}) - N x_1 x_2 \cdots x_m \Big|.$$

To be precise, this is called the *star discrepancy*, and it measures the worst-case distance between the ideal and the actual number of points included in a rectangle of the type specified by Eq. (9.1). In Section 4.3.2 we have considered the Kolmogorov–Smirnov measure of fit for probability distributions: Star discrepancy can be interpreted as such a measure of fit for a multidimensional uniform distribution with independent marginals.

When computing a multidimensional integral on the unit hypercube, it is natural to look for low-discrepancy sequences. Some theoretical results suggest that low-discrepancy sequences may perform better than pseudorandom sequences obtained through an LCG or its variations. From the theory of confidence intervals, we know that the estimation error with Monte Carlo sampling is something like

$$O\big(\frac{1}{\sqrt{N}}\big),$$

where N is the sample size. With certain point sequences, it can be shown that the error is something like

$$O\left(\frac{(\log N)^m}{N} \right), \qquad (9.2)$$

where m is the dimension of the space in which we are integrating.[1] It is conjectured that the expression in Eq. (9.2) is a lower bound on what can be achieved, and the term "low-discrepancy sequence" is used when referring to sequences achieving this bound. Various such sequences have been proposed in the literature, and some are available in R. In the following, we illustrate the basic ideas behind two of them, namely, Halton and Sobol sequences.

---

[1]This is stated in a slightly more precise form in Section 9.4. See [6] for a detailed and rigorous account of these results.

9.2 Halton sequences

The Halton sequence was the first sequence for which the attainment of the low-discrepancy bound of Eq. (9.2) was shown. This sequence relies on a generalization of the Van der Corput one-dimensional sequence, which is based on the following simple recipe:

- Represent an integer number n in a base b, where b is a prime number:

$$n = (\cdots d_4 d_3 d_2 d_1 d_0)_b.$$

- Reflect the digits and add a radix point to obtain a number within the unit interval:

$$h = (0.d_0 d_1 d_2 d_3 d_4 \cdots)_b.$$

More formally, if we represent an integer number n as

$$n = \sum_{k=0}^{m} d_k b^k,$$

the nth number in the Van der Corput sequence with base b is

$$V(n, b) = \sum_{k=0}^{m} d_k b^{-(k+1)}.$$

▣ Example 9.1 Van der Corput sequence in base 2

The generation of a sequence with base 2 is quite instructive. Let us start with $n = 1$, which may be represented in base 2 as

$$n = 1 = 1 \times 2^0 + 0 \times 2^1 + 0 \times 2^2 + \cdots = (\ldots 0001)_2.$$

Reflecting this number we find

$$V(1, 2) = (0.1000\ldots)_2 = 1 \times 2^{-1} = 0.5.$$

Applying the same drill to $n = 2$ yields

$$n = 2 = 0 \times 2^0 + 1 \times 2^1 + 0 \times 2^2 + \cdots = (\ldots 0010)_2$$
$$\Rightarrow (0.0100\ldots)_2 = 0 \times 2^{-1} + 1 \times 2^{-2} = 0.25.$$

Now the pattern should be clear enough:

$$n = 3 = 1 \times 2^0 + 1 \times 2^1 + 0 \times 2^2 + \cdots = (\ldots 0011)_2$$
$$\Rightarrow (0.1100\ldots)_2 = 1 \times 2^{-1} + 1 \times 2^{-2} = 0.75,$$
$$n = 4 = 1 \times 2^0 + 0 \times 2^1 + 1 \times 2^2 + \cdots = (\ldots 0100)_2$$
$$\Rightarrow (0.0010\ldots)_2 = 0 \times 2^{-1} + 0 \times 2^{-2} + 1 \times 2^{-3} = 0.125.$$

This sequence may be generated in R using the `halton` function from the `randtoolbox` package:

```
> library(randtoolbox)
> halton(10, dim = 1)
  [1] 0.5000 0.2500 0.7500 0.1250 0.6250 0.3750
      0.8750 0.0625 0.5625 0.3125
```

We see from the example how a Van der Corput sequence works: It fills the unit interval by jumping around in such a way to maintain a balanced coverage. A Halton sequence in multiple dimensions is obtained by associating a Van der Corput sequence with each dimension, using prime numbers for each base. For instance, the snapshot

```
> halton(10, dim = 3)
          [,1]         [,2] [,3]
  [1,] 0.5000 0.33333333 0.20
  [2,] 0.2500 0.66666667 0.40
  [3,] 0.7500 0.11111111 0.60
  [4,] 0.1250 0.44444444 0.80
  [5,] 0.6250 0.77777778 0.04
  [6,] 0.3750 0.22222222 0.24
  [7,] 0.8750 0.55555556 0.44
  [8,] 0.0625 0.88888889 0.64
  [9,] 0.5625 0.03703704 0.84
 [10,] 0.3125 0.37037037 0.08
```

shows that the base $b = 2$ is associated with dimension 1, whereas bases $b = 3$ and $b = 5$ are associated with dimensions 2 and 3, respectively.

It is instructive to compare how a pseudorandom sample and a Halton sequence cover the unit square $(0, 1) \times (0, 1)$. The following snapshot produces the plots in Figs. 9.1 and 9.2:

```
> set.seed(55555)
> haltonSeq <- halton(1000, dim = 2)
> windows()
> plot(haltonSeq[,1],haltonSeq[,2],pch=20)
> grid(nx=10,col='black')
> windows()
> plot(runif(1000),runif(1000),pch=20)
> grid(nx=10,col='black')
```

The judgment is a bit subjective here, but it could be argued that the coverage of the Halton sequence is more even. What is certainly true is that using non-prime numbers as bases may result in quite unsatisfactory patterns, such as the one

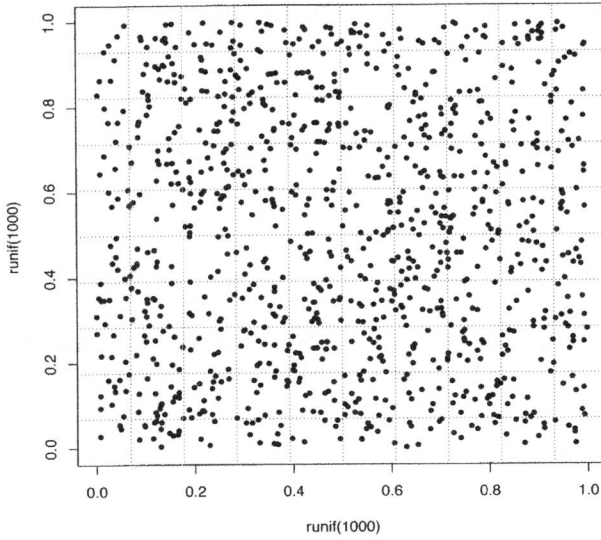

FIGURE 9.1 **Pseudorandom sample in two dimensions.**

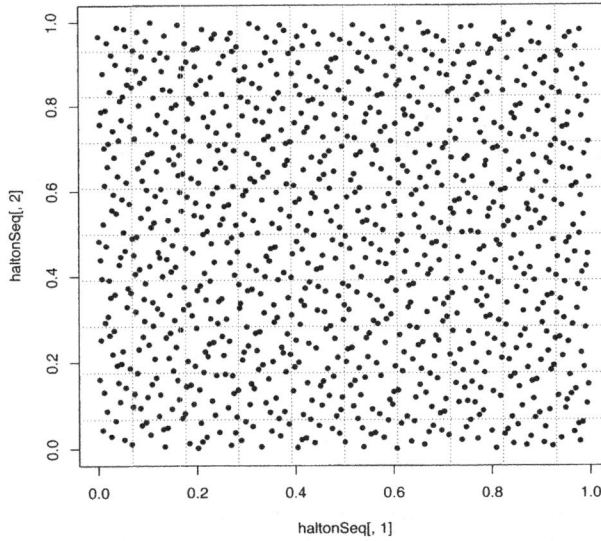

FIGURE 9.2 **Covering the bidimensional unit square with Halton sequences.**

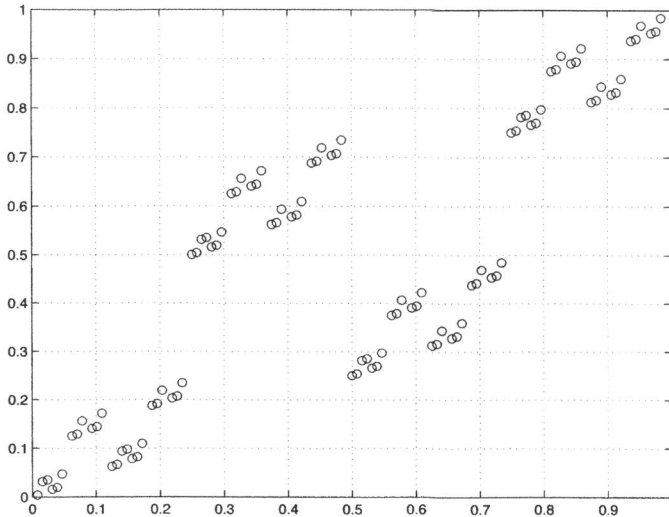

FIGURE 9.3 **Bad choice of bases in Halton sequences.**

shown in Fig. 9.3. This plot has been obtained using Van der Corput sequences with bases 2 and 4.

Example 9.2 **Numerical integration with Halton sequences**

Let us explore the use of Halton low-discrepancy sequences in a bidimensional integration context. Suppose that we want to compute

$$\int_0^1 \int_0^1 e^{-xy} \left(\sin 6\pi x + \cos 8\pi y \right) \, dx \, dy.$$

The surface corresponding to this function is illustrated in Fig. 9.4 and features a nasty oscillatory behavior. In R, we first write a function to compute the integrand and then use the `adaptIntegrate` function from the `cubature` package to get an estimate by traditional quadrature formulas:

```
> library(cubature)
> f <- function(x) exp(-x[1]*x[2])*
      (sin(6*pi*x[1])+cos(8*pi*x[2]))
> adaptIntegrate(f,rep(0,2),rep(1,2),tol=1e-4)
$integral
[1] 0.01986377
$error
[1] 1.983034e-06
```

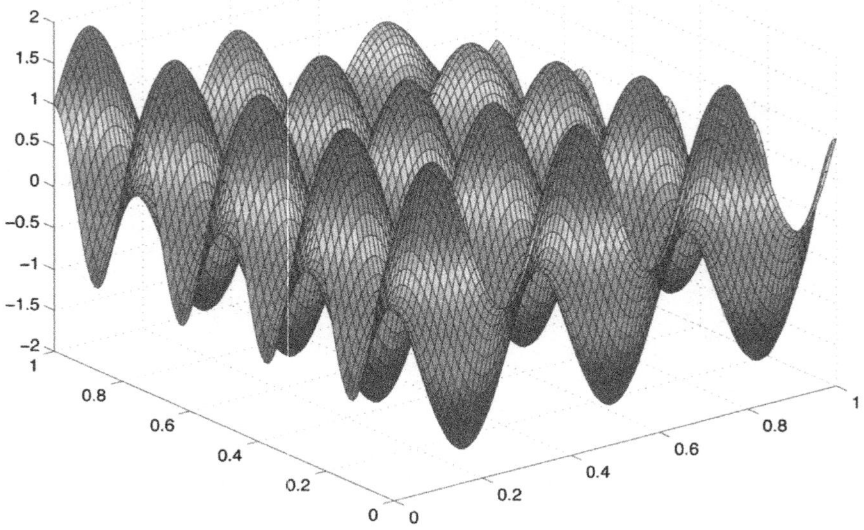

FIGURE 9.4 **Plot of the integrand function in Example 9.2.**

```
$functionEvaluations
[1] 13957
$returnCode
[1] 0
```

It is easy to see that Monte Carlo estimates based on a sample of 10,000 pseudorandom points are not reliable:

```
> set.seed(55555)
> mean(apply(matrix(runif(20000),ncol=2),1,f))
[1] 0.02368898
> mean(apply(matrix(runif(20000),ncol=2),1,f))
[1] 0.01311604
> mean(apply(matrix(runif(20000),ncol=2),1,f))
[1] 0.01870101
> mean(apply(matrix(runif(20000),ncol=2),1,f))
[1] 0.01796109
> mean(apply(matrix(runif(20000),ncol=2),1,f))
[1] 0.01680492
```

> In the above snapshot we sample 20,000 observations from the uni-
> form distribution, but they are arranged into a two-column matrix and
> correspond to 10,000 points; also note the use of `apply`. On the con-
> trary, the result obtained by the bidimensional Halton sequence seems
> fairly satisfactory:
>
> ```
> > library(randtoolbox)
> > mean(apply(halton(10000,dim=2),1,f))
> [1] 0.01987514
> ```

The example above looks quite promising. Nevertheless, there are issues in
using Halton sequences for high-dimensional problems. To see why, let us plot
the Van der Corput sequence for a large base:

```
> X <- halton(n=1000,dim=30)
> plot(X[,30],pch=20)
> 1/X[1,30]
[1] 113
```

Here we generate a Halton sequence with dimension 30 and plot the last column
of the output matrix, which corresponds to a Van der Corput sequence with
base $b = 113$. The resulting plot, shown in Fig. 9.5, features long monotonic
subsequences. This is not surprising, given the large base: For instance, we
have to move from $1/113$ to $112/113$ before breaking the first subsequence.
This results in a very bad coverage of the unit square when we use Halton
sequences with high bases associated with high dimensions. The following
command produces the plot in Fig. 9.6:

```
> plot(X[,30],X[,29],pch=20)
```

This is just a projection of the 30-dimensional sequence on two margins, but it
is reasonable to expect that such a pattern may have a strong adverse effect on
the results.

9.3 Sobol low-discrepancy sequences

With Halton sequences, we have to use Van der Corput sequences associated
with large prime numbers for high-dimensional problems, and as we have seen
this may result in poor performance. A way to overcome this issue is to generate
only Van der Corput sequences associated with a suitably small base. This idea
has been pursued and resulted in two more low-discrepancy sequences:

- In Faure sequences, only one base b is used for all dimensions. It is
  required that the prime number b is larger than the problem dimensionality

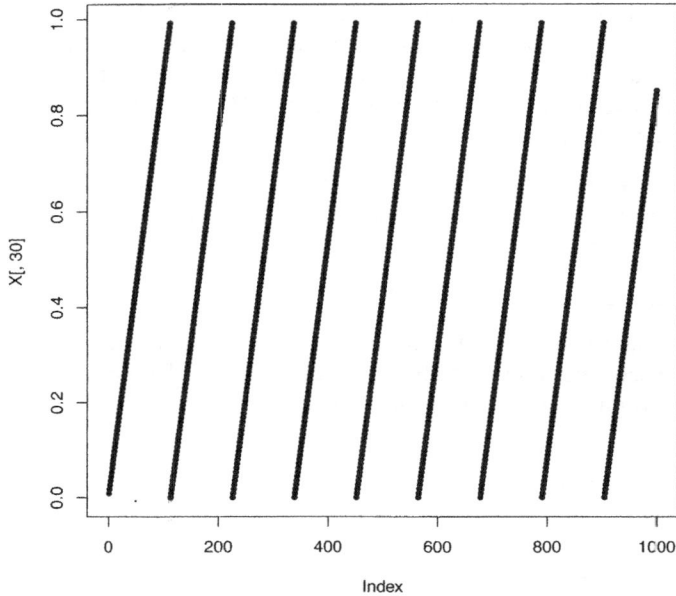

FIGURE 9.5 **Long monotonic subsequences in a Van der Corput sequence with a large base.**

m; then, different permutations of the resulting Van der Corput sequence are associated with each dimension.

- In Sobol sequences, only the smallest base, $b = 2$, is used. Again, suitable permutations of the corresponding Van der Corput sequence are associated with each dimension.

Here we only outline the basic ideas behind Sobol sequences, which rely on more sophisticated concepts in the algebra of polynomials on a binary field. Before doing so, however, it is useful to get acquainted with this kind of sequence by using the function `sobol` provided by the package `randtoolbox`:

```
> sobol(n=10,dim=4)
          [,1]    [,2]    [,3]    [,4]
  [1,]  0.5000  0.5000  0.5000  0.5000
  [2,]  0.7500  0.2500  0.7500  0.2500
  [3,]  0.2500  0.7500  0.2500  0.7500
  [4,]  0.3750  0.3750  0.6250  0.1250
  [5,]  0.8750  0.8750  0.1250  0.6250
  [6,]  0.6250  0.1250  0.3750  0.3750
  [7,]  0.1250  0.6250  0.8750  0.8750
```

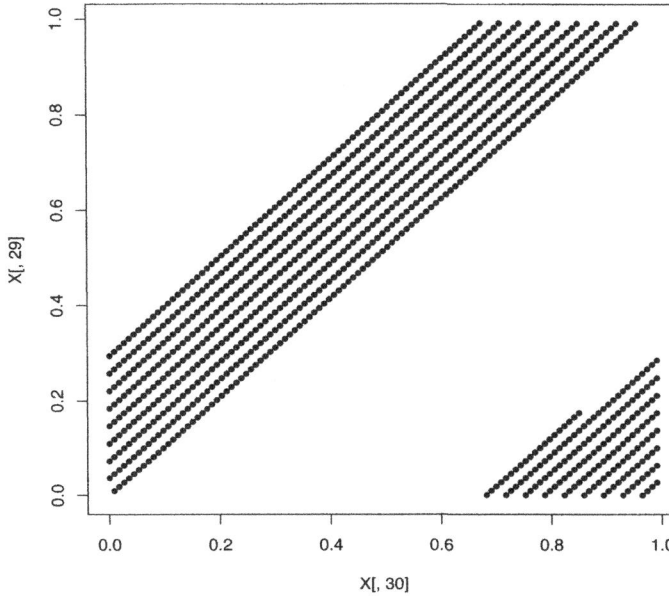

FIGURE 9.6 **Poor coverage of the unit square when large bases are used in Halton sequences.**

```
 [8,]  0.1875  0.3125  0.3125  0.6875
 [9,]  0.6875  0.8125  0.8125  0.1875
[10,]  0.9375  0.0625  0.5625  0.9375
```

It is easy to see that the four columns are all permutations of the Van der Corput sequence with base $b = 2$.

9.3.1 SOBOL SEQUENCES AND THE ALGEBRA OF POLYNOMIALS

The uninterested reader may safely skip this section, which is more technical and not needed in the rest of the book. We will only give some hints at the theory behind Sobol sequences; the section can also be seen as an excuse to illustrate some R functions oriented toward bit manipulation. Let us focus on the generation of a one-dimensional sequence

$$\left(x^{(1)}, x^{(2)}, \ldots, x^{(n)}, \ldots\right)$$

in the unit interval $[0, 1]$. A Sobol sequence is generated on the basis of a set of *direction numbers* $v_1, v_2, \ldots$; we will see shortly how direction numbers are selected, but for now just think of them as numbers which are less than 1. To get the nth number in the Sobol sequence, consider the binary representation of the integer n:[2]

$$n = (\dots b_3 b_2 b_1)_2.$$

The corresponding number in the sequence is obtained by computing the bitwise exclusive OR of the direction numbers v_i for which $b_i \neq 0$:

$$x^{(n)} = b_1 v_1 \oplus b_2 v_2 \oplus \cdots . \tag{9.3}$$

If direction numbers are chosen properly, a low-discrepancy sequence will be generated [8]. A direction number may be thought as a binary fraction:

$$v_i = (0.v_{i1} v_{i2} v_{i3} \dots)_2,$$

or as

$$v_i = \frac{m_i}{2^i},$$

where $m_i < 2^i$ is an odd integer. To generate direction numbers, we exploit primitive polynomials over the field $\mathbb{Z}_2$, i.e., polynomials with binary coefficients:

$$P = x^d + a_1 x^{d-1} + \cdots + a_{d-1} x + 1, \qquad a_k \in \{0, 1\}.$$

Irreducible polynomials are those polynomials which cannot be factored; primitive polynomials are a subset of the irreducible polynomials and are strongly linked to the theory of error-correcting codes, which is beyond the scope of this book. Some irreducible polynomials over the field $\mathbb{Z}_2$ are listed, e.g., in [7, Chapter 7], to which the reader is referred for further information. Given a primitive polynomial of degree d, the procedure for generating direction numbers is based on the recurrence formula:

$$v_i = a_1 v_{i-1} \oplus a_2 v_{i-2} \oplus \cdots \oplus a_{d-1} v_{i-d+1} \oplus v_{i-d} \oplus [v_{i-d}/2^d], \qquad i > d.$$

This is better implemented in integer arithmetic as

$$m_i = 2a_1 m_{i-1} \oplus 2^2 a_2 m_{i-2} \oplus \cdots$$
$$\oplus 2^{d-1} a_{d-1} m_{i-d+1} \oplus 2^d m_{i-d} \oplus m_{i-d}. \tag{9.4}$$

Some numbers $m_1, \dots, m_d$ are needed to initialize the recursion. They may be chosen arbitrarily, provided that each m_i is odd and $m_i < 2^i$.

---

[2]In this section, the rightmost digit in the (binary) representation of an integer number is denoted by b_1 instead of d_0; this is just a matter of convenience, as we associate bit positions with direction numbers rather than powers of a generic base.

▣ Example 9.3 Direction numbers

As an example, let us build the set of direction numbers on the basis of the primitive polynomial

$$x^3 + x + 1.$$

In this polynomial we have coefficients $a_1 = 0$ and $a_2 = 1$. Therefore, the recursive scheme of Eq. (9.4) runs as follows:

$$m_1 = 4m_{i-2} \oplus 8m_{i-3} \oplus m_{i-3},$$

which may be initialized with $m_1 = 1$, $m_2 = 3$, $m_3 = 7$. The reasons why this may be a good choice are given in [2]. We may carry out the necessary computations step by step in R, using the `bitXor` function.

```
> library(bitops)
> m <- c(1, 3, 7)
> i <- 4
> m[i] <- bitXor(4*m[i-2],bitXor(8*m[i-3],m[i-3]))
> i <- 5
> m[i] <- bitXor(4*m[i-2],bitXor(8*m[i-3],m[i-3]))
> i <- 6
> m[i] <- bitXor(4*m[i-2],bitXor(8*m[i-3],m[i-3]))
> m
[1]  1  3  7  5  7 43
```

Given the integer numbers m_i, we may build the direction numbers v_i. To implement the generation of direction numbers, we may use a function like `GetDirNumbers`, which is given in Fig. 9.7. The function requires a primitive polynomial p, a vector of initial numbers m, and the number n of direction numbers we want to generate. On exit, we obtain the direction numbers v and the integer numbers m.

```
> p <- c(1,0,1,1)
> m0 <- c(1,3,7)
> GetDirNumbers(p,m0,6)
$v
[1] 0.500000 0.750000 0.875000 0.312500
    0.218750 0.671875
$m
[1]  1  3  7  5  7 43
```

The code is not optimized; for instance, the first and last coefficients of the input polynomial should be 1 by default, and no check is done on the congruence in size of the input vectors.

```
GetDirNumbers <- function(p,m0,n){
  degree <- length(p)-1
  p <- p[2:degree]
  m <- c(m0, numeric(n-degree))
  for (i in ((degree+1):n)){
    m[i] <- bitXor(m[i-degree], 2^degree * m[i-degree])
    for (j in 1:(degree-1))
      m[i] <- bitXor(m[i], 2^j * p[j]*m[i-j])
  }
  v <- m/(2^(1:length(m)))
  return(list(v=v,m=m))
}
```

FIGURE 9.7 **R code to generate direction numbers for Sobol sequences.**

After computing the direction numbers, we could generate a Sobol sequence according to Eq. (9.3). However, an improved method was proposed by Antonov and Saleev [1], who proved that the discrepancy is not changed by using the Gray code representation of n. Gray codes are discussed, e.g., in [7, Chapter 20]; all we need to know is the following:

1. A Gray code is a function mapping an integer i to a corresponding binary representation $G(i)$; the function, for a given integer N, is one-to-one for $0 \leq i \leq 2^N - 1$.

2. A Gray code representation for the integer n is obtained from its binary representation by computing

$$\ldots g_3 g_2 g_1 = (\ldots b_3 b_2 b_1)_2 \oplus (\ldots b_4 b_3 b_2)_2.$$

3. The main feature of such a code is that the codes for consecutive numbers n and $n+1$ differ only in one position.

Example 9.4 Gray code

Computing a Gray code is easily accomplished in R. For instance, we may define a `gray` function and compute the Gray codes for the numbers $i = 0, 1, \ldots, 15$ as follows:

```
> gray <- function (x) bitXor(x,bitShiftR(x,b=1))
> codes <- matrix(0,nrow=16,ncol=4)
> for (i in 1:16) codes[i,]
    <- as.numeric(intToBits(gray(i-1)))[c(4,3,2,1)]
> codes
```

```
         [,1]  [,2]  [,3]  [,4]
   [1,]    0     0     0     0
   [2,]    0     0     0     1
   [3,]    0     0     1     1
   [4,]    0     0     1     0
   [5,]    0     1     1     0
   [6,]    0     1     1     1
   [7,]    0     1     0     1
   [8,]    0     1     0     0
   [9,]    1     1     0     0
  [10,]    1     1     0     1
  [11,]    1     1     1     1
  [12,]    1     1     1     0
  [13,]    1     0     1     0
  [14,]    1     0     1     1
  [15,]    1     0     0     1
  [16,]    1     0     0     0
```

We have used the function `bitShiftR` of the package `bitops` to shift the binary representation of x one position to the right and the function `intToBits` to obtain the binary representation of a number. We see that indeed the Gray codes for consecutive numbers i and $i+1$ differ in one position; that position corresponds to the rightmost zero bit in the binary representation of i (adding leading zeros if necessary).

Using the feature of Gray codes, we may streamline the generation of a Sobol sequence. Given $x^{(n)}$, we have $x^{(n+1)} = x^{(n)} \oplus v_c$, where c is the index of the rightmost zero bit b_c in the binary representation of n.

9.4 Randomized and scrambled low-discrepancy sequences

In Chapter 7 we have seen how confidence intervals may be used to measure the reliability of Monte Carlo estimates. With low-discrepancy sequences, we cannot apply that idea, as there is nothing random in quasi–Monte Carlo simulation. Actually, there is a result, known as the Koksma–Hlawka theorem, which provides us with a bound on the integration error we incur when approximating an integral on the unit hypercube:

$$\left| \frac{1}{N} \sum_{k=1}^{M} f\left(\mathbf{x}^{(k)}\right) - \int_{(0,1)^m} f(\mathbf{u}) \, d\mathbf{u} \right| \le V(f) \cdot D^*\left(\mathbf{x}^{(1)}, \dots, \mathbf{x}^{(N)}\right),$$

where $D^*\left(\mathbf{x}^{(1)},\ldots,\mathbf{x}^{(N)}\right)$ is the star discrepancy of the point sequence $\mathcal{L}_N = \left(\mathbf{x}^{(1)},\ldots,\mathbf{x}^{(N)}\right)$, and $V(f)$ is a quantity related to the integrand function, known as the Hardy–Krause variation. We will refrain from defining this form of variation, but let us say that, in general, variations try to capture the variability of a function, as this clearly affects the difficulty in approximating its integral numerically. Unfortunately, this bound is not quite useful, unless we have an operational way to evaluate the variation $V(f)$. Another issue with such bounds is that they can be very weak. We may compare the above bound with the corresponding Monte Carlo probabilistic bound obtained by a pseudorandom sequence $\{\mathbf{U}^{(k)}\}$ in the unit hypercube:

$$\left|\frac{1}{N}\sum_{k=1}^{M} f\left(\mathbf{U}^{(k)}\right) - \int_{(0,1)^m} f(\mathbf{u})\,d\mathbf{u}\right| \le z_{1-\alpha/2}\frac{\sigma_f}{\sqrt{n}},$$

with probability $1-\alpha/2$, where $\sigma_f^2 = \mathrm{Var}[f(\mathbf{U})]$. It is true that the discrepancy bound in Eq. (9.2) is more appealing than the $1/\sqrt{n}$ factor in Monte Carlo sampling, but the latter is related to a quite operational measure of estimate quality.

An array of methods have been designed both to allow the construction of confidence intervals and to improve the quality of point sequences:

Random shift. This is simple idea that lends itself to application with Korobov point sets (see Section 2.4.3). Given a point set $\{\mathbf{u}^{(i)}\}$, a uniform random vector $\mathbf{V}$ in $[0,1)^m$ is sampled and used to shift the point set to yield a sequence consisting of

$$\widetilde{\mathbf{U}}^{(i)} = \left(\mathbf{u}^{(i)} + \mathbf{V}\right)\ \mathrm{mod}\ 1.$$

Digital shift. The digital shift differs from the random shift in how the shift is applied, i.e., in an arithmetic modulo b, where b is the base used to represent the numbers in the unit interval. Consider the expansion in base b of the coordinate j of a point $\mathbf{u}^{(i)}$ in $[0,1)^m$:

$$u_j^{(i)} = \sum_{k=1}^{\infty} u_{jk}^{(i)}\, b^{-k}.$$

A random shift $\mathbf{V}$ in $[0,1)^m$ is sampled, with coordinates

$$V_j = \sum_{k=1}^{\infty} V_{jk} b^{-k},$$

and shifted points are generated as

$$\widetilde{\mathbf{U}}^{(i)} = \sum_{k=1}^{\infty}\left[\left(u_{jk}^{(i)} + V_{jk}\right)\mathrm{mod}\,b\right] b^{-k}.$$

Scrambling. Scrambling is a deeper readjustment of points in the sequence. A scrambling procedure was originally proposed by Tezuka to improve Faure sequences, and is therefore known as Faure–Tezuka scrambling; an alternative approach is due to Owen. Both of these scrambling approaches are available in R and can be applied to Sobol sequences.

Given a randomized shifting, we may generate a sample of sequences and come up with a confidence interval. The approach is justified by distributional results that we skip here. We just want to illustrate how scrambling of Sobol sequences is applied in R by setting the optional parameter `scrambling` in the function `sobol`:

```
> sobol(n=10,dim=4,scrambling=1)
           [,1]       [,2]       [,3]       [,4]
 [1,] 0.8897523 0.13510869 0.99200249 0.06669691
 [2,] 0.5694336 0.58325994 0.73452467 0.99942088
 [3,] 0.2349515 0.27649581 0.07017102 0.36903149
 [4,] 0.0633796 0.68916714 0.76618928 0.52431285
 [5,] 0.7416083 0.39046729 0.41350642 0.15471716
 [6,] 0.8119298 0.82890844 0.15676318 0.78641582
 [7,] 0.3836831 0.02996351 0.52293140 0.40991741
 [8,] 0.4830195 0.50730205 0.05096463 0.24351555
 [9,] 0.8369830 0.32138658 0.62871468 0.62484527
[10,] 0.6416652 0.89649355 0.87904912 0.44233483
```

By setting `scrambling` to 1, Owen scrambling is selected, whereas 2 corresponds to Faure–Tezuka scrambling. We may check the effect of scrambling on the integration problem of Example 9.2:

```
> adaptIntegrate(f,rep(0,2),rep(1,2),tol=1e-6)$integral
[1] 0.01986377
> mean(apply(sobol(10000,dim=2),1,f))
[1] 0.01976074
> mean(apply(sobol(10000,dim=2,scrambling=1),1,f))
[1] 0.01981418
> mean(apply(sobol(10000,dim=2,scrambling=2),1,f))
[1] 0.01963135
```

In this specific example, Owen scrambling seems to work well.

9.5   Sample path generation with low-discrepancy sequences

Whatever low-discrepancy sequence we use, we need to transform the "uniform" numbers into something else. Since normal variables are ubiquitous in financial applications, let us consider how we may generate standard normal variables. To generate normal variates, we may either use the Box–Muller method, which we described in Section 5.5.1, or the inverse transform method. We *cannot* apply polar rejection, because when using low-discrepancy sequences we must integrate over a space with a well-defined dimensionality. We must know exactly how many quasi-random numbers we need, whereas with rejection-based methods we cannot anticipate that.

We recall the Box–Muller algorithm here for the sake of convenience. To generate two independent standard normal variates, we should first generate two independent random numbers U_1 and U_2, and then set

$$X = \sqrt{-2 \ln U_1} \cos(2\pi U_2),$$
$$Y = \sqrt{-2 \ln U_1} \sin(2\pi U_2).$$

Rather than generating pseudorandom numbers, we may use a bidimensional low-discrepancy sequence. Given the potentially weird effects of the Box–Muller transformation, which we have illustrated in Fig. 5.12, one could argue that the inverse transform method is a safer approach. Since fast and accurate methods to invert the CDF of the standard normal have been devised, we will only pursue the latter approach.

▣ Example 9.5 Pricing a vanilla call option

> As an illustration, let us price a vanilla call option using Halton and Sobol sequences. The code in Fig. 9.8 receives a sequence of uniform numbers, possibly obtained by a pseudorandom generator, and transforms them to standard normals by inversion. The snapshot below shows that low-discrepancy sequences yield much more accurate answers than naive Monte Carlo, for a relatively small sample size.

```
> S0 <- 50; K <- 50; sigma <- 0.4
> r <- 0.05; T <- 5/12
> EuVanillaCall(S0,K,T,r,sigma)
[1] 5.614995
> set.seed(55555)
> BlsLowD(S0,K,T,r,sigma,runif(5000))
[1] 5.673162
> BlsLowD(S0,K,T,r,sigma,halton(5000,dim=1))
[1] 5.592316
> BlsLowD(S0,K,T,r,sigma,sobol(5000,dim=1))
[1] 5.605094
> BlsLowD(S0,K,T,r,sigma,
          sobol(5000,dim=1,scrambling=1))
[1] 5.612801
```

> In this case, too, it seems that scrambling a Sobol sequence works fairly well.

```
BlsLowD <- function(S0,K,T,r,sigma,LDsequence){
  nuT <- (r - 0.5*sigma^2)*T
  sigmaT <- sigma*sqrt(T)
  vetEps <- qnorm(LDsequence)
  ST <- S0*exp(nuT+sigmaT*vetEps)
  dpayoff <- exp(-r*T)*pmax(0,ST-K)
  return(mean(dpayoff))
}
```

FIGURE 9.8 **Pricing a vanilla call option with low-discrepancy sequences.**

A final important note concerns the correct use of low-discrepancy sequences for sample path generation. Generating sample paths for geometric Brownian motion is rather easy, as we only need standard normal variables. Consider the following matrix of uniform numbers:

```
> halton(10,dim=5)
          [,1]       [,2] [,3]       [,4]       [,5]
 [1,] 0.5000 0.33333333 0.20 0.14285714 0.09090909
 [2,] 0.2500 0.66666667 0.40 0.28571429 0.18181818
 [3,] 0.7500 0.11111111 0.60 0.42857143 0.27272727
 [4,] 0.1250 0.44444444 0.80 0.57142857 0.36363636
 [5,] 0.6250 0.77777778 0.04 0.71428571 0.45454545
 [6,] 0.3750 0.22222222 0.24 0.85714286 0.54545455
 [7,] 0.8750 0.55555556 0.44 0.02040816 0.63636364
 [8,] 0.0625 0.88888889 0.64 0.16326531 0.72727273
 [9,] 0.5625 0.03703704 0.84 0.30612245 0.81818182
[10,] 0.3125 0.37037037 0.08 0.44897959 0.90909091
```

It must be realized that with these numbers we may generate ten sample paths with five time instants along each path (not including the initial value of the process). Each dimension of the sequence (number of columns) must correspond to a time instant, and each row corresponds to a sample path. A common mistake is to associate a dimension with a sample path, but by doing so we lose the intertemporal independence property of the standard Wiener processes.

The function in Fig. 9.9 is quite similar to the vectorized code of Fig. 6.3. It can use either Halton or Sobol sequences and, for the latter ones, we may also select the type of scrambling. As a first check, let us generate and plot five sample paths, without any scrambling:

```
> S0 <- 50; mu <- 0.1; sigma <- 0.4;
> T <- 1/12; numSteps <- 30; numRepl <- 5
> pathsH <- lowDiscrGBM(S0,mu,sigma,T,numSteps,
                  numRepl,type="halton")
> pathsS <- lowDiscrGBM(S0,mu,sigma,T,numSteps,numRepl)
> par(mfrow=c(1,2))
```

```
lowDiscrGBM <- function (S0,mu,sigma,T,numSteps,numRepl,
                         type="sobol",scrambling=0){
  dt <- T/numSteps
  nuT <- (mu-sigma^2/2)*dt
  sigmaT <- sqrt(dt)*sigma
  if (toupper(type) == "HALTON")
    uSeq <- halton(numRepl,dim=numSteps)
  else
    uSeq <- sobol(numRepl,dim=numSteps,scrambling=scrambling)
  normMatrix <- nuT + sigmaT*qnorm(uSeq)
  logIncrements <- cbind(log(S0),normMatrix)
  logPath <- t(apply(logIncrements,1,cumsum))
  return(exp(logPath))
}
```

FIGURE 9.9 **Generating sample paths of GBM by low-discrepancy sequences.**

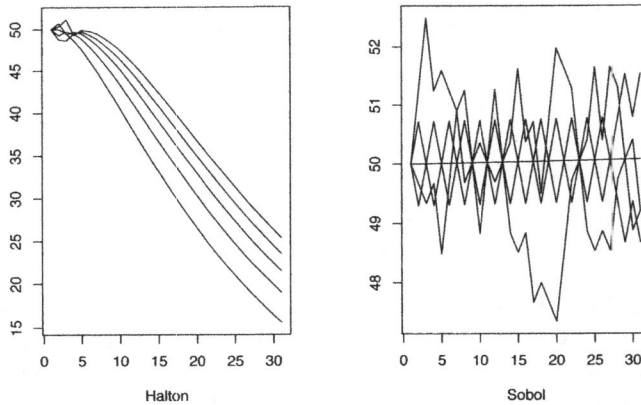

FIGURE 9.10 **Five GBM paths by Halton and Sobol sequences.**

```
> plot(pathsH[1,],type='l',ylim=range(pathsH),
                 xlab="Halton",ylab="")
> for (k in 2:numRepl) lines(pathsH[k,])
> plot(pathsS[1,],type='l',ylim=range(pathsS),
                 xlab="Sobol",ylab="")
> for (k in 2:numRepl) lines(pathsS[k,])
> par(mfrow=c(1,1))
```

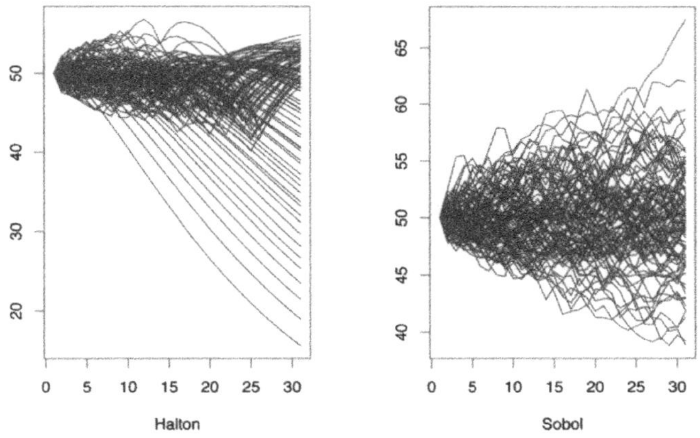

FIGURE 9.11 **More GBM paths by Halton and Sobol sequences.**

The sample paths, displayed in Fig. 9.10 do not look quite satisfactory. The case of Halton sequence shows decreasing paths for latter time periods; the case of Sobol sequence shows one sample path that looks like a straight horizontal line. A bit of reflection explains why we observe such weird patterns. We have already noted that, for large bases, Van der Corput sequences suffer from long monotonically increasing subsequences. Thus, they start with small numbers, which correspond to negative normal shocks and a monotonic decrease in the stochastic process, at least in the first steps. On the contrary, the first part of the sample paths seems less problematic, as there we are using small bases. As to the Sobol sequences, since the first point consists of a repetition of values 0.5, which correspond to zero normal shocks, the first part of the sample path may look like a straight line, even though it is actually the initial part of an exponential.

If we sample more paths, as shown in Fig. 9.11, the results look a bit more reasonable. Indeed, skipping the initial parts of a low-discrepancy sequence is a strategy that is sometimes adopted, as well as the Brownian bridge, in order to break some nasty patterns. Another possibility, of course, is to use scrambling. The ten sample paths of Fig. 9.12 have been produced by Sobol sequences with Owen scrambling, and they definitely look more realistic.

For further reading

- We have introduced low-discrepancy sequences in a rather informal way, without referring to fundamental concepts such as the digital nets. A rig-

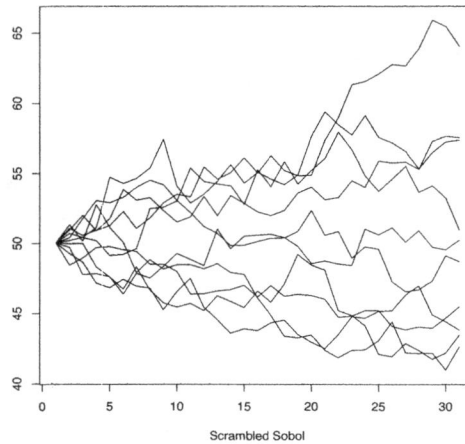

FIGURE 9.12 **GBM paths by scrambled Sobol sequences.**

orous account can be found in [6], whereas a less extensive but readable treatment can be found in [3, Chapter 5].

- You may also see [5], where Chapter 5, among other things, deals with quality measures, and Chapter 6 analyzes questions related to decomposition and effective dimensions.

- Another interesting reference is [4, Chapter 8], which is more specifically aimed at financial applications.

References

1 I.A. Antonov and V.M. Saleev.   An economic method of computing lp_τ sequences. *USSR Computational Mathematics and Mathematical Physics*, 19:252–256, 1979.

2 P. Bratley and B.L. Fox. Algorithm 659: Implementing Sobol's quasirandom sequence generator. *ACM Transactions on Mathematical Software*, 14:88–100, 1988.

3 P. Glasserman. *Monte Carlo Methods in Financial Engineering.* Springer, New York, 2004.

4 P. Jaeckel. *Monte Carlo Methods in Finance.* Wiley, Chichester, 2002.

5 C. Lemieux. *Monte Carlo and Quasi–Monte Carlo Sampling.* Springer, New York, 2009.

6 H. Niederreiter. *Random Number Generation and Quasi–Monte Carlo Methods*. Society for Industrial and Applied Mathematics, Philadelphia, PA, 1992.

7 W.H. Press, S.A. Teukolsky, W.T. Vetterling, and B.P. Flannery. *Numerical Recipes in C* (2nd ed.). Cambridge University Press, Cambridge, 1992.

8 I.M. Sobol. On the distribution of points in a cube and the approximate evaluation of integrals. *USSR Computational Mathematics and Mathematical Physics*, 7:86–112, 1967.

## Part Five

Miscellaneous Applications

## Chapter Ten

Optimization

There is a multifaceted interaction between optimization and Monte Carlo simulation, which is illustrated by the rich connections between this and other chapters of this book. Monte Carlo methods may be used when dealing with stochastic optimization models, i.e., when some of the parameters characterizing a decision problem are uncertain. This is clearly relevant in finance, and there is an array of approaches depending on the problem structure and on our purpose:

- In stochastic programming (SP) models we generalize deterministic mathematical programming models in order to deal with uncertainty that is typically represented by a scenario tree. The scenario tree may be generated by crude Monte Carlo sampling, or we may adopt other approaches like Gaussian quadrature (see Chapter 2), variance reduction (see Chapter 8), or low-discrepancy sequences (see Chapter 9). The stochastic programming models that we consider here can be extended to account for risk aversion, as illustrated in Chapter 13.

- Another approach to decision making under uncertainty relies on dynamic programming (DP). We will illustrate the advantages and the disadvantages of DP with respect to SP models. The principle behind DP is embodied in the Bellman recursive equation and is extremely powerful; nevertheless, DP is plagued by the so-called curse of dimensionality. There are approaches to overcome this issue and to make DP a viable computational tool. On the one hand, we have numerical dynamic programming methods based on relatively straightforward tools from numerical analysis; on the other hand we have approximate dynamic programming (ADP) methods, which rely more heavily on learning by Monte Carlo simulation. ADP methods are also used in Chapter 11 to price options with early exercise features.

- Both SP and classical DP approaches rely on a formal model of the system, in the form of explicit mathematical expressions. However, we may have to cope with systems that are too complicated for analytical modeling. In such a case, we may rely on simulation-based optimization, where there is also a clear role for Monte Carlo methods. The idea is also appealing when we want to reduce the complexity of optimization under uncertainty by making decisions on the basis of parameterized decision

405

rules. Learning by search and simulation is a possible strategy in such cases. Sometimes, we may also wish to take advantage of simulation to estimate the sensitivity of performance with respect to some parameters, in order to guide the search process. Sensitivity estimation is the topic of Chapter 12.

A further use of Monte Carlo methods is to check the out-of-sample robustness of a solution. Within an optimization model, we may only afford a limited representation of uncertainty; the in-sample value of the objective function may be a misleading indicator of quality. Thus, it may be useful to verify the performance of the selected solution in a much wider set of scenarios.

Stochastic optimization in all of its forms does not tell the whole story about the interaction between optimization and Monte Carlo methods. Stochastic search strategies are also used to solve *deterministic* nonconvex optimization problems. There is an array of useful strategies, including some fancy ones mimicking biological systems, such as genetic algorithms or particle swarm optimization. Such methods are typically derivative-free, i.e., unlike classical nonlinear programming methods, they do not rely on information provided by the gradient or the Hessian matrix of the objective function. This allows to use them when the objective is not guaranteed to be differentiable or even continuous. Furthermore, by blending exploitation with exploration behavior, they may be able to overcome the local optimality issues that make nonconvex optimization difficult. The optimization of nonconvex functions may arise, e.g., in model calibration or parameter estimation by maximum likelihood (see Chapter 4), when the underlying model is complicated enough.

As the reader can imagine, there is an enormous range of models and methods involved, which cannot be covered in a single chapter. Nevertheless, we offer here a set of basic examples in order to get a taste for this wide and quite interesting application field. In Section 10.1 we outline a classification of optimization models, for the sake of the uninitiated. Then, we provide a few concrete examples in Section 10.2, with due emphasis on economic and financial applications. Given our limited aim and scope, we will *not* go through traditional concepts like Lagrange multipliers in nonlinear programming or the simplex method for linear programming.[1] When necessary, we will just use R functions from suitable packages implementing these methods. Emphasis is partly on modeling and partly on alternative methods, which are illustrated in Section 10.3 for global (nonconvex) optimization. These methods can also be applied to simulation-based optimization, which is the subject of Section 10.4. In Section 10.5 we illustrate model building in stochastic programming. We cannot cover specialized solution methods for this class of models, but we insist on scenario generation in Section 10.5.3, illustrating important concepts such as in-sample and out-of-sample stability. The last three sections are devoted to stochastic dynamic programming. The cornerstone principle of dynamic programming, the recursive functional equation due to Bellman, is introduced in Section 10.6,

---

[1] An elementary introduction can be found in [8, Chapter 12].

where we also compare this approach against stochastic programming with recourse. In Section 10.7 we describe "standard" numerical methods for dynamic programming, which are well-suited to problems with a rather moderate size, whereas in Section 10.8 we describe ADP strategies that are aimed at larger scale problems and feature interesting links with machine learning.

10.1 Classification of optimization problems

A fairly general statement of an optimization problem is

$$\min_{\mathbf{x} \in S} f(\mathbf{x}), \tag{10.1}$$

which highlights the following three building blocks:

- A vector of decision variables $\mathbf{x}$, representing a solution of the problem in mathematical form.
- A feasible set S, also called feasible region, to which $\mathbf{x}$ must belong; in this book we consider $S \subseteq \mathbb{R}^n$, i.e., decisions are represented by an n-dimensional tuple of real numbers.
- An objective function $f(\cdot)$, mapping each feasible solution $\mathbf{x} \in S$ to a performance measure $f(\mathbf{x}) \in \mathbb{R}$, which we are interested in optimizing.

This rather dry and abstract statement encompasses an incredible variety of decision problems. Note that there is no loss of generality in considering minimization problems, as a maximization problem can be easily converted into a minimization one:

$$\max f(\mathbf{x}) \quad \Rightarrow \quad -\min\left[-f(\mathbf{x})\right].$$

Indeed, many software tools assume that the problem is either in minimization or in maximization form and, if necessary, leave the burden of changing the sign of the objective function to the user. In other cases, a flag is used to characterize the kind of problem.

In this book we only deal with finite-dimensional optimization problems, involving a finite number of decision variables. In continuous-time optimal control problems, where system dynamics is described by differential equations, decisions may be represented by functions $u(t)$ where $t \in [t_0, T]$; hence, such problems are infinite-dimensional. We do consider dynamic optimization problems in this chapter, but we assume that time has been discretized; this results in a finite-dimensional problem when the time horizon is finite. When the time horizon is infinite, in principle we do have an infinite-dimensional problem (though countably so); however, if the system is stationary, we may look for a stationary policy by dynamic programming. Dynamic programming relies on functional equations whose unknown are, in fact, infinite-dimensional objects if the underlying state space is continuous. As we shall see, we always boil down the overall problem to a sequence of finite-dimensional subproblems by suitable discretizations. Hence, what we really deal with are mathematical programming problems, but we will use this term and the more general "optimization problem" interchangeably.

When $S \equiv \mathbb{R}^n$ we speak of an *unconstrained* optimization problem; otherwise we have a *constrained* optimization problem. The feasible region S, in concrete terms, is typically described by a set of of equality and inequality constraints, resulting in the mathematical program:

$$\begin{aligned} \min \quad & f(\mathbf{x}) && (10.2) \\ \text{s.t.} \quad & h_i(\mathbf{x}) = 0, \quad i \in E, \\ & g_i(\mathbf{x}) \leq 0, \quad i \in I, \\ & \mathbf{x} \in \mathcal{X} \subset \mathbb{R}^n. \end{aligned}$$

The condition $\mathbf{x} \in \mathcal{X}$ may include additional restrictions, such as the integrality of some decision variables. As we shall see, the most common form of this restriction concerns binary decision variables, i.e., $x_j \in \{0, 1\}$; this trick of the trade is useful to model logical decisions like "we do it" vs. "we do not." Simple lower and upper bounds, describing box constraints like $l_j \leq x_j \leq u_j$, are usually considered apart from general inequalities for the sake of computational efficiency. So, we might also have $\mathcal{X} = \{\mathbf{x} \in \mathbb{R}^n \mid \mathbf{l} \leq \mathbf{x} \leq \mathbf{u}\}$, where vectors $\mathbf{l}$ and $\mathbf{u}$ collect the lower and upper bounds on decision variables, respectively.

An optimization problem can be very simple or on the verge of intractability, depending on some essential features of its building blocks. Such features are often more important than the sheer size of the problem. We will consider concrete modeling examples later, shedding light on such features and showing the astonishing variety of problems that we may address, but an essential classification framework involves the following dimensions:

- Convex vs. nonconvex problems
- Linear vs. nonlinear problems
- Deterministic vs. stochastic problems

Convex vs. nonconvex optimization problems

By far, the most important property of an optimization problem is convexity. Convexity is a property of sets, in particular of the feasible set S, which can be extended to a property of functions, in our case, the objective function.

DEFINITION 10.1 (Convex sets) *A set $S \subseteq \mathbb{R}^n$ is a* convex *if*

$$\mathbf{x}, \mathbf{y} \in S \Rightarrow \lambda \mathbf{x} + (1 - \lambda)\mathbf{y} \in S \qquad \forall \lambda \in [0, 1].$$

Convexity can be grasped intuitively by observing that points of the form $\lambda \mathbf{x} + (1 - \lambda)\mathbf{y}$, where $0 \leq \lambda \leq 1$, are simply the points on the straight-line segment joining $\mathbf{x}$ and $\mathbf{y}$. So, a set S is convex if the line segment joining any pair of points $\mathbf{x}, \mathbf{y} \in S$ is contained in S. This is illustrated in Fig. 10.1: S_1

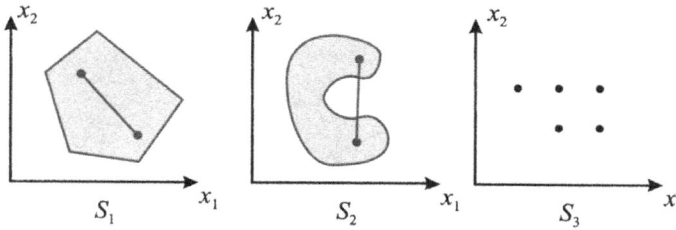

FIGURE 10.1 **An illustration of set convexity.**

is convex, but S_2 is not. S_3 is a discrete set and it is not convex; this fact has important consequences for discrete optimization problems. It is easy to see that the intersection of convex sets is a convex set. This does not necessarily apply to set union.

Example 10.1 Polyhedra are convex sets

Polyhedral sets are essential in describing the feasible set of a linear programming problem. A hyperplane $\mathbf{a}_i^\mathsf{T}\mathbf{x} = b_i$, where $b_i \in \mathbb{R}$ and $\mathbf{a}_i, \mathbf{x} \in \mathbb{R}^n$, divides $\mathbb{R}^n$ into two *half-spaces* expressed by the linear inequalities $\mathbf{a}_i^\mathsf{T}\mathbf{x} \leq b_i$ and $\mathbf{a}_i^\mathsf{T}\mathbf{x} \geq b_i$. Note that we always assume column vectors, so we use transposition to specify a row vector $\mathbf{a}_i^\mathsf{T}$. A *polyhedron* $P \subseteq \mathbb{R}^n$ is a set of points satisfying a finite collection of linear inequalities, i.e.,

$$P = \{\mathbf{x} \in \mathbb{R}^n \mid \mathbf{A}\mathbf{x} \geq \mathbf{b}\},$$

where matrix $\mathbf{A} \in \mathbb{R}^{m,n}$ collects row vectors $\mathbf{a}_i^\mathsf{T}$, $i = 1, \ldots, m$, and column vector $\mathbf{b} \in \mathbb{R}^m$ collects the right-hand sides b_i. Note that a linear equality can be rewritten as two inequalities with opposite senses. A polyhedron is therefore the intersection of a finite collection of half-spaces and is convex. As we will discuss later (see Eqs. (10.4) and (10.5)), the feasible set of a linear programming problem can be described, e.g., as $S = \{\mathbf{x} \in \mathbb{R}^n \mid \mathbf{A}\mathbf{x} = \mathbf{b}, \mathbf{x} \geq \mathbf{0}\}$ or $S = \{\mathbf{x} \in \mathbb{R}^n \mid \mathbf{A}\mathbf{x} \geq \mathbf{b}\}$, and it is a polyhedron.

Example 10.2 The union of convex sets need not be convex

The closed intervals $[2, 5]$ and $[6, 10]$ are clearly convex sets. However, their union $[2, 5] \cup [6, 10]$ is not. As a more practically motivated

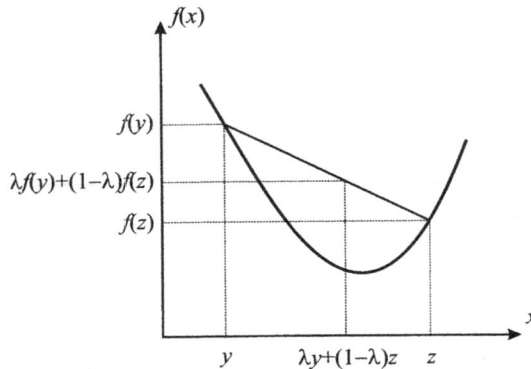

FIGURE 10.2 **An illustration of function convexity.**

example, consider an activity whose level is measured by the decision variable x_j; in concrete, x_j may represent the weight of an asset in a financial portfolio. A possible restriction is

$$x_j \in \{0\} \cup [l_j, u_j],$$

meaning that we may not invest in that asset, i.e., $x_j = 0$, but if we do, there are a lower bound l_j and an upper bound u_j on its weight. This restriction specifies a nonconvex set, and it should not be confused with the box constraint $l_j \leq x_j \leq u_j$, which corresponds to a convex set. Later, we shall see how we can represent such a nonconvexity by the introduction of binary decision variables.

Set convexity can be extended to function convexity as follows.

DEFINITION 10.2 (Convex functions) *A function* $f \colon \mathbb{R}^n \to \mathbb{R}$, *defined over a convex set* $S \subseteq \mathbb{R}^n$, *is a* convex function *on* S *if, for any* **y** *and* **z** *in* S, *and for any* $\lambda \in [0, 1]$, *we have*

$$f\big(\lambda \mathbf{y} + (1 - \lambda)\mathbf{z}\big) \leq \lambda f(\mathbf{y}) + (1 - \lambda)f(\mathbf{z}). \qquad (10.3)$$

The definition can be interpreted by looking at Fig. 10.2. If we join any two points on the function graph with a line segment, all of the segment lies above the function graph. In other words, a function is convex if its epigraph, i.e., the region above the function graph, is a convex set. If the condition (10.3) is satisfied with strict inequality for all $\mathbf{y} \neq \mathbf{z}$, the function is *strictly* convex. We illustrate examples of convex and nonconvex functions in Fig. 10.3. The first function is convex, whereas the second is not. The third function is a *polyhedral* convex function, and it is kinky. This example shows that a convex function

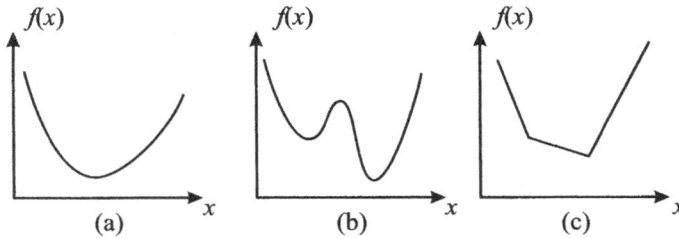

FIGURE 10.3 **Convex and nonconvex functions.**

need not be differentiable everywhere. Note that in the second case we have a local minimum that is not a global one. Indeed, convexity is so relevant in minimization problems because it rules out local minima.

🔲 Example 10.3 **Local minima in polynomial optimization**

Consider the following unconstrained problem:

$$\min f(x) = x^4 - 10.5x^3 + 39x^2 - 59.5x + 30.$$

The objective function is a polynomial, which can be dealt with in R by the `polynom` package, as illustrated by the script in Fig. 10.4. The script produces the following output:

```
out1 <- optim(par=4,fn=f,method="BFGS")
> out1$par; out1$value
[1] 3.643656
[1] -0.6934635
out2 <- optim(par=1,fn=f,method="BFGS")
> out2$par; out2$value
[1] 1.488273
[1] -1.875737
```

showing that we end up with different solutions, depending on the initial point. The script also produces the plot in Fig. 10.5: We observe that the polynomial features two local minimizers, one of which is also the global one, as well as one local maximizer. The function goes to infinity when x is large in absolute value, so there is no global maximizer. With a high-degree polynomial, there are potentially many local maxima and minima; in fact, multivariable polynomial optimization is, in general, a difficult nonconvex problem.

```
library(polynom)
p <- c(30, -59.5, 39, -10.5, 1)
f <- as.function(polynomial(p))
x <- seq(from=1,to=4,by=0.01)
plot(x,f(x),type='l')
out1 <- optim(par=4,fn=f,method="BFGS")
out1$par; out1$value
out2 <- optim(par=1,fn=f,method="BFGS")
out2$par; out2$value
xx <- seq(from=1,to=4,by=0.01)
plot(xx,f(xx),type='l')
```

R programming notes:

1. The polynomial is represented by a vector collecting coefficients in *increasing* order.

2. The vector is transformed into a polynomial object by the function polynomial, and this in turn is converted into a function by using as.function.

3. The function optim implements some very basic optimization algorithms, requiring an initial solution par and a function fn; its output consists of the minimizer and the corresponding value.

FIGURE 10.4 **Local minima in polynomial optimization.**

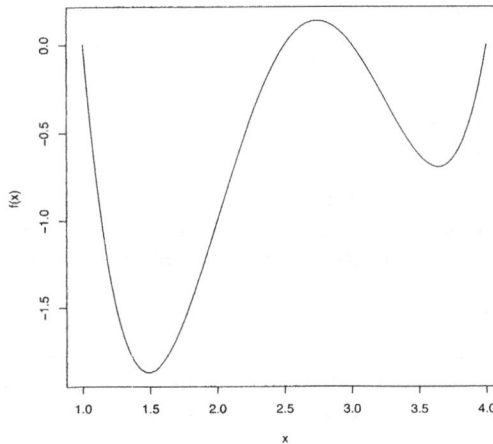

FIGURE 10.5 **Global and local optima of a polynomial.**

When solving a minimization problem, it is nice to have a convex objective function. When the objective has to be maximized, it is nice to deal with a concave function.

DEFINITION 10.3 (Concave function) *A function f is* concave *if* $(-f)$ *is convex.*

Thus, a concave function is just a convex function turned upside down. We have observed that a convex function features a convex epigraph; hence, function convexity relies on set convexity. A further link between convex sets and convex functions is that the set $S = \{\mathbf{x} \in \mathbb{R}^n \mid g(\mathbf{x}) \leq 0\}$ is convex if g is a convex function.

◾ Example 10.4   Variations on a circle

> Consider the function $g(\mathbf{x}) = x_1^2 + x_2^2 - 1$. The constraint $g(\mathbf{x}) \leq 0$ describes a circle of radius 1, which is clearly a convex set. Indeed, $g(\mathbf{x})$ is a convex function. However, the constraint $g(\mathbf{x}) \geq 0$ describes the *exterior* of the circle, which is clearly nonconvex. Indeed, the latter constraint is equivalent to $-g(\mathbf{x}) \leq 0$, where the function describing the set is concave. The equality constraint $g(\mathbf{x}) = 0$ describes the boundary of the circle, i.e., the circumference, which again is nonconvex.

In general, equality constraints like

$$h(\mathbf{x}) = 0$$

do not describe convex sets. We may understand why by rewriting the equality as two inequalities:

$$h(\mathbf{x}) \leq 0,$$
$$-h(\mathbf{x}) \geq 0.$$

There is no way in which $h(\cdot)$ and $-h(\cdot)$ can be both convex, unless the involved function is affine:

$$h(\mathbf{x}) = \mathbf{a}^\mathsf{T}\mathbf{x} - b.$$

It is pretty intuitive that convexity of the objective function makes an unconstrained problem relatively easy, since local minima are ruled out. But what about convexity of the feasible set of a constrained problem? Actually, convexity must be ensured in both the objective function and the feasible set, as shown by the next example.

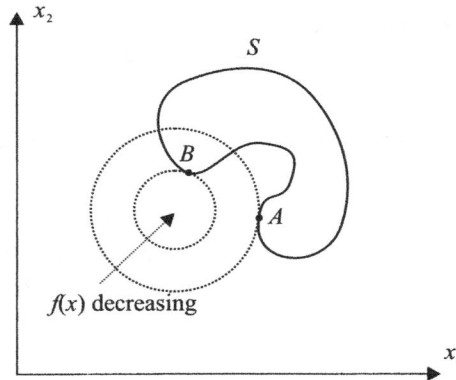

FIGURE 10.6 **Local optima due to a nonconvex feasible set.**

Example 10.5 **Nonconvexities in the feasible set**

Even if the objective function is convex, nonconvexities in the feasible set may make the problem difficult to solve. In Fig. 10.6 we observe two level curves of a convex objective function like $f(x_1, x_2) = (x_1 - \alpha)^2 + (x_2 - \beta)^2$. The feasible set is a sort of bean and is nonconvex. We may notice that B is the global minimum, as there the feasible set is tangent to the lowest level curve. However, we also have a local optimum at point A: In its neighborhood there is no better feasible solution.

DEFINITION 10.4 (Convex optimization problem) *The minimization problem* $\min_{\mathbf{x} \in S} f(\mathbf{x})$ *is convex if both S and $f(\cdot)$ are convex. The maximization problem* $\max_{\mathbf{x} \in S} f(\mathbf{x})$ *is convex if S is convex and f is concave.*

Note that we also speak of a convex problem when we are maximizing a concave objective. We speak of a concave problem when we *minimize* a concave objective on a convex set. Concave problems are not as nice as convex problems, as the following example illustrates.

Example 10.6 **A concave problem**

Consider the following one-dimensional problem:

$$\begin{aligned} \min \quad & -(x-2)^2 + 3 \\ \text{s.t.} \quad & 1 \le x \le 4 \end{aligned}$$

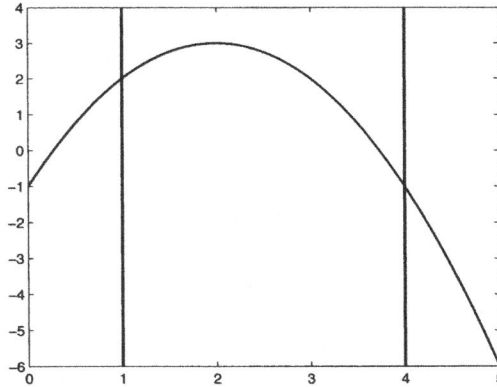

FIGURE 10.7 **A concave problem may have local optima, but they lie on the boundary of the convex feasible set.**

This is a concave problem, since the leading term in the quadratic polynomial defining the objective is negative, so that the second-order derivative is negative everywhere. In Fig. 10.7 we show the objective function and the feasible set. The stationarity point $x = 2$ is of no use to us, since it is a maximizer. We see that local minimizers are located at the boundary of the feasible set. A local minimizer lies at the left boundary, $x = 1$, and the global minimizer is located at the right boundary, $x = 4$.

It turns out that the property illustrated in the example, i.e., there is an optimal solution on the boundary of the (convex) feasible set, does apply to concave problems in general. This kind of structure may be exploited in global optimization procedures.

Linear vs. nonlinear optimization problems

An optimization model in the concrete form (10.2) can be linear or nonlinear. We have a linear programming (LP) problem if all of the involved functions are linear affine, as in the following model:

$$\begin{aligned} \min \quad & \mathbf{c}^{\mathsf{T}}\mathbf{x} \\ \text{s.t.} \quad & \mathbf{A}\mathbf{x} = \mathbf{b}, \\ & \mathbf{x} \geq \mathbf{0}, \end{aligned} \tag{10.4}$$

where $\mathbf{x} \in \mathbb{R}^n$, $\mathbf{c} \in \mathbb{R}^n$, $\mathbf{b} \in \mathbb{R}^m$, and $\mathbf{A} \in \mathbb{R}^{m,n}$. Clearly, such a problem makes sense if $m < n$. The above form is an LP in standard form, involving only equality constraints and non-negative decision variables. Another way to state the problem is the canonical form

$$\min \quad \mathbf{c}^\mathsf{T}\mathbf{x} \qquad\qquad (10.5)$$
$$\text{s.t.} \quad \mathbf{A}\mathbf{x} \geq \mathbf{b}.$$

Using tricks of the trade, it is possible to convert any LP in one of the two forms above. The important point is that LP is both a convex and concave programming problem. This implies that any local optimum is also a global one, and that we may just bother with points on the boundary of the feasible region. More precisely, we have an optimal solution corresponding to a vertex of the feasible region, which is polyhedral. This is exploited in the celebrated simplex algorithm, which is available in plenty of commercial packages. There are also alternative strategies based on interior point methods.

Example 10.7 The `linprog` package

> Say that we want to solve the following LP:
>
> $$\begin{aligned} \max \quad & 45x_1 + 60x_2 \\ \text{s.t.} \quad & 15x_1 + 35x_2 \leq 2400, \\ & 25x_1 + 15x_2 \leq 2400, \\ & 0 \leq x_1 \leq 100, \\ & 0 \leq x_2 \leq 50. \end{aligned}$$
>
> We may resort to the `linprog` package, which provides us with the function `solveLP` to solve LPs by the simplex algorithm. The following script solves the problem:
>
> ```
> library(linprog)
> Amat <- rbind(c(15,35),c(25,15),c(1,0),c(0,1))
> bvet <- c(2400,2400,100,50)
> cvet <- c(45,60)
> result <- solveLP(cvet,bvet,Amat,maximum=TRUE)
> cat('optimal obj ', result$opt, '\n')
> cat('optimal x ', result$solution, '\n')
> ```
>
> Note that we have to include the upper bounds in the matrix of generic linear constraints; better solvers allow us to keep bounds apart. The resulting output is
>
> ```
> optimal obj  5538.462
> optimal x  73.84615 36.92308
> ```

If even one function involved in problem (10.2) is nonlinear, be it the objective or a function defining an equality or inequality constraint, we step into the domain of nonlinear programming. Contrary to what one may imagine, the real difficulty boundary is not marked by the linear/nonlinear dichotomy. The defining property is convexity.

For instance, we lose convexity in LP if we restrict a subset of variables to take only integer values: We get a mixed-integer LP model, MILP for short, which is nonconvex and requires much more complicated solution procedures. In principle, since there is no easy optimality condition to check for MILPs, enumeration of solution is the only viable approach; partial enumeration strategies based on branch and bound methods are widely available, even though they may be quite expensive in terms of computational effort. On the contrary, there are nonlinear programming problems that are rather easy to solve. One such example is convex quadratic programming (QP):

$$\min \quad \tfrac{1}{2}\mathbf{x}^\mathsf{T}\mathbf{Q}\mathbf{x} + \mathbf{h}^\mathsf{T}\mathbf{x} \tag{10.6}$$
$$\text{s.t.} \quad \mathbf{A}\mathbf{x} = \mathbf{b},$$
$$\mathbf{x} \geq \mathbf{0}.$$

In QPs, we have an objective involving a quadratic form, whereas all of the constraints are linear. The problem is easy if the quadratic form is convex, which is the case if the matrix $\mathbf{Q}$ is positive semidefinite. It is worth noting that a quite relevant example of positive semidefinite matrix in financial applications is a covariance matrix; see Section 10.2.1.

📑 Example 10.8 The `quadprog` **package**

Say that we want to solve the following QP:

$$\min \quad x_1^2 + x_2^2$$
$$\text{s.t.} \quad x_1 + x_2 = 4$$

with obvious solution $\mathbf{x}^* = [2,2]^\mathsf{T}$. We may resort to the `quadprog` package. The following script solves the problem:

```
library(quadprog)
Qmat <- rbind(c(2,0),c(0,2))
bvet <- c(4)
Amat <- matrix(c(1,1),nrow=2)
result <- solve.QP(Qmat, c(0,0),Amat,bvet,meq=1)
cat('optimal obj ', result$value, '\n')
cat('optimal x ', result$solution, '\n')
```

The resulting output is

```
optimal obj  8
optimal x   2 2
```

We note a few peculiarities of `solve.QP`:

- We have to double the matrix $\mathbf{Q}$, since the objective function is assumed in the form shown in Eq. (10.6), with a leading $\frac{1}{2}$.
- We have to specify that the first `meq=1` constraints are equalities, whereas all of the rest of them (none, here) are assumed to be inequalities.
- We have to transpose the matrix $\mathbf{A}$, as for some reason `solve.QP` assumes constraints in the forms $\mathbf{A}^\mathsf{T}\mathbf{x} \sim \mathbf{b}$, where $\sim$ can be $=$ and/or $\geq$.

As we show later, QPs have a long tradition in finance. Another class of relevant problems is quadratically constrained QP (QCQP), which involves convex quadratic forms in the constraints as well. There are other classes of nonlinear, yet convex problems for which quite efficient solution procedures are available. The more general nonlinear programming problems are typically harder but not intractable, provided that convexity applies. On the contrary, nonconvex problems require global optimization methods and are quite challenging. This is where Monte Carlo methods come into play.

Deterministic vs. stochastic optimization problems

In most interesting and realistic problems we have to cope with some form of uncertainty in the data. This is certainly the case in financial problems, where return on investments is uncertain. We may have even problems with an uncertain time horizon, such as pension planning. A rather abstract form of a stochastic optimization problem is

$$\min_{\mathbf{x}\in S} \mathrm{E}_{\boldsymbol{\xi}}\left[f(\mathbf{x},\boldsymbol{\xi})\right], \tag{10.7}$$

which is an obvious generalization of the abstract problem (10.1). Here the objective function depends on decision variables $\mathbf{x}$ and random variables $\boldsymbol{\xi}$. Taking the expectation eliminates $\boldsymbol{\xi}$, yielding a function of $\mathbf{x}$. Assuming that the random variables are continuous with joint density $g_{\boldsymbol{\xi}}(\mathbf{0})$, we have

$$\mathrm{E}_{\boldsymbol{\xi}}\left[f(\mathbf{x},\boldsymbol{\xi})\right] = \int_{\Xi} f(\mathbf{x},\mathbf{z})g_{\boldsymbol{\xi}}(\mathbf{z})\,d\mathbf{z} \equiv H(\mathbf{x}).$$

The random variables represent uncertainty that is realized *after* we make the decision $\mathbf{x}$, as illustrated in the following simple example.

■ Example 10.9 **The newsvendor problem**

The newsvendor problem is a prototypical problem in supply chain management and concerns production or purchasing decisions for items subject to perishability or obsolescence risk, like fashion goods. There is a limited time window for sales, and we must decide the order quantity q before knowing the realization of a random demand D during the time window. Items are bought (or produced) at unit cost c and sold at the full retail price $p > c$ within the sales time window; any unsold item after the time window must be marked down and sold at price $p_u < c$. If the case $q < D$ occurs, we have an opportunity cost corresponding to the lost profit margin $m \equiv p - c$ for each unit of unsatisfied demand. On the other hand, when $q > D$ we incur a cost $c_u \equiv c - p_u$ for each unsold item. The newsvendor problem seems quite unrelated with financial problems. Nevertheless, to see a useful connection, imagine a retail bank that has to decide on the amount of cash required to meet withdrawal requests; ideally, we would like to set apart exactly the cash needed, and any surplus or shortfall is associated with a cost. Furthermore, the newsvendor problem is a simple decision problem with an analytical solution, which we use later to illustrate issues in scenario generation.

Let us use the notation $X^+ \equiv \max\{0, X\}$ and express profit as:

$$
\begin{aligned}
\pi(q, D) &= -cq + p \min\{q, D\} + p_u \max\{0, q - D\} \\
&= -cq + pD - p(D - q)^+ + p_u(q - D)^+ \\
&\quad + m(D - q)^+ - m(D - q)^+ \\
&= -cq + pD - p(D - q)^+ + p_u(q - D)^+ \\
&\quad + p(D - q)^+ - c(D - q)^+ - m(D - q)^+ \\
&= pD - c[q + (D - q)^+] + p_u(q - D)^+ - m(D - q)^+ \\
&= pD - cD - c(q - D)^+ + p_u(q - D)^+ - m(D - q)^+ \\
&= mD - c_u(q - D)^+ - m(D - q)^+.
\end{aligned}
\tag{10.8}
$$

Our problem is the maximization of expected profit,

$$
\max_q E_D\left[\pi(q, D)\right],
\tag{10.9}
$$

which can also be recast in terms of cost minimization:

$$
\min_q E_D\left[c_u(q - D)^+ + m(D - q)^+\right].
\tag{10.10}
$$

To see the equivalence of models (10.9) and (10.10), observe that the expectation of Eq. (10.8) involves the the expected demand, which is a given constant and does not play any role. In the second form

(10.10), we associate a cost c_u with each unit of surplus and a cost m with each unit of shortfall with respect to demand. In the case of a continuous demand distribution, it is easy to show that the optimal solution solves the equation

$$F_D(q^*) = \frac{m}{m + c_u}, \qquad (10.11)$$

where $F_D(x) \equiv P\{D \leq x\}$ is the cumulative distribution function (CDF) of demand. In other words, the profit margin m and the cost of unsold items c_u define a quantile of the demand distribution. If demand is normal, $D \sim \mathsf{N}(\mu, \sigma^2)$, then $q^* = \mu + z\sigma$, where z is the standard normal quantile corresponding to a probability level given by the ratio in Eq. (10.11). Note that, in general, the optimal solution is *not* to order the expected value of demand. If uncertainty is significant, we should tilt our decision toward larger values if the critical ratio is larger than 0.5, and toward smaller values when it is smaller than 0.5. The important takeaway is that, when faced with an optimization problem involving uncertain data, we should *not* solve a deterministic problem based on their expected values.

The formulation (10.7) actually encompasses cases in which we are not really able to formulate the problem as a mathematical program explicitly. It may be the case that the form of the objective function is not known, but we can only estimate its value by a simulation model. At first, this may sound weird, but there are indeed optimization problems in which we are not quite able to evaluate the objective function exactly. In fact, the objective function could be related to the performance of a complex dynamical system, such that there is no closed-form expression of the objective function. Under uncertainty, the decisions themselves may not be explicit, but given in the form of a decision rule to be applied on the basis of the current state and the previous realization of uncertain parameters. In such a setting, we have to resort to simulation-based optimization approaches, where Monte Carlo methods play a prominent role.

While the formulation of the model (10.7) is general, it obscures two different sides of the coin:

- Uncertainty in optimality
- Uncertainty in feasibility[2]

---

[2] In principle, we may stipulate that $H(\mathbf{x}) = +\infty$ for an infeasible solution, but this is not very helpful computationally.

Infeasibility is not an issue in the newsvendor model above, but it is in general. We will be more specific on how stochastic optimization can be addressed in Section 10.5.

10.2 Optimization model building

The ability of solving an optimization problem is of no use without the ability of building a sensible optimization model. Model building is a difficult art to learn, possibly a bit out of the scope of this book. Nevertheless, it is useful to illustrate a few relevant examples to get a glimpse of the resulting model classes:

- Financial portfolio optimization, possibly including qualitative constraints on portfolio composition that require the introduction of logical decision variables
- Asset pricing problems, in which emphasis is on the value of the objective function rather than the decision variables
- Parameter estimation and model calibration problems

10.2.1 MEAN–VARIANCE PORTFOLIO OPTIMIZATION

Mean–variance optimization is the prototypical *static* portfolio optimization model. The decision concerns setting the weights w_i, $i = 1, \ldots, n$, of a set of n assets, in such as way to strike a satisfactory trade-off between expected return and risk. Let

- $\boldsymbol{\mu} = \left[\mu_1, \ldots, \mu_n \right]^\mathsf{T}$ be a vector collecting the expected return of each asset.
- μ_T be a target expected return that we wish to achieve.
- $\boldsymbol{\Sigma} \in \mathbb{R}^{n,n}$ be the covariance matrix collecting the covariances σ_{ij} between the return of assets $i, j = 1, \ldots, n$.

If we assume that risk can be measured by the standard deviation of portfolio return, one sensible model for the asset allocation decision is:

$$\min \quad \mathbf{w}^\mathsf{T} \boldsymbol{\Sigma} \mathbf{w} = \sum_{i=1}^{n} \sum_{j=1}^{n} w_i \sigma_{ij} w_j$$

$$\text{s.t.} \quad \boldsymbol{\mu}^\mathsf{T} \mathbf{w} = \mu_T, \tag{10.12}$$

$$\sum_{i=1}^{n} w_i = 1,$$

$$w_i \geq 0.$$

Here we minimize variance, rather than standard deviation, since this yields an equivalent, but highly tractable convex quadratic programming problem. The constraint (10.12) enforces the achievement of the target expected return; then we also require that portfolio weights add up to 1, and the non-negativity requirements forbid short-selling.

From a computational point of view, this is a trivial problem that can be solved quite efficiently. However, the model itself has a few weak points:

1. Using standard deviation as a risk measure might make sense for symmetric distributions, such as a multivariate normal; however, asset returns are not quite symmetric, especially if financial derivatives are involved. Alternative risk measures, such as conditional value-at-risk can be adopted; see Section 13.5.

2. The model is static and does not allow for dynamic rebalancing. We may formulate a full-fledged multiperiod model, as we do in Section 10.5.2; alternatively, we may simplify the problem by enforcing some structure on the investment policy, as we do in Section 10.2.4.

3. Even if we stay within the easier domain of static optimization models, we may wish to enforce additional constraints on the portfolio, such as an upper bound on the number of stock names in our portfolio, as we do in Section 10.2.2 below.

These variations on the theme lead to nonconvex and/or stochastic optimization models, which may be quite challenging to formulate and solve; Monte Carlo methods may play a pivotal role in tackling them, in one way or another.

10.2.2 MODELING WITH LOGICAL DECISION VARIABLES: OPTIMAL PORTFOLIO TRACKING

One of the features that may make an optimization problem nonconvex and much harder to solve is the presence of decision variables restricted to integer values. By far, the most common case involves logical (or binary) decision variables, taking values in the set $\{0, 1\}$. We first outline a few standard modeling tricks involving such variables, and then we describe a practically relevant portfolio optimization model requiring their introduction.

Modeling fixed charges

Consider an economic activity whose level is measured by a continuous decision variable $x \geq 0$. The total cost $\mathrm{TC}(x)$ related to the activity may consist of two terms:

- A variable cost, which is represented by a term like cx, where c is the unit variable cost.
- A fixed charge f, which is only paid if $x > 0$, i.e., if the activity is carried out at a positive level.

In principle, we could introduce a step function such as

$$\gamma(x) = \begin{cases} 1, & \text{if } x > 0, \\ 0, & \text{if } x = 0, \end{cases}$$

and express total cost as

$$\text{TC}(x) = cx + f\gamma(x).$$

Unfortunately, the step function is nonlinear and discontinuous at the origin, as it jumps from 0 to 1. To avoid the resulting technical difficulty, we may resort to an alternative representation by introducing the following binary decision variable:

$$\delta = \begin{cases} 1, & \text{if } x > 0, \\ 0, & \text{otherwise.} \end{cases}$$

Then, we link x and δ by the so-called *big-M constraint*:

$$x \leq M\delta, \tag{10.13}$$

where M is a suitably large constant. To see how Eq. (10.13) works, imagine that $\delta = 0$; then, the constraint reads $x \leq 0$, whatever the value of M is. This constraint, together with the standard non-negativity restriction $x \geq 0$, enforces $x = 0$. If $\delta = 1$, the constraint is $x \leq M$, which is nonbinding if M is large enough; in practice, M should be a sensible upper bound on the level of activity x.

Semicontinuous variables

A possible requirement on the level of an activity is that, if it is undertaken, its level should be in the interval $[m_i, M_i]$. Note that this is *not* equivalent to requiring that $m_i \leq x_i \leq M_i$. Rather, we want something like

$$x_i \in \{0\} \cup [m_i, M_i],$$

which involves a nonconvex set (recall that the union of convex sets need not be convex). For instance, we may use a constraint like this to avoid the inclusion of assets with negligible weights in a portfolio. Using the big-M trick as above, we may write

$$x_i \geq m_i\delta_i, \qquad x_i \leq M_i\delta_i.$$

These constraints define a *semicontinuous decision variable*.

Portfolio tracking and compression

Active portfolio managers aim at achieving superior return by tacking advantage of skill or analytical capabilities. On the contrary, passive portfolio managers aim at replicating an index or tracking a benchmark portfolio as close as possible. There is some theoretical support for the passive view, as the efficient market hypothesis, in its various forms, essentially states that there is no way to beat the market. The capital asset pricing model, too, supports the view that the market portfolio is the optimal risky component in an overall portfolio. The matter is highly controversial, but one thing can be taken for granted: A passive portfolio must have a very low management cost. One way to achieve this is to restrict the number of asset names included in the portfolio, i.e., to enforce a cardinality constraint.

To formulate the problem, let us consider a universe of n assets and a benchmark portfolio with given weights w_i^b, $i, \ldots, n$. We want to find the weights $w_i \geq 0$ of a tracking portfolio, in such a way to minimize a measure of distance between the target and the tracking portfolio. An additional restriction is that at most $C_{\max}$ assets can be included, i.e., at most $C_{\max}$ weights can be strictly positive (we assume that short-selling is forbidden). In order to express the cardinality constraint, we introduce a set of binary variables δ_i, one for each asset, modeling the inclusion of asset i in the tracking portfolio:

$$\delta_i = \begin{cases} 1, & \text{if asset } i \text{ is included in the tracking portfolio,} \\ 0, & \text{otherwise.} \end{cases}$$

The binary variables δ_i and the continuous variables w_i are linked by the big-M constraint

$$w_i \leq M\delta_i, \tag{10.14}$$

and the cardinality constraint reads as

$$\sum_{i=1}^{n} \delta_i \leq C_{\max}.$$

Since a weight cannot be larger than 1, when we rule out short-selling, we can set $M = 1$ in Eq. (10.14); the big-M constraints can be tightened by setting $M = \overline{w}_i$, where $\overline{w}_i$ is some suitable upper bound on the weight of asset i.[3]

Now the problem is how to measure the portfolio tracking error. A trivial distance metric can be defined by an L_1 norm:

$$\sum_{i=1}^{n} \left| w_i - w_i^b \right|.$$

By proper model formulation, this metric yields an MILP model. However, this distance metric does not take any return model into account. To see the point, let us consider the following weights for two assets:

$$w_1^b = 0.1, \qquad w_2^b = 0.2,$$
$$w_1 = 0.3, \qquad w_2 = 0.$$

---

[3]Tightening such bounds will speed up branch and bound methods, which are the standard solution methods for mixed-integer programs.

According to the L_1 norm, the distance between the target and the tracking portfolio, considering only these two assets, is

$$|0.3 - 0.1| + |0 - 0.2| = 0.4.$$

However, what is the distance if the two assets are perfectly correlated? Of course, the actual distance would be zero, in such a case. More generally, the actual distance depends on the link between the two returns. An alternative distance metric is TEV (tracking error variance), defined as

$$\text{TEV} \equiv \text{Var}(R_p - R_b),$$

where R_p and R_b are the return of the tracking portfolio and the benchmark, respectively. Then, TEV can be expressed as

$$\text{Var}\left(\sum_{i=1}^{n} w_i R_i - \sum_{i=1}^{n} w_i^b R_i\right) = \text{Var}\left[\sum_{i=1}^{n} \left(w_i - w_i^b\right) R_i\right]$$
$$= \sum_{i=1}^{n} \sum_{j=1}^{n} \left(w_i - w_i^b\right) \sigma_{ij} \left(w_j - w_j^b\right),$$

where σ_{ij} is the covariance between the returns of assets i and j. Straightforward minimization of TEV, subject to a cardinality constraint, yields the following model:

$$\min \quad \sum_{i=1}^{n} \sum_{j=1}^{n} \left(w_i - w_i^b\right) \sigma_{ij} \left(w_j - w_j^b\right) \tag{10.15}$$

$$\text{s.t.} \quad \sum_{i=1}^{n} w_i = 1,$$

$$\sum_{i=1}^{n} \delta_i \leq C_{\max},$$

$$w_i \leq \overline{w}_i \delta_i \qquad \forall i,$$

$$w_i \geq 0, \qquad \delta_i \in \{0, 1\} \qquad \forall i.$$

This model is a mixed-integer quadratic programming problem with a convex relaxation. Commercial solvers are able to tackle this model structure. Nevertheless, some stochastic search methods, such as the sampling based algorithms described in Section 10.3, have been proposed as alternatives for large-scale problem instances.

10.2.3 A SCENARIO-BASED MODEL FOR THE NEWSVENDOR PROBLEM

We have introduced the newsvendor problem in Example 10.9. The basic version of the problem can be solved analytically, and the same applies to many interesting variants. Nevertheless, it is instructive, as well useful for what follows, to consider a scenario-based model formulation. The idea is to approximate the probability distribution for demand by a set of S scenarios, characterized by a probability π_s and a demand realization D_s, $s = 1, \ldots, S$. The main decision variable is the ordered amount $q \geq 0$ and, in order to express the objective in cost minimization form, we introduce two non-negative deviations with respect to realized demand:

$$q - D_s = y_s^+ - y_s^- \qquad \forall s, \tag{10.16}$$

where y_s^+ is a surplus and y_s^- is a shortfall. The, the model can be formulated as an LP model:

$$\min \quad \sum_{s=1}^{S} \pi_s \left(c_u y_s^+ + m y_s^- \right) \tag{10.17}$$

$$\text{s.t.} \quad q = D_s + y_s^+ - y_s^- \qquad \forall s,$$

$$q, y_s^+, y_s^- \geq 0.$$

As we shall discover later, this can be considered as a very simple example of two-stage stochastic LP model. From a practical point of view, it is a basis for more interesting models allowing for interactions among multiple products, e.g., by demand substitution.

10.2.4 FIXED-MIX ASSET ALLOCATION

In Section 10.2.1 we have considered a static, single-period asset allocation model. Later, in Section 10.5.2, we will consider a full-fledged multiperiod model, allowing for dynamic portfolio adjustments. An intermediate strategy can be devised by specifying a fixed-mix, i.e., by keeping portfolio weights w_i constants over multiple periods. Wealth will change along the way, and to keep the relative weights constant, we will sell high-price assets, and buy low-price assets. Thus, the strategy is intrinsically a contrarian one, rather than a trend-following one. Let us also assume that we only care about mean and variance, as we did before, and that our mean–risk objective is the expected value of terminal wealth penalized by its variance. Unlike the static case, expressing variance of the terminal wealth is not easily accomplished; hence, we resort to a set of S return scenarios. Each scenario is a history of total[4] asset returns R_{it}^s, over a planning horizon consisting of time periods $t = 1, \ldots, T$. Each scenario is associated with a probability π_s, $s = 1 \ldots, S$. The model we describe here is due to [32], to which we refer the reader for further information and computational experiments, and is basically an extension of the mean–variance framework.

Let W_0 be the initial wealth. Then, wealth at the end of time period 1 in scenario s will be

$$W_1^s = W_0 \sum_{i=1}^{n} R_{i1}^s w_i.$$

---

[4]For the sake of notational simplicity, here we denote by R the *total* return, whereas when R is the *rate* of return, total return is $1 + R$.

Note that wealth is scenario-dependent, but the asset allocation is not. In general, when we consider two consecutive time periods, we have

$$W_t^s = W_{t-1}^s \sum_{i=1}^n R_{it}^s w_i \qquad \forall t, s.$$

Unfolding the recursion, we see that wealth at the end of the planning horizon is

$$W_T^s = W_0 \prod_{t=1}^T \left(\sum_{i=1}^n R_{it}^s w_i \right) \qquad \forall s.$$

Within a mean–variance framework, we may build a quadratic utility function depending on the terminal wealth. Given a parameter λ related to risk aversion, the objective function is

$$\max \quad \mathrm{E}[W_T] - \lambda \operatorname{Var}(W_T).$$

To express the objective function, we recall that $\operatorname{Var}(X) = \mathrm{E}[X^2] - \mathrm{E}^2[X]$ and write the model as

$$
\begin{aligned}
\max \quad & W_0 \sum_{s=1}^S \pi_s \left[\prod_{t=1}^T \left(\sum_{i=1}^n R_{it}^s w_i \right) \right] \\
& + \lambda W_0^2 \left\{ \left[\sum_{s=1}^S p^s \left[\prod_{t=1}^T \left(\sum_{i=1}^n R_{it}^s w_i \right) \right] \right]^2 \right. \\
& \qquad\qquad \left. - \sum_{s=1}^S p^s \left[\prod_{t=1}^T \left(\sum_{i=1}^n R_{it}^s w_i \right) \right]^2 \right\} \\
\text{s.t.} \quad & \sum_{i=1}^n w_i = 1, \\
& 0 \le w_i \le 1.
\end{aligned}
$$

This looks like a very complex problem; however, while the objective function is a bit messy, the constraints are quite simple. The real difficulty is that this is a nonconvex problem. To see why, just note that the objective turns out to be a polynomial in the decision variables; since polynomials may have many minima and maxima (see Example 10.3), we face a nonlinear nonconvex problem. Once again, stochastic search algorithms, based on Monte Carlo sampling, can be used to cope with these problems.

10.2.5 ASSET PRICING

Asset pricing is a central problem in financial engineering, as well as in financial economics and corporate finance. We know from Section 3.9.2 that the price of a European-style option can be expressed as the discounted expected payoff, under the risk-neutral measure. Assuming a constant risk-free rate r, with continuous compounding, we have

$$f(S_0, 0) = e^{-rT} \mathrm{E}^{\mathbb{Q}} \big[\Phi(S_T) \big], \qquad (10.18)$$

where S_0 is the underlying asset price now and $\Phi(S_T)$ is the option payoff, depending on the underlying asset price S_T at maturity. In principle, pricing a European-style option by Monte Carlo methods is easily accomplished, provided that we can generate sample paths efficiently; if we do not consider hedging issues, we have just to sit down, cross our fingers, and wait until maturity. However, there are assets whose value depends on our choices along the way. Generally speaking, the value of a resource depends on how the resource itself is used. American-style options, unlike their European-style counterparts, can be exercised at any date prior to expiration. One easy conclusion is that an American-style option cannot cost less than the corresponding European one, as it offers more opportunities. However, early exercise opportunities make pricing much harder, as we should consider an optimal exercise strategy, which in turn calls for the solution of a stochastic dynamic optimization problem. At each exercise opportunity, if the option is in-the-money, we have to compare the *intrinsic value* of the option, i.e., the immediate payoff that we may earn from exercising the option early, and the *continuation value*, i.e., the value of postponing exercise and waiting for better opportunities.

Formally, the price of an American option can be written as

$$\max_{\tau} \mathrm{E}^{\mathbb{Q}} \big[e^{-r\tau} \Phi(S_\tau) \big]. \qquad (10.19)$$

The difference with Eq. (10.18) is that now we have to solve an optimization problem with respect to a stopping time τ. The term "stopping time" has a very precise meaning in the theory of stochastic processes, but here we may simply interpret it as the time at which we exercise the option, according to some well-defined strategy.

Example 10.10 An American-style put option

A put option is in-the-money at time t if $S(t) < K$. Immediate exercise in this case yields a payoff $K - S(t)$. Clearly, early exercise will not occur if the option is not in-the-money, but even if $S(t) < K$ it may be better to keep the option alive and wait. This means that early exercise will occur only if the option is "enough" in-the-money. By how much, it will generally depend on time to expiration. When we are close to option expiration, there are less opportunities, and we are more inclined to exercise; but when we are far from expiration, we are more inclined to wait. Formally, we have to find a boundary separating the early exercise region, where the intrinsic value is larger

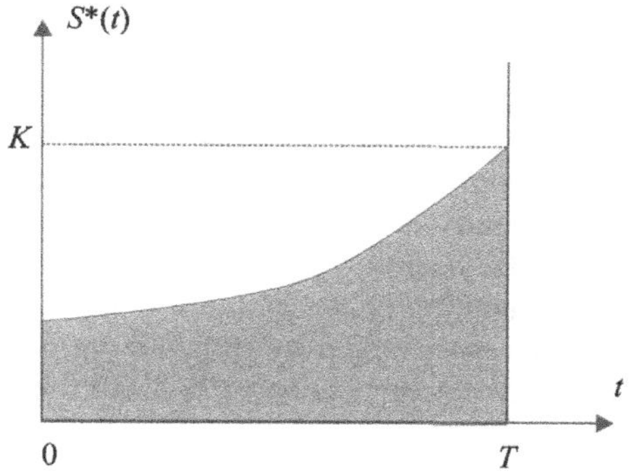

FIGURE 10.8 **Qualitative sketch of the early exercise boundary for a vanilla American put. The option is exercised within the shaded area.**

than the continuation value, from the continuation region. Qualitatively, for an American put option we would expect an early exercise boundary like the one depicted in Fig. 10.8. This boundary specifies a stock price $S^*(t)$ such that if $S(t) < S^*(t)$, i.e., if the option is deeply enough in-the-money, then we exercise. If we are above the boundary, we are in the continuation region, and we keep the option alive. Finding the boundary is what makes the problem difficult; for a detailed treatment of the exercise boundary for American options see, e.g., [31, Chapter 4].

From the last example, we understand that the time at which the option will be exercised is indeed a random variable, in this case the time at which we cross the exercise boundary, if we cross it. More complicated strategies are conceivable and must be applied to exotic and multidimensional options, and this is what the stopping time captures. The stopping time is related to any conceivable strategy, provided that it is nonanticipative; we may only use information gathered from time 0 up to time t. From a practical perspective, in order to price an option with early exercise features, we have to solve a possibly nontrivial optimization problem. For low-dimensional options, numerical methods based on suitable discretizations of the state space, like binomial or trinomial lattices, or finite-difference methods to solve partial differential equations work very well. For high-dimensional or path-dependent options, we have to resort to Monte Carlo methods. In the past, this was considered impossible, but by framing the problem within a stochastic dynamic programming framework, we may tackle it, as we shall see in Section 11.3.

10.2.6 PARAMETER ESTIMATION AND MODEL CALIBRATION

We have introduced maximum-likelihood estimation in Section 4.2.3. Parameter values are chosen in such a way as to maximize a likelihood function related to a selected probability distribution. In very simple cases, the solution is easily found analytically. In nontrivial cases, numerical optimization methods are needed for the solution of the resulting problems, possibly because of the complexity of the likelihood function or the addition of constraints on parameter values, on the basis of some additional information we might have. Furthermore, the resulting problem may be nonconvex and call for global optimization strategies.

A similar kind of problem is encountered when we have to calibrate a pricing model. For instance:

- Given a set of bond prices, we might want to infer the term structure of interest rates, i.e., the yield curve.

- Given a set of quoted option prices written on the same underlying asset, we might want to estimate a volatility surface.

Example 10.11 Fitting the yield curve

At time t we observe a set of bond prices $P_k(t)$, $k = 1, \ldots, n$. Let us assume for the moment that the cash flows associated with these bonds, i.e., the payment of coupons along the way and the face value at bond maturity, occur at a set of common time instants T_i, $i = 1, \ldots, m$. Let $r(t, T)$ denote the continuously compounded interest rate applying over the time interval $[t, T]$. Based on observed bond prices, we want to estimate $r(t, T)$ for a range of maturities. According to straightforward bond pricing formulas, the bond prices should be given by

$$\hat{P}_k(t) = \sum_{i=1}^{m} c_{k,i} e^{-r(t,T_i)(T_i - t)}, \qquad k = 1, \ldots, n,$$

where $c_{k,i}$ is the cash flow from bond k at time T_i. These predicted prices are compared against the observed prices $P_k^o(t)$. If $m = n$, the number of unknown parameters and prices are the same, and we may solve a system of nonlinear equations to find the interest rates for maturities T_i, where we assume $\hat{P}_k(t) = P_k^o(t)$. A textbook approach, known as bootstrapping, assumes that exactly one bond will mature at each time instant T_i; then, the system of equations has a "triangular" structure and we can solve one equation at a time, finding one rate per equation. In practice, it may be difficult to find a suitable set of bonds lending itself to such a simplified solution strategy. Liquidity and other issues may affect bond prices, introducing noise and some irregularity in the resulting yield curve.

To smooth those effects, it is preferable to choose $m > n$ and minimize an error distance between observed and predicted prices, with respect to rates:

$$\min \sum_{k=1}^{n} \left[\hat{P}_k(t) - P_k^o(t) \right]^2.$$

This is a nonlinear least-squares problem that can be solved by numerical optimization methods. However, it is typically desirable not to consider a restricted set of cash flow times T_i, in order to widen the set of bonds we may use for calibration. Furthermore, we would like to fit a term structure with some reasonable degree of smoothness, in such a way that we can estimate rates for maturities different from the aforementioned cash flow times T_i. One way to accomplish this is by spline interpolation, but due to possible pricing anomalies exact interpolation may yield unreasonable results, and least-squares approximation may be preferred. More generally, we may consider a set of basis functions $\phi_j(T)$, $j = 1, \ldots, J$, whose linear combination

$$\sum_{j=1}^{J} \alpha_j \phi_j(T)$$

gives the yield curve as a function of maturity T. Here, the decision variables of the optimization problem are the coefficients α_j, and we might also consider estimating the discount curve directly, which is then recast as the yield curve. The use of basis functions to boil down an infinite dimensional problem to a finite dimensional one is very common, and we will take advantage of the idea later, when dealing with numerical dynamic programming.

The calibration of the yield curve on the basis of quoted prices is an example of an *inverse problem*. Model calibration is also a cornerstone of option pricing in incomplete markets. As we have seen in Section 3.9.3, market incompleteness arises whenever we step out of the safe BSM domain based on GBM process with very simple drift and volatility terms. If volatility is stochastic or the underlying variable is nontradable, we may need a more sophisticated approach. In Eq. (3.126) we have pointed out the existence of an unknown function $m^*(\cdot, \cdot)$ serving as a drift in an Itô differential equation, under a pricing measure. We report the equation here for the sake of convenience, recasting it in terms of a generic state variable X_t:

$$dX_t = m^*(X_t, t)\, dt + s(X_t, t)\, dW_t.$$

Once again, note that for pricing purposes the drift term is related to a risk-neutral measure, and it cannot be estimated on the basis of a time series of X_t. By a similar token, we may not want to fit the volatility function on the basis of historical values of volatility; rather, we want to extract the current market view, which is implicit in asset prices. Hence, we are faced with an inverse problem. Depending on the kind of dependence on time, the functions $m^*(\cdot, \cdot)$ and $s(\cdot, \cdot)$ may depend on a finite or infinite set of parameters. The second case occurs when these terms explicitly depend on time, in order to improve their matching performance against quoted prices. Whatever the case, by a suitable discretization mechanism, possibly based on basis functions, interpolation, and whatnot, we end up with a pricing model depending on a vector of parameters $\boldsymbol{\alpha}$. The pricing model predicts a price $\hat{P}_k(\boldsymbol{\alpha})$ for a set of securities with quoted (observed) price P_k^o, and it calibrated by solving the optimization problem

$$\min_{\boldsymbol{\alpha}} \sum_{k=1}^{n} \left[\hat{P}_k(\boldsymbol{\alpha}) - P_k^o\right]^2.$$

One might wonder why we should calibrate a model by fitting a set of unobservable parameters against quoted prices of exchange-traded instruments. Clearly, the aim is not to recover quoted prices per se, but to fit a model in order to price over-the-counter assets and to assess option price sensitivity with respect to a set of risk factors.

Depending on the complexity of the pricing model, we may end up with relatively simple optimization models, amenable to classical nonlinear programming procedures (most notably, the Levenberg–Marquardt algorithm), or an extremely difficult global optimization problem. In the latter case, the Monte Carlo based global optimization methods that we describe in the next section might be the only viable solution method.

10.3 Monte Carlo methods for global optimization

When dealing with a convex optimization problem featuring a well-defined objective function, we may take advantage of useful information provided by its gradient and Hessian matrix. Unfortunately, it may be the case that the objective function is badly behaved, i.e., nonconvex, possibly discontinuous, or even not known in analytical form. Then, we may resort to algorithms that rely only on function evaluations, possibly obtained by stochastic simulation, and a search mechanism to explore the solution space randomly. Monte Carlo methods may be used at both levels:

1. In this section we consider stochastic search methods for global optimization. The optimization problem, per se, is a purely deterministic one, but

we take advantage of random exploration to overcome local optimality issues.

2. Later, in Section 10.4, we consider simulation-based optimization, where Monte Carlo simulation is needed to estimate the objective function itself, and a deterministic search strategy is used to explore the solution space.

As the reader can imagine, there is no clear-cut line separating the two ideas, as random sampling may be employed at both levels.

For any specific decision problem it is possible to devise an ad hoc heuristic method providing us with a good, possibly near-optimal, solution. Here we only consider fairly general strategies, which may be customized to the specific problem at hand. Indeed, a wide class of such methods are known under the name of *metaheuristics*, which emphasizes their generality and flexibility. Indeed, they are able to cope with both continuous nonconvex problems as well as discrete optimization problems, like optimal portfolio tracking, where the nonconvexity arises from the feasible set, which includes binary decision variables. We first introduce local search and other metaheuristic principles in Section 10.3.1. Then, in Section 10.3.2 we describe simulated annealing, a simple stochastic search method inspired by an interesting physical analogy. Other methods, such as genetic algorithms (Section 10.3.3) and particle swarm optimization (Section 10.3.4) rely on analogies with biological systems and, unlike simulated annealing, work on a population of solutions, rather than a single one.

10.3.1 LOCAL SEARCH AND OTHER METAHEURISTICS

Local search is a wide class of simple exploration strategies based on the idea of generating a sequence of solutions exploring the neighborhood of the current one. Consider a generic optimization problem

$$\min_{x \in S} f(x),$$

defined over a feasible set S. The basic idea is to improve a given feasible solution by the application of a set of local perturbations. We associate a feasible solution x with a neighborhood $\mathcal{N}(x)$, which is defined by applying a set of simple perturbations to x. Different perturbations yield different *neighborhood structures*. The simplest local search algorithm is *local improvement*: Given a current (incumbent) solution $\mathbf{x}^0$, an alternative (candidate) solution $\mathbf{x}'$ is searched for in the neighborhood of the incumbent:

$$f(\mathbf{x}') = \min_{\mathbf{x} \in \mathcal{N}(\mathbf{x}^0)} f(\mathbf{x}).$$

If the candidate solution improves the objective function, i.e., $f(\mathbf{x}') < f(\mathbf{x}^0)$, then the incumbent solution is replaced by the candidate and the process is repeated. The search is terminated when we cannot find any improving solution, and the incumbent solution is returned. Clearly, there is no guarantee that an optimal solution is found, as the process may get trapped in a bad locally optimal solution. Furthermore the solution is locally optimal with respect to the selected neighborhood structure, which must be simple enough to allow for an efficient search, but not so simplistic that the exploration process is ineffective. If the candidate solution is the optimal solution in the neighborhood, we speak of a *best-improving* local search; if we stop exploring the neighborhood as soon as an improving solution is found, we speak of a *first-improving* local search.

Example 10.12   Neighborhood structures for local search

Neighborhood structures may be devised for both discrete and continuous optimization problems. If all the components of $\mathbf{x} \in \mathbb{R}^n$ are restricted to binary values, a simple neighborhood structure is obtained by complementing each variable in turn, i.e., for each $i = 1, \ldots, n$ we set

$$x_i' = \begin{cases} 1 & \text{if } x_i^0 = 0, \\ 0 & \text{otherwise.} \end{cases}$$

The neighborhood consists of n points and it is easy to explore, but possibly very poor. We may consider also complementing pairs of variables, as well as swapping values of variables having different values.

The case of continuous variables adds a degree of difficulty, as it is not clear by how much we should perturb each component. We may define steps δ_i for each component and generate $2n$ alternative solutions by applying the perturbation in both directions:

$$\mathbf{x}^0 \pm \mathbf{e}_i \delta_i,$$

where $\mathbf{e}_i$ is a unit vector with 1 in position i and 0 in the other positions. This is the simplest pattern for searching the neighborhood, but some mechanism should be devised to dynamically adjust the step lengths.

Alternatively, we may generate a random candidate at a time, by sampling randomly in the neighborhood of the incumbent solution using a suitable probability distribution. For instance, we may adopt a multivariate normal distribution with expected value $\mathbf{x}^0$ and a diagonal covariance matrix $\boldsymbol{\Sigma}$. This approach lends itself to first-improving strategy.

Actually, devising a clever and effective neighborhood is not trivial at all. A further complication is that due attention must be paid to constraints. In some cases we might just discard solutions violating them; however, this is not easily accomplished when dealing with equality constraints, as staying inside the feasible region might be difficult. An alternative strategy is to relax constraints and augment the objective function by a corresponding penalty function. For instance, the constrained problem

$$
\begin{aligned}
\min \quad & f(\mathbf{x}) \\
\text{s.t.} \quad & h_i(\mathbf{x}) = 0, \quad i \in E, \\
& g_i(\mathbf{x}) \le 0, \quad i \in I,
\end{aligned}
$$

may be transformed into the unconstrained problem

$$
\min \quad f(\mathbf{x}) + \sum_{i \in E} \omega_i \big[h_i(\mathbf{x}) \big]^2 + \sum_{i \in I} \omega_i \max \big\{ 0, g_i(\mathbf{x}) \big\},
$$

for suitably large penalty coefficients ω_i, possibly expressing the relative importance of constraints. The penalty coefficients might also be dynamically adjusted, allowing the search process to occasionally leave the feasible region and getting back into it at another point. This kind of approach, called strategic oscillation, may be used, along with other trick of the trade, to balance exploitation (the need to stick to a good solution and explore its neighborhood) and exploration (the need to explore other subregions of the feasible set).

Local improvement is not suitable to global optimization. However, one may relax the requirement that the new solution in the search path is an improving one. By accepting nonimproving candidates, we may hope to escape from local optima. Two such strategies are:

- Simulated annealing, whereby the neighborhood is randomly sampled and a nonimproving solution is accepted with a suitably chosen probability.

- Tabu search, whereby we explore the neighborhood and always accept the best solution found, even if it is nonimproving. Since this may result in repeating a cycle of solutions, tabu attributes are used to control the search process and forbid revisiting solutions. Tabu search, in its basic form, is a deterministic search approach and we will not deal with it.

Both simulated annealing and tabu search work with a single incumbent solution. In order to find a better trade-off between exploration and exploitation, one may work with a *population* of solutions. Examples of population-based approaches are genetic algorithms and particle swarm optimization.

10.3.2 SIMULATED ANNEALING

Simulated annealing owes its name to an analogy between cost minimization in optimization and energy minimization in physical systems. The local improvement strategy behaves much like physical systems do according to classical mechanics. It is impossible for a system to have a certain energy level at a certain time instant then and to increase it without external input: If you place a ball in a hole, it will stay there. This is not true in statistical mechanics; according to these physical models, at a temperature above absolute zero, thermal noise makes an increase in the energy of a system possible. This applies to changes

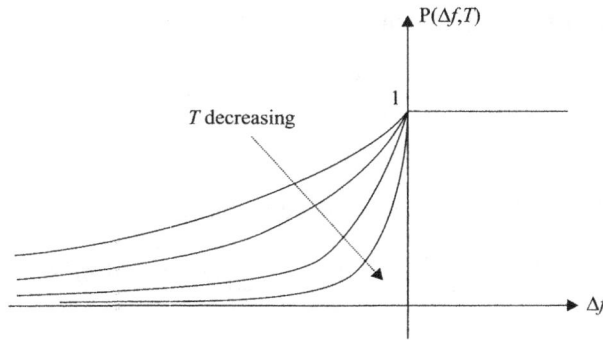

FIGURE 10.9 **Acceptance probabilities as a function of cost increase for different temperatures in simulated annealing.**

at a microscopic level, and an increase in energy is more likely to occur at high temperatures. The probability P of this upward jump depends on the increase in energy ΔE and the temperature T, according to the Boltzmann distribution

$$P(\Delta E, T) = \exp\left(-\frac{\Delta E}{KT}\right),$$

where K is the Boltzmann constant. Annealing is a metallurgical process by which a melted material is slowly cooled in order to obtain a good (low-energy) solid-state configuration. If the temperature is decreased too fast, the system gets trapped in a local energy minimum, and a glass is produced. But if the process is slow enough, random kinetic fluctuations due to thermal noise allow the system to escape from local minima, reaching a point very close to the global optimum.

In simulated annealing we consider the change in the objective function when moving from the incumbent solution to a candidate solution:

$$\Delta f = f(\mathbf{x}') - f(\mathbf{x}^0).$$

The candidate solution is accepted with probability

$$P(\Delta f, T) = \min\left\{\exp\left(-\frac{\Delta f}{T}\right), 1\right\}.$$

An improving candidate yields $\Delta f < 0$ and is always accepted. A nonimproving candidate is randomly accepted, according to a probability depending on the amount of cost increase Δf and the control parameter T, as shown in Fig. 10.9. At very high temperatures, most candidate solutions are accepted, whereas at intermediate temperatures candidate solutions that increase cost are accepted with lower probability if the deterioration is significant. For $T \to 0$ the acceptance probability collapses into a step function, and the method behaves like local improvement. In Section 14.4 we will interpret simulated annealing within the framework of Markov chain Monte Carlo methods, as a way to sample from a density that is concentrated on the set of globally optimal solutions. However, to reach that set without getting trapped in local optima, the temperature should be decreased according to a *cooling schedule*. The simplest cooling schedule is

$$T_k = \beta T_{k-1}, \qquad 0 < \beta < 1.$$

In practice, it is advisable to keep the temperature constant for a certain number of steps, in order to reach a thermodynamic equilibrium before changing the control parameter. More sophisticated adaptive cooling strategies have been proposed. Furthermore, we have to establish a termination criterion. The simplest idea is to stop the algorithm after a given number of iterations; alternatively, we may stop after a maximum number of iterations without accepting a candidate.

We show an extremely naive implementation of simulated annealing in Fig. 10.10, for illustrative purpose only. We may test it on the Rastrigin function[5]

$$f(\mathbf{x}) = \sum_{i=1}^{n} \left[x_i^2 + 10(1 - \cos(2\pi x_i)) \right]^2,$$

which has a global optimum for $\mathbf{x} = \mathbf{0}$:

```
> Rastrigin<-function(x) sum(x^2-10*cos(2*pi*x))+10*length(x)
> dim <- 20
> lb <- rep(-100,dim)
> ub <- rep(100,dim)
> set.seed(55555)
> x0 <- runif(dim,min=lb,max=ub)
> out <- optim(par=x0,fn=Rastrigin,lower=lb,upper=ub,
+              method="L-BFGS-B")
> out[c("value","counts")]
$value
[1] 1361.003
$counts
function gradient
      43       43
>
> set.seed(55555)
> out <- NaiveSA(fn=Rastrigin,par=x0,temp0=100,
+                beta=0.9999,maxit=200000,sigma=1)
> out[c("value","counts")]
$value
[1] 125.437
$counts
[1] 2e+05
```

We use $n = 20$ and start search from a randomly selected point within a box. For compatibility reasons with other R functions that we want to use for comparison

---

```
NaiveSA <- function(fn,par,temp0,beta,maxit,sigma){
  xNow <- par
  fNow <- fn(xNow)
  xBest <- xNow
  fBest <- fNow
  temp <- temp0
  dim <- length(x0)
  for (k in 1:maxit){
    # generate candidate
    xNew <- xNow+rnorm(dim,sd=sigma)
    fNew <- fn(xNew)
    if (fNew<fNow){
      xNow <- xNew; fNow <- fNew
      if (fNew < fBest){
        xBest <- xNew; fBest <- fNew
      }
    }
    else{
      if (runif(1) <= exp((fNow-fNew)/temp)){
        xNow <- xNew; fNow <- fNew
      }
    }
    temp <- beta*temp
  }
  return(list(value=fBest,par=xBest,counts=maxit))
}
```

R programming notes:

1. The function minimizes `fn` starting from the initial solution `par`.

2. The initial temperature is `temp0` and is decreased using the cooling parameter `beta`.

3. The algorithm is terminated after `maxit` iterations, and candidates are selected using a multivariate normal distribution $N(x_0, \sigma^2 I)$, where x_0 is the current distribution and $\sigma^2 I$ is a diagonal covariance matrix with entries σ^2.

4. We return the value of the objective, the minimizer, and the (obvious in this case) number of function evaluations. This conforms to standard R usage.

FIGURE 10.10 **A naive implementation of simulated annealing.**

purposes, we constrain all decision variables in a box with lower bound -100 and upper bound 100. A classical nonlinear programming method, available as an option in the standard R function `optim` gets quickly stuck in a bad solution, whereas the naive annealing does better. Actually, we may take advantage of a better implementation of simulated annealing offered by `optim`:

```
> set.seed(55555)
> out <- optim(par=x0,fn=Rastrigin,method="SANN",
+                 control=list(temp=100,tmax=10,maxit=200000))
> out[c("value","counts")]
$value
[1] 81.33197
$counts
function gradient
   200000      NA
```

Here, `tmax` is the maximum number of iterations without decreasing the temperature. Still better results are obtained by function `GenSA` offered in the package having the same name:

```
> library(GenSA)
> set.seed(55555)
> out <- GenSA(fn = Rastrigin,lower=lb,upper=ub,par=x0)
> out[c("value","par","counts")]
$value
[1] 0
$par
  1.149916e-11   8.491049e-12 -1.670109e-11 -3.363457e-11
 -2.144013e-11  -4.907911e-13 -3.319979e-11 -3.287746e-11
  6.624013e-12  -2.850014e-11  3.173777e-11 -1.580739e-11
 -2.003915e-13  -2.192858e-09  2.288534e-11  3.180884e-11
  1.039823e-11  -7.188424e-12  3.151424e-11  1.477490e-11
$counts
[1] 257528
```

This function implements a generalized simulated annealing algorithm; see [44] for more information.

10.3.3   GENETIC ALGORITHMS

Unlike simulated annealing, genetic algorithms work with a set of solutions rather than a single point. The idea is based on the survival-of-the-fittest mechanism of biological evolution, whereby a population of individuals evolves by a combination of the following elements:

- Selection according to a fitness criterion, which in an optimization problem is related to the objective function.

- Mutation, whereby some features of an individual are randomly changed.

- Crossover, whereby two individuals are selected for reproduction and generate offsprings with features resulting from a mix of their parents' features.

There is a considerable freedom in the way the above idea is translated into a concrete algorithm. A critical issue is how to balance exploration and exploitation. On the one hand, biasing the selection of the new population toward

```
library(GA)
Rastrigin<-function(x){sum(x^2-10*cos(2*pi*x))+10*length(x)}
dim <- 20
lb <- rep(-100,dim)
ub <- rep(100,dim)
Fitness <- function(x) -Rastrigin(x)
set.seed(55555)
out <- ga(type="real-valued",fitness=Fitness,min=lb,max=ub,
          popSize=2000,maxiter=200)
summary(out)
```

FIGURE 10.11 **Using the GA package to minimize the Rastrigin function.**

high quality individuals may lead to premature termination; on the other hand, unless a certain degree of elitism is used to keep good individuals into the population, the algorihtm might wander around and skip the opportunity of exploring promising subregions of the solution space.

A key ingredient is the encoding of a feasible solution as a string of features, corresponding to the genes of a chromosome. For continuous optimization in $\mathbb{R}^n$ the choice of representing each solution by the corresponding vector $\mathbf{x}$ is fairly natural.[6] Let us consider the simplest form of crossover: Given two individuals $\mathbf{x}$ and $\mathbf{y}$ in the current pool, a "breakpoint" position $k \in \{1, 2, \ldots, n\}$ is randomly selected and two offsprings are generated as follows:

$$\left\{ \begin{array}{l} x_1, x_2, \ldots, x_k, x_{k+1}, \ldots, x_n \\ y_1, y_2, \ldots, y_k, y_{k+1}, \ldots, y_n \end{array} \right\} \Rightarrow \left\{ \begin{array}{l} x_1, x_2, \ldots, x_k, y_{k+1}, \ldots, y_n \\ y_1, y_2, \ldots, y_k, x_{k+1}, \ldots, x_n \end{array} \right\}.$$

Variations on the theme are possible; for instance, a double crossover may be exploited, in which two breakpoints are selected for the crossover. The two new solutions are clearly feasible for an unconstrained optimization problem in $\mathbb{R}^n$; when constraints are added, this guarantee is lost in general. As we have already mentioned, one can enforce hard constraints by eliminating noncomplying individuals. A usually better alternative is a relaxation by a suitable penalty function.

There are a few packages implementing genetic algorithms in R. Figure 10.11 shows how to use one of them, GA, to minimize the Rastrigin function. Note that we have to set the "real-valued" option to specify that we are dealing with a continuous optimization, and that there are other settings aimed at discrete optimization. Furthermore, genetic algorithms typically maximize a fitness function; hence, we have to change the sign of the Rastrigin function. We obtain the following (edited) output:

---

[6]The choice is not that obvious for many problems in combinatorial optimization, e.g., when solutions correspond to permutations of objects.

```
> summary(out)
+-----------------------------------+
|            Genetic Algorithm      |
+-----------------------------------+

GA settings:
Type                   =  real-valued
Population size        =  2000
Number of generations  =  200
Elitism                =
Crossover probability  =  0.8
Mutation probability   =  0.1

GA results:
Iterations             = 200
Fitness function value = -30.58069
Solution               =
        x1            x2            x3          x4            x5
  0.9207202  -0.02426609   -0.9619024   2.046103   -0.009269147
        x6            x7            x8          x9           x10
 -0.1023605     1.02851   -0.0135309  0.08916896   0.06605338
       x11           x12          x13         x14           x15
  0.9552585   0.04914588   -0.9231975  -0.9239431   -0.8744139
       x16           x17          x18         x19           x20
 0.01952913   -2.000306  -0.07129762  -0.9101591  0.002521083
```

We notice that some options related to crossover and mutation probabilities, as well as the degree of elitism, have been set to default values. We may also provide the `ga` function with customized functions to carry out mutation, crossover, etc. Genetic algorithms offer plenty of customization options, which is a mixed blessing, as it may be awkward to find a robust and satisfactory setting. It is also possible to collect the best solutions found along the way and to refine them by a classical nonlinear programming method, finding a local minimizer nearby.

10.3.4   PARTICLE SWARM OPTIMIZATION

Particle swarm optimization (PSO) methods are an alternative class of stochastic search algorithms, based on a population of m particles exploring the space of solutions, which we assume is a subset of $\mathbb{R}^n$. The position of each particle j in the swarm is a vector

$$\mathbf{x}_j(t) = \left[x_{1j}(t), x_{2j}(t), \ldots, x_{nj}(t), \right]^{\mathsf{T}} \in \mathbb{R}^n,$$

which changes in discrete time ($t = 1, 2, 3, \ldots$) according to three factors:

1. *Inertia.* Each particle is also associated with a velocity vector $\mathbf{v}_j(t)$, which tends to be maintained.

2. *Cognitive factor.* Each particle tends to move toward its personal best, i.e., the best point $\mathbf{p}_j^*$ that it has visited so far.

3. *Social factor.* Each particle tends to move toward the global best of the swarm, i.e., the best point $\mathbf{g}^*$ that has been visited so far by the whole set of particles.

The idea is to mimic the interactions of members of a swarm looking for food, and a classical statement of the algorithm is the following:

$$\mathbf{v}_j(t+1) = \mathbf{v}_j(t) + c_1 r_{1j}(t) \left[\mathbf{p}_j^*(t) - \mathbf{x}_j(t)\right]$$
$$+ c_2 r_{2j}(t) \left[\mathbf{g}_j^*(t) - \mathbf{x}_j(t)\right], \tag{10.20}$$
$$\mathbf{x}_j(t+1) = \mathbf{x}_j(t) + \mathbf{v}_j(t+1). \tag{10.21}$$

Equation (10.20) governs the evolution of each component $i = 1, \ldots, n$ of the velocity of each particle $j = 1, \ldots, m$. The new velocity depends on:

- The previous velocity, i.e., the inertia factor.
- The difference between the personal best and the current position, i.e., the cognitive factor, scaled by a coefficient c_1 and multiplied by a random variable $r_{1j}(t)$; a typical choice is a uniform distribution $U(0,1)$.
- The difference between the global best and the current position, i.e., the social factor, scaled by a coefficient c_2 and multiplied by a random variable $r_{2j}(t)$; in this case, too, a typical choice is a uniform distribution $U(0,1)$.

Equation (10.21) simply changes each component of the current position according to the new velocity. At each iteration, the personal and global bests are updated when necessary, and other adjustments are used in order to keep velocity within a given range, as well as the positions, if bounds on the variables are specified.

PSO can be regarded as another Monte Carlo search approach, and several variants have been proposed (see the end of chapter references):

- In small world methods, the global best position for neighboring particles is used, rather than the whole swarm.
- In quantum PSO different rules are used to evolve particle positions. A version of the algorithm generates a new particle position as

$$\mathbf{x}_j(t+1) = \mathbf{p}_j(t) \pm \beta \cdot \left|\mathbf{p}_j(t) - \mathbf{x}_j(t)\right| \cdot \log(1/U),$$

where U is a uniformly distributed random variable, β is a coefficient to be chosen, and $\mathbf{p}_j(t)$ is a random combination of the personal and the global best. The $\pm$ is resolved by the flip of a fair coin, i.e., $+$ and $-$ have both 50% probability.
- Some algorithms use Lévy flights, i.e., the shocks to current positions are generated by using heavy-tailed distributions, which is contrasted against the normal distribution underlying geometric Brownian motion and standard random walks.
- In the firefly algorithm, the quality of each solution corresponds to the light intensity of a firefly attracting other particles.

```
library(pso)
Rastrigin<-function(x){sum(x^2-10*cos(2*pi*x))+10*length(x)}
dim <- 20
lb <- rep(-100,dim)
ub <- rep(100,dim)
set.seed(55555)
x0 <- runif(dim,min=lb,max=ub)
out <- psoptim(par=x0,fn=Rastrigin,lower=lb,upper=ub,
                control=list(maxit=10000,trace=1,REPORT=1000,
                                hybrid=F))
out <- psoptim(par=x0,fn=Rastrigin,lower=lb,upper=ub,
                control=list(maxit=1000,trace=1,REPORT=100,
                                hybrid=T))
```

FIGURE 10.12 **Using the** pso **package to minimize the Rastrigin function.**

Some of these algorithms have a physical motivation, whereas some have a biological motivation; they differ in the number of coefficients to be set and the corresponding complexity of fine tuning, though some self-adaptive methods are available. As the reader can imagine, it is easy to get lost because of the sheer number of proposals and combinations. The key is to experiment with some of these variants and find the best combination for the specific problem at hand. A welcome tool is the pso R package, which implements some variants of the approach. The code in Fig. 10.12 shows how to use the psoptim function to minimize the Rastrigin function. The function has a similar interface as the standard optim function, plus several control parameters controlling the maximum number of iterations (maxit), the tracing of execution (trace and REPORT), as well as the possibility of carrying out a local optimization for each particle using a standard nonlinear optimizer (if hybrid is set). The code produces the following output:

```
> out <- psoptim(par=x0,fn=Rastrigin,lower=lb,upper=ub,
+           control=list(maxit=10000,trace=1,REPORT=1000,
+                           hybrid=F))
S=18, K=3, p=0.1576, w0=0.7213, w1=0.7213, c.p=1.193,
c.g=1.193, v.max=NA, d=894.4, vectorize=FALSE, hybrid=off
It 1000: fitness=27.86
It 2000: fitness=27.86
It 3000: fitness=27.86
It 4000: fitness=27.86
It 5000: fitness=27.86
It 6000: fitness=27.86
It 7000: fitness=27.86
It 8000: fitness=27.86
It 9000: fitness=27.86
It 10000: fitness=27.86
```

```
Maximal number of iterations reached
> out <- psoptim(par=x0,fn=Rastrigin,lower=lb,upper=ub,
+               control=list(maxit=1000,trace=1,REPORT=100,
+                            hybrid=T))
S=18, K=3, p=0.1576, w0=0.7213, w1=0.7213, c.p=1.193,
c.g=1.193, v.max=NA, d=894.4, vectorize=FALSE, hybrid=on
It 100: fitness=35.82
It 200: fitness=28.85
It 300: fitness=24.87
It 400: fitness=21.89
It 500: fitness=18.9
It 600: fitness=14.92
It 700: fitness=14.92
It 800: fitness=14.92
It 900: fitness=8.955
It 1000: fitness=6.965
Maximal number of iterations reached
```

Indeed, we notice that the hybrid algorithm is more effective, at the cost of a possibly larger computational effort. Some experimentation with control parameters may be needed to fine tune the algorithm performance.

10.4   Direct search and simulation-based optimization methods

In the previous section we have considered Monte Carlo-based stochastic search approaches to optimize a deterministic objective function. The difficulty was in the nonconvexity of the objective function, but not in its evaluation. When dealing with optimization under uncertainty, as in the case

$$\min_{\mathbf{x}\in S} E_{\boldsymbol{\xi}}\left[f(\mathbf{x},\boldsymbol{\xi})\right],$$

the function evaluation itself may be difficult. Thus, we may be forced to adopt Monte Carlo sampling to obtain noisy estimates of the objective function. In well-structured problems, we may build a formal optimization model, which can be tackled by stochastic programming with recourse or stochastic dynamic programming, to be discussed later. However, sometimes the system is so complicated that we may only estimate the objective function by some stochastic simulator acting as a black box: We feed the black box with x and we obtain an estimate of the expected performance. As a consequence, some nice properties such as continuity and differentiability may not be always guaranteed. Population-based approaches, like genetic algorithms and particle swarm optimization may help, since they do not place any restriction on the objective function; furthermore, since they work on a population, they tend to be less vulnerable to estimation errors. We may also keep a pool of the most promising solutions found, which could be simulated to a greater degree of accuracy to yield the optimal solution.

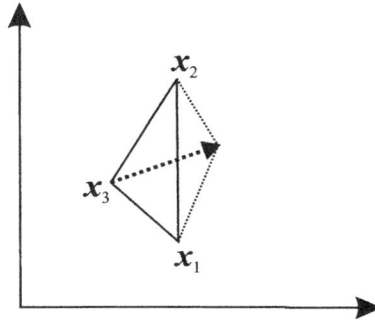

FIGURE 10.13 **Reflection of the worst value point in the Nelder–Mead simplex search procedure.**

Here we discuss two approaches which may be used for simulation-based optimization:

- *Simplex search*, not to be confused with the simplex algorithm for linear programming, is a sort of population-based algorithm for local optimization of possibly discontinuous functions.
- *Metamodeling* aims at building a suitable approximation of the true objective function by regression strategies. This higher-level model may be tackled by traditional nonlinear programming techniques.

We should note that these methods are not aimed at global optimization, unlike those discussed in the previous section. Nevertheless they may be more effective and efficient in specific settings.

10.4.1 SIMPLEX SEARCH

The simplex search algorithm, also known as Nelder–Mead method, is a deterministic search strategy relying on functions evaluations only. Rather than working with a single point, it uses a simplex of $n+1$ points in $\mathbb{R}^n$. A simplex in $\mathbb{R}^n$ is the convex hull of a set of $n+1$ affinely independent points $\mathbf{x}_1, \ldots, \mathbf{x}_{n+1}$.[7] The convex hull of a set of points is just the set of points that may be obtained by taking convex combinations (linear combinations whose weights are nonnegative and add up to 1) of the elements of the set. For instance, in two dimensions, a simplex is simply a triangle, whereas in three dimensions it is a tetrahedron.

The rationale behind the method is illustrated in Fig. 10.13 for a minimization problem in $\mathbb{R}^2$. In this case, the simplex search generates a sequence of

[7]Affine independence in $\mathbb{R}^n$ means that the vectors $(\mathbf{x}_2 - \mathbf{x}_1), \ldots, (\mathbf{x}_{n+1} - \mathbf{x}_1)$ are linearly independent. For $n = 2$ this means that the three points do not lie on the same line. For $n = 3$ this means that the four points do not lie on the same plane.

sets consisting of three points, in which the worst one is discarded and replaced by a new one. For instance, let us assume that x_3 is associated with the worst objective value; then, it seems reasonable to move away from x_3 by reflecting it through the center of the face formed by the other two points, as shown in the figure. A new simplex is obtained and the process is repeated. The generation of the new point is easily accomplished algebraically. If x_{n+1} is the worst point, we compute the centroid of the other n points as

$$\mathbf{c} = \frac{1}{n} \sum_{i=1}^{n} \mathbf{x}_i,$$

and we try a new point of the form

$$\mathbf{x}_r = \mathbf{c} + \alpha(\mathbf{c} - \mathbf{x}_{n+1}).$$

Clearly, the key issue is finding the right reflection coefficient $\alpha > 0$. If $\mathbf{x}_r$ turns out to be even worse than $\mathbf{x}_{n+1}$, we may argue that the step was too long, and the simplex should be contracted. If $\mathbf{x}_r$ turns out to be the new best point, we have found a good direction and the simplex should be expanded. Different strategies have been devised in order to improve the convergence of the method.[8] Although the simplex search procedure has its merit, it does not overcome the possible difficulties due to the nonconvexity of the objective function or the discrete character of some decision parameters. In fact, it was originally developed for experimental, response surface optimization.

The Nelder–Mead algorithm is offered as an option in the basic `optim` function. Furthermore, there are some R packages implementing variants of this and other derivative-free optimization methods, such as `neldermead` and `dfoptim`. The package `optimsimplex` offers a framework to implement specific variants of simplex search.

10.4.2 METAMODELING

We have discussed several derivative-free algorithms. Avoiding reliance on information provided by the gradient and the Hessian matrix of the objective function is certainly a good idea when we are dealing with a badly behaved function. However, in other cases, we may miss an opportunity to better drive the search process. Metamodeling aims to build an approximation of the objective function

$$h(\mathbf{x}) \equiv \mathrm{E}_{\boldsymbol{\xi}}\big[f(\mathbf{x}, \boldsymbol{\xi})\big]$$

on the basis of simulation experiments. Hence, we use a simulation model as a black box, whose response surface is modeled by a suitably chosen approximating function. This kind of higher-level model explains the name of the overall approach.

---

[8]See, e.g., [21].

If we want to find the gradient $\nabla h(\mathbf{x}_k)$ at some point $\mathbf{x}_k$, we may build a local linear approximator like

$$\hat{h}(\mathbf{x}) = \alpha + \sum_{i=1}^{n} \beta_i x_i = \alpha + \boldsymbol{\beta}^\mathsf{T}\mathbf{x}.$$

Such a model can be easily found by linear regression. Then, we just set

$$\nabla h(\mathbf{x}_k) \approx \boldsymbol{\beta}.$$

Apparently, the gradient does not depend on $\mathbf{x}_k$. However, the linear metamodel does depend on that point, as the simulation experiments that we use in linear regression have to be carried out in its neighborhood. They should not be too close, otherwise the noise in the evaluation of $f(\cdot,\cdot)$ will be overwhelming. However, they should not be placed too far, as the linear approximation is likely to be valid only on a neighborhood of $\mathbf{x}_k$. Statistical methods for design of experiments can also be put to good use to limit the number of simulation runs needed to feed the linear regression model. Also variance reduction strategies, like common random numbers may help.

The estimated gradient may be used within standard nonlinear programming algorithms to generate a sequence of points like

$$\mathbf{x}_{k+1} = \mathbf{x}_k - \alpha_k \nabla h(\mathbf{x}_k)$$

for a step length $\alpha_k > 0$. This is the steepest descent (or gradient) method, which suffers from poor convergence. One way to improve performance is to build a quadratic metamodel

$$\hat{h}(\mathbf{x}) = \alpha + \boldsymbol{\beta}^\mathsf{T}\mathbf{x} + \tfrac{1}{2}\mathbf{x}^\mathsf{T}\boldsymbol{\Gamma}\mathbf{x}.$$

If the square matrix $\boldsymbol{\Gamma}$ is positive semidefinite, the above quadratic form is convex and it is easily minimized by solving a system of linear equations, yielding a new approximation $\mathbf{x}_{k+1}$ of the optimal solution. A new quadratic approximation is built, centered on the new point, and the procedure is repeated until some convergence condition is met. This kind of approach is related to sequential quadratic programming methods for traditional nonlinear programming. It may be advisable to use a linear metamodel at the early stages of the process, since this may result in a faster movement toward the minimizer; then, when convergence of the gradient method becomes problematic, the procedure may switch to the quadratic metamodel.

A noteworthy feature of metamodeling is that it can cope with constraints on the decision variables, since we solve a sequence of nonlinear programming models, whereas alternative approaches have to resort to penalty functions. An obvious disadvantage is that it is likely to be quite expensive in computational terms, if many simulation experiments are needed to filter random noise and find a good metamodel. Alternative methods, such as perturbation analysis, have been proposed to estimate sensitivities with a single simulation run. We will consider a few simple application examples to financial derivatives in Chapter 12.

10.5   Stochastic programming models

Earlier in the chapter we have introduced decision models under uncertainty. In this section we describe one possible modeling framework, stochastic programming with recourse. For the sake of simplicity, we will just deal with linear programming (LP) models under uncertainty, but the approach can be extended to nonlinear optimization models. We begin with two-stage models in Section 10.5.1. Then, we generalize the approach to multistage problems in Section 10.5.2. Finally, we stress important issues related to scenario generation in Section 10.5.3. An alternative approach to dynamic decision making under uncertainty, stochastic dynamic programming, will be the subject of the last sections of the chapter.

10.5.1   TWO-STAGE STOCHASTIC LINEAR PROGRAMMING WITH RECOURSE

To get a clear picture on how uncertainty can be introduced into an optimization model, let us consider an uncertain LP problem:

$$\begin{aligned}\min \quad & \mathbf{c}(\boldsymbol{\xi})^{\mathsf{T}}\mathbf{x} \\ \text{s.t.} \quad & \mathbf{A}(\boldsymbol{\xi})\,\mathbf{x} = \mathbf{b}(\boldsymbol{\xi}), \\ & \mathbf{x} \geq \mathbf{0},\end{aligned}$$

where all problem data depend on random variables $\boldsymbol{\xi}$ with support $\boldsymbol{\Xi}$. Stated as such, the model is not even posed in a sensible way, since the objective function is a random variable, i.e., a function of the underlying events, and the minimization is not well defined. When we select a decision, we only select the probability distribution of cost or profit, and in order to assign a proper meaning to the problem, we must specify how probability distributions are ranked. The most natural approach is to take the expected value of the objective, if we ignore risk aversion.[9] However, even if we settle for some sensible optimality criterion, there are serious issues with *feasibility*. We should wonder whether we can require feasibility for any possible contingency, i.e.,

$$\mathbf{A}(\boldsymbol{\xi})\,\mathbf{x} = \mathbf{b}(\boldsymbol{\xi}) \qquad \forall \boldsymbol{\xi} \in \boldsymbol{\Xi}.$$

This is clearly not possible in general, as we get an inconsistent set of equations. The case of inequality constraints is a bit easier. Nevertheless, requiring

$$\mathbf{A}(\boldsymbol{\xi})\,\mathbf{x} \geq \mathbf{b}(\boldsymbol{\xi}) \qquad \forall \boldsymbol{\xi} \in \boldsymbol{\Xi},$$

yields a *fat* solution that may be overly expensive, assuming that such a feasible solution exists. More often than not, we must accept a controlled violation

---

[9]See Chapter 13 to see how risk measures may be included in an optimization model.

of constraints. One possibility is to settle for a solution with a probabilistic guarantee:

$$P\{A(\xi)x \geq b(\xi)\} \geq 1 - \alpha,$$

for a suitably small α. This idea leads to chance-constrained models, which have a sensible interpretation in terms of solution reliability. Chance-constrained models are indeed useful in some cases, but they suffer from definite limitations:

- Per se, they do not model the flow of information and the sequence of decision stages; in real life, we make a decision here and now, but then we revise and adjust it along the way, when uncertainty is progressively resolved.

- They can lead to quite risky solutions, as nothing is said about the effect of a constraint violation, even though this has low probability α.[10]

- Even if we choose a very small α, can we trust our ability to estimate very small probabilities?

- In general, chance-constrained models do not result in convex optimization problems. The essential reason for this unpleasing state of the matter is that the union of convex sets need not be convex.[11]

An alternative modeling framework, to which one may resort to overcome these issues, is stochastic programming with recourse. Let us consider the stochastic constraints

$$T(\xi)x = h(\xi).$$

As we have pointed out, it is impossible to satisfy these constraints with an assignment of x that is valid for any realization of the random variable ξ. However, we may think of "adjusting" the solution, *after* observing uncertainty, by some action represented by decision variables $y(\xi)$:

$$Wy(\xi) + T(\xi)x = h(\xi).$$

These variables are called recourse variables, and they are random in the sense that they are adapted to the random variable ξ, as they are made after uncertainty is resolved. They are, therefore, second-stage decisions corresponding to contingency plans, whereas the x are first-stage, here-and-now, decisions. The matrix W is called recourse matrix and represent the "technology" that we use to adjust first-stage decisions. Its structure has a remarkable impact on the difficulty of the problem, and it is worth noting that we are assuming deterministic recourse, as the matrix W is not random. Problems with stochastic recourse are generally more difficult and are subject to some pathologies.

---

[10]This is related to some issues with risk measures like value-at-risk; see Chapter 13.

[11]Convexity of chance-constrained problems is discussed in [25].

■ Example 10.13 Simple recourse in the newsvendor model

In Section 10.2.3 we have formulated the newsvendor model within a scenario-based framework. The first-stage decision q, the amount we produce, should ideally match demand D^s in every scenario, which is clearly impossible. In the model of Eq. (10.17) we just penalize deviations from the target:

$$q = D_s + y_s^+ - y_s^- \qquad \forall s = 1, \ldots, S.$$

Thus, we immediately see that in this model the structure of the recourse matrix is

$$\mathbf{W} = [\mathbf{I}, -\mathbf{I}],$$

where $\mathbf{I}$ is the identity matrix of order S. This case is referred to as *simple* recourse, as the second-stage variables are just used for bookkeeping and cost evaluation, and are not really linked to decisions. However, we could apply more clever decisions, such as asking a third-party to produce or sell us additional items, possibly using express shipping, or trying to persuade the customer to accept a substitute item. When we include these recourse actions in the model, and we account for multiple items, we really introduce second-stage decisions.

Clearly, these adjustments are not for free and incur some cost. We should minimize the total expected cost of both our initial and recourse decisions, which yields the following optimization model:

$$\min \quad \mathbf{c}^\mathsf{T}\mathbf{x} + \mathrm{E}\left[\mathbf{q}(\boldsymbol{\xi})^\mathsf{T}\mathbf{y}(\boldsymbol{\xi})\right] \tag{10.22}$$

$$\text{s.t.} \quad \mathbf{A}\mathbf{x} = \mathbf{b}, \tag{10.23}$$

$$\mathbf{W}\mathbf{y}(\boldsymbol{\xi}) + \mathbf{T}(\boldsymbol{\xi})\,\mathbf{x} = \mathbf{h}(\boldsymbol{\xi}) \qquad \forall \boldsymbol{\xi} \in \Xi, \tag{10.24}$$

$$\mathbf{x}, \mathbf{y}(\boldsymbol{\xi}) \geq \mathbf{0}. \tag{10.25}$$

This is a two-stage stochastic LP model with recourse and can be interpreted as follows:

- The first-stage decisions $\mathbf{x}$ must satisfy the deterministic constraints (10.23) and are associated with the deterministic first-stage cost $\mathbf{c}^\mathsf{T}\mathbf{x}$.
- At the second stage, a random event occurs, associated with random data depending on $\boldsymbol{\xi}$. Given this information, a set of second-stage (recourse) actions $\mathbf{y}(\boldsymbol{\xi}) \geq \mathbf{0}$ are taken.
- The second-stage decisions $\mathbf{y}(\boldsymbol{\xi})$ are related to first-stage decisions by constraints (10.24).
- The second-stage decisions result in a cost $\mathbf{q}(\boldsymbol{\xi})^\mathsf{T}\mathbf{y}(\boldsymbol{\xi})$.

- In (10.22) we minimize the sum of the first-stage cost and the expected value of second-stage cost.

We may also introduce the recourse function $\mathcal{Q}(\mathbf{x})$ and rewrite the model as the deterministic equivalent

$$\begin{aligned}\min \quad & \mathbf{c}^\mathsf{T}\mathbf{x} + \mathcal{Q}(\mathbf{x}) \\ \text{s.t.} \quad & \mathbf{A}\mathbf{x} = \mathbf{b}, \\ & \mathbf{x} \geq \mathbf{0},\end{aligned}$$

where

$$\mathcal{Q}(\mathbf{x}) \equiv \mathrm{E}\left[Q(\mathbf{x}, \boldsymbol{\xi})\right],$$

and

$$Q(\mathbf{x}, \boldsymbol{\xi}) \equiv \min_{\mathbf{y}} \left\{ \mathbf{q}(\boldsymbol{\xi})^\mathsf{T}\mathbf{y} \mid \mathbf{W}\mathbf{y} = \mathbf{h}(\boldsymbol{\xi}) - \mathbf{T}(\boldsymbol{\xi})\,\mathbf{x}, \ \mathbf{y} \geq \mathbf{0} \right\}.$$

This formulation shows that a stochastic linear program is, in general, a non-linear programming problem. Nonlinear problems are usually less pleasing to solve than linear ones, but this is not the really bad news. In fact, the recourse function $\mathcal{Q}(\mathbf{x})$ looks like a "hopeless" function to deal with:

- It is an expectation, with respect to the joint distribution of $\boldsymbol{\xi}$; hence, it is a multidimensional integral, if random variables are continuous. This integral may require Monte Carlo methods for its approximation.

- As if a multidimensional integral were not bad enough, it involves a function that we do not really know, as it is implicitly defined by an optimization problem.

Luckily, in many cases of practical interest, we can prove interesting properties of the recourse function, most notably convexity, which implies continuity (on an open domain). In some cases, $\mathcal{Q}(\mathbf{x})$ is differentiable; in other cases it is polyhedral. This does not imply that it is easy to evaluate the recourse function, but we may resort to statistical sampling (scenario generation) and take advantage of both convexity and problem structure.

We may represent uncertainty by a discrete probability distribution, where $\mathcal{S}$ is the set of scenarios and π_s is the probability of scenario $s \in \mathcal{S}$. We obtain the scenario tree (fan) in Fig. 10.14, where $\boldsymbol{\xi}_s$ is the realization of random variables corresponding to scenario s. One obvious way to sample a scenario tree is by Monte Carlo, in which case the scenarios have the same probability $\pi_s = 1/|\mathcal{S}|$, but other scenario generation strategies are available. This yields the following LP:

$$\begin{aligned}\min \quad & \mathbf{c}^\mathsf{T}\mathbf{x} + \sum_{s \in \mathcal{S}} \pi_s \mathbf{q}_s^\mathsf{T}\mathbf{y}_s \\ \text{s.t.} \quad & \mathbf{A}\mathbf{x} = \mathbf{b}, \\ & \mathbf{W}\mathbf{y}_s + \mathbf{T}_s\mathbf{x} = \mathbf{h}_s \qquad \forall s \in \mathcal{S}, \\ & \mathbf{x}, \mathbf{y}_s \geq \mathbf{0}.\end{aligned}$$

Future
scenarios

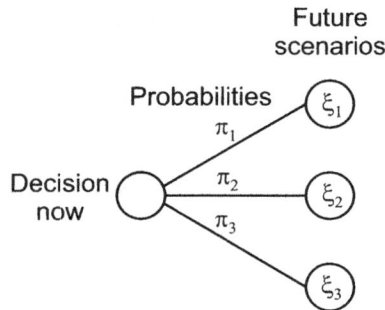

FIGURE 10.14 **A scenario fan.**

Thus we are left with the task of solving a plain LP, even though a possibly large-scale one. A crucial point is the ability to generate a good and robust solution on the basis of a limited number of representative scenarios.

10.5.2 A MULTISTAGE MODEL FOR PORTFOLIO MANAGEMENT

Multistage stochastic programming formulations arise naturally as a generalization of two-stage models. At each stage, we gather new information and we make decisions accordingly, taking into account immediate costs and expected future recourse cost. The resulting decision process may be summarized as follows:[12]

- At the beginning of the first time period (at time $t = 0$) we select the decision vector $\mathbf{x}_0$; this decision has a deterministic immediate cost $\mathbf{c}_0^\mathsf{T}\mathbf{x}_0$ and must satisfy the constraint

$$\mathbf{A}_{00}\mathbf{x}_0 = \mathbf{b}_0.$$

- At the beginning of the second time period we observe random data ($\mathbf{A}_{10}$, $\mathbf{A}_{11}$, $\mathbf{c}_1$, $\mathbf{b}_1$); then, on the basis of this information, we make decision $\mathbf{x}_1$; this second decision has an immediate cost $\mathbf{c}_1^\mathsf{T}\mathbf{x}_1$ and must satisfy the constraint

$$\mathbf{A}_{10}\mathbf{x}_0 + \mathbf{A}_{11}\mathbf{x}_1 = \mathbf{b}_1.$$

Note that these data are not known at time $t = 0$, but only at time $t = 1$; the new decision depends on the realization of these random variables and is also affected by the previous decision.

- We repeat the same scheme as above for time periods up to $H - 1$, where H is our planning horizon.

- Finally, at the beginning of the last time period H, we observe random data ($\mathbf{A}_{H,H-1}$, $\mathbf{A}_{HH}$, $\mathbf{c}_H$, $\mathbf{b}_H$); then, on the basis of this information we

---

[12]See, e.g., [39] for a more detailed discussion.

make decision x_H, which has an immediate cost $c_H^T x_H$ and must satisfy the constraint

$$A_{H,H-1} x_{H-1} + A_{HH} x_H = b_H.$$

From the perspective of time period $t = 0$, the decisions $x_1, \ldots, x_H$ are random variables, as they will be adapted to the realization of the underlying stochastic process. However, the only information that we may use in making each decision consists on the observed history so far. The resulting structure of the dynamic decision process can be appreciated by the following recursive formulation of the multistage problem:

$$
\min_{\substack{A_{00} x_0 = b_0 \\ x_0 \geq 0}} c_0^T x_0 + E \left[\min_{\substack{A_{10} x_0 + A_{11} x_1 = b_1 \\ x_1 \geq 0}} c_1^T x_1 \right. \tag{10.26}
$$

$$
\left. + E \left[\cdots + E \left[\min_{\substack{A_{H,H-1} x_{H-1} + A_{HH} x_H = b_H \\ x_H \geq 0}} c_H^T x_H \right] \right] \right].
$$

In this formulation, we observe that a decision x_t depends directly only on the previous decision x_{t-1}. In general, decisions may depend on all of the past history, leading to a slightly more complicated model. However, we may often introduce additional state variables, in such a way that the above formulation applies. What the above formulation hides is the dependence structure of the random variables ξ_t, $t = 1, \ldots, H$, which are the underlying risk factors of the random matrices and vectors A, b, and c:

- In the easiest case, they are mutually independent.
- In the Markovian case, ξ_t depends only on ξ_{t-1}.
- In the most general case, a possibly complicated path dependency should be accounted.

From the viewpoint of stochastic programming, all of the above cases may be tackled. On the contrary, the stochastic dynamic approach that we consider later relies on a Markovian structure and is more limited. We shall also see that dynamic programming takes advantage of a recursive formulation, which is implicit in the above multistage model. On the other hand, a definite advantage of the dynamic programming framework is that decisions are given in feedback form, as a function of the current system state. Indeed, it should be noted that, in practice, the relevant output of a stochastic programming model is the set of immediate decisions x_0. The remaining decision variables could be regarded as contingency plans, which should be implemented as time goes by; however, it is more likely that the model will be solved again and again according to a rolling-horizon logic.

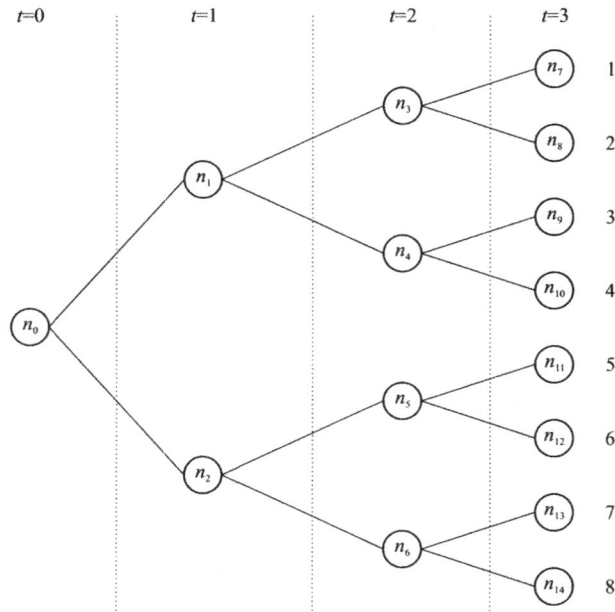

FIGURE 10.15 Scenario tree for a simple asset–liability management problem.

While this formulation points out the dynamic optimization nature of multistage problems, we usually resort to deterministic equivalents based on discretized scenario trees, like the one depicted in Fig. 10.15. Each node n_k in the tree corresponds to an event, where we should make a decision. We have an initial node n_0 corresponding to time $t = 0$. Then, for each event node, we branch descendant nodes. In the figure, each node has two successors, but the branching factors are arbitrary. The terminal nodes (the leaves) have no successor and their number gives the number of scenarios in the tree. A scenario is a path from the root node to a leaf node. For instance, scenario 2 consists of the node sequence (n_0, n_1, n_3, n_8). Each branch is associated with a conditional probability of occurrence $P(n_k \mid n_i)$, where $n_i = a(n_k)$ is the immediate predecessor of node n_k. If scenarios are sampled by plain Monte Carlo, this probability is related to the branching factor. For instance, we may assume that all conditional probabilities in Fig. 10.15 are 0.5. The probability of a scenario is the product of the conditional probabilities of the branches we meet on the path. Hence, in this example we have 8 scenarios with probability $\frac{1}{8}$.

Decision variables are associated with each node of the tree, as we show in the next section. It is important to realize that the tree structure enforces nonanticipativity of decisions. For instance, at node n_1 we do not know if we are on scenario 1, 2, 3, or 4. The decisions $\mathbf{x}(n_1)$ at node n_1 will be the same for each of those 4 scenarios, which cannot be distinguished at node n_1. Alternatively, we could associate decisions $\mathbf{x}(s, t)$ with scenarios and time steps. Therefore, rather than second-stage decisions $\mathbf{x}(n_1)$ and $\mathbf{x}(n_1)$, we would include decisions $\mathbf{x}(s, 1)$, for $s = 1, 2, \ldots, 8$. However, we should include nonanticipativity constraints explicitly:[13]

$$\mathbf{x}(1, 1) = \mathbf{x}(2, 1) = \mathbf{x}(3, 1) = \mathbf{x}(4, 1); \quad \mathbf{x}(5, 1) = \mathbf{x}(6, 1) = \mathbf{x}(7, 1) = \mathbf{x}(8, 1).$$

In the next section we illustrate a simple financial model where decisions are associated with nodes, rather than scenarios, resulting in a more compact model formulation. The choice between the two approaches, however, depends on the solution algorithm that we choose for solving a possibly large-scale optimization model. In fact, we realize that if the branching structure is not kept under control, the number of nodes will grow exponentially.

10.5.2.1 A multistage financial planning model

To illustrate how multistage stochastic programming can be used to tackle financial problems, let us consider a simple asset–liability management problem.

- We have a stream of stochastic liabilities that should be funded by a portfolio of assets. We represent uncertainty by a scenario tree, and L^n is the liability we have to meet in node $n \in \mathcal{N}$, where $\mathcal{N}$ is the set of nodes in the tree.

- We denote the root node of the tree by n_0; each node $n \in \mathcal{N} \setminus \{n_0\}$ has a unique predecessor denoted by $a(n)$.[14]

- We may invest in assets indexed by $i \in \mathcal{I}$. Of course, asset prices are random, and let P_i^n is the price for asset i at node n.

- We are given a set of initial holdings $\overline{h}_i^{n_0}$ for each asset $i \in \mathcal{I}$ at the root node.

- The initial portfolio can be rebalanced at node n_0, as well as at other nodes of the tree.

- For the sake of simplicity, we do not consider the possibility of receiving new cash along the way, as it would be the case, e.g., for a pension fund receiving new contributions. The only way to raise cash is by selling assets.

- When we buy or sell assets, we incur proportional (linear) transaction costs; the transaction cost is a percentage c of the traded value, for both buying and selling any asset. It is easy to introduce proportional transaction costs depending on the liquidity of each single asset, as well as on the sign of the trade.

- We want to maximize the expected utility of the terminal wealth at the leaves of the tree. Let $\mathcal{S}$ be the set of terminal nodes in the tree, where

---

[13]Readers with a background on measure-theoretic probability will immediately recognize that this is a measurability condition with respect to a filtration.

[14]We denote set difference by $\setminus$.

we assume to liquidate the portfolio and pay the last liability. Actually, this need not be really the case, as the model is solved in a rolling-horizon framework.

As we have already pointed out, a scenario in the tree is a path from the root node n_0 to a terminal node $s \in \mathcal{S}$; let π_s be the corresponding probability. We introduce the following decision variables:

- z_i^n, the amount of asset i purchased at node n
- y_i^n, the amount of asset i sold at node n
- x_i^n, the amount of asset i held at node n, after rebalancing
- W^s, the wealth at terminal node $s \in \mathcal{S}$

Variables z_i^n, y_i^n, and x_i^n are only defined at nodes $n \in \mathcal{N} \setminus \mathcal{S}$, as we assume that no rebalancing occurs at terminal nodes, where the portfolio is liquidated, we pay the liability L^s, and we measure the terminal wealth W^s. Let $u(w)$ is the utility for wealth w; this function is used to express utility of terminal wealth.

On the basis of this notation, we may write the following model:

$$\max \quad \sum_{s \in \mathcal{S}} \pi^s u(W^s) \tag{10.27}$$

$$\text{s.t.} \quad x_i^{n_0} = \overline{h}_i^{n_0} + z_i^{n_0} - y_i^{n_0} \quad \forall i \in \mathcal{I}, \tag{10.28}$$

$$x_i^n = x_i^{a(n)} + z_i^n - y_i^n \quad \forall i \in \mathcal{I}, \forall n \in \mathcal{N} \setminus (\{n_0\} \cup \mathcal{S}), \tag{10.29}$$

$$(1 - c) \sum_{i=1}^{I} P_i^n y_i^n - (1 + c) \sum_{i=1}^{I} P_i^n z_i^n = L^n \quad \forall n \in \mathcal{N} \setminus \mathcal{S}, \tag{10.30}$$

$$W^s = (1 - c) \sum_{i=1}^{I} P_i^s x_i^{a(s)} - L^s \quad \forall s \in \mathcal{S}, \tag{10.31}$$

$$x_i^n, z_i^n, y_i^n, W^s \geq 0. \tag{10.32}$$

The objective (10.27) is the expected utility of terminal wealth. Equation (10.28) expresses the initial asset balance, taking the current holdings into account; the asset balance at intermediate trading dates, i.e., at all nodes with exception of root and leaves, is taken into account by Eq. (10.29). Equation (10.30) ensures that enough cash is generated by selling assets in order to meet the liabilities; we may also reinvest the proceeds of what we sell in new asset holdings. Note how proportional transaction costs are expressed for selling and purchasing. Equation (10.31) is used to evaluate terminal wealth at leaf nodes, after portfolio liquidation and payment of the last liability. In practice, we would repeatedly solve the model on a rolling-horizon basis, so the exact expression of the objective function is a bit debatable. The role of terminal utility is just to ensure that we are left in a good position at the end of the planning horizon, in order to avoid nasty end-of-horizon effects.

10.5.3 SCENARIO GENERATION AND STABILITY IN STOCHASTIC PROGRAMMING

The key ingredient in any stochastic programming model is a scenario tree, as the quality of the solution depends critically on how well uncertainty is captured. In principle, given a model of the underlying stochastic process, straight-forward Monte Carlo simulation could be used to generate a scenario tree, on the basis of the sampling and path generation methods of Chapters 5 and 6. Unfortunately, a suitably rich tree may be needed to capture uncertainty and, given the exponential growth of nodes, this is not practical in the multistage case. An additional difficulty in financial problems is that the tree should be arbitrage-free, which places additional requirements on the branching structure.

One possible way out is a radical change in the solution paradigm. Rather than generate the tree and then solve a large-scale optimization problem, we may interleave optimization and sampling and stop when a convergence criterion is met; this kind of approach relies on sophisticated decomposition strategies that are beyond the scope of this book,[15] and it requires the development of specific software. If we want to stick to commercial solvers, we may resort to the following ideas for clever scenario generation:

- Gaussian quadrature
- Variance reduction methods or low-discrepancy sequences
- Optimized scenario generation

In the next sections we outline these ideas. Whatever scenario generation strategy we select, it is important to check its effect on solution stability, as we discuss in Section 10.5.3.4. We will use the simple newsvendor model to illustrate these concepts in Section 10.5.3.5.

We also remark a couple of possibly useful guidelines:

1. If the model aims to yield robust first-stage solutions, it may be advisable to use a rich branching factor at the first-stage, and a limited branching factor at later stages.

2. The number of stages can also be limited by using non-uniform time steps. For instance, the first two time steps could correspond to one month, and the later ones to one year.

3. A clever way to limit the number of stages is to include in the objective function a term measuring the quality of the terminal state. By doing so, we may avoid myopic decisions and reduce the size of the tree.

4. Sometimes, we may reduce the effective dimensionality of the problem by applying data reduction techniques such as principal component analysis; see Section 3.8.1. However, we must be able to sample the few principal

---

[15]Two approaches in this vein are stochastic decomposition [19] and stochastic dual dynamic programming [35].

components accurately and generate corresponding scenarios for all the random variables involved in the model.

10.5.3.1 Gaussian quadrature

We have introduced Gaussian quadrature in Section 2.2 as a tool for numerical integration. When the integral represents the expected value of a function $g(\mathbf{x}, \boldsymbol{\xi})$ depending on continuous random variables $\boldsymbol{\xi}$ with joint density $f_{\boldsymbol{\xi}}(\cdot)$, a quadrature formula represents a discretization of the underlying continuous distribution:

$$\mathrm{E}_{\boldsymbol{\xi}}\left[g(\mathbf{x}, \boldsymbol{\xi})\right] = \int_{\Xi} g(\mathbf{x}, \mathbf{y}) f_{\boldsymbol{\xi}}(\mathbf{y})\, d\mathbf{y} \approx \sum_{s=1}^{S} \pi_s g(\mathbf{x}, \boldsymbol{\xi}_s), \qquad (10.33)$$

where the weights π_s correspond to scenario probabilities, and the integration nodes $\boldsymbol{\xi}_s$ are the realizations of the random variables in the discrete approximation. Note that, unlike Monte Carlo sampling, different probabilities are associated with different scenarios. Since Gaussian quadrature formulas ensure exact integration of polynomials up to a given order, this ensures moment matching. As we have remarked, this approach can be applied up to a limited dimensionality, however. Nevertheless, it may work very well for two-stage models.

10.5.3.2 Variance reduction and low-discrepancy sequences

Low-discrepancy sequences are another tool for numerical integration that can be used to generate scenarios; see [29]. There is also some room for variance reduction strategies:

- Common random numbers are used in the stochastic decomposition approach [19].
- For an application of importance sampling see, e.g., [22].
- Several variance reduction strategies are used in [18] to improve the accuracy of objective function evaluation in stochastic programming.
- Antithetic sampling and randomized quasi-Monte Carlo are used in [28] to hedge contingent claims.

See also [40, Chapter 4] to appreciate the role of the aforementioned techniques for statistical inference in stochastic programming.

10.5.3.3 Optimized scenario generation

Arguably, the most sophisticated approaches to scenario generation are those based on the solution of an optimization problem. We have already appreciated the role of moment matching in Section 2.5. The idea can be applied to scenario generation as well, since we aim at finding a good discretization of a continuous distribution.

To illustrate the approach, consider a variable distributed as a multivariate normal, $\mathbf{X} \sim \mathsf{N}(\boldsymbol{\mu}, \boldsymbol{\Sigma})$. The expected values and covariances of the discrete approximation should match those of the continuous distribution. Furthermore, since we are dealing with a normal distribution, we know that skewness and kurtosis for each component should be 0 and 3, respectively.

Let us denote by x_i^s the realization of X_i, $i = 1, \ldots, n$, in scenario s, $s = 1, \ldots, S$. Natural requirements are:

$$\frac{1}{S} \sum_{s=1}^{S} x_i^s \approx \mu_i \qquad \forall i,$$

$$\frac{1}{S} \sum_{s=1}^{S} (x_i^s - \mu_i)(x_j^s - \mu_j) \approx \sigma_{ij} \qquad \forall i, j,$$

$$\frac{1}{S} \sum_{s=1}^{S} \frac{(x_i^s - \mu_i)^3}{\sigma_i^3} \approx 0 \qquad \forall i,$$

$$\frac{1}{S} \sum_{s=1}^{S} \frac{(x_i^s - \mu_i)^4}{\sigma_i^4} \approx 3 \qquad \forall i.$$

Approximate moment matching is obtained by minimizing the following squared error:

$$w_1 \sum_{i=1}^{n} \left[\frac{1}{S} \sum_{s=1}^{S} x_i^s - \mu_i \right]^2$$

$$+ w_2 \sum_{i=1}^{n} \sum_{j=1}^{n} \left[\frac{1}{S} \sum_{s=1}^{S} (x_i^s - \mu_i)(x_j^s - \mu_j) - \sigma_{ij} \right]^2$$

$$+ w_3 \sum_{i=1}^{n} \left[\frac{1}{S} \sum_{s=1}^{S} \left(\frac{x_i^s - \mu_i}{\sigma_i} \right)^3 \right]^2$$

$$+ w_4 \sum_{i=1}^{n} \left[\frac{1}{S} \sum_{s=1}^{S} \left(\frac{x_i^s - \mu_i}{\sigma_i} \right)^4 - 3 \right]^2. \qquad (10.34)$$

The objective function includes four weights w_k which may be used to fine tune performance. It should be mentioned that the resulting scenario optimization problem need not be convex. However, if we manage to find any solution with a low value of the "error" objective function, this is arguably a satisfactory solution, even though it is not necessarily the globally optimal one. The idea can be generalized to "property matching," since one can match other features that are not necessarily related to moments; see, e.g., [20].

Moment (or property) matching has been criticized, since it is possible to build counterexamples, i.e., pairs of quite different distributions that share the first few moments. An alternative idea is to rely on metrics fully capturing the distance between probability distributions. Thus, given a scenario tree topology, we should find the assignment of values and probabilities minimizing some distance with respect to the "true" distribution. Alternatively, given a large scenario tree, we should try to reduce it to a more manageable size by optimal scenario reduction. From a formal point of view, there are different ways to define a distance between two probability measures $\mathbb{P}$ and $\mathbb{Q}$. One possibility has its roots in the old Monge transportation problem, which consists of finding the optimal way of transporting mass (e.g., soil, when we are building a road). The problem has a natural probabilistic interpretation, which was pointed out by Kantorovich, when we interpret mass in a probabilistic sense (see [38], for more details). Thus, we may define a transportation functional:

$$\mu_c(\mathbb{P}, \mathbb{Q}) \equiv \inf \left\{ \int_{\Xi \times \Xi} c(\boldsymbol{\xi}, \tilde{\boldsymbol{\xi}}) \eta(d\boldsymbol{\xi}, d\tilde{\boldsymbol{\xi}}) : \pi_1 \eta = \mathbb{P}, \ \pi_2 \eta = \mathbb{Q} \right\}.$$

Here $c(\cdot, \cdot)$ is a suitably chosen cost function; the problem calls for finding the minimum of the integral over all joint measures η, defined on the Cartesian product $\Xi \times \Xi$, where Ξ is the support of random variable $\boldsymbol{\xi}$, whose marginals coincide with $\mathbb{P}$ and $\mathbb{Q}$, respectively (π_1 and π_2 represent projection operators).

As the reader can imagine, we are threading on difficult ground, requiring some knowledge of functional analysis. Nevertheless, the idea is not too difficult to grasp if we deal with discrete distributions, where mass is concentrated on points (atoms). The Kantorovich distance, in the discrete case, boils down to

$$\mu_c(\mathbb{P}, \mathbb{Q}) = \inf \ \sum_{i=1}^{S} \sum_{j=1}^{\tilde{S}} \eta_{ij} c(\boldsymbol{\xi}^i, \tilde{\boldsymbol{\xi}}^j)$$

$$\text{s.t.} \ \sum_{i=1}^{S} \eta_{ij} = q_j \qquad \forall j,$$

$$\sum_{j=1}^{\tilde{S}} \eta_{ij} = p_i \qquad \forall i,$$

$$\eta_{ij} \geq 0,$$

where

$$c(\boldsymbol{\xi}^i, \tilde{\boldsymbol{\xi}}^j) = \sum_{\tau=1}^{T} \left| \boldsymbol{\xi}_\tau^i - \tilde{\boldsymbol{\xi}}_\tau^j \right|.$$

Therefore, the evaluation of distance requires the solution of a transportation problem, which is a well-known instance of a linear programming problem. Given this distance, it is possible to build heuristics in which scenarios are successively eliminated, redistributing probability mass in an optimal way, after having prescribed the scenario cardinality or the quality of the approximation. See, e.g., [17] for more details.

We should mention that a counterargument to proponents of the optimal approximation approach is that we are not really interested in the distance between the ideal distribution and the scenario tree. What really matters is the quality of the solution. For instance, if we are considering a mean–variance portfolio optimization problem along the lines of Section 10.2.1, it can be argued that matching the first two moments is all we need. See, e.g., [26, Chapter 4] for a discussion of these issues.

10.5.3.4 In- and out-of-sample stability

Whatever scenario reduction strategy we employ, it is important to check the stability of the resulting solution. There are quite sophisticated studies on the stability of stochastic optimization, relying on formal distance concepts. Here we just consider a down-to-earth approach along the lines of [26]. The exact problem is

$$\min_{\mathbf{x}} \mathrm{E}^{\mathbb{P}}[f(\mathbf{x}, \boldsymbol{\xi})],$$

which involves the expectation of the objective under the true measure $\mathbb{P}$. The actual problem we solve is based on an approximate scenario tree $\mathcal{T}$, possibly resulting from random sampling:

$$\min_{\mathbf{x}} \hat{f}(\mathbf{x}; \mathcal{T}),$$

where with the notation $\hat{f}$ we point out that we are just estimating the true expected value of the objective, given the generated scenario tree. Each scenario tree induces an optimal solution. Thus, if we sample trees $\mathcal{T}_i$ and $\mathcal{T}_j$, we obtain solutions $\mathbf{x}_i^*$ and $\mathbf{x}_j^*$, respectively. If the tree generation mechanism is reliable, we should see some stability in what we get. We may consider stability in the solution itself, but it is easier, and perhaps more relevant, to check stability in the value of the objective function. After all, if we want to minimize cost or maximize profit, this is what matters most.

There are two concepts of stability. *In-sample stability* means that if we sample two scenario trees, the values of the solutions that we obtain are not too different:

$$\hat{f}(\mathbf{x}_i^*; \mathcal{T}_i) \approx \hat{f}(\mathbf{x}_j^*; \mathcal{T}_j). \tag{10.35}$$

This definition does not apply directly if the scenario tree is generated deterministically; in that case, we might compare trees with slightly different branching structures to see if the tree structure that we are using is rich enough to ensure solution robustness. This concept of stability is called in-sample, as we evaluate a solution using the same tree we have used to find it. But since the tree is only a limited representation of uncertainty, we should wonder what happens when we apply the solution in the real world, where a different scenario may unfold. This leads to *out-of-sample stability*, where we compare the objective value from the optimization model against the actual expected performance of the solution. If the trees are reliable, we should not notice a significant difference in performance:

$$\mathrm{E}^{\mathbb{P}}[f(\mathbf{x}_i^*, \boldsymbol{\xi})] \approx \mathrm{E}^{\mathbb{P}}[f(\mathbf{x}_j^*, \boldsymbol{\xi})]. \qquad (10.36)$$

This kind of check is not too hard in two stage models, as we should just plug the first-stage solution into several second-stage problems, sampling a large number of solutions. Typically, solving a large number of second-stage problems is not too demanding, especially if they are just linear programs; as a rule, it takes more time to solve one large stochastic LP than a large number of deterministic LPs. Unfortunately, this is not so easy in a multistage setting. To see why, consider the second-stage decisions. They are associated with a realization of random variables, which will be different from the out-of-sample realization, and there is no obvious way to adapt the solution.[16] Some ideas can be devised of course, but a realistic evaluation should rely on a rolling-horizon simulation whereby we solve a sequence of multistage stochastic programming problems and apply only the first-stage solution. This is technically feasible but quite expensive computationally. A possible (cheaper) alternative is to check solutions on different trees to see if

$$\hat{f}(\mathbf{x}_i^*; \mathcal{T}_j) \approx \hat{f}(\mathbf{x}_j^*; \mathcal{T}_i).$$

In the next section we apply these ideas on a very simple problem, for the purpose of illustration.

10.5.3.5 Scenario generation and the newsvendor problem

We have introduced the newsvendor model in Example 10.9, as a basic optimization model under uncertainty. Given its structure, we are able to find the optimal solution in explicit form. If we assume that demand is normally distributed, $D \sim \mathsf{N}(\mu, \sigma^2)$, then we know that the optimal decision is to make/buy a quantity

$$q^* = \mu + z_\beta \sigma$$

of items, where z_β is the quantile of the standard normal for the probability level

$$\beta = \frac{m}{m + c_u},$$

defined by the profit margin m (underage cost) and by the cost of unsold items c_u (overage cost). We have also seen that maximization of expected profit is equivalent to minimization of an expected deviation cost

$$\mathrm{E}_D\left[c_u(q - D)^+ + m(D - q)^+\right].$$

---

[16]As we shall see later in the chapter, dynamic programming provides us with decisions in feedback form, which makes it very suitable for out-of-sample simulation and evaluation of policies.

In Section 10.2.3 we have formulated the problem as a scenario-based LP, and it is interesting to check the accuracy of scenario generation mechanisms on this prototypical example, as well as investigating stability. To this end, we should also express the objective function analytically, which is not too difficult to accomplish for normally distributed demand,

Let us denote the normal PDF of demand by $f_D(x)$ and express the expected deviation cost in the case of normal demand:

$$
\begin{aligned}
\mathrm{E}_D\big[C\,(q,D)\,\big] &= c_u \int_{-\infty}^{q}(q-x)f_D(x)\,dx + m\int_{q}^{+\infty}(x-q)f_D(x)\,dx \\
&= c_u \int_{-\infty}^{q}(q-x)f_D(x)\,dx + c_u\int_{q}^{+\infty}(q-x)f_D(x)\,dx \\
&\quad - c_u \int_{q}^{+\infty}(q-x)f_D(x)\,dx + m\int_{q}^{+\infty}(x-q)f_D(x)\,dx \\
&= c_u \int_{-\infty}^{+\infty}(q-x)f_D(x)\,dx + (c_u+m)\int_{q}^{+\infty}(x-q)f_D(x)\,dx \\
&= c_u(q-\mu) + (c_u+m)\int_{q}^{+\infty}(x-q)f_D(x)\,dx.
\end{aligned}
$$

Since we have assumed a normal distribution, the integration domain starts from $-\infty$, which looks unreasonable for a non-negative demand. The assumption is plausible only if the probability of a negative demand is practically negligible. The last integral represents expected lost sales as a function of q and is related to the *loss function*. We digress a little to illustrate the essential properties of the loss function in the normal case.[17]

Technical note: The loss function. We define the standard loss function for a standard normal variable as

$$
L(z) \equiv \int_{z}^{+\infty}(x-z)\phi(x)\,dx,
$$

where $\phi(x)$ is the standard normal density. For a standard normal, the loss function can be expressed in terms of the PDF $\phi(x)$ and the corresponding CDF $\Phi(x)$:

$$
\begin{aligned}
L(z) &= \int_{z}^{+\infty} x\phi(x)\,dx - z\int_{z}^{+\infty}\phi(x)\,dx \\
&= \frac{1}{\sqrt{2\pi}}\int_{z}^{+\infty} xe^{-x^2/2}\,dx - z\left(1 - \int_{-\infty}^{z}\phi(x)\,dx\right).
\end{aligned}
$$

[17]As we shall see in Chapter 13, the same reasoning can be used to evaluate conditional value-at-risk in a normal case.

The first integral can be evaluated by substitution of variable:

$$t = x^2/2 \quad \Rightarrow \quad dt = x\,dx.$$

Changing the integration limits accordingly we find

$$\frac{1}{\sqrt{2\pi}} \int_z^{+\infty} x e^{-x^2/2}\,dx = \frac{1}{\sqrt{2\pi}} \int_{z^2/2}^{+\infty} e^{-t}\,dt = \frac{1}{\sqrt{2\pi}} e^{-z^2/2} = \phi(z).$$

Putting all of it together, we end up with

$$L(z) = \phi(z) - z(1 - \Phi(z)). \tag{10.37}$$

Given the standard loss function, we may express lost sales in the case of a generic normal random variable:

$$\int_q^{+\infty} (x - q)\frac{1}{\sigma}\phi\left(\frac{x-\mu}{\sigma}\right) dx = \sigma \int_q^{+\infty} \left(\frac{x-\mu}{\sigma} - \frac{q-\mu}{\sigma}\right)\phi\left(\frac{x-\mu}{\sigma}\right) dx$$

$$= \sigma \int_{(q-\mu)/\sigma}^{+\infty} \left(t - \frac{q-\mu}{\sigma}\right)\phi(t)\,dt$$

$$= \sigma L\left(\frac{q-\mu}{\sigma}\right),$$

where we have used the substitution of variable $t = (x - \mu)/\sigma$.

Using the loss function of a generic normal, we may express the deviation cost for a normal demand explicitly:

$$E_D[C(q,D)] = c_u(q - \mu) + (c_u + m)\sigma L\left(\frac{q-\mu}{\sigma}\right).$$

The R code of Fig. 10.16 takes advantage of this knowledge and yields the optimal decision and the related cost, given purchase cost, selling price, markdown price, as well as expected value and standard deviation of demand:

```
> pur <- 10
> sell <- 12
> mdown <- 5
> mu <- 100
> sigma <- 30
> newsOptimal(pur,sell,mdown,mu,sigma)
optimal q  83.02154
optimal cost  71.38016
```

In a practical setting, we would probably be forced to round the solution when demand is restricted to integer values, in which case the normal distribution would be an approximation. By the way, since $\mu > 3\sigma$ we may neglect the embarrassing possibility of negative demand.

```
newsOptimal <- function(pur,sell,mdown,mu,sigma){
  # define standard loss function
  lossFun <- function(z) dnorm(z) - z*(1-pnorm(z))
  # get economics
  m <- sell - pur    # profit margin
  cu <- pur - mdown  # cost of unsold items
  # get quantile and optimal q
  z <- qnorm(m/(m+cu))
  q <- mu + z*sigma
  # print solution and cost
  cat('optimal q ', q, '\n')
  cost <- cu*(q-mu)+(cu+m)*sigma*lossFun((q-mu)/sigma)
  cat('optimal cost ', cost, '\n')
}
```

FIGURE 10.16 **Solving the newsvendor problem optimally in the normal case.**

Given this analytical benchmark, we may use the scenario-based LP model of Section 10.2.3 to check the accuracy of the solution and the estimated expected cost. If we use straightforward Monte Carlo for scenario generation, then the scenarios are equiprobable: $\pi_s = 1/S$, $s = 1,\ldots,S$. We may use the function `solveLP` of package `linprog` to solve the model, but this requires building the technology matrix explicitly, which is in this case quite simple as the recourse matrix is just $\mathbf{W} = [\mathbf{I}, -\mathbf{I}]$. The resulting code is shown in Fig. 10.17. Demand scenarios and probabilities are an input to the procedure, so that we may check the performance of Monte Carlo scenario generation against other strategies. The figure also includes code to check in- and out-of-sample stability. Note that in scenario generation we clip demand to non-negative values, as we may not rule out generation of a negative demand, especially with a large sample. In order check in-sample stability, let us repeat the generation of 50 random scenarios five times:

```
obj   66.801   q   82.7
obj   58.2748  q   79.6
obj   77.0129  q   82.3
obj   68.2175  q   80.9
obj   59.6849  q   85.3
```

We observe some variability, especially in the objective function. As this casts some doubts on the quality of the solution, we may repeat the procedure with 1,000 scenarios, with a corresponding increase in computational effort, which yields

```
obj   71.4185  q   81.3
obj   70.1544  q   82.4
obj   70.801   q   82.5
```

```
newsLP <- function(m,cu,demand,probs){
  library(linprog)
  numScen <- length(demand)
  Amat <- cbind(rep(1,numScen),diag(numScen),-diag(numScen))
  cvet <- c(0, m*probs, cu*probs)
  eqs <- rep("=",numScen)
  # solve and print solution
  result <- solveLP(cvet,demand,Amat,const.dir=eqs,
                    lpSolve=TRUE)
  return(list(cost=result$opt,q=result$solution[1]))
}

# use naive MC and check in-sample stability
m <- sell-pur
cu <- pur-mdown
set.seed(55555)
numScen <- 50
probs <- rep(1/numScen,numScen)
qv <- numeric(5)
for (k in 1:5){
  demand <- pmax(rnorm(numScen,mu,sigma),0)
  out <- newsLP(m,cu,demand,probs)
  qv[k] <- out$q
  cat('obj ',round(out$cost,4),' q ',round(out$q,1),'\n')
}

# Check out-of-sample stability
outScost <- function(q,m,cu,demand){
  mean(cu*pmax(q-demand,0)+m*pmax(demand-q,0))}
set.seed(55555)
outsDemand <- pmax(rnorm(100000,mu,sigma),0)
for (k in 1:5){
  trueCost <- outScost(qv[k],m,cu,outsDemand)
  cat('true cost ',round(trueCost,4),'\n')
}
```

FIGURE 10.17 **Solving the newsvendor problem by stochastic LP and checking in- and out-of-sample stability.**

```
obj   71.3892   q   84.7
obj   72.8487   q   84.3
```

Now we see that the solution is more stable, and fairly close to the true optimum. As a further check, we may assess out-of-sample stability, by generating a large number of scenarios (one million) and estimating the cost of each solu-

tion generated by the stochastic LP model. A surprising fact is that the solution
obtained by using 50 scenarios proves to be a rather good one:

```
true cost   71.4412
true cost   71.8959
true cost   71.4542
true cost   71.6127
true cost   71.6429
```

The same happens with the solutions obtained with 1,000 scenarios:

```
true cost   71.5528
true cost   71.4514
true cost   71.4477
true cost   71.5522
true cost   71.5014
```

This finding can be explained by observing that in some cases the true objective function is rather flat around the optimal solution, which makes the model
rather robust. However, this need not be the case in other settings. For instance,
in financial applications the model might take full advantage of the faulty information included in badly generated scenarios, yielding a very unstable solution.
Furthermore, here we may afford a large number of scenarios because this is a
very simple model, but a parsimonious approach must be pursued in general.
Thus, it is also interesting to see what happens if we generate scenarios using
Gaussian quadrature:[18]

```
> library(statmod)
> quad <- gauss.quad.prob(10,dist="normal",mu=mu,sigma=sigma)
> out <- newsLP(m,cu,pmax(quad$nodes,0),quad$weights)
> cat('obj ',round(out$cost,4),' q ',round(out$q,1),'\n')
obj  65.5008   q   85.5
> trueCost <- outScost(out$q,m,cu,outsDemand)
> cat('true cost ',round(trueCost,4),'\n')
true cost   71.6757
```

With only ten scenarios, we get a solution that proves to be good out-of-sample,
even though the in-sample evaluation of the objective function does not look too
satisfactory. A less pleasing feature of Gaussian quadrature for this example is
that, in order to capture the tails of the distribution, it generates negative as well
as rather extreme demand values:

```
> quad$nodes
-45.783885  -7.454705    25.470225   56.020327   85.451929
114.548071  143.979673  174.529775 207.454705 245.783885
> quad$weights
4.310653e-06 7.580709e-04 1.911158e-02 1.354837e-01
3.446423e-01 3.446423e-01 1.354837e-01 1.911158e-02
```

---

[18]See Section 2.2.2.

```
7.580709e-04 4.310653e-06
```

Even though negative values have small probability, this shows a potential issue, which is actually related to the fact that we are using a normal distribution to model an intrinsically non-negative variable. A similar issue could arise when modeling rates of return of stock shares, which cannot be less than -100%.

In Gaussian quadrature we determine both nodes and weights together. We may also pursue a simple approach in which probabilities are fixed to a uniform value, and we look for nodes. One such approach can be devised by integrating random variate generation by the inverse transform method with a sort of "deterministic stratification." The generation of a standard normal random variable by the inverse transform method, as we have seen in Chapter 5, requires the generation of a uniform random variable $U \sim U(0, 1)$ and the inversion of the CDF of the standard normal, $Z = \Phi^{-1}(U)$. We may partition the unit interval uniformly, define a representative point of each subinterval, and generate the corresponding standard normal. For instance, if we use five intervals, their end points are

$$0, \quad 0.2, \quad 0.4, \quad 0.6, \quad 0.8, \quad 1.$$

Each subinterval has probability 0.2, and we might consider their midpoints as their representative points. If so, the nodes would be

$$\Phi^{-1}(0.1), \quad \Phi^{-1}(0.3), \quad \Phi^{-1}(0.5), \quad \Phi^{-1}(0.7), \quad \Phi^{-1}(0.9).$$

We note that, in the stated form, this is at best a simple heuristic; in order to apply stratification we should find the conditional expectation of the standard normal on each interval. This may actually be done, but it is left as an exercise for the reader, and we just check the performance of this simple approach:

```
> numVals <- 50
> aux <- (0:numVals)/numVals
> U <- (aux[-1]+aux[-(numVals+1)])/2
> demand <- pmax(0, mu+sigma*qnorm(U))
> probs <- rep(1/numVals,numVals)
> out <- newsLP(m,cu,demand,probs)
> cat('obj ',round(out$cost,4),' q ',round(out$q,1),'\n')
obj 70.8987   q 83.4
> trueCost <- outScost(out$q,m,cu,outsDemand)
> cat('true cost ',round(trueCost,4),'\n')
true cost  71.4444
```

Needless to say, the performance is arguably satisfactory because of the simple and well-behaved problem we are tackling. Still, the idea might be worth considering, and it can be applied to small-dimensional cases (where we build hyper-rectangular cells, rather than subintervals), possibly with some preliminary Monte Carlo runs to find the representative point of each cell. It is interesting to note that in this case we do not generate "extreme" node values:

```
> range(demand)
[1]  30.20956 169.79044
```

Whether this is a strenght or a weakness depends on the application. As a final example, let us resort to low-discrepancy sequences and try scrambled Sobol sequences:[19]

```
> library(randtoolbox)
> numVals <- 50
> eps <- sobol(numVals,dim=1,normal=TRUE,scrambling=2)
> demand <- pmax(0, mu+sigma*eps)
> probs <- rep(1/numVals,numVals)
> out <- newsLP(m,cu,demand,probs)
> cat('obj ',round(out$cost,4),' q ',round(out$q,1),'\n')
obj  73.4044  q  82
> trueCost <- outScost(out$q,m,cu,outsDemand)
> cat('true cost ',round(trueCost,4),'\n')
true cost  71.4798
```

Once again, the result does not look bad, but the reader is warned against drawing any conclusion on the basis of this extremely limited experiment. Anyway, we observe that both Monte Carlo and quasi-Monte Carlo methods can be applied to relatively high-dimensional problems, whereas both Gaussian quadrature and stratification are more limited. Unfortunately, the limitation of R as an optimization tool does not allow to cope with more interesting examples. However, we should mention that state-of-the-art commercial solvers, such as IBM CPLEX and Gurobi do offer an interface with R.

10.6 Stochastic dynamic programming

Dynamic programming (DP) is, arguably, the single most powerful optimization principle that we have at our disposal. In fact, it can be used to tackle a wide variety of problems:

- Continuous as well discrete problems
- Discrete- as well as continuous-time problems
- Deterministic as well as stochastic problems
- Finite- as well as infinite-horizon problems

Actually, for some problem families, dynamic programming is the only viable approach. Since there is no free lunch, power and generality must come at some price. On the one hand, dynamic programming is a principle, rather than a well defined optimization algorithm. Thus, unlike the simplex method for linear programming, there is no on-the-shelves software package, and a considerable degree of customization and implementation effort may be needed to make it work. On the other hand, dynamic programming is associated with the so-called curse of dimensionality. As we shall see, the curse actually takes different

---

[19]See Section 9.4.

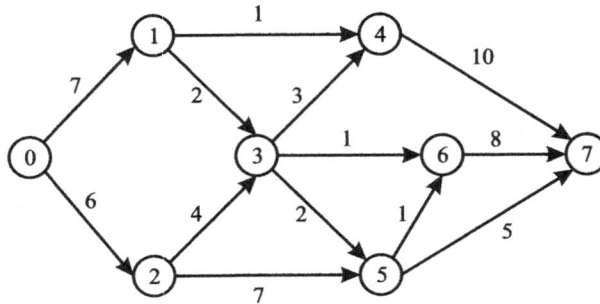

FIGURE 10.18 **A shortest path problem.**

forms, but the essence is that, as a rule, conventional dynamic programming can only be applied to rather small-scale problems. These limitations can be overcome, at least in part, by resorting to numerical approximations where Monte Carlo methods play a fundamental role. Here we are interested in discrete-time decision models under uncertainty; hence, there is a considerable connection with stochastic programming with recourse.

To introduce the dynamic programming concept in the simplest way, we show how it can be used to solve a deterministic shortest path problem in Section 10.6.1. Then, in Section 10.6.2, we formalize the principle in the form of the recursive functional equation originally introduced by Bellman; this form allows us to cope with stochastic optimization problems. We outline the adaption of the approach to infinite-horizon problems in Section 10.6.3. Finally, in Section 10.6.4, we briefly compare stochastic programming with recourse and stochastic dynamic programming; as expected, the two approaches have somewhat complementary strengths and weaknesses, as well as many intersections.

10.6.1 THE SHORTEST PATH PROBLEM

The easiest way to introduce dynamic programming is by considering one of its most natural applications, i.e., finding the shortest path in a network. Graph and network optimization are not customary topics in finance or economics, but a quick look at Fig. 10.18 is enough to understand what we are talking about. A network consists of a set of nodes (numbered from 0 to 7 in our toy example) and a set of arcs joining pairs of nodes. Arcs are labeled by a number which can be interpreted as the arc length (or cost). Our purpose is finding a path in the network, starting from node 0 and leading to node 7, such that the path has total minimal length. For instance, by summing the arc lengths that are followed on the path $(0, 1, 4, 7)$, we see that its total length is 18, whereas path $(0, 1, 3, 5, 7)$ has length 16. At each node, we must choose the next node to visit. We may immediately appreciate that this problem bears some resemblance to dynamic decision making; given some state we are in, we should decide what to do in order to optimize an outcome that depends on the whole path. A greedy decision need not be the optimal one; for instance, the closest node to the starting point 0 in our network is node 2, but there is no guarantee that this arc is on an optimal path.

Of course, we could simply enumerate all of the possible paths to spot the optimal one; here we have just a finite set of alternatives and there is no uncertainty involved, so the idea is conceptually feasible. However, this approach becomes quickly infeasible in practice, as the network size increases. Furthermore, we need a better idea in view of an extension to a stochastic case: If arc costs are random, running a Monte Carlo simulation for each possible path to estimate its expected length is out of the question. So we must come up with some clever way to avoid exhaustive enumeration. Dynamic programming is one possible approach to accomplish this aim. It is worth noting that more efficient algorithms are available for the shortest path problem, but the idea we illustrate here can be extended to problems featuring infinite state spaces (countable or not) and uncertain data. Let $\mathcal{N} = \{0, 1, 2, \ldots, N\}$ be the node set and $\mathcal{A}$ be the arc set; let the start and final nodes be 0 and N, respectively. For simplicity, we assume that the network is acyclic and that the arc lengths c_{ij}, $i, j \in \mathcal{N}$, are non-negative: If we had the possibility of getting trapped in a loop of negative length arcs, the optimal cost would be $-\infty$, and we do not want to consider such pathological cases.

The starting point is to find a *characterization* of the optimal solution, which can be translated into a constructive algorithm. Let V_i be the length of the shortest path from node $i \in \mathcal{N}$ to node N (denoted by $i \overset{*}{\to} N$). Assume that, for a specific $i \in \mathcal{N}$, node j lies on the optimal path $i \overset{*}{\to} N$. Then the following property holds:

$$j \overset{*}{\to} N \text{ is a subpath of } i \overset{*}{\to} N.$$

In other words, the optimal solution for a problem is obtained by assembling optimal solutions for smaller subproblems, which suggests a useful decomposition strategy. To understand why, consider the decomposition of $i \overset{*}{\to} N$ into the subpaths $i \to j$ and $j \to N$. The length of $i \overset{*}{\to} N$ is the sum of the lengths of the two subpaths:

$$V_i = L(i \to j) + L(j \to N). \tag{10.38}$$

Note that the second subpath is not affected by *how* we go from i to j. This is strongly related to the concept of state in Markovian dynamic systems: How we get to state j has no influence on the future. Now, assume that the subpath $j \to N$ is not the optimal path from j to N. Then we could improve the second term of Eq. (10.38) by assembling a path consisting of $i \to j$ followed by $j \overset{*}{\to} N$. The length of this new path would be

$$L(i \to j) + L(j \overset{*}{\to} N) < L(i \to j) + L(j \to N) = V_i,$$

which is a contradiction, as we have assumed that V_i is the optimal path length.

This observation leads to the following recursive equation for the shortest path from a generic node i to the terminal node N:

$$V_i = \min_{(i,j)\in\mathcal{A}} \{c_{ij} + V_j\} \qquad \forall j \in \mathcal{N}. \qquad (10.39)$$

In other words, to find the optimal path from node i to node N, we should consider the immediate cost c_{ij} of going from i to all of its immediate successors j, plus the optimal cost of going from j to the terminal node. Note that we do not only consider the immediate cost, as in a greedy decision rule. We also add the future cost of the optimal sequence of decisions starting from each state j that we can visit next; this is what makes the approach nonmyopic. The function V_i is called *cost-to-go* or *value function* and is defined recursively by Eq. (10.39). The value function, for each point in the state space, tells us what the future optimal cost would be, if we reach that state and proceed with an optimal policy. This kind of recursive equation, whose exact form depends on the problem at hand, is the heart of dynamic programming and is an example of a *functional equation*. In the shortest path problem, we have a finite set of states, and the value function is a vector; in a continuous-state model, the value function is an infinite-dimensional object.

Solving the problem requires finding the value function V_0 for the initial node, and to do that we should go backward from the terminal node.[20] We can associate a terminal condition $V_N = 0$ with our functional equation, as the cost for going from N to N is zero. Then we unfold the recursion by considering the immediate predecessors i of the terminal node N; for each of them, finding the optimal path length is trivial, as this is just c_{iN}. Then we proceed backward, labeling each node with the corresponding value function. In this unstructured network, we may label a node only when all of its successors have been labeled; we can always find a correct node ordering in acyclic networks.

Example 10.14 A shortest path

> Let us find the shortest path for the network depicted in Fig. 10.18. We have the terminal condition $V_7 = 0$ for the terminal node, and we look for its immediate predecessors 4 and 6 (we cannot label node 5 yet, because node 6 is one of its successors). We have
>
> $$V_4 = c_{47} + V_7 = 10 + 0 = 10,$$
> $$V_6 = c_{67} + V_7 = 8 + 0 = 8.$$

---

[20]We are considering here only the *backward* version of dynamic programming. For the shortest path and other deterministic combinatorial optimization problems, we could also apply a *forward* equation (see, e.g., [9, Appendix D]). The natural formulation of stochastic problems is backward, but forward DP plays a major role in some approximate methods, as we shall see later.

Now we may label node 5:

$$V_5 = \min \left\{ \begin{array}{c} c_{56} + V_6 \\ c_{57} + V_7 \end{array} \right\} = \min \left\{ \begin{array}{c} 1 + 8 \\ 5 + 0 \end{array} \right\} = 5.$$

Then we consider node 3 and its immediate successors 4, 5, and 6:

$$V_3 = \min \left\{ \begin{array}{c} c_{34} + V_4 \\ c_{35} + V_5 \\ c_{36} + V_6 \end{array} \right\} = \min \left\{ \begin{array}{c} 3 + 10 \\ 2 + 5 \\ 1 + 8 \end{array} \right\} = 7.$$

By the same token we have

$$V_1 = \min \left\{ \begin{array}{c} c_{13} + V_3 \\ c_{14} + V_4 \end{array} \right\} = \min \left\{ \begin{array}{c} 2 + 7 \\ 1 + 10 \end{array} \right\} = 9,$$

$$V_2 = \min \left\{ \begin{array}{c} c_{23} + V_3 \\ c_{25} + V_5 \end{array} \right\} = \min \left\{ \begin{array}{c} 4 + 7 \\ 7 + 5 \end{array} \right\} = 11,$$

$$V_0 = \min \left\{ \begin{array}{c} c_{01} + V_1 \\ c_{02} + V_2 \end{array} \right\} = \min \left\{ \begin{array}{c} 7 + 9 \\ 6 + 11 \end{array} \right\} = 16.$$

Apart from getting the optimal length, which is 16, we may find the optimal path by looking for the nodes optimizing each single decision, starting from node 0:

$$0 \to 1 \to 3 \to 5 \to 7.$$

10.6.2 THE FUNCTIONAL EQUATION OF DYNAMIC PROGRAMMING

The shortest path problem is one of the prototypical examples of dynamic programming, yet it lacks the structure of most problems of interest in finance and economics, which are dynamic problems where uncertainty unfolds over time. We have introduced the three essential families of dynamic models in Chapter 3. Dynamic programming can be applied to continuous-time and discrete-event models, too, but here we only consider discrete-time models, which are relatively simple but do hide some potential traps. In fact, when discretizing time

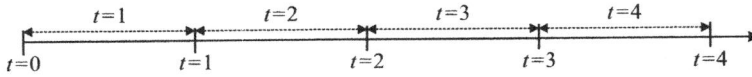

FIGURE 10.19 **Illustrating time conventions.**

it is necessary to be very clear about the meaning of time instants and time intervals:[21]

- We consider *time instants* indexed by $t = 0, 1, 2, \ldots$. At these time instants we observe the system state and make a decision.

- By a *time interval* t we mean the time interval between time instants $t - 1$ and t. After the decision at time instant $t - 1$, the system evolves and a new state is reached at time t. During the time interval, some random "disturbance" will be realized, influencing the transition to the new state.

These definitions are illustrated in Fig. 10.19. Note that, with this timing convention, we emphasize the fact that noise is realized *during* time interval t, *after* making the decision at time instant $t - 1$.

Let us introduce a bit of notation:

- The vector of state variables at time instant t is denoted by $\mathbf{s}_t$; with the time convention above $\mathbf{s}_t$ is the value of the state variables at the end of time period t, as a result of what happened between time instants $t - 1$ and t, i.e., during time interval t.

- The vector of decision variables is denoted by $\mathbf{x}_t$; these decisions are based on the knowledge of $\mathbf{s}_t$. Here we assume that states are perfectly observable, but the framework can be applied to inexact observations as well.

- The state $\mathbf{s}_{t+1}$ at time instant $t + 1$ depends on $\mathbf{s}_t$ and $\mathbf{x}_t$, but also on possible random perturbations that take place during time interval $t + 1$, i.e., between time instants t, when we made the last decision, and $t + 1$, when we have to make another one. We represent these perturbations by ϵ_{t+1}; the subscript $t + 1$ emphasizes the fact this is *new information*.

Then, the system dynamics is represented by a state equation like

$$\mathbf{s}_{t+1} = S_{t+1}\left(\mathbf{s}_t, \mathbf{x}_t, \epsilon_{t+1}\right), \tag{10.40}$$

where S_{t+1} is the state transition function over time period $t + 1$; the subscript may be omitted if the dynamics do not change over time, i.e., if the system is stationary. The latter is the common assumption when dealing with infinite-horizon problems.

---

[21]For an extensive treatment of modeling issues in DP, see [37, Chapter 5].

By making the decision $\mathbf{x}_t$ in state $\mathbf{s}_t$ we incur an immediate cost $C_t(\mathbf{s}_t, \mathbf{x}_t)$. We would like to minimize the expected cost incurred over a given planning horizon,

$$\min \mathrm{E}\left[\sum_{t=0}^{T} \gamma^t C_t(\mathbf{s}_t, \mathbf{x}_t)\right], \qquad (10.41)$$

where $\gamma \leq 1$ is a suitable discount factor. Clearly, when dealing with profits or utility functions, we rewrite the problem in terms of a maximization. However, in the statement above we are not stating precisely what kind of decision process we are dealing with. In a deterministic setting, we would minimize with respect to decision variables $\mathbf{x}_t$, and we would come up with a plan specifying all of the future decisions. However, we have already seen in multistage stochastic programming with recourse that decisions should be adapted to the occurred contingencies. As a consequence, apart from the first-stage ones, decisions are in fact random variables. In stochastic dynamic programming we make this even clearer, as decisions are obtained by the application of a feedback policy.[22] We define a policy, denoted by π, whereby an action is selected on the basis of the current state:

$$\mathbf{x}_t = A_t^\pi(\mathbf{s}_t). \qquad (10.42)$$

Hence, a more precise statement of the problem is

$$\min_{\pi \in \Pi} \mathrm{E}\left[\sum_{t=0}^{T} \gamma^t C_t\big(\mathbf{s}_t, A_t^\pi(\mathbf{s}_t)\big)\right], \qquad (10.43)$$

where Π is the set of feasible policies. By feasible we mean that each decision $\mathbf{x}_t$ is constrained to be in $X_t(\mathbf{s}_t)$, the set of feasible decisions at time t if we are in state $\mathbf{s}_t$.

Stated as such, the problem does not look quite tractable. However, the shortest path example suggests a suitable decomposition strategy. When in state $\mathbf{s}_t$ at time t, we should select the decision $\mathbf{x}_t \in X_t(\mathbf{s}_t)$ that minimizes the sum of the immediate cost, and the expected value of cost-to-go that we will incur from time period $t + 1$ onward:

$$V_t(\mathbf{s}_t) = \min_{\mathbf{x}_t \in X_t(\mathbf{s}_t)} \left\{C_t(\mathbf{s}_t, \mathbf{x}_t) + \gamma \mathrm{E}\big[V_{t+1}\big(S_{t+1}(\mathbf{s}_t, \mathbf{x}_t, \epsilon_{t+1})\big)\big|\mathbf{s}_t\big]\right\}. \quad (10.44)$$

More generally, we talk of a value function, as the term is better suited to both minimization and maximization problems. The conditioning with respect to $\mathbf{s}_t$ stresses the fact that we are taking an expectation conditional on the current state; it might even be the case that uncertainty depends on our decisions, and if so we should also condition with respect to $\mathbf{x}_t$. We will not analyze the conditions under which Eq. (10.44) can be used to solve problem (10.43).

---

[22]We are cutting several corners here, as we assume that an optimal decision in feedback form can be found. Actually, this depends on technical conditions that we take for granted, referring the reader to the literature for a rigorous treatment.

The intuition gained from the shortest path problem suggests that the stochastic process ϵ_t must be Markovian. With more general path dependency, the decomposition argument behind dynamic programming fails. Sometimes, a model reformulation based on additional state variables can be used to transform a non-Markovian model into a Markovian one, at the price of an increase in complexity. The careful reader will immediately notice a similarity between this modeling framework and stochastic programming with recourse, where the recourse cost plays the role of the value function. The recursive structure of a multistage problem, which is pretty evident in Eq. (10.26), is also visible here. However, multistage stochastic programming with recourse does not require the Markov property. On the other hand, we should note that standard stochastic programming models with recourse assume exogenous uncertainty.

Equation (10.44) is the functional equation of dynamic programming, also called the optimality equation, and it requires to find the value function for each time instant. Typically, the functional equation is solved backward in time, taking advantage of the fact that the problem in the last time period is a myopic choice. At the last decision time instant, $t = T$, we should solve the problem

$$V_T(\mathbf{s}_T) = \min_{\mathbf{x}_T \in X_T(\mathbf{s}_T)} \left\{ C_T(\mathbf{s}_T, \mathbf{x}_T) + \gamma \mathrm{E}\left[V_{T+1}\left(S_{T+1}(\mathbf{s}_T, \mathbf{x}_T, \epsilon_{T+1}) \right) \middle| \mathbf{s}_T \right] \right\},$$

(10.45)

for every possible state $\mathbf{s}_T$. The value function $V_{T+1}(\mathbf{s}_{T+1})$ is given by a boundary condition. For problem (10.41), the boundary condition is

$$V_{T+1}(\mathbf{s}_{T+1}) = 0 \qquad \forall \mathbf{s}_{T+1},$$

since the objective function includes only a trajectory cost. Sometimes, a cost (or reward) $\Phi_{T+1}(\mathbf{s}_{T+1})$ is associated with the terminal state, and the problem is

$$\min \mathrm{E}\left[\sum_{t=0}^{T} \gamma^t C_t(\mathbf{s}_t, \mathbf{x}_t) + \gamma^{T+1} \Phi_{T+1}(\mathbf{s}_{T+1}) \right].$$

(10.46)

For instance, in life-cycle consumption–saving models for pension economics, the objective function could include both a utility function depending on consumption and some utility from bequest. In this case, we impose the boundary condition

$$V_{T+1}(\mathbf{s}_{T+1}) = \Phi_{T+1}(\mathbf{s}_{T+1}) \qquad \forall \mathbf{s}_{T+1}.$$

Whatever boundary condition we select, problem (10.45) is a *static* optimization model. By solving it for every state $\mathbf{s}_T$, we build the value function $V_T(\mathbf{s}_T)$ at time $t = T$. Then, unfolding the recursion backward in time, we find the value functions $V_{T-1}(\mathbf{s}_{T-1})$, $V_{T-2}(\mathbf{s}_{T-2})$, and so on, down to $V_1(\mathbf{s}_1)$. Finally, given the initial state $\mathbf{s}_0$, we find the next optimal decision by solving the static problem

$$\min_{\mathbf{x}_0 \in X_0(\mathbf{s}_0)} \left\{ C_0(\mathbf{s}_0, \mathbf{x}_0) + \gamma \mathrm{E}\left[V_1\left(S_1(\mathbf{s}_0, \mathbf{x}_0, \epsilon_1) \right) \middle| \mathbf{s}_0 \right] \right\}.$$

(10.47)

Unlike stochastic programming with recourse, we have a clear way to make also the next decisions, as the full set of value functions allows us to solve each problem over time, finding the decision x_t as a function of the current state s_t. Thus, we find the optimal policy in an implicit feedback form, and the dynamic problem is decomposed into a sequence of static problems.

The skeptical reader will wonder what is the price to pay for this extremely powerful approach. The price is actually pretty evident in Eq. (10.44). We should solve an optimization problem for each possible value of the state variable s_t. If we are dealing with a continuous state space, this is impossible, unless we may take advantage of some specific problems structure (we shall see an example in Section 10.7.1). But even if the state space is discrete, when the dimensionality is large, i.e., when s_t is a vector with several components, the sheer computational burden is prohibitive. This is called the *curse of dimensionality*. Actually, a close look at Eq. (10.44) shows that we face a three-fold curse of dimensionality:

1. The dimension of the state space, which is related to the state variable s_t.

2. The need to evaluate, or approximate, the expected value of the value function in the next state, which is related to the disturbance ϵ_{t+1} and may call for expensive numerical integration.

3. The need to solve a possibly large-scale optimization problem with respect to a high-dimensional decision vector x_t.

The shortest path problem is so easy because it is deterministic, it features a discrete state space, and the of possible decisions is discrete as well. In general the application of DP is not trivial, and calls for the integration of several numerical tricks of the trade, as we show in Section 10.7. Even so, DP in its standard form can be tackled only for moderately sized problems. However, recent trends in approximate dynamic programming, which we outline in Section 10.8, integrate statistical learning with optimization, paving the way for a much wider array of applications. Typically, these approaches rely on Monte Carlo simulation.

10.6.3 INFINITE-HORIZON STOCHASTIC OPTIMIZATION

The recursive form of Eq. (10.44) needs some adjustment when coping with an infinite-horizon problem like

$$\min E \left[\sum_{t=0}^{+\infty} \gamma^t C(s_t, x_t) \right], \qquad (10.48)$$

where we assume that immediate costs are bounded and $\gamma < 1$, so that the series converges to a finite value.[23] In this case, we typically drop the subscript t from

---

[23] In some engineering applications, the average cost, rather than the discounted one is of interest. However, discounted DP is the rule in finance and economics, so we will stick to it.

the immediate cost, as well as from the state-transition equation

$$\mathbf{s}_{t+1} = S\left(\mathbf{s}_t, \mathbf{x}_t, \boldsymbol{\epsilon}_{t+1}\right),$$

i.e., we assume a stationary model. An infinite-horizon model may be of interest per se, or it can be a trick to avoid the need to specify a terminal state value function. In this case, the functional equation boils down to

$$V(\mathbf{s}) = \min_{\mathbf{x} \in X(\mathbf{s})} \left\{ C(\mathbf{s}, \mathbf{x}) + \gamma \mathrm{E}\left[V\left(S(\mathbf{s}, \mathbf{x}, \boldsymbol{\epsilon})\right)\right] \right\}, \tag{10.49}$$

where $X(\mathbf{s})$ is the set of feasible decisions when we are in state s. The good news is that we need only one value function, rather than a value function for each time instant. The bad news is that now we have a value function defined as the fixed point of a possibly complicated operator, as we have $V(\mathbf{s})$ on both sides of the equation. In general, an iterative method is needed to solve Eq. (10.49).

10.6.4 STOCHASTIC PROGRAMMING WITH RECOURSE VS. DYNAMIC PROGRAMMING

Before we delve into numerical solution methods for stochastic dynamic programming, it is useful to pause and compare this approach with stochastic programming with recourse, since there is a clear connection between the two.

As a general consideration, stochastic programming models are mostly favored within the operations research community, whereas dynamic programming is favored by economists.[24] One reason is that stochastic programming is intrinsically operational: What we care about is the first-stage decision, which is what we actually implement. The next stages are included to find a nonmyopic and robust solution, but they are not really contingency plans to be implemented. When uncertainty is progressively resolved, we typically recalibrate and resolve the model, within a rolling-horizon framework. On the contrary, the use of a model in economics is generally quite different. Consider a life-cycle, strategic asset allocation model.[25] Usually, a very limited number of assets is included in the model, possibly only two, risk-free and risky. Indeed, the aim there is not really to manage a portfolio of a pension fund. The aim is to investigate the behavior of individual workers in order to see what drives their decisions (precautionary savings, habit formation, income path over time, etc.). To simulate consumption–saving and asset allocation decisions over several years, dynamic programming is definitely more convenient, as once we have found the sequence of value functions, we may simulate the sequence of optimal decisions, which calls for the solution of a sequence of static optimization models, without the

[24]Of course, this is not an absolute rule, as DP is also widely used by operations researchers, too. On the contrary, stochastic programming with recourse is less common in economics.
[25]We will see a simplified model in this vein in Section 10.7.3. For an extensive treatment of these kind of model see, e.g., [10].

need of solving a dynamic model repeatedly. The result is a pattern in terms of consumption–saving decisions, as well as an asset allocation between the risky and risk-free assets. The pattern may be checked against empirical evidence in order to assess the validity of the model and the assumptions behind it.

There is also a clear connection between recourse functions in stochastic programming and value functions in dynamic programming. However, a recourse function is instrumental in finding a first-stage decision. Decisions at latter stages are made after observing the new state, which need not be included in the scenario tree; however, there is no obvious way to adjust a recourse decision when an unplanned scenario is realized. The value function is essential in finding decisions in feedback form; hence, they may be used for any scenario. This is a definite advantage of dynamic programming, which comes at a price. First, we have to rely on a Markovian structure, whereas stochastic programming may cope with more general path dependence. Second, stochastic programming may cope with more complicated, less stylized decision models.

Another feature of dynamic programs is that, as a rule, they deal with multistage models where the same kind of decision must be made over and over. For instance, in a consumption–saving problem, at each time step we have to decide how much to consume and how much to save. However, there are decision problems in which the decisions ate different stages are completely different. For instance, consider the design of a logistic network. At one stage, we cope with the fixed costs of installing warehouses and other facilities; hence, the decision variables define the network structure, which must be designed under demand uncertainty. At later stages, we have to cope with variable transportation costs; hence, the second-stage decision are related to the optimal routing of items, in order to satisfy a known demand, with constraints imposed by the selected network structure. Clearly, this latter kind of model is less common in finance or economics, but it contributes to explain with stochastic programming is more common in operational decision problems.

Another noteworthy point is that, so far, we have assumed that the important output of the optimization model is a set of decisions, possibly in feedback form. However, there are problems in which it is the value of the objective function that is relevant. One important class of such problems is the pricing of American-style options. As we have seen in Section 10.2.5, pricing options with early exercise features requires the solution of a stochastic optimization problem. Since our ability to represent uncertainty by scenario trees is limited, the validity of a price found by stochastic programs is questionable. Armed with a sequence of value functions, possibly approximated ones, we may simulate the application of an early exercise policy with a large number of scenarios. Due to approximations in the solution of the dynamic programming problems, we will get a lower bound on the option price, but at least we may check its reliability, as we shall see in Chapter 11.

Finally, we observe that DP is well-suited to cope with infinite horizon problems, whereas stochastic programming with recourse is not. However,the two approaches may be integrated, since the value function for an infinite-horizon program may be used to avoid end-of-horizon effects in a finite-horizon stochastic program. Indeed, there are important connections between the approaches, especially when it comes to devising sophisticated decomposition algorithms to cope with large-scale problems. Thus, both approaches should be included in our bag of tools to tackle a wide class of relevant problems.

10.7   Numerical dynamic programming

The decomposition principle behind dynamic programming is extremely powerful, but it requires a set of value functions, which are infinite-dimensional objects when the state space is continuous. Solving a sequence of functional equations involves a mix of numerical methods for integration, optimization, function approximation, and equation solving. In this section we illustrate the building blocks of the mix, by applying dynamic programming to three examples. In Section 10.7.1 we deal with a deterministic budget allocation problem with a continuous-state space. The problem structure allows for an analytical solution, which we use as a benchmark for a numerical method based on the discretization of the state space and the approximation of value functions by cubic splines. Then, in Section 10.7.2 we tackle a simple infinite-horizon, stochastic problem. The example is very simple, as it features a discrete state space and a discrete set of decisions, nevertheless it illustrates the concept of value iteration, which is a key ingredient in approximate dynamic programming. Finally, we tackle a simple consumption–saving problem with income uncertainty in Section 10.7.3. Even though quite simplified with respect to real-life problems, this example is challenging, as it involves mixed states, both continuous and discrete, continuous disturbances, as well as continuous decisions.

10.7.1   APPROXIMATING THE VALUE FUNCTION: A DETERMINISTIC EXAMPLE

Let us consider a simple resource allocation problem. We have a resource budget B that can be allocated to a set of n activities; let $x_i \geq 0$ be the amount of resource allocated to activity i. If the contribution to profit from activity i is represented by $C_i(x_i)$, our problem is

$$\max \quad \sum_{k=1}^{n} C_k(x_k)$$

$$\text{s.t.} \quad \sum_{k=1}^{n} x_k \leq B,$$

$$x_k \geq 0.$$

A typical choice for the contribution function is a concave function like $C(x) = \sqrt{x}$, to model decreasing marginal returns. Note that we assume that a continuous allocation of resources is feasible. If $C(\cdot)$ is increasing, then we may replace the inequality by an equality constraint, as the budget will be fully used for sure. Clearly, there is no need for dynamic programming to solve the above problem, which may be easily tackled as an ordinary nonlinear program. By the way, if the contribution functions are the same for each activity and is concave, it is easy to see the in the optimal solution the budget is equally allocated to each activity. Nevertheless, using dynamic programming could be valuable in the case of different contribution functions and discrete decisions, i.e., when resources are not infinitely divisible; in such a case, the above problem would be a potentially nasty nonlinear integer programming problem. Furthermore, it is instructive to see how a problem where time does not play any role can be recast as a sequential decision process. Since the structure of the problem is simple, we may be able to use dynamic programming and solve the problem analytically by finding the value functions explicitly. Then, we may compare the analytical solution with a numerical solution based on the discretization of the state space and the approximation of the value functions.

To recast the problem as a dynamic program, let us imagine a sequence of decisions. We start with an initial budget $b_1 = B$. Then we allocate an amount of resource $x_1 \leq b_1$ to the first activity, which yields an immediate reward $C_1(x_1)$ and leaves us with a residual budget $b_2 = b_1 - x_1$. After deciding x_2, we have a residual budget $b_3 = b_2 - x_2$, and so on. It is easy to see that the residual budget b_k, before making decision x_k, is the natural state variable, with state transition function $b_{k+1} = b_k - x_k$. Furthermore, we define the value function $V_k(b_k)$, the optimal profit that we obtain by optimally allocating the residual budget b_k to activities $k, k+1, \ldots, n$. The value function satisfies the optimality equation

$$V_k(b_k) = \max_{0 \leq x_k \leq b_k} \left\{ C_k(x_k) + V_{k+1}(b_k - x_k) \right\}. \qquad (10.50)$$

It is a fairly simple exercise[26] to solve the equation analytically by backward induction. Let us assume that $C_k(x_k) = \sqrt{x_k}$, for $k = 1, \ldots, n$. At the last stage, we have

$$V_n(b_n) = \max_{0 \leq x_n \leq b_n} \sqrt{x_n} = \sqrt{b_n},$$

since at the last stage the obvious decision is to allocate whatever is left to the last activity. Plugging this value function into Eq. (10.50) for $k = n - 1$, we find

$$V_{n-1}(b_{n-1}) = \max_{0 \leq x_{n-1} \leq b_{n-1}} \left\{ \sqrt{x_{n-1}} + \sqrt{b_{n-1} - x_{n-1}} \right\},$$

whose optimal solution is $x_{n-1}^* = b_{n-1}/2$. Plugging the optimal decision into this objective function, we find the value function

$$V_{n-1}(b_{n-1}) = 2\sqrt{\frac{b_{n-1}}{2}}.$$

---

[26]See, e.g., [37, pp. 29–31].

```
x <- 1:5
y = c(8,5,3,7,4)
f <- splinefun(x,y)
plot(x,y)
pts <- seq(from=1,to=5,by=0.01)
lines(pts,f(pts))
```

FIGURE 10.20 **A script to illustrate cubic splines.**

If we repeat the argument going backward with respect to index $k = n - 2, n - 3, \ldots, 1$, and we also apply the principle of mathematical induction, we obtain

$$V_k(b_k) = (n - k + 1)\sqrt{\frac{b_k}{n - k + 1}}$$

and

$$x_k^* = \frac{b_k}{n - k + 1}.$$

Note that when we consider subscript k, it means that we are about to make a choice for activity k; thus, we have already allocated a share of budget to activities from 1 to $k - 1$, and there are still $n - (k - 1)$ to go. The last condition, in fact, tells us that the residual budget should be shared in fair slices.

Of course, since the profit contributions are identical and concave, the conclusion is far from surprising, and we have arguably used a pretty huge sledgehammer to crack a tiny nut. The message is that dynamic programming can be also applied when we have different profit contributions, discrete units of budget, and uncertainty as well. Furthermore, we are now in the position to experiment with a numerical method, in order to get acquainted with and to gain confidence in the solution strategies that are needed in more complicated cases. The big issue here is that each value function is an infinite dimensional object, since we are dealing with a continuous state variable. In the discrete-state case, we may just tabulate the value function; in the continuous case, we must somehow discretize it. Therefore, we may define a grid of budget values, and solve a finite number of problems of the form (10.50) for points on the grid. Each optimization is solved by a numerical method. The numerical method to find $V_k(b_k)$ will often need values $V_{k+1}(b_{k+1})$ outside the grid. Values for state values outside the grid can be found by an interpolation strategy; a common interpolation strategy relies on cubic splines. A cubic spline is a piecewise polynomial function passing through a specified set of points: Each polynomial piece is of order three, and some essential continuity and differentiability properties are guaranteed. Given a set of points, the R function `splinefun` returns a function that may be used to interpolate outside the grid. An example if its use is given in the script in Fig. 10.20, which produces the spline depicted in Fig. 10.21.

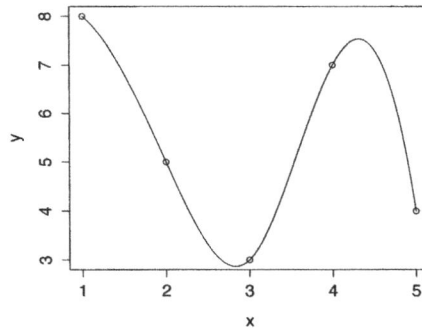

FIGURE 10.21 A cubic spline.

Figure 10.22 shows a script to solve the budget allocation problem using state space discretization and interpolation by cubic splines. The function receives the following inputs:

- The available `budget`.

- The list `funcList` containing the profit function for each activity; we need not use the same function for each activity.

- The number of points `numPoints` on the grid; the grid spacing is uniform and the same discretization is used for each value function.

Then, the function allocates a matrix of function values `valFunMatrix`, which will store the values of each value function on the grid. These values are used to specify one cubic spline for each decision stage; the list of cubic splines is contained in the list `splinesList`. The core of the procedure consists of two nested `for` loops, one over the decision stages and one over the points on the grid. For each point on the grid, we build the objective function as

```
objFun<-function(z) funcList[[t]](z)+splinesList[[t+1]](b-z)
```

Note that this objective function consists of the sum of the profit function for the current activity, if carried out at level `z`, and the next value function evaluated for the remaining budget `b-z`. The optimization is carried out numerically by the function `optimize`. This function can maximize over the bounded interval $[0, b]$, but it returns an error if $b = 0$; the code takes care of this case. After finding the whole sequence of value functions, we have a second part of the procedure, where we step forward and apply the optimal decisions stage by stage. The output consists of a list including the optimal solution and the overall profit, as well as the value functions on the grid. We may try and check the procedure by the following R script:

```
> budget <- 20
```

```
> numPoints <- 50
> f <- function(x) sqrt(x)
> funcList <- list(f,f,f)
> out <- Budgeting(budget,funcList,numPoints)
> out$par
[1]  6.666658  6.666673  6.666669
> out$value
[1]  7.745967
```

In this trivial case, the optimal solution and the corresponding profit are, trivially:

$$x_1^* = x_2^* = x_3^* = \frac{20}{3} \approx 6.666667, \quad \sum_{k=1}^{3} \sqrt{x_k^*} = 3 \times \sqrt{\frac{20}{3}} \approx 7.745967.$$

As we can see, we do find a sensible approximation of the optimal solution. Approximating the value function by splines is only one possibility. We may also use approaches based on collocation methods or the projection onto a space of elementary functions, such as polynomials, by linear regression. In a stochastic problem, function approximation must be coupled with a strategy to approximate the expectation, as we shall see later.

10.7.2 VALUE ITERATION FOR INFINITE-HORIZON PROBLEMS

When we deal with a finite-horizon problem, we may build a sequence of value functions iteratively, starting from a boundary condition. However, if we tackle an infinite-horizon problem, we face the genuine functional equation

$$V(\mathbf{s}) = \min_{\mathbf{x} \in X(\mathbf{s})} \left\{ C(\mathbf{s}, \mathbf{x}) + \gamma \mathrm{E}\big[V\big(S(\mathbf{s}, \mathbf{x}, \epsilon)\big)\big] \right\}. \tag{10.51}$$

We notice that this equation defines the value function $V(\mathbf{s})$ as a fixed point of an operator that involves the solution of an optimization problem. The basic approach to solve the above equation is value iteration. The approach is conceptually very simple to apply to problems with a finite state space and a discrete set of actions. Such problems are referred to as Markov decision processes (MDP), and we consider a toy example in the next section. Then we outline an extension to continuous-state problems and some ideas to speed up value iteration.

10.7.2.1 Value iteration for Markov decision processes

If the state space is finite, we may think of the value function as a vector $\mathbf{V} \in \mathbb{R}^n$, where n is the size of the state space. If the set $\mathcal{A}_i$ of possible decisions in each state $i = 1, \ldots, n$ is finite as well, we may associate each decision $a \in \mathcal{A}_i$ with a transition probability matrix P_{ij}^a, which gives the probability of moving from state i to state j if we choose action a. In this context, decisions

```
Budgeting <- function(budget,funcList,numPoints){
  ### INITIALIZATION: set up a simple grid
  budgetGrid <- seq(0,budget,length.out=numPoints)
  # prepare matrix of function values for each time step
  numSteps <- length(funcList)
  valFunMatrix <- matrix(0,nrow=numPoints,ncol=numSteps)
  splinesList <- vector("list",numSteps)
  # start from last step (increasing contribution functions)
  valFunMatrix[,numSteps] <- funcList[[numSteps]](budgetGrid)
  # now step backward in time
  for (t in seq(from=numSteps-1,to=1,by=-1)) {
    # build function adding immediate contribution and
    # value function t plus 1, at the next stage
    splinesList[[t+1]] <- splinefun(budgetGrid,
                             valFunMatrix[,t+1])
    for (k in 1:numPoints){
      b <- budgetGrid[k]
      objFun <- function(z) funcList[[t]](z)+
                          splinesList[[t+1]](b-z)
      if (b>0){
        outOptim <- optimize(objFun,lower=0,upper=b,
                          maximum=TRUE)
        valFunMatrix[k,t] <- outOptim$objective}
      else  valFunMatrix[k,t] <- objFun(0)
    }
  }
  # Now optimize forward in time and return values
  residualB <- budget;  objVal <- 0
  x <- numeric(numSteps)
  for (t in 1:(numSteps-1)){
    objFun <- function(z) funcList[[t]](z)+
                        splinesList[[t+1]](residualB-z)
    if (residualB > 0){
      outOptim <- optimize(objFun,lower=0,upper=residualB,
                          maximum=TRUE)
      x[t] <- outOptim$maximum}
    else  x[t] < 0
    objVal <- objVal + funcList[[t]](x[t])
    residualB <- residualB - x[t]
  }
  x[numSteps] <- residualB
  objVal <- objVal + funcList[[numSteps]](x[numSteps])
  # assemble output
  out <- list(par=x,value=objVal,valFuns=valFunMatrix)
}
```

FIGURE 10.22 **Using state space discretization and splines to approximate the value functions of the budget allocation problem.**

are typically referred to as *actions*. The transition probability matrix defines a Markov chain, and it is easy to compute the expectation in the functional equation. Solving the optimization problem is easy as well (unless we are dealing with a huge state space), as it may be tackled by explicit enumeration.

The value iteration algorithm for such a problem can be stated as follows, where V_i^n, collected into vector $\mathbf{V}^n$, is the value function approximation for state $i \in \mathcal{S}$ at iteration n:

Step 0. Initialization:

- Set $V_i^0 = 0$ for all states $i \in \mathcal{S}$.
- Choose a tolerance parameter ϵ.
- Set the iteration counter $n = 1$.

Step 1. For each state i compute

$$V_i^n = \min_{a \in \mathcal{A}_i} \left\{ C(i, a) + \gamma \sum_{j \in \mathcal{S}} P_{ij}^a V_j^{n-1} \right\} \qquad (10.52)$$

Step 2. If the termination condition

$$\| \mathbf{V}^n - \mathbf{V}^{n-1} \| < \epsilon \frac{1 - \gamma}{2\gamma}$$

is met, stop and return the policy that solves problem (10.52), as well as the approximation $\mathbf{V}^n$; otherwise, increment the iteration counter $n = n + 1$ and repeat step 1.

The value iteration algorithm is best illustrated by an example. Figure 10.23 depicts a toy problem that serves well as an illustration.[27] The state space consists of four states:

- When we are at state 1, we stay there with probability 0.7, and we will move to state 2 with probability 0.3. In state 1, there is no decision to make.
- In state 2, we have a choice between two actions. We may "flush" the state and move back to state 1, earning an immediate reward of 10; if we do not flush, the next state is random: At the next time instant we may still be there, or move on to state 3, with probabilities 0.8 and 0.2, respectively.
- In state 3, a similar consideration applies, but if we flush, we get a larger immediate reward of 15; the probability of moving on to a better state is 0.1, which is lower for state 3 than for state 2.

---

[27]The example is borrowed from [37, p. 107]; since it is problem 3.11 in that book, we will add the PW311 prefix to R functions to remark this fact.

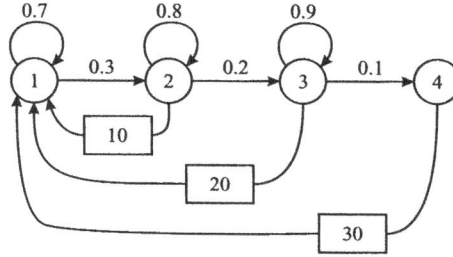

FIGURE 10.23 **A simple infinite-horizon MDP.**

- If we are patient enough and we reach state 4, then we may move back to square 1 and earn an immediate reward of 20.

Clearly, there is a trade-off between flushing the state immediately and waiting for better opportunities, which is not too different from early exercise decisions in American-style derivatives, where we have to compare the immediate payoff from early exercise against the continuation value. Since this is an infinite-horizon problem, we also need to define a discount factor, say, $\gamma = 0.8$.

Ths script of Fig. 10.24 implements value iteration for this toy example. In this case, we modify value iteration by maximizing the value with respect to the available actions, rather than minimizing it. The following snapshot shows the optimal solution that we find by its application:

```
> probs = c(.7,.8,.9,0)
> payoff = c(0,10,20,30)
> discount=.8
> out <- PW311_ValueIteration(probs, payoff, discount)
> out
$stateValues
[1]   9.677415 17.741931 27.741931 37.741931
$policy
[1]  0 1 1 0
```

When the action is 0, we wait and hope to move to a better state; if the action is 1, we grab the reward and move back to square 1. Note that, for the sake of simplicity, we do not use full state transition probabilities as the input; we only use the probability of remaining in each state if we wait and see. We observe that we should always flush the state, when we have to make a decision. A good check on the values of states 2, 3, and 4 is that they differ by 10, which is exactly the increment in the payoff if we start in a better state and apply the optimal policy. With the above data, the best policy is greedy, in the sense that we should always go for the immediate payoff, without waiting for better opportunities. This depends on two factors: the low probability of moving to a better state and the heavy discount factor. Let us play with these numbers to check our intuition:

```
PW311_ValueIteration <- function(probs, payoff, discount,
                            eps=0.00001, maxIter=1000){
  # Initialization
  numStates <- length(payoff)
  oldV <- numeric(numStates)
  newV <- numeric(numStates)
  myeps <- eps*(1-discount)/2/discount
  stopit <- FALSE
  count <- 0
  # precompute invariant discounted probs
  dprobStay <- discount*probs
  dprobMove <- discount*(1-probs)
  # we have to decide only in interior states
  indexVet <- 2:(numStates-1)
  norm2 <- function(v) {return(sqrt(sum(v^2)))}
  # value iteration
  while (!stopit){
    count <- count + 1
    # update: first and last state are apart
    newV[1] <- dprobStay[1]*oldV[1]+dprobMove[1]*oldV[2]
    valueFlush <- payoff[indexVet] + discount*oldV[1]
    valueWait <- dprobStay[indexVet]*oldV[indexVet]+
                 dprobMove[indexVet]*oldV[indexVet+1]
    newV[indexVet] <- pmax(valueWait,valueFlush )
    newV[numStates] <- payoff[numStates] + discount*oldV[1]
    if ((max(newV-oldV)<myeps) || (count > maxIter))
      stopit <- TRUE
    else
      oldV <- newV
  }
  # return state values and policy vector (1->flush is true)
  return(list(stateValues=newV,
              policy=c(0,valueFlush>valueWait,1)))
}
```

FIGURE 10.24 **Numerical value iteration for a simple MDP.**

```
> probs = c(.6,.6,.6,0)
> payoff = c(0,10,20,30)
> discount=.95
> out <- PW311_ValueIteration(probs, payoff, discount)
> out
$stateValues
[1] 60.51969 68.48281 77.49371 87.49371
$policy
[1] 0 0 1 0
```

```
> probs = c(.6,.6,.6,0)
> payoff = c(0,10,20,30)
> discount=.99
> out <- PW311_ValueIteration(probs, payoff, discount)
> out
$stateValues
[1] 342.1122 350.7518 359.6096 368.6910
$policy
[1] 0 0 0 0
```

It is important to notice that, in this very simple setting, we may easily compute the expectation, since it involves a finite sum with very few terms. In other settings, this may be very hard to do, possibly because of the sheer size of the state space, or because the dynamics are complicated and it is very difficult to find the transition probability matrix. Later, we will consider approximate dynamic programming approaches that rely on Monte Carlo simulation and do not require an explicit transition probability matrix. Monte Carlo simulation is also quite useful to simulate a policy which is implicit in the value function. The code in Fig. 10.25 receives the problem data, as well as the initial state and the value function, and it estimates the value of the optimal policy over a given number of steps. Due to discounting, there is little point to simulate a large number of steps. For instance $0.8^{40} = 0.0001329228$. Thus, `numRepl` complete replications should be run, which may also be used to come up with a confidence interval.

10.7.2.2   Value iteration for continuous-state problems

When dealing with a continuous-state problem, we have to adopt a suitable discretization strategy for the state space, but also for the value function. Cubic splines are one possible choice, but a more general idea is to approximate the value function by a linear combination of simple building blocks:

$$V(\mathbf{s}) \approx \sum_{j=1}^{n} c_j \phi_j(\mathbf{s}).$$

The functions $\phi_j(\cdot)$ are called basis functions, and they allow us to boil down the infinite-dimensional functional equation (10.51) to the finite-dimensional problem of finding the coefficients c_j, $j = 1, \ldots, n$. This kind of approach is quite common in numerical methods for functional equations, including partial differential equations and integral equations, and it is called the *collocation method*. In this context, the points in the discretized state space are called collocation nodes. Given collocation nodes $\mathbf{s}_1, \ldots, \mathbf{s}_n$, where now the subscript refers to the discretization and not time, the Bellman equation for each state $\mathbf{s}_i$

```
PW311_Simulate <- function(probs,payoff,discount,startState,
                           numRepl,numSteps,policy){
  numStates <- length(probs)
  sumRewards <- numeric(numRepl)
  for (k in 1:numRepl){
      state <- startState
      dRewards <- numeric(numSteps)
      discf <- 1
      for (t in 1:numSteps){
          if (state == 1){
              # state 1, no choice
            if (runif(1) > probs[state])
              state <- 2
          }
          else if (state == numStates){
              # last state, get reward and back to square 1
              state <- 1
              dRewards[t] <- discf*payoff[numStates]
          }
          else {
              # apply policy
              if (policy[state] == 1){
                  # get reward and go back to square 1
                  dRewards[t] <- discf*payoff[state]
                  state <- 1
              }
              else {
                  # random move
                  if (runif() > probs[state])
                      state <- state + 1
              }
          } # end if on states
          discf <- discf * discount
      } # end single replication
      sumRewards[k] <- sum(dRewards)
  } # end replications
  # return point estimate and confidence interval (95%)
  aux <- t.test(sumRewards)
  value <- as.numeric(aux$estimate)
  ci <- as.numeric(aux$conf.int)
  return(list(estimate=value,conf.int=ci))
}
```

FIGURE 10.25 **Simulating a policy for a simple MDP.**

reads:[28]

$$\sum_{j=1}^{n} c_j \phi_j(\mathbf{s}_i) = \min_{\mathbf{x} \in X(\mathbf{s}_i)} \left\{ C(\mathbf{s}_i, \mathbf{x}) + \gamma \, \mathrm{E}\left[\sum_{j=1}^{n} c_j \phi_j\big(S(\mathbf{s}_i, \mathbf{x}, \epsilon) \big) \right] \right\},$$

$$i = 1, \ldots, n. \quad (10.53)$$

The reader may notice that if the number of collocation nodes and the number of basis functions are indeed the same, Eq. (10.53) is a set of n nonlinear equations in the n unknown coefficients c_j. We may rewrite it as

$$\mathbf{\Phi} \mathbf{c} = \boldsymbol{\nu}(\mathbf{c}),$$

where $\mathbf{\Phi}$ is the collocation matrix with elements $\Phi_{ij} = \phi_j(\mathbf{s}_i)$, and the collocation function $\boldsymbol{\nu}(\mathbf{c})$ is a vector function with components

$$\nu_i(\mathbf{c}) = \min_{\mathbf{x} \in X(\mathbf{s}_i)} \left\{ C(\mathbf{s}_i, \mathbf{x}) + \gamma \, \mathrm{E}\left[\sum_{j=1}^{n} c_j \phi_j\big(S(\mathbf{s}_i, \mathbf{x}, \epsilon) \big) \right] \right\}.$$

To solve the nonlinear equations, we may use the Newton's method. When dealing with a single equation $f(x) = 0$, where $x \in \mathbb{R}$, the method consists of the following iterative scheme:

$$x^{(k+1)} = x^{(k)} - \frac{f\big(x^{(k)}\big)}{f'\big(x^{(k)}\big)}.$$

When dealing with a system of linear equations, which can be represented as $\mathbf{F}(\mathbf{x}) = \mathbf{0}$, where $\mathbf{F} : \mathbb{R}^n \longrightarrow \mathbb{R}^n$, the division by the first-order derivative is replaced by the premultiplication by the inverse of the Jacobian matrix:[29]

$$\mathbf{x}^{(k+1)} = \mathbf{x}^{(k)} - \left[\mathbf{J}\big(\mathbf{x}^{(k)}\big) \right]^{-1} \mathbf{F}\big(x^{(k)}\big).$$

We recall that the entries of the Jacobian matrix are the first-order derivatives of each component of the vector function $\mathbf{F}(\cdot)$ with respect to each element of the vector variable $\mathbf{x}$. In our case, we must take a derivative of a function defined by an optimization operator. To this end, we need a result in sensitivity analysis of optimization problems, known as the envelope theorem.[30] Consider an optimization problem depending on a vector $\boldsymbol{\alpha} \in \mathbb{R}^m$ of parameters:

$$\min_{\mathbf{x} \in X} f(\mathbf{x}; \boldsymbol{\alpha}).$$

---

[28]The treatment here follows [34], to which we refer for economic and financial application examples.

[29]In practice, the Jacobian matrix is not inverted, and a system of linear equations is solved; see, e.g., [7].

[30]See, e.g., [43].

The solution of the problem yields an optimal solution $\mathbf{x}^*(\boldsymbol{\alpha})$ depending on the parameters, as well as an optimal cost

$$g(\boldsymbol{\alpha}) \equiv f(\mathbf{x}^*(\boldsymbol{\alpha}); \boldsymbol{\alpha}).$$

We may be interested in investigating the sensitivity of $g(\cdot)$ with respect to the parameters. The envelope theorem, under some suitable conditions, yields a surprisingly simple result:

$$\frac{\partial g}{\partial \alpha_j} = \frac{\partial f}{\partial \alpha_j}(\mathbf{x}^*(\boldsymbol{\alpha}); \boldsymbol{\alpha}), \qquad j = 1, \ldots, m.$$

By putting all of this together, we come up the iteration scheme

$$\mathbf{c} \leftarrow \mathbf{c} - \big[\boldsymbol{\Phi} - \boldsymbol{\Theta}(\mathbf{c})\big]^{-1}\big[\boldsymbol{\Phi}\mathbf{c} - \boldsymbol{\nu}(\mathbf{c})\big],$$

where the Jacobian matrix $\boldsymbol{\Theta}(\mathbf{c})$ may be computed by resorting to the envelope theorem,

$$\theta_{ij}(\mathbf{c}) = \frac{\partial \nu_i}{\partial c_j}(\mathbf{c}) = \gamma\,\mathrm{E}\Big[\phi_j\big(\mathbf{g}(\mathbf{s}_i, \mathbf{x}_i^*, \epsilon)\big)\Big],$$

and $\mathbf{x}_i^*$ is the optimal decision for the optimization problem that we solve to find the value of the collocation function $\nu_i(\mathbf{c})$ corresponding to state $\mathbf{s}_i$.

We are left with one issue: How can we approximate the expectation, assuming the shock ϵ is a continuous random variable? We may use the scenario generation approaches that we have considered in Section 10.5.3. For instance, we may use a Gaussian quadrature formula with weights (probabilities) π_k and nodes (discretized shock values) ϵ_k, respectively, $k = 1, \ldots, K$. Therefore, in order to evaluate the collocation function, for each collocation node $\mathbf{s}_i$ we have to solve the following optimization problem:

$$\nu_i(\mathbf{c}) = \min_{\mathbf{x} \in X(\mathbf{s}_i)} \left\{ C(\mathbf{s}_i, \mathbf{x}) + \gamma \sum_{k=1}^{K}\sum_{j=1}^{n} \pi_k c_j \phi_j\big(S(\mathbf{s}_i, \mathbf{x}, \epsilon_k)\big) \right\},$$

with Jacobian

$$\theta_{ij}(c) = \gamma \sum_{k=1}^{K} \pi_k \phi_j\big(S(\mathbf{s}_i, \mathbf{x}_i, \epsilon_k)\big).$$

We will not delve into this approach any further, but the message is that, by choosing a suitable set of basis functions, we may approximate the value function. The idea can also be applied when the dimensionality of the random shocks precludes the application of quadrature formulas and we have to resort to Monte Carlo sampling.

10.7.2.3 Speeding up value iteration

Value iteration is a fairly straightforward approach, whose convergence is not too difficult to prove in theory. However, it may suffer from slow convergence in practice. There are a few ways to speed convergence up, and we refer to the end of chapter references. Here we just want to point out one intrinsic weakness of value iteration. The value function is instrumental in determining the optimal policy in feedback form. It may happen that we do find the optimal policy relatively early, but the value function takes many more steps to converge, as it may take several iterations to assess the value of the optimal policy. Thus, we may try to improve convergence by a better assessment of the value of a policy. This leads to policy iteration approaches. Due to space limitations, we cannot illustrate policy iteration, but we may mention that value and policy iteration can be integrated within approximate dynamic programming strategies.

10.7.3 A NUMERICAL APPROACH TO CONSUMPTION–SAVING

In Section 1.4.3 we have used a simple life-cycle consumption–saving problem to motivate stochastic DP. Here we consider a little variation on the theme in order to illustrate the application of numerical methods to a simple but interesting problem. The additional feature that we introduce is a stylized model of income uncertainty based on a discrete-time Markov chain. Let us summarize our modeling assumptions:

- Time is discretized by time instants indexed by $t = 0, 1, \ldots, T$, where the time horizon T is assumed deterministic. In more realistic models, quite common in pension economics, the time horizon is a random variable whose distribution is given by mortality tables.

- At time $t = 0$, the decision maker (or agent) is endowed with an initial wealth W_0 and has to decide the consumed amount C_0. It is not possible to borrow money, so $0 \leq C_0 \leq W_0$. What is not consumed is the saved amount, $S_0 = W_0 - C_0$.

- A second decision is the allocation of S_0 between a risky and a risk-free asset; let $\alpha_0 \in [0, 1]$ be the fraction allocated to the risky asset, R_1 the rate of return of the risky asset, and r_f the rate of return of the risk-free asset. Note that we use R_1 rather than R_0 to emphasize that this piece of information is not known when the decision is made; in other words, return R_1 is realized over the time period $t = 1$, defined by the pair of time instants $(0, 1)$ (see Fig. 10.19). For the sake of simplicity, both the risk-free rate and the distribution of risky return are assumed constant over time. There is no intertemporal dependence between risky returns, which are assumed independent.

- At time instant $t = 1$, the available wealth W_1 is the sum of capital and labor (noncapital) income. The capital depends on the realized portfolio return:

$$S_0 \left[\alpha_0(1 + R_1) + (1 - \alpha_0)(1 + r_f) \right]$$
$$= (W_0 - C_0) \left[1 + r_f + \alpha_0(R_1 - r_f) \right].$$

The labor income is L_1, which may take one of three values depending on the state of employment. Again, our notational choice emphasizes that this piece of information is not known when the decision is made at $t = 0$.

- The state of employment at time t, denoted by λ_t, may take three values in the set $\mathcal{L} = \{\alpha, \beta, \eta\}$, and $L_\alpha > L_\beta > L_\eta$. We may interpret η as "unemployed," α as "fully-employed," and β as an intermediate situation. The dynamics of this state is modeled by a matrix of time-independent transition probabilities with elements

$$\pi_{ij} = \mathrm{P}\left\{\lambda_{t+1} = j \mid \lambda_t = i\right\}, \qquad i, j \in \mathcal{L}. \tag{10.54}$$

The initial employment state λ_0 is given, and the corresponding income is already included in the initial wealth W_0.

- An immediate utility $u(C_t)$ is obtained from consumption.
- All of the above holds, with the natural adjustment of time subscripts, for time periods up to time $t = T - 1$. At time $t = T$, the terminal wealth W_T is fully consumed; the model could be extended by accounting for some utility from bequest. This implies that terminal wealth is completely consumed, $C_T = W_T$.
- The overall objective is to maximize total expected utility

$$\max \mathrm{E}\left[\sum_{t=0}^{T} \gamma^t u(C_t)\right],$$

where $\gamma \in (0, 1)$ is a subjective discount factor.

In order to tackle the above problem within a dynamic programming framework, we have to specify the state variables. In our case, the natural choice is the pair

$$\mathbf{s}_t = (W_t, \lambda_t).$$

The peculiarity of this model is the mixed nature of the state, which is continuous with respect to the first component and discrete with respect to the second one. Dynamic programming requires to find the set of value functions

$$V_t(W_t, \lambda_t), \qquad t = 1, 2, \ldots, T - 1,$$

with the boundary condition

$$V_T(W_T, \lambda_T) = u(W_T).$$

The general form of the functional equation is

$$V_t(W_t, \lambda_t) = \max_{0 \le C_t \le W_t, 0 \le \alpha_t \le 1} \left\{u(C_t) + \gamma\, \mathrm{E}_t\left[V_{t+1}(W_{t+1}, \lambda_{t+1})\right]\right\}, \tag{10.55}$$

where the notation $\mathrm{E}_t[\cdot]$ points out that expectation is conditional on the current state. The dynamic equations for state transitions are

$$\lambda_{t+1} = M_\Pi(\lambda_t), \tag{10.56}$$

$$W_{t+1} = L_{t+1}(\lambda_{t+1}) + (W_t - C_t)\left[1 + r_f + \alpha_t(R_{t+1} - r_f)\right], \tag{10.57}$$

where M_Π in Eq. (10.56) represents the stochastic evolution of the employment state according to the matrix Π collecting the transition probabilities of Eq. (10.54), and the evolution of wealth in Eq. (10.57) depends on the random rate of return of the risky asset and the labor income L_{t+1} which is a function of employment state λ_{t+1}. Note that we are also assuming that the employment state evolution and the return from the risky asset are independent. If we assume that both financial returns and employment depend on underlying macroeconomic factors, we should account for their correlation.

Now, in order to solve the functional equation (10.55) numerically, we have to specify the following three pieces of the puzzle:

1. The discretization of the state space
2. The approximation of the value function at each time instant
3. The discretization of the conditional expectation

Then, we will state the corresponding optimization model to be solved at each point in the grid and prepare R code to solve the DP recursion and simulate the optimal policy.

State space discretization

We have to discretize only the wealth component of the state variable. The simplest idea, as we have seen in Section 10.7.1, is to use a uniform grid that does not depend on time. This may be accomplished by setting up a grid of wealth values $W^{(k)}$, $k = 0, 1, \ldots, K$, where $W^{(0)} = 0$ and $W^{(K)} = \overline{W}$ is some sensible upper bound on wealth. One might wonder if such a bound exists, if the probability distribution of return of the risky asset has infinite support; depending on how the value function is approximated, we may need a way to *extrapolate* outside the grid, whereas in the budget allocation problem we only have to interpolate inside the grid. Another issue is if we should really use the same bound for all time periods, as wealth might increase; adaptive grid may be used to improve accuracy and/or efficiency. However, in our case, the agent consumes wealth along the way and unlike pension models, we have no accumulation–decumulation phase. Hence, for the sake of simplicity, we refrain from using adaptive grids and just use a time-invariant grid of states:

$$ \mathbf{s}^{k,i} = \left(W^{(k)}, i \right), \qquad k = 0, 1, \ldots, K; \; i \in \{\alpha, \beta, \eta\}, $$

where we suppress dependence on time to ease notation.

Approximation of the value function

To approximate the value function, we use cubic splines, as we did in Section 10.7.1. The value function depends on two state variables, but the second one is discrete and can only take three values. Hence, we associate three cubic splines

$$ v_{t,\alpha}(w), \quad v_{t,\beta}(w), \quad v_{t,\eta}(w) $$

with each time bucket, depending only on wealth. The nodes for these splines are defined by the discretized grid $\mathbf{s}^{k,i}$. As we have already pointed out, we may need to extrapolate the splines for large values of wealth. Whatever choice we take, there is a degree of arbitrariness, but some standard choices are available in R, in the form of options for use with the function `splinefun`:

- If the `type` option is set to `natural`, a natural cubic spline is built, and extrapolation is linear, using the slope of nearest point on the grid.
- If the choice is `monoH.FC`, interpolation based on Hermite polynomials is carried out, ensuring the monotonicity in the data, if present.
- The `hyman` is similar to the previous one, but cubic polynomials are selected.

Discretization of the conditional expectation

We have to discretize only the dynamics described by Eq. (10.57), which in turn requires generating suitable return scenarios for the risky asset. Let us assume that the price of the risky asset is modeled by geometric Brownian motion. If we assume that time periods correspond to one year, this means that consecutive prices of the asset are related by

$$P_{t+1} = P_t \exp\left[\left(\mu - \frac{\sigma^2}{2}\right) + \sigma\epsilon\right],$$

where $\epsilon \sim \mathsf{N}(0, 1)$. This in turn implies that the annual rate of return is

$$R_{t+1} = \frac{P_{t+1} - P_t}{P_t} = \exp\left[\left(\mu - \frac{\sigma^2}{2}\right) + \sigma\epsilon\right] - 1.$$

All we need, then, is to discretize the standard normal distribution, e.g., by Gaussian quadrature, or by any other scenario generation strategy that we have discussed in Section 10.5.3.5. Whatever approach we select, we will define scenarios characterized by realizations ϵ_l, with corresponding rate of return R_l, and probabilities q_l, $l = 1, \ldots, L$. Note that the subscript l has nothing to do with time, as we assume that the rates of return of the risky asset for different time periods are independent and identically distributed. Note that μ is related to the continuously compounded return on the risky asset, but the risk-free rate r_f must be expressed with annual compounding for consistency in Eq. (10.57).

The optimization model

After all of the above choices, we are finally ready to state the discretized optimization model corresponding to the single-step problem (10.55). At time t, we have one such problem for each point in the state grid $\mathbf{s}^{k,i}$, where wealth available is $W_t = W^{(k)}$ and the current employment state is $\lambda_t = i \in \mathcal{L}$. We have discretized the return of the risky asset with L scenarios, associated with probability q_l and rate of return R_l, and there are three possible employment states; hence, after making decisions C_t and α_t, the next state in the optimization model is one of $L \times 3$ future scenarios, indexed by (lj) and characterized by available wealth $W_{t+1}^{(lj)}$. Since the two risk factors are assumed independent, each probability is given by $q_l \times \pi_{ij}$. Hence, the optimization model $\mathcal{P}(\mathbf{s}^{k,i})$, conditional on the state at time t, reads as follows:

$$\max \quad u(C_t) + \beta \sum_{l=1}^{L} q_l \left[\sum_{j \in \mathcal{L}} \pi_{ij} \, v_{t+1,j} \left(W_{t+1}^{(lj)} \right) \right] \tag{10.58}$$

$$\text{s.t.} \quad W_{t+1}^{(lj)} = L_j + \left(W^{(k)} - C_t \right) \left[1 + r_f + \alpha_t (R_l - r_f) \right],$$

$$l = 1, \dots, L; \, j \in \mathcal{L}, \tag{10.59}$$

$$0 \le C_t \le W^{(k)},$$

$$0 \le \alpha_t \le 1,$$

where the objective function requires knowledge of the spline functions $v_{t+1,j}(\cdot)$ for the next time period. At the terminal time instant T the available wealth is fully consumed. Therefore, the boundary conditions are

$$V_T \left(W^{(k)}, i \right) = u \left(W^{(k)} \right) \qquad \forall \mathbf{s}^{k,i}.$$

We solve the functional equation backward to find the approximate value functions for each time period $t = T - 1, T - 2, \dots, 1$. To find the initial decision, we solve the above problem at time instant $t = 0$, with initial wealth W_0 and current employment state λ_0; in the solution we use the splines $v_{1,j}$.

Implementation in R

The code implementing numerical DP for the consumption–saving problem is shown in Fig. 10.26. The function `DPSaving` receives the following input arguments:

- The initial states `W0` and `empl0`.
- The utility function `utilFun` and the discount factor `gamma`.
- The number of time periods `numSteps`.
- The specification of the wealth grid, consisting of `numPoints` points ranging from `Wmin` to `Wmax`.
- The vector `income` containing the income for each state, and the corresponding transition matrix `transMatrix`.
- The return scenarios `retScenarios` and the corresponding probabilities `probs` for the risky asset, as well as the risk free rate `rf`.
- The `splineOption` parameter specifying the type of spline, which is natural by default, but could be monotonicity preserving.

```
DPSaving <- function(W0,empl0,utilFun,gamma,numSteps,
                numPoints,Wmin,Wmax,income,transMatrix,
                retScenarios,probs,rf,splineOption="natural"){
  # initialization
  numScen <- length(probs)
  wealthValues <- seq(Wmin,Wmax,length.out=numPoints)
  valFunMatrix <- matrix(0,nrow=numPoints,ncol=3)
  valFunsList <- vector("list",numSteps)
  # initialize last time period
  auxFunsList <- vector("list",3)
  for (i in 1:3) auxFunsList[[i]] <- utilFun
  valFunsList[[numSteps]] <- auxFunsList
  # backward recursion for each time period
  for (t in seq(from=numSteps-1,to=1,by=-1)) {
    # prepare nested list of splinefuns
    auxFunsList <- vector("list",3)
    valFunsList[[t]] <- auxFunsList
    # nested for loop on the grid
    for (i in 1:3){
      for (k in 1:numPoints){
        # build objective function
        objFun <- function(x){
            C <- x[1]; alpha <- x[2]; out <- 0
            for (j in 1:3){
              for (l in 1:numScen){
                futWealth <- income[j]+(wealthValues[k]-C)*
                  (1+rf+alpha*(retScenarios[l]-rf))
                out <- out+probs[l]*transMatrix[i,j]*
                  valFunsList[[t+1]][[j]](futWealth)}}
            out <- utilFun(C)+gamma*out
            return(-out)}
        # solve optimization problem
        x0 <- c(wealthValues[k]/2,0.5)
        outOptim <- nlminb(start=x0,objective=objFun,
                    lower=c(0,0),upper=c(wealthValues[k],1))
        valFunMatrix[k,i] <- -outOptim$objective
      }
      # use function values and build spline
      valFunsList[[t]][[i]] <- splinefun(wealthValues,
                      valFunMatrix[,i],method=splineOption)
    } # end for on wealth values
  } # end for loop on employment states
  return(valFunsList)
}
```

FIGURE 10.26 **Numerical DP for the consumption–saving problem.**

The output of the function is the list valFunsList. This is a nested list containing, for each time period, a list of three spline functions, one for each employment state. To generate the value functions, the code does the following:

- It prepares the vector `wealthValues` containing the wealth values on the grid.
- It allocates the matrix `valFunMatrix`, which stores the value functions at each wealth value on the grid, for each employment state; these values are used to build the spline functions.
- `valFunsList[[numSteps]]` is the list of value functions for the last time instant, and it consists of a three-element list referring to the utility function (at the last stage, whatever the employment state, the agent consumes the avaliable wealth).
- Then, the main loop goes backward in time preparing the objective function at each time step, according to the optimization moodel (10.58); the objective function `objFun` depends on the two decision variables and is the sum of immediate utility and the discounted expected value of the next stage value function.
- The model is solved by `nlminb`, which minimizes a nonlinear function subject to bounds; note that we change the sign of `objFun`, as it has to be maximized; the initial solution for the optimization solver is arbitrarily chosen (we consume 50% the available wealth and 50% of saving is allocated to the risky asset).
- `valFunMatrix`, on exit from the loop for each time stage, stores the values of each state; this is used to build the three splines, one per employment state, by the selected method (natural spline by default); the result is assigned to `valFunsList[[t]][[i]]`.

The code in Fig. 10.27 is used to run the numerical DP. The function `MakeScenariosGQ` generates scenarios of risky asset returns using Gaussian quadrature. We use 10 quadrature nodes, and the state grid consists of 30 points. It is important to plot some value functions to check that they make sense, which also depends on our selection of grid boundaries and the type of spline. Note that the natural spline can be used to extrapolate outside the grid, with a slope corresponding to the closest boundary point on the grid. Figure 10.28 shows some value functions in the case of a logarithmic utility function. The functions look concave and increasing, which is what we expect. We may also consider a power utility function

$$u(C) = \frac{C^{1-\rho}}{1-\rho}.$$

The degree of risk aversion of the power utility function depends on the coefficient ρ. When $\rho = 1$, the power utility boils down to the logarithmic one. A moderate degree of risk aversion is associated to $\rho = 5$.[31] In Fig. 10.29 we show

---

[31] See [10, Chapter 7].

```
MakeScenariosGQ <- function(mu,sigma,numScenarios){
  require(statmod)
  aux <- gauss.quad.prob(numScenarios,dist="normal",
                         mu=mu-sigma^2/2,sigma=sigma)
  probRets <- aux$weights
  riskyRets <- exp(aux$nodes)-1
  out <- list(retScenarios=riskyRets,probs=probRets)
}

utilFun <- function(x) log(x); gamma <- 0.97
numSteps <- 10; W0 <- 100
mu <- 0.07; sigma <- 0.2; rf <-0.03
empl0 <- 2; income <- c(5,20,40)
transMatrix <- matrix(
  c(0.6,0.3,0.1,0.2,0.5,0.3,0.2,0.1,0.7),
  nrow=3,byrow=TRUE)
Wmin <- 1; Wmax <- 500; numPoints <- 30
numScenarios <- 11
scen <- MakeScenariosGQ(mu,sigma,numScenarios)
retScenarios <- scen$retScenarios
probs <- scen$probs
valFuns <- DPSaving(W0,empl0,utilFun,gamma,numSteps,
               numPoints,Wmin,Wmax,income,transMatrix,
               retScenarios,probs,rf)

x=Wmin:Wmax
par(mfrow=c(2,2))
plot(x,valFuns[[10]][[2]](x),type='l',ylab='time=10')
plot(x,valFuns[[9]][[2]](x),type='l',ylab='time=9')
plot(x,valFuns[[5]][[2]](x),type='l',ylab='time=5')
plot(x,valFuns[[1]][[2]](x),type='l',ylab='time=1')
par(mfrow=c(1,1))
```

FIGURE 10.27 Running the function DPSaving and plotting selected value functions.

the impact of the choice of spline function, for a power utility with $\rho = 5$. The value functions refer to $t = 9, 10$ and employment state β, and have been built using both the natural and the monotonicity preserving splines, corresponding to spline option monoH.F. We notice that the power utility at the end of the time horizon $t = 10$ displays larger risk aversion than the logarithmic utility. When we step back to $t = 9$, the natural spline yields an unreasonable function featuring some spurious oscillations. The monotonicity preserving spline, on the contrary, avoids this behavior. This shows that some care must be taken when choosing an interpolation strategy, especially with long time horizons, since errors tend to cumulate when we proceed backward in time.

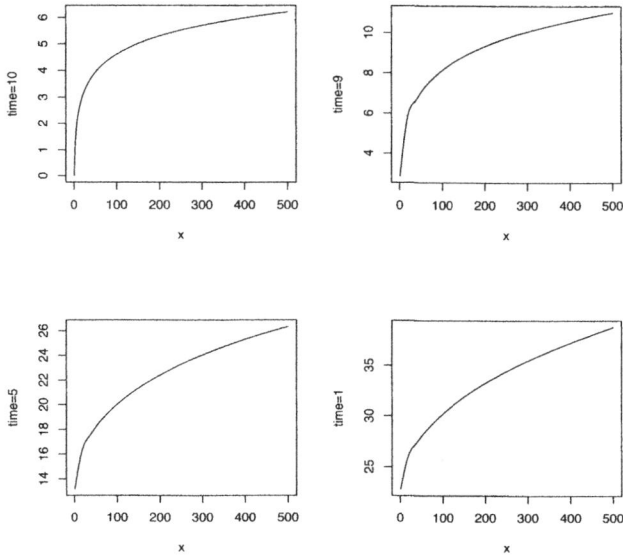

FIGURE 10.28 **Value functions for logarithmic utility.**

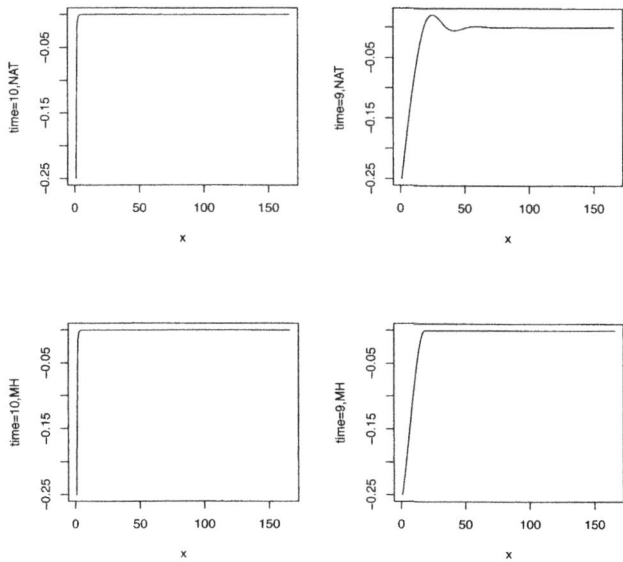

FIGURE 10.29 **Value functions for power utility.**

Simulation of the resulting policy

As we have remarked before, a notable advantage of stochastic DP is that we
obtain a set of value functions that can be used to simulate a sequence of decisions out of sample, e.g., in scenarios that are different from those used to learn
a policy. We might even check robustness against uncertainty in the probability distribution itself. In this section we use Monte Carlo runs to compare the
performance of the DP policy against a simple rule based on fixed decisions.
According to the fixed rule, at each time instant from $t = 0$ to $t = T - 1$ a
fixed percentage of the available wealth is consumed, and a fixed percentage
of savings is allocated to the risky asset; at the end of the planning horizon,
the whole wealth is consumed. The first thing we need is a piece of code to
generate scenarios of risky asset returns and employment states. This is accomplished by the code in Fig. 10.30. The returns are generated by sampling
a geometric Brownian motion, and the stochastic process for the employment
state is a discrete-time Markov chain.[32] The output sample paths are collected
in a list consisting of two matrices with `numRepl` rows and `numSteps` columns.
The code in Fig. 10.31 receives these scenarios and returns a list including the
discounted utility, as well as the corresponding sample paths for α_t, C_t, and W_t.
Clearly, α_t is fixed here, but we prefer to return the same output as the code in
Fig. 10.32, which solves an optimization model at each step.

Figure 10.33 reports a script to compare the optimal DP and the fixed policy, using the same out-of-sample scenarios (100 scenarios are generated to this
purpose). These out-of-sample scenarios should not be confused with the scenarios that are used to solve the DP recursion, as well as each single-stage decision problem. We may compare the average utility achieved, for the logarithmic
utility function as follows:

```
> mean(utilValsFixed)
[1] 31.28172
> mean(utilValsDP)
[1] 32.3301
> t.test(utilValsDP,utilValsFixed)
Welch Two Sample t-test
t = 2.9255, df = 196.126, p-value = 0.003845
alternative hypothesis: true difference in means
is not equal to 0
```

The difference does not look huge, even though a paired t-test suggests that it is
statistically significant. Actually, comparing utility functions may be misleading, as they are meant as ordinal, rather than cardinal utility. If we repeat the
experiment for the power utility function, with $\rho = 5$, we obtain an apparently
different picture:

```
> mean(utilValsFixed)
[1] -6.53674e-05
```

---

[32] See Section 14.2 for more details.

```
sampleEmployment <- function(empl0,cumProbs,numSteps){
  # cumProbs are cumulated probs for efficiency
  state <- c(empl0, numeric(numSteps))
  for (t in 1:numSteps){
    u <- runif(1)
    state[t+1] <- sum(u > cumProbs[state[t],])+1
  }
  return(state[-1])
}

MakeSamplePaths <- function(empl0,transMatrix,mu,sigma,
                            numSteps,numRepl){
  # cumulate transition probs once
  cumProbs <- t(apply(transMatrix,1,cumsum))
  emplPaths <- matrix(0,nrow=numRepl,ncol=numSteps)
  retPaths <- matrix(0,nrow=numRepl,ncol=numSteps)
  for (k in 1:numRepl){
    # sample employment scenario
    emplPaths[k,] <-
            sampleEmployment(empl0,cumProbs,numSteps)
    # sample return scenario
    retPaths[k,] <-
            exp(rnorm(numSteps,mu-sigma^2/2,sigma))-1
  }
  return(list(emplPaths=emplPaths,retPaths=retPaths))
}
```

R programming notes:

1. The function `MakeSamplePaths` uses `sampleEmployment` as an auxiliary function to generate a single sample path of employment states, not including the initial one.

2. Given the current state, we need to sample the next one on the basis of a row of the transition matrix. We use the same trick as in Fig. 5.7 to sample from an empirical discrete distribution, which needs the cumulative probabilities.

3. To avoid cumulating the same probabilities over and over, we compute them once in the function `MakeSamplePaths` by applying `cumsum` along the rows of the matrix `transMatrix`.

FIGURE 10.30 **Sample path generation for the life-cycle problem.**

```
SimulateFixedSaving <- function(W0,empl0,utilFun,gamma,
                          income,transMatrix,mu,sigma,rf,
                          scenarios,alpha,fractCons){
  # allocate data structures
  dims <- dim(scenarios$retPaths)
  numRepl <- dims[1]
  numSteps <- dims[2]
  alphaPaths <- matrix(alpha,nrow=numRepl,ncol=numSteps)
  CPaths <- matrix(0,nrow=numRepl,ncol=numSteps+1)
  wealthPaths <- matrix(0,nrow=numRepl,ncol=numSteps+1)
  utilVals <- numeric(numRepl)
  # start replications
  for (k in 1:numRepl){
    # get employment and return scenario
    emplPath <- scenarios$emplPaths[k,]
    riskyRets <- scenarios$retPaths[k,]
    wealthNow <- W0
    stateNow <- empl0
    for (t in 0:(numSteps-1)){
      # collect path info
      C <- fractCons*wealthNow
      CPaths[k,t+1] <- C
      wealthPaths[k,t+1] <- wealthNow
      utilVals[k] <- utilVals[k] + gamma^(t)*utilFun(C)
      # update wealth according to actual scenario
      wealthNow <- income[emplPath[t+1]]+(wealthNow-C)*
        (1+rf+alpha*(riskyRets[t+1]-rf))
    }
    # at last step, just consume available wealth
    wealthPaths[k,numSteps+1] <- wealthNow
    CPaths[k,numSteps+1] <- wealthNow
    utilVals[k] <- utilVals[k] +
                   gamma^(numSteps)*utilFun(wealthNow)
  }
  out <- list(alphaPaths=alphaPaths, CPaths=CPaths,
              wealthPaths=wealthPaths, utilVals=utilVals)
  return(out)
}
```

FIGURE 10.31 **Simulating a fixed decision policy.**

```
SimulateDPSaving <- function(W0,empl0,utilFun,gamma,
                valFunsList,income,transMatrix,inRetScenarios,
                probs,rf,outScenarios){
  dims <- dim(outScenarios$retPaths)
  numRepl <- dims[1]; numSteps <- dims[2]
  alphaPaths <- matrix(0,nrow=numRepl,ncol=numSteps)
  CPaths <- matrix(0,nrow=numRepl,ncol=numSteps+1)
  wealthPaths <- matrix(0,nrow=numRepl,ncol=numSteps+1)
  utilVals<-numeric(numRepl);numInScenarios<-length(probs)
  for (k in 1:numRepl){  # start replications
    emplOutScenario <- outScenarios$emplPaths[k,]
    retOutScenario <- outScenarios$retPaths[k,]
    wealthNow <- W0  # initialize state
    stateNow <- empl0
    for (t in 0:(numSteps-1)){
      objFun <- function(x){ # build objective function
        C <- x[1]; alpha <- x[2]; out <- 0
        for (j in 1:3){
          for (l in 1:numInScenarios){
            futWealth <- income[j]+(wealthNow-C)*
            (1+rf+alpha*(retScenarios[l]-rf))
            out <- out+probs[l]*transMatrix[stateNow,j]*
            valFunsList[[t+1]][[j]](futWealth)}}
        out <- utilFun(C)+gamma*out
        return(-out)}
      x0 <- c(wealthNow/2,0.5) # solve optim. problem
      outOptim <- nlminb(start=x0,objective=objFun,
                    lower=c(0,0),upper=c(wealthNow,1))
      # collect path info
      C <- outOptim$par[1]; alpha <- outOptim$par[2]
      CPaths[k,t+1] <- C; alphaPaths[k,t+1] <- alpha
      wealthPaths[k,t+1] <- wealthNow
      utilVals[k] <- utilVals[k] + gamma^(t)*utilFun(C)
      wealthNow<-income[emplOutScenario[t+1]]+(wealthNow-C)*
      (1+rf+alpha*(retOutScenario[t+1]-rf))
    }
    wealthPaths[k,numSteps+1] <- wealthNow
    CPaths[k,numSteps+1] <- wealthNow
    utilVals[k] <- utilVals[k] +
            gamma^(numSteps)*utilFun(wealthNow)
  }
  out <- list(alphaPaths=alphaPaths, CPaths=CPaths,
            wealthPaths=wealthPaths, utilVals=utilVals)
  return(out)
}
```

FIGURE 10.32 **Simulating the optimal DP policy.**

```
numRepl <- 100
set.seed(55555)
scenarios <- MakeSamplePaths(empl0,transMatrix,mu,sigma,
                             numSteps,numRepl)
alpha <- 0.5
fractCons <- 0.5
out <- SimulateFixedSaving(W0,empl0,utilFun,gamma,
                           income,transMatrix,mu,sigma,rf,
                           scenarios,alpha, fractCons)
utilValsFixed <- out$utilVals
out <- SimulateDPSaving(W0,empl0,utilFun,gamma,valFuns,
                        income,transMatrix,retScenarios,
                        probs,rf,scenarios)
utilValsDP <- out$utilVals
```

FIGURE 10.33 **Comparing the optimal DP and a fixed policy.**

```
> mean(utilValsDP)
[1] -2.547044e-05
> t.test(utilValsDP,utilValsFixed)
Welch Two Sample t-test
t = 3.4024, df = 124.686, p-value = 0.0008983
alternative hypothesis: true difference in means
is not equal to 0
```

Now the difference looks more relevant, but once again the comparison should just be taken as an illustration. Furthermore, we could use the simulation-based optimization strategies of Section 10.4 to find the best settings for the fixed policy. We urge the reader to further investigate these policies. For instance, by checking the paths of α_t, which is stored in the output matrix `out$alphaPaths` by the simulation function, we find that with the logarithmic utility we should always invest 100% in the risky asset, whereas the allocation is more conservative for the power utility with $\rho = 5$.

10.8 Approximate dynamic programming

The numerical methods for stochastic dynamic programming that we have discussed in the previous section are certainly a useful tool for tackling some dynamic optimization problems under uncertainty, but they are not a radical antidote against the curses of dimensionality. Over the years, an array of techniques, collectively labeled as approximate dynamic programming (ADP) have been proposed. They rely more exensively on Monte Carlo methods to learn an approximation of the value function $V_t(s_t)$ at time instant t for state s_t, by an iterative process. Let $\overline{V}_t^n(\mathbf{s}_t)$ be such an approximation at iteration n. The iteration superscript n should not be confused with the time subscript t, which is relevant for finite-horizon problems; for infinite-horizon problems, the notation may be replaced by $\overline{V}^n(\mathbf{s})$. The idea is to use Monte Carlo methods to generate observations of the value functions at different states, by applying approximate policies for a sequence of sample paths. After having observed n sample paths, the current approximation is $\overline{V}_t^n(\mathbf{s}_t)$.

ADP is related to reinforcement learning and borders with artificial intelligence and machine learning. It is also worth mentioning that some model-free forms of ADP have been devised for applications in which there is no model of the system, and one can only learn by trial-and-error experimentation with the real system. Needless to say, we cannot offer a full account of ADP methods,[33] but we can point out some of their most relevant features:

- In Section 10.8.1 we outline a basic version of ADP, stressing the role of Monte Carlo sampling to learn a sequence of approximated value functions.

- In Section 10.8.2 we introduce a powerful device in ADP: the post-decision state variable.

- Finally, in Section 10.8.3, we revisit the toy MDP of Section 10.7.2.1 to illustrate the application of Q-learning, which can be interpreted as an ADP approach based on post-decision state variables.

- We will see another, more significant, application of ADP in Section 11.3.3, where we show how the idea can be used to price American-style options by approximating value functions with linear regression and a set of basis functions.

10.8.1 A BASIC VERSION OF ADP

Figure 10.34 outlines a very basic version of ADP. This procedure is just the statement of a principle, rather than a working algorithm. Nevertheless, we may observe one common ingredient of many ADP methods: the absence of the usual backward recursion. Indeed, the algorithm generates a sequence of states and decisions, based on the approximation of the value function for iteration $n - 1$, by going *forward* in time. The solution of the optimization problem (10.60) provides us with an observation of the value of state $\mathbf{s}_t^n$, which is used to improve the approximation of the value function. How this is accomplished exactly depends on the kind of state space. If we are dealing with a small and finite state space, like the Markov decision process of Fig. 10.23, the value function is just a vector listing the value for each state. Then, we might update

---

[33]The book by Warren Powell [37] is the recommended reading for a general presentation of ADP, and it also provided the basis for much of the following presentation.

Step 0. Initialization

- Initialize the approximated value function $\overline{V}_t^0(\mathbf{s}_t)$ for all states.
- Select the total number of iterations N and set the iteration counter $n = 1$.

Step 1. Generate a sample path of the driving random variables $\{\epsilon_t^n\}$, $t = 1, \ldots, T$, and choose the initial state $\mathbf{s}_0^n$.

Step 2. For t in $\{0, 1, 2, \ldots, T\}$:

- Solve the optimization problem

$$\min_{\mathbf{x}_t \in X_t} \left\{ C_t(\mathbf{s}_t^n, \mathbf{x}_t) + \gamma \mathrm{E}\left[\overline{V}_{t+1}^{n-1}\big(S_{t+1}\left(\mathbf{s}_t^n, \mathbf{x}_t\right)\big) \right] \right\} \tag{10.60}$$

to find a new observation $\overline{v}_t^n$ of the approximate value function in $\mathbf{s}_t^n$ and the optimal decision $\mathbf{x}_t^*$.
- Use $\overline{v}_t^n$ to update $\overline{V}_t^n(\mathbf{s}_t^n)$.
- Generate the next state $\mathbf{s}_{t+1}^n$ on the basis of the current state $\mathbf{s}_t^n$, the optimal decision $\mathbf{x}_t^*$, and the disturbance ϵ_{t+1}^n.

Step 3. Increment the iteration counter, $n = n + 1$; if $n \leq N$, go to step 1.

FIGURE 10.34 **A simple ADP approach.**

the value function for the visited state $\mathbf{s}_t^n$ as follows:

$$\overline{V}_t^n(\mathbf{s}_t^n) = \alpha^n \, \overline{v}_t^n + (1 - \alpha^n) \, \overline{V}_t^{n-1}(\mathbf{s}_t^n), \tag{10.61}$$

where the smoothing coefficient $\alpha^n \in (0, 1)$ mixes the new and the old information. When looking at Eq. (10.61), it is important to realize that we are *not* estimating the expected value of a random variable by sampling a sequence of i.i.d. observations: The decisions we make depend on the current approximation of the value function and, since the approximations themselves change over the iterations, we are chasing a moving target. An effective ADP algorithm, among other things, must strike a difficult balance between conflicting requirements:

- *Speed and convergence.* If the smoothing coefficient α^n is close to zero, much weight is given to the previous aproximation of the value function, possibly resulting in painfully slow learning. However, if it is too large, erratic behavior may arise. It is possible to adapt the coefficient dynamically, starting with a relatively large value in order to quickly improve the approximation at early stages, and then reducing it in order ensure convergence at later stages. We also mention that convergence may be improved if we start with a sensible initialization of the value function, possibly found by simulating the application of some good heuristic policy.

- *Exploitation and exploration.* When solving the problem (10.60) at each iteration, we rely on the current approximation of the value function. It may happen that a good state looks like a very bad one because of a poor value function approximation. The problem is that we may make decisions in such a way to steer away from that state, preventing the possibility of learning that it is actually a good one. Hence, we should allow for a significant amount of exploration in order to learn where good states are located. Unfortunately, this conflicts with the need for exploitation, i.e., using the knowledge we have about good states in order to produce good policies.

The above observations suggest that there are indeed quite some challenges in devising a good ADP algorithm. However, from the pseudocode of Fig. 10.34 it may be difficult to see any definite advantage with respect to more classical numerical approaches in dealing with the three curses of dimensionality:

- One curse of dimensionality concerns the difficulty in solving each optimization model, when the decision variables x_t are continuous and high-dimensional. From this point of view, we may rely on sophisticated optimization technology, but there is really no difference between ADP and classical numerical DP.

- The traditional curse of dimensionality concerns the dimensionality of the state space, and it is clearly impossible to update the approximation of value functions using Eq. (10.61) for a continuous state space. Discretizing the state space using a regular grid is only feasible for low-dimensional spaces, but we may approximate value functions like we have done with the collocation method of Section 10.7.2.2, relying on a set of basis functions $\phi_j(\cdot)$, $j = 1, \ldots, m$:

$$\overline{V}_t(\mathbf{s}_t) \approx \sum_{j=1}^{m} c_{jt} \phi_j(\mathbf{s}_t). \tag{10.62}$$

Here we assume that the set of basis functions is the same for all time instants, whereas the coefficients c_{jt} can depend on time. Clearly, finding a suitable set of basis function capturing the features of good states is non trivial. Nevertheless, we may estimate the coefficients by linear regression, based on the observations provided by Eq. (10.60). Since, as we have mentioned, we are chasing a moving target, it may be advisable to use weighted least squares, in order to assign more weights to recent observations. Furthermore, rather than using least squares at each step, recursive procedures can be used to estimate the coefficients iteratively, rather than classical batch estimation where all the observations are used.[34]

- A last curse of dimensionality, however, is still present in the ADP procedure of Fig. 10.34, as we need the expectation of the of the value function for the next state at the next time step. For high-dimensional random variables ϵ_t, this is an overwhelming task. A powerful trick to get rid of the expectation is to resort to a different way of writing the recursive Bellman equation, based on post-decision state variables, which we discuss in the next section.

10.8.2 POST-DECISION STATE VARIABLES IN ADP

The classical way of writing the Bellman equation involves a thorny expectation. When dealing with a stochastic optimization problem, there is no way to avoid having to deal with an expectation somewhere. Nevertheless, we may try to move it to a more convenient place, by rewriting the functional equation around post-decision state variables.[35] Before introducing the concept, it is useful to illustrate the rationale behind the approach in a simple context.

Example 10.15 Forecasting by linear regression models

Consider a multiple linear regression model

$$Y_t = \alpha + \sum_{j=1}^{m} \beta_j f_{jt} + \epsilon_t,$$

where m factors are used as regressors (explanatory variables, if you prefer). Now, imagine that we want to use this model to forecast Y_t. If the regressors are stochastic, there is no way to use a model like this, unless we are able to forecast the factors themselves. An alternative model could be

$$Y_t = \alpha + \sum_{j=1}^{m} \beta_j f_{j,t-1} + \epsilon_t.$$

Now we are relating the future value of Y_t to *past* observations of the factors, which may be more useful for forecasting purposes.

---

[34] In the linear regression based approach that we discuss in Section 11.3.3, a batch procedure is used.

[35] This way of rewriting the optimality equations is associated with end-of-period value functions in [24, Chapter 12]. We prefer the terminology and the approach of [37].

The idea of the above example can also be applied to ADP, by breaking down the transition dynamics

$$\mathbf{s}_{t+1} = S_{t+1}\left(\mathbf{s}_t, \mathbf{x}_t, \epsilon_{t+1}\right)$$

in two steps. The sequence of events is:

1. We observe the state $\mathbf{s}_t$.
2. We make the decision $\mathbf{x}_t$.
3. The random variable ϵ_{t+1} is realized, leading to a new state $\mathbf{s}_{t+1}$.

This results in the following history of the stochastic decision process:

$$\left(\mathbf{s}_0, \mathbf{x}_0, \epsilon_1, \mathbf{s}_1, \mathbf{x}_1, \epsilon_2, \ldots, \mathbf{s}_{t-1}, \mathbf{x}_{t-1}, \epsilon_t, \ldots\right).$$

We may introduce a post-decision state variable $\mathbf{s}_t^x$, which represents the system state after the decision $\mathbf{x}_t$, but *before* the realization of the disturbance ϵ_{t+1}. The state evolution is modified as

$$\left(\mathbf{s}_0, \mathbf{x}_0, \mathbf{s}_0^x, \epsilon_1, \mathbf{s}_1, \mathbf{x}_1, \mathbf{s}_1^x, \epsilon_2, \ldots, \mathbf{s}_{t-1}, \mathbf{x}_{t-1}, \mathbf{s}_{t-1}^x, \epsilon_t, \ldots\right).$$

Let us illustrate the idea by a simple example.

Example 10.16 Post-decision state in the life-cycle problem

> In Section 10.7.3 we have defined the state as the pair $\mathbf{s}_t = (W_t, \lambda_t)$, consisting of available wealth and the current employment state. This is a *pre-decision* state variable, as we still have to make the consumption and allocation decision. After making the decision, we may introduce the post-decision state variable $\mathbf{s}_t^x = (Q_t, B_t, \lambda_t)$, where Q_t and B_t are the amounts invested in the risky and risk-free asset, respectively, and λ_t has the same interpretation as before (in fact, the employments state is not affected by our decision).

In the above example, we introduce one more component in the state variable. In general, if we have n assets and there are no transaction costs, the essential state variable for an asset allocation problem is just available wealth. If we introduce post-decision variables, we need n variables, which increases complexity a lot. However, if we include transaction costs into the model, there is no increase in complexity, as we must account for each asset individually, anyway. Whether the post-decision state is more or less complex than the pre-decision state depends on the problem structure.[36] However, the significant

---

[36]We shall consider pricing of American-style options by Monte Carlo methods in Section 11.3.3. In that case, the post-decision state variable is related to the continuation value of the option, with no increase in complexity.

advantage of the former is apparent when we rewrite the optimality equations in terms of $V_t^x(\mathbf{s}_t^x)$, the value of being in the post-decision state $\mathbf{s}_t^x$, rather than in terms of $V_t(\mathbf{s}_t)$. We have:

$$V_{t-1}^x(\mathbf{s}_{t-1}^x) = \mathrm{E}\left[V_t(\mathbf{s}_t) \mid \mathbf{s}_{t-1}^x\right] \tag{10.63}$$

$$V_t(\mathbf{s}_t) = \min_{\mathbf{x}_t \in X_t} \left\{ C_t(\mathbf{s}_t, \mathbf{x}_t) + \gamma V_t^x(\mathbf{s}_t^x) \right\} \tag{10.64}$$

$$V_t^x(\mathbf{s}_t^x) = \mathrm{E}\left[V_{t+1}(\mathbf{s}_{t+1}) \mid \mathbf{s}_t^x\right]. \tag{10.65}$$

Equation (10.63) relates the value function for the post-decision state variable at time $t-1$ with the value function for the pre-decision state variable at time t; the expectation is with respect to ϵ_t, the disturbance term used in the transition from $\mathbf{s}_{t-1}^x$ to $\mathbf{s}_t$. The most relevant point is the form of Eq. (10.64). This is the optimality equation written using the post-decision value function: This equation does not involve any expectation and requires the solution of a *deterministic* optimization problem. Equation (10.65) is just Eq. (10.63) moved forward by one time period. Clearly, we obtain the standard Bellman equation by plugging (10.65) into (10.64):

$$V_t(\mathbf{s}_t) = \min_{\mathbf{x}_t \in X_t} \left\{ C_t(\mathbf{s}_t, \mathbf{x}_t) + \gamma \mathrm{E}\left[V_{t+1}(\mathbf{s}_{t+1})\right] \right\}.$$

But if we substitute (10.64) into (10.63) we find a different form of the optimality equation:

$$V_{t-1}^x(\mathbf{s}_{t-1}^x) = \mathrm{E}\left\{ \min_{\mathbf{x}_t \in X_t} \left[C_t(\mathbf{s}_t, \mathbf{x}_t) + \gamma V_t^x(\mathbf{s}_t^x) \right] \right\}. \tag{10.66}$$

Now, the expectation is outside the minimization. Obviously, we did not eliminate the expectation altogether, but the outside expectation may be dealt with by statistical learning based on Monte Carlo methods. With this alternative DP form, we get rid of one curse of dimensionality, which may have a significant impact on some problems.

If we adopt post-decision state variables, the basic ADP algorithm of Fig. 10.34 is modified as shown in Fig. 10.35. A few comments are in order:

- $\overline{V}_t^{x,n}$ is the approximation at iteration n of the post-decision value function at time instant t.

- We denote by $S_t^x(\cdot, \cdot)$ the transition function from the current pre-decision to the next post-decision state. This transition function is deterministic.

- An important point is how we use the observation $\overline{v}_t^n$ to update the post-decision value function. When tackling the optimization problem (10.67), we are at pre-decision state $\mathbf{s}_t^n$, and the optimal decision leads us to the post-decision state $\mathbf{s}_t^{x,n}$. However, we do *not* want to update the value of $\mathbf{s}_t^n$, as in order to use this knowledge in the next iterations we should use an expectation, and the related curse of dimensionality would creep back into our algorithmic strategy. We update the value of the post-decision

Step 0. Initialization

- Initialize the approximated value function $\overline{V}_t^{x,0}(\mathbf{s}_t)$ for all states.
- Select the total number of iterations N and set the iteration counter $n = 1$.

Step 1. Generate a sample path of the driving random variables $\{\epsilon_t^n\}$, $t = 1, \ldots, T$, and choose the initial state $\mathbf{s}_0^n$.

Step 2. For t in $\{0, 1, 2, \ldots, T\}$:

- Solve the optimization problem

$$\min_{\mathbf{x}_t \in X_t} \left\{ C_t(\mathbf{s}_t^n, \mathbf{x}_t) + \gamma \overline{V}_t^{x,n-1}\big(S_t^x(\mathbf{s}_t^n, \mathbf{x}_t)\big) \right\} \qquad (10.67)$$

to find a new observation $\overline{v}_t^n$ of the approximate value function in $\mathbf{s}_{t-1}^{x,n}$ and the optimal decision $\mathbf{x}_t^*$.
- Use $\overline{v}_t^n$ to update $\overline{V}_{t-1}^{x,n}(\mathbf{s}_{t-1}^{x,n})$.
- Generate the next post-decision state $\mathbf{s}_t^{x,n}$, based on the current state $\mathbf{s}_t^n$ and the optimal decision $\mathbf{x}_t^*$; generate the next pre-decision state $\mathbf{s}_{t+1}^n$, based on $\mathbf{s}_t^{x,n}$ and the disturbance ϵ_{t+1}^n.

Step 3. Increment the iteration counter, $n = n + 1$; if $n \leq N$, go to step 1.

FIGURE 10.35 **ADP based on the post-decision state variable.**

state $\mathbf{s}_{t-1}^{x,n}$ that was visited before reaching the current pre-decision state $\mathbf{s}_t^n$. In other words, we use the optimal value of the objective function in problem (10.67) to estimate the value of the post-decision state that lead us to the current post-decision state $\mathbf{s}_t^n$, via the realization of the random variable ϵ_t^n. Hence, we take $\overline{v}_t^n$ as an observation of the random variable whose expectation occurs in the optimality equation (10.66).

Needless to say, the procedure of Fig. 10.35 is just an outline, and many critical details must be filled to come up with in a working algorithm. We refer to the extensive treatment in [37] for several examples. The idea is illustrated in the next section for the toy Markov decision process of Section 10.7.2.1.

10.8.3 *Q*-LEARNING FOR A SIMPLE MDP

We refer again to the MDP depicted in Fig. 10.23. In this very small problem instance computing an expectation is quite easy and poses no challenge at all; however, we may use this example to investigate how we can adapt the idea of post-decision states. For instance, imagine that we are in state 2. If we decide to take the immediate payoff, we will be at state 1 at the next step; otherwise, a flip of a (biased) coin will decide if we will stay at state 2 or move on to state 3. Let $s \in \{1, 2, 3, 4\}$ be the current pre-decision state and $a \in \{0, 1\}$ be the selected action (we assume that 1 corresponds to taking the immediate payoff and getting back to state 1). The post-decision state, for this problem structure, is the pair (s, a).[37] In general, the action a selects a transition probability matrix, which is used explicitly in value iteration. We can estimate the value of taking action a in state s, denoted by $Q(s, a)$, by a strategy known as Q-learning, which may be applied to discrete-state, discrete-action problems with an infinite horizon. The values $Q(s, a)$ are called Q-factors, and we learn them iteratively in this approach. Let $\overline{Q}^{n-1}(s, a)$ be the estimate of the factor at iteration $n - 1$. Then, if we are in the pre-decision state s^n at iteration n, we select the next action as

$$a^n = \arg\max_{a \in \mathcal{A}} \overline{Q}^{n-1}(s^n, a). \tag{10.68}$$

Note that we do not need any expectation to solve this optimization problem. After selecting the optimal action, we evaluate the immediate cost (or reward) and observe the next pre-decision state s^{n+1}. Note that we associate the next state with the next iteration $n + 1$, rather than the next time instant, as we are dealing with an infinite-horizon problem solved iteratively. When using Monte Carlo simulation, the next state may be generated by a suitable sampling strategy, and we do not need to know the transition matrix explicitly.[38] Now we may update the Q-factor on the basis of the new observation[39]

$$\hat{q}^n = C(s^n, a^n) + \gamma \max_{a' \in \mathcal{A}} \overline{Q}^{n-1}(s^{n+1}, a'), \tag{10.69}$$

using the smoothing equation

$$\overline{Q}^n(s^n, a^n) = \alpha^{n-1}\,\hat{q}^n + (1 - \alpha^{n-1})\,\overline{Q}^{n-1}(s^n, a^n),$$

where $\alpha^{n-1} \in (0, 1)$ is a smoothing coefficient. To see the connection with value functions, consider the estimate of the value of state s^n,

$$\overline{V}^n(s^n) = \max_{a \in \mathcal{A}} \overline{Q}^n(s^n, a),$$

and rewrite Eq. (10.69) as

$$\hat{q}^n = C(s^n, a^n) + \gamma \overline{V}^{n-1}(s^{n+1}). \tag{10.70}$$

---

[37] In this section we denote the state by s, rather than by boldface s, as we assume that states can be enumerated.

[38] In some artificial intelligence or control engineering problems, we could even apply the strategy online to the actual system.

[39] We assume that the immediate cost is deterministic, but it could well be random; in this case, we should use the notation $\hat{C}(s^n, a^n)$, as we are dealing with an observation of a random variable.

Indeed, the essential advantage of the approach is that no expectation is involved. The disadvantage is that a lot of Q-factors are needed if the state and the action spaces are large.

The R code of Fig. 10.36 shows how Q-learning may be applied to the simple MDP of Section 10.7.2.1. The code is rather straightforward, and the only detail worth noting is that action indexes in R are incremented by 1 when accessing vectors and matrices. In fact, we associate action $a = 1$ with the decision of collecting the immediate payoff, and action $a = 0$ with the decision of waiting. However, vector/matrix indexing starts from 1 in R, so those actions correspond to indexes 2 and 1, respectively. We also recall that we have to make a decision only in the interior states 2 and 3; in the boundary state 1 we cannot earn any immediate reward, and in the boundary state 4 we just collect the reward and move back to state 1.

Let us compare the result of Q-learning with the result of value iteration. The Q-factors have to be initialized. For action $a = 0$, we initialize the factors to 50; for action $a = 1$, the factors are initialized with the payoffs (zero for state 1). We select the smoothing coefficient $\alpha = 0.1$ and run 10,000 iterations, obtaining the following output:

```
> probs <- c(.7,.8,.9,0); payoff <- c(0,10,20,30)
> discount <- 0.8
> PW311_ValueIteration(probs, payoff, discount)
$stateValues
[1]  9.677415 17.741931 27.741931 37.741931
$policy
[1] 0 1 1 1


> numSteps <- 10000; startState <- 1; alpha <- 0.1
> startQFactors <- matrix(c(rep(50,4),payoff),ncol=2)
> set.seed(55555)
> PW311_QLearning(probs,payoff,discount,numSteps,
          startState,startQFactors,alpha)
$QFactors
           [,1]       [,2]
[1,]   9.877252  0.00000
[2,]   9.851703 18.63811
[3,]  19.704227 25.67057
[4,]  50.000000 40.12486
$policy
[1] 0 1 1 1
```

If we compare the factors corresponding to the optimal decisions with the state values obtained by value iteration, we find them in fairly good agreement. What really, matters, however, is that the state value estimates are good enough to induce the optimal policy. This may look reassuring, but what if we change the initialization a bit? In the following run, we initialize the factors for action $a = 0$ to 500:

```
> startQFactors <- matrix(c(rep(500,4),payoff),ncol=2)
```

```r
PW311_QLearning <- function(probs,payoff,discount,numSteps,
                            startState,startQFactors,alpha){
  # NOTE: action indexes in R are incremented by 1
  # when accessing vectors and matrices
  QFactors <- startQFactors
  state <- startState
  numStates <- length(probs)
  for (k in 1:numSteps){
    if (state == 1){ # state 1, no choice
      newState <- ifelse(runif(1)>probs[state],2,1)
      action <- 0
      reward <- 0
    }
    else if (state == numStates){
      # last state, get reward and go back to square 1
      newState <- 1
      action <- 1
      reward <- payoff[state]
    }
    else { # Choose optimal action
      action <- which.max(QFactors[state,]) - 1
      if (action == 1){# get reward and go back to square 1
        newState <- 1
        reward <- payoff[state]
      }
      else { # random transition
        if (runif(1) > probs[state]){
          newState <- newState + 1
          reward <- 0
        }
      }
    }
    } # end if on states
    qhat <- reward+discount*max(QFactors[newState,])
    QFactors[state,action+1] <-
      alpha*qhat+(1-alpha)*QFactors[state,action+1]
    state <- newState
  } # end for
  # find optimal policy (for interior states)
  indexVet <- 2:(numStates-1)
  policy <- c(0,QFactors[indexVet,2]>QFactors[indexVet,1],1)
  out <- list(QFactors=QFactors,policy=policy)
  return(out)
}
```

FIGURE 10.36 **Tackling the MDP of Section 10.7.2.1 by *Q*-learning.**

```
> set.seed(55555)
> PW311_QLearning(probs,payoff,discount,numSteps,
         startState,startQFactors,alpha)
$QFactors
          [,1]       [,2]
[1,]   50.24164   0.00000
[2,]   92.49797  10.00000
[3,]  194.02190  20.00000
[4,]  500.00000  70.24866
$policy
[1] 0 0 0 1
```

With this initialization, we assign a vary large value to the option of waiting, and after 10,000 iterations we fail to find the optimal policy. We may observe that the value of the action $a = 0$ in states 2 and 3 is lower than its initial value, and we may argue that after several steps we will end up with the right choices. This is also a case for adopting a larger value of α, at least in the initial steps, in order to accelerate learning.[40] However, the next run points out a more disturbing facet of value function initialization:

```
> startQFactors <- matrix(c(rep(1,4),rep(0,4)),ncol=2)
> set.seed(55555)
> PW311_QLearning(probs,payoff,discount,numSteps,
         startState,startQFactors,alpha)
$QFactors
            [,1]       [,2]
[1,]    3.256359   0.00000
[2,]    6.009678   0.00000
[3,]   12.634094   0.00000
[4,]    1.000000  32.60307
$policy
[1] 0 0 0 1
```

This is a case of bad initialization, too. The initial value of action $a = 0$ in states 1, 2, and 3 seems to converge from below to the correct value. However, the corresponding value of the action $a = 1$ is stuck to zero, and it will remain there. With the above version of Q-learning, we will never choose action $a = 1$ in those states, and we will never learn its true value. This is a general concern related to the balance between exploitation and exploration. On the one hand, we want to exploit a good policy to optimize the objective function; on the other hand, we want to explore actions and states in order to learn a possibly better policy.

One way to strike a better balance between these conflicting objectives is to select a random action every now and then, without solving the optimization problem (10.68). The code in Fig. 10.37 applies this idea, which is known as epsilon-greedy policy. The function is essentially the same as the Q-learning

---

[40]The whole Chapter 11 of [37] is devoted to selecting adaptive stepsizes.

```
PW311_QepsLearning <- function(probs,payoff,discount,
                numSteps,startState,startQFactors,
                alpha,eps){
  .........
  for (k in 1:numSteps){
    if (state == 1){
    ......... }
    else if (state == numStates){
    ......... }
    else {
      #-------------- Code modified here !
      if (runif(1) > eps) # Choose optimal action
        action <- which.max(QFactors[state,]) - 1
      else # Choose random action
        action <- ifelse(runif(1)<0.5,0,1)
      #--------------
      if (action == 1){
      ......... }
    } # end if on states
  .........
  out <- list(QFactors=QFactors,policy=policy)
  return(out)
}
```

FIGURE 10.37 **Adapting the *Q*-learning code of Fig. 10.36 to improve exploration (abridged code).**

version of Fig. 10.36, but it takes one additional parameter, eps, which is the probability of choosing the next action randomly; when we are in interior states 2 and 3, a coin is flipped, a random action is taken with probability ϵ, and the currently best decision is taken with probability $1 - \epsilon$. As we may see from the snapshot below, now we do find the optimal policy, even with the bad initialization:

```
> startQFactors <- matrix(c(rep(1,4),rep(0,4)),ncol=2)
> eps <- 0.1
> set.seed(55555)
> PW311_QepsLearning(probs,payoff,discount,numSteps,
         startState,startQFactors,alpha,eps)
$QFactors
          [,1]     [,2]
[1,] 10.423041  0.00000
[2,] 15.748525 18.87826
[3,]  4.153173 23.19135
[4,]  1.000000 17.75869
$policy
```

```
[1]  0  1  1  1
```

We may also observe that the state values are not exact, but this is enough to find the optimal policy. Clearly, balancing exploration and exploitation is not trivial with large state spaces, and a successful strategy requires a lot of problem-specific adaptation. Another useful observation is that with discrete-state, discrete-action systems we do not necessarily need a very precise assessment of value functions in order to find the optimal policy. Indeed, alternative policy iteration strategies use a policy to assess state values and aim at improving the policy itself.

In continuous-state problems, the accuracy of the value function may be more or less relevant, depending on the case. For instance, in the newsvendor problem of Section 10.2.3, we may not need a very accurate representation of uncertainty, which has an impact on the approximation of the objective function, as even a solution reasonably close to the optimal one performs rather well out of sample. However, a poor representation of uncertainty may prove fatal in other applications; for instance, badly generated scenarios may contain arbitrage opportunities in portfolio optimization, which leads to unreasonable solutions. Finally, we note that there are problems where the main output of the optimization is the value of the objective function, rather than the decision variables or the policy. One example is pricing options with early exercise features. The regression-based ADP approach that we discuss later in Section 11.3.3 bears some similarity with Q-learning and the example we have discussed here, as we must compare the value of waiting, the continuation value of the option, against the value of immediate exercise, the immediate payoff. These are not Q-factors associated with discrete states, and after learning the policy it is important to simulate its application out-of-sample, on a large set of scenarios, in order to improve the estimate. Furthermore, bias issues should be carefully considered. As we shall see, a suboptimal policy yields a *lower* bound on the option value, which need to be complemented by an upper bound.

For further reading

- The most important skill in optimization is model building, rather than model solving. Readers interested in sharpening their model building skills will find plenty of enlightening examples in [46]. For more specific examples of optimization modeling in finance see, e.g., [13].

- An elementary treatment of optimization algorithms can be found in [8], where some model building tricks of the trade are also illustrated. If you look for deeper technical treatments of optimization methods, [1] and [47] are excellent references for nonlinear programming and integer programming, respectively.

- Model calibration problems are often solved as nonlinear programs; see, e.g., [12] or [23, Chapter 15]. An example of application of integer programming to financial optimization can be found in [5].

- Several Monte Carlo methods for optimization problems are discussed in [30], including some approaches that we did not mention, like the cross-entropy method.

- There is a huge literature on metaheuristics and related approaches. We did not cover tabu search; see, e.g., [16] for a financial application. Genetic algorithms are covered in [33]. For the original references on particle swarm optimization, see [14] and [36]. Several variants of swarm optimization have been proposed; see, e.g., [48] for the firefly variant and the use of Levy flights, or [42] for the quantum variant.

- A general reference on stochastic optimization is [41]. Simulation-based optimization is dealt with in [15]; see [27] for an extensive treatment of metamodeling.

- Modeling with stochastic programming is illustrated in [25]. An extensive treatment of theoretical and algorithmic issues can be found in [6] or [26].

- The key reference for approximate dynamic programming is [37]. Other useful references are [2] and [3]. The former illustrates dynamic programming in general, and the latter is more focused on ADP. See also [4] for the use of neural networks as tools for the approximation of value functions. *Q*-learning was introduced in the Ph.D. thesis by Watkins; see [45].

- Several applications of dynamic programming to economics can be found in [34], which also describes numerical DP methods implemented in a MATLAB toolbox. Several numerical DP tricks are described in [11].

- R offers several optimization packages; see `http://cran.r-project.org/web/views/Optimization.html` for a listing. However, if you need to solve really tough optimization problems, you may need to resort to state-of-the-art solvers like CPLEX (see `http://www-01.ibm.com/software/info/ilog/` or `www.ilog.com`) and Gurobi (see `http://www.gurobi.com/`). Both CPLEX and Gurobi offer an interface with R.

References

1 M.S. Bazaraa, H.D. Sherali, and C.M. Shetty. *Nonlinear Programming. Theory and Algorithms* (2nd ed.). Wiley, Chichester, West Sussex, England, 1993.

2 D.P. Bertsekas. *Dynamic Programming and Optimal Control Vol. 1* (3rd ed.). Athena Scientific, Belmont, MA, 2005.

3 D.P. Bertsekas. *Dynamic Programming and Optimal Control Vol. 2* (4th ed.). Athena Scientific, Belmont, MA, 2012.

4 D.P. Bertsekas and J.N. Tsitsiklis. *Neuro-Dynamic Programming*. Athena Scientific, Belmont, MA, 1996.

5 D. Bertsimas, C. Darnell, and R. Stoucy. Portfolio construction through

mixed-integer programming at Grantham, Mayo, Van Otterloo and Company. *Interfaces*, 29:49–66, 1999.

6 J.R. Birge and F. Louveaux. *Introduction to Stochastic Programming.* Springer, New York, 1997.

7 P. Brandimarte. *Numerical Methods in Finance and Economics: A MATLAB-Based Introduction* (2nd ed.). Wiley, Hoboken, NJ, 2006.

8 P. Brandimarte. *Quantitative Methods: An Introduction for Business Management.* Wiley, Hoboken, NJ, 2011.

9 P. Brandimarte and A. Villa. *Advanced Models for Manufacturing Systems Management.* CRC Press, Boca Raton, FL, 1995.

10 J.Y. Campbell and L.M. Viceira. *Strategic Asset Allocation.* Oxford University Press, Oxford, 2002.

11 C. Carroll. Solving microeconomic dynamic stochastic optimization problems. Lecture notes downloadable from `http://www.econ.jhu.edu/people/ccarroll/index.html`.

12 T.F. Coleman, Y. Li, and A. Verma. Reconstructing the unknown volatility function. *Journal of Computational Finance*, 2:77–102, 1999.

13 G. Cornuejols and R. Tütüncü. *Optimization Methods in Finance.* Cambridge University Press, New York, 2007.

14 R.C. Eberhart and J. Kennedy. Particle swarm optimization. In *Proceedings of the IEEE International Conference on Neural Networks*, p. 1942Ũ1948, 1995.

15 M.C. Fu. Optimization by simulation: A review. *Annals of Operations Research*, 53:199–247, 1994.

16 F. Glover, J.M. Mulvey, and K. Hoyland. Solving dynamic stochastic control problems in finance using tabu search with variable scaling. In I.H. Osman and J.P. Kelly, editors, *Meta-Heuristics: Theory and Applications*, pp. 429–448. Kluwer Academic, Dordrecht, The Netherlands, 1996.

17 H. Heitsch and W. Roemisch. Scenario reduction algorithms in stochastic programming. *Computational Optimization and Applications*, 24:187–206, 2003.

18 J.L. Higle. Variance reduction and objective function evaluation in stochastic linear programs. *INFORMS Journal of Computing*, 10:236–247, 1998.

19 J.L. Higle and S. Sen. *Stochastic Decomposition.* Kluwer Academic Publishers, Dordrecht, 1996.

20 K. Hoyland and S.W. Wallace. Generating scenario trees for multistage decision problems. *Management Science*, 47:296–307, 2001.

21 D.G. Humphrey and J.R. Wilson. A revised simplex search procedure for stochastic simulation response surface optimization. *INFORMS Journal on Computing*, 12:272–283, 2000.

22 G. Infanger. *Planning under Uncertainty: Solving Large-Scale Stochastic*

Linear Programs. Boyd and Fraser, Danvers, MA, 1994.

23 J. James and N. Webber. *Interest Rate Modelling.* Wiley, Chichester, 2000.

24 K.L. Judd. *Numerical Methods in Economics.* MIT Press, Cambridge, MA, 1998.

25 P. Kall and S.W. Wallace. *Stochastic Programming.* Wiley, Chichester, 1994.

26 A.J. King and S.W. Wallace. *Modeling with Stochastic Programming.* Springer, Berlin, 2012.

27 J.P.C. Kleijnen. *Design and Analysis of Simulation Experiments.* Springer, Berlin, 2008.

28 M. Koivu. Variance reduction in sample approximations of stochastic programs. *Mathematical Programming,* 103:463–485, 2005.

29 M. Koivu and T. Pennanen. Epi-convergent discretizations of stochastic programs via integration quadratures. *Numerische Mathematik,* 100:141–163, 2005.

30 D.P. Kroese, T. Taimre, and Z.I. Botev. *Handbook of Monte Carlo Methods.* Wiley, Hoboken, NJ, 2011.

31 Y.K. Kwok. *Mathematical Models of Financial Derivatives.* Springer, Berlin, 1998.

32 C.D. Maranas, I.P. Androulakis, C.A. Floudas, A.J. Berger, and J.M. Mulvey. Solving long-term financial planning problems via global optimization. *Journal of Economic Dynamics and Control,* 21:1405–1425, 1997.

33 Z. Michalewicz. *Genetic Algorithms + Data Structures = Evolution Programs.* Springer, Berlin, 1996.

34 M.J. Miranda and P.L. Fackler. *Applied Computational Economics and Finance.* MIT Press, Cambridge, MA, 2002.

35 L.M.V.G. Pinto M.V.F. Pereira. Multi-stage stochastic optimization applied to energy planning. *Mathematical Programming,* 52:359–375, 1991.

36 R. Poli, J. Kennedy, and T. Blackwell. Particle swarm optimization: An overview. *Swarm Intelligence,* 1:33–57, 2007.

37 W.B. Powell. *Approximate Dynamic Programming: Solving the Curses of Dimensionality* (2nd ed.). Wiley, Hoboken, NJ, 2011.

38 S.T. Rachev. *Probability Metrics and the Stability of Stochastic Models.* Wiley, New York, 1991.

39 A. Ruszczyński and A. Shapiro. Stochastic programming models. In A. Ruszczyński and A. Shapiro, editors, *Stochastic Programming.* Elsevier, Amsterdam, 2003.

40 A. Shapiro, D. Dentcheva, and A. Ruszczyński. *Lectures on Stochastic Programming: Modeling and Theory.* SIAM, Philadelphia, PA, 2009.

41 J.C. Spall. *Introduction to Stochastic Search and Optimization: Estimation, Simulation, and Control.* Wiley, Hoboken, NJ, 2003.

42 J. Sun, C.-H. Lai, and X.-J. Wu. *Particle Swarm Optimisation: Classical and Quantum Perspectives*. CRC Press, Boca Raton, FL, 2011.

43 A. Takayama. *Analytical Methods in Economics*. Harvester Wheatsheaf, New York, 1994.

44 C. Tsallis and D.A. Stariolo. Generalized simulated annealing. *Physica A*, 233:395–406, 1996.

45 C.J.C.H. Watkins and P. Dayan. Q-learning. *Machine Learning*, 8:279–292, 1992.

46 H.P. Williams. *Model Building in Mathematical Programming* (4th ed.). Wiley, Chichester, 1999.

47 L.A. Wolsey. *Integer Programming*. Wiley, New York, 1998.

48 X.-S. Yang. Firefly algorithm, Lévy flights and global optimization. In M. Bramer, R. Ellis, and M. Petridis, editors, *Research and Development in Intelligent Systems XXVI*, pp. 209–218. Springer, 2010.

## Chapter Eleven

Option Pricing

Option pricing is one of the most important application fields for Monte Carlo methods, since option prices may be expressed as expected values under a suitable probability measure. Monte Carlo sampling is a quite flexible strategy that can be adopted when alternative numerical methods, like binomial/trinomial lattices or finite difference methods for solving partial differential equations, are inefficient or impossible to apply. For low-dimensional and path-independent options, Monte Carlo methods are not quite competitive with these alternatives, but for multidimensional or path-dependent options they are often the only viable computational strategy. Even more so when one has to deal with more complicated processes than an innocent geometric Brownian motion (GBM), or even a full-fledged term structure of interest rates. In the past, it was claimed that a significant limitation of Monte Carlo methods was their inability to cope with early exercise features. More recently, ideas from stochastic optimization and stochastic dynamic programming have been adapted to option pricing, paving the way to some Monte Carlo methods for American- and Bermudan-style options.

In this chapter we rely on the path generation methods of Chapter 6, where we have first seen how to price a vanilla call option by random sampling in Section 6.2.1. We will also make good use of the variance reduction strategies of Chapter 8, which were also illustrated in the simple case of a vanilla call option. Finally, when dealing with early exercise features, we use concepts of stochastic programming, like scenario trees, and approximate dynamic programming, which we introduced in Sections 10.5 and 10.8.

To make things a bit more interesting and realistic, we will consider simple examples of multidimensional and path-dependent options, as well as extremely basic interest rate derivatives. As a first example, we consider bidimensional spread options in Section 11.1, where we apply both naive Monte Carlo and stratified sampling. Then, in Section 11.2 we consider path-dependent derivatives, like barrier and arithmetic average Asian options. In these first two sections we consider European-style options, which lend themselves very well to sampling methods, since only the estimation of an expectation is required. The case of American- or Bermudan-style options is more challenging, as their early exercise features require the solution of an optimization problem. In Section 11.3 we show how certain optimization concepts can be adapted to option pricing. We introduce these fundamental concepts within the easy and comfortable

BSM (Black–Scholes–Merton) world. This is a sensible choice from a pedagogical point of view, and sometimes allows us to compare the results obtained
by sampling methods against exact prices. However, the BSM world does not
capture all of the intricacies of financial markets. In Section 11.4 we illustrate
an example of option pricing with stochastic volatility. Finally, we consider
extremely simple interest rate derivatives Section 11.5, i.e., zero coupon bonds
and vanilla options on them, in the cases of the Vasicek and Cox–Ingersoll–Ross
models.

We only consider option pricing in this chapter, but related issues are dealt
with in Chapters 12, where we estimate option greeks, and 13, where we measure the risk of a portfolio of options. Some background about risk-neutral
pricing is provided in Section 3.9.

11.1   European-style multidimensional options in the BSM world

Let us consider what is arguably the simplest option depending on two risk
factors, namely, a European-style *spread option*. This is an option written on
two stocks, whose price dynamics under the risk-neutral measure are modeled
by the following GBMs:

$$dU(t) = rU(t)\, dt + \sigma_U U(t)\, dW_U(t),$$
$$dV(t) = rV(t)\, dt + \sigma_V V(t)\, dW_V(t),$$

where the two Wiener processes have instantaneous correlation ρ. Thus, the
driving process is a bidimensional geometric Brownian motion, for which we
can generate sample paths as shown in Section 6.2.2. The option payoff is

$$\max\big\{V(T) - U(T) - K, 0\big\},$$

i.e., there is a positive payoff when the spread between the two prices exceeds
the threshold K. When $K = 0$ the option is also called *exchange option*. To
see where the name comes from, consider the value at maturity of a portfolio
consisting of one share of U and one option:

$$U(T) + \max\big\{V(T) - U(T), 0\big\} = \max\big\{V(T), U(T)\big\}.$$

Hence, the option allows us to exchange one asset for the other at maturity.
The exchange option is actually fairly simple to deal with, and indeed there is
an analytical pricing formula which is a fairly straightforward generalization of
the Black–Scholes–Merton formula:

```
EuExchange <- function(V0,U0,sigmaV,sigmaU,rho,T,r){
  sigmahat <- sqrt(sigmaU^2+sigmaV^2-2*rho*sigmaU*sigmaV)
  d1 <- (log(V0/U0) + 0.5*T*sigmahat^2)/(sigmahat*sqrt(T))
  d2 <- d1 - sigmahat*sqrt(T)
  return(V0*pnorm(d1) - U0*pnorm(d2))
}
```

FIGURE 11.1 **Exact pricing for an exchange option.**

$$P = V_0\Phi(d_1) - U_0\Phi(d_2),$$

$$d_1 = \frac{\ln(V_0/U_0) + \hat{\sigma}^2 T/2}{\hat{\sigma}\sqrt{T}},$$

$$d_2 = d_1 - \hat{\sigma}\sqrt{T},$$

$$\hat{\sigma} = \sqrt{\sigma_V^2 + \sigma_U^2 - 2\rho\sigma_V\sigma_U},$$

where $\Phi(\cdot)$ is the CDF of the standard normal distribution, and $V_0 = V(0)$, $U_0 = U(0)$. The reason why we get this type of formula is that the payoff has a homogeneous form, which allows to simplify the corresponding partial differential equation by considering the ratio V/U of the two prices.[1] R code implementing this formula is shown in Fig. 11.1.

In this simple case, we have a benchmark against which to assess the quality of Monte Carlo estimates. In order to apply Monte Carlo, we need a path generation strategy, which is rather simple in this case. We recall from Section 6.2.2 that, to generate sample paths for two correlated Wiener processes, we need correlated standard normals. One possible approach relies on the Cholesky factor for the covariance matrix, which in this case is actually a correlation matrix:

$$\Sigma = \begin{bmatrix} 1 & \rho \\ \rho & 1 \end{bmatrix}.$$

It may be verified by straightforward matrix multiplication that $\Sigma = \mathbf{L}\mathbf{L}^\mathsf{T}$, where

$$\mathbf{L} = \begin{bmatrix} 1 & 0 \\ \rho & \sqrt{1-\rho^2} \end{bmatrix}.$$

---

[1] See, e.g., [1, pp. 184–188] for a proof.

```
EuSpreadNaiveMC <- function(V0,U0,K,sigmaV,sigmaU,rho,T,r,
                            numRepl,c.level=0.95){
   # generate sample paths
   S0Vet <- c(V0,U0); muVet <- rep(r,2)
   sigmaVet <- c(sigmaV,sigmaU)
   corrMatrix <- matrix(c(1,rho,rho,1),nrow=2)
   numSteps <- 1
   pathArray <- simMVGBM(S0Vet,muVet,sigmaVet,corrMatrix,
                         T,numSteps,numRepl)
   # collect discounted  payoffs
   dpayoff <- exp(-r*T)*
              pmax(0,pathArray[,2,1]-pathArray[,2,2]-K)
   # in pathArray, the first index refers to sample paths,
   # the second one to time, and the last one to the asset
   aux <- t.test(dpayoff,conf.level=c.level)
   value <- as.numeric(aux$estimate)
   ci <- as.numeric(aux$conf.int)
   return(list(estimate=value,conf.int=ci))
}
```

FIGURE 11.2 **Pricing a spread option by naive Monte Carlo.**

Hence, to simulate bidimensional correlated Wiener processes, we must generate two independent standard normal variates Z_1 and Z_2 and use

$$\epsilon_1 = Z_1, \tag{11.1}$$

$$\epsilon_2 = \rho Z_1 + \sqrt{1 - \rho^2} Z_2, \tag{11.2}$$

to drive path generation. However, for the sake of generality, let us take advantage of the function simMVGBM that we developed in Chapter 6. The resulting R code is displayed in Fig. 11.2 for a spread option. The code is fairly simple, and the only thing worth mentioning is that the array storing the sample paths is tridimensional. To price the exchange option and check the results against the exact formula, we just need to select $K = 0$:

```
> V0 <- 50; U0 <- 60; sigmaV <- 0.3
> sigmaU <- 0.4; rho <- 0.7; T <- 5/12
> r <- 0.05; K <- 0
> EuExchange(V0,U0,sigmaV,sigmaU,rho,T,r)
[1] 0.8633059
> numRepl <- 200000
> set.seed(55555)
> EuSpreadNaiveMC(V0,U0,0,sigmaV,sigmaU,rho,T,r,numRepl)
$estimate
[1] 0.8536415
$conf.int
```

[1] 0.8427841 0.8644989

Now, can we improve the quality of the above estimate? In this bidimensional case, it is fairly easy to resort to stratification with respect to both random variables. However, for illustration purposes, let us pursue both single and double stratification. By single stratification, we mean stratifying with respect to one random variable, e.g., the standard normal ϵ_1 that drives the price of the first stock. Given Eqs. (11.1) and (11.2), we immediately see that the conditional distribution of ϵ_2 given ϵ_1 is normal with

$$\mathrm{E}[\epsilon_2 \,|\, \epsilon_1] = \rho\epsilon_1, \qquad \mathrm{Var}(\epsilon_2 \,|\, \epsilon_1) = (1 - \rho^2)\mathrm{Var}(Z_2) = (1 - \rho^2).$$

However, we can obtain the same result by using a more general result about conditional distributions for the multivariate normal, which was given in Section 3.3.4.1 and can be applied in more general settings. The details are given in Example 3.7, and we urge the reader to check that we indeed obtain the same results. When stratifying a standard normal, it may be convenient to stratify the underlying uniform variable. As we have seen in Example 8.4, if we have m strata, indexed by $j = 1, \ldots, m$, and N replications, we may generate $N_j = N/m$ observations. Each stratum is an interval of the form

$$\left[\frac{j-1}{m}, \frac{j}{m}\right).$$

So, our procedure requires:

1. To generate uniform random variables U_k, $k = 1, \ldots, N_j$, for each stratum j, and to transform them into

$$V_{jk} = \frac{U_k + j - 1}{m} = \frac{j-1}{m} + \frac{U_k}{m},$$

the observation k within stratum j.

2. To invert the CDF of the standard normal and generate

$$\epsilon_{1jk} = \Phi^{-1}(V_{jk}).$$

3. To sample $\epsilon_{2jk} \sim \mathsf{N}(\rho\epsilon_{1jk}, 1 - \rho^2)$.

4. To generate the corresponding lognormal variables and evaluate the option payoff.

All of this is accomplished by the R code in Fig. 11.3. Note that, in order to compute the overall estimate and the confidence interval, according to Eq. (8.13), we have to compute the sample mean and the sample variance within each stratum.

Double stratification requires to generate two uniform variables in a subsquare of the unit square:

$$\left[\frac{j-1}{m}, \frac{j}{m}\right) \times \left[\frac{l-1}{m}, \frac{l}{m}\right), \qquad j, l = 1, \ldots, m.$$

```
EuSpreadStrat1 <- function(V0,U0,K,sigmaV,sigmaU,rho,T,r,
                           numRepl,numStrat,c.level=0.95){
  nuTV <- (r - 0.5*sigmaV^2)*T
  sigmaTV <- sigmaV*sqrt(T)
  nuTU <- (r - 0.5*sigmaU^2)*T
  sigmaTU <- sigmaU*sqrt(T)
  rho2 <- sqrt(1-rho^2)
  numPerStrat <- ceiling(numRepl/numStrats)
  meanStrat <- numeric(numStrats)
  varStrat <- numeric(numStrats)
  for (k in 1:numStrats) {
    U <- (k-1)/numStrats+runif(numPerStrat)/numStrats
    epsV <- qnorm(U)
    epsU <- rnorm(numPerStrat,rho*epsV,rho2)
    VT <- V0*exp(nuTV+sigmaTV*epsV)
    UT <- U0*exp(nuTU+sigmaTU*epsU)
    dPayoff <- exp(-r*T)*pmax(0,VT-UT-K)
    meanStrat[k] <- mean(dPayoff)
    varStrat[k] <- var(dPayoff)
  }
  meanTot <- mean(meanStrat)
  varTot <- mean(varStrat)
  alpha <- (100-c.level)/100
  half <- qnorm(1-alpha/2)*sqrt(varTot)
  ci <- c(meanTot-half,meanTot+half)
  return(list(estimate=meanTot,conf.int=ci))
}
```

FIGURE 11.3 **Pricing a spread option by single stratification.**

The rest of the procedure is quite similar to the one for single stratification and can be implemented as shown in Fig. 11.4.

Now we may compare the accuracy of naive Monte Carlo against the two stratifications. We choose $K = 0$ in order to use the exact formula for the exchange option, and we also note that, for a fair comparison, if we set the number of strata to 25 in simple stratification, we should set it to 5 in double stratification, since the parameter numStrat specifies the number of strata along each dimension:

```
> V0 <- 50; U0 <- 45; sigmaV <- 0.2
> sigmaU <- 0.3; rho <- 0.5; T <- 1
> r <- 0.05; K <- 0
> EuExchange(V0,U0,sigmaV,sigmaU,rho,T,r)
[1] 7.887551
> numRepl <- 10000
> set.seed(55555)
```

```
EuSpreadStrat2 <- function(V0,U0,K,sigmaV,sigmaU,rho,T,r,
                           numRepl,numStrat,c.level=0.95){
  nuTV <- (r - 0.5*sigmaV^2)*T
  sigmaTV <- sigmaV*sqrt(T)
  nuTU <- (r - 0.5*sigmaU^2)*T
  sigmaTU <- sigmaU*sqrt(T)
  rho2 <- sqrt(1-rho^2)
  numPerStrat <- ceiling(numRepl/numStrats)
  meanStrat <- matrix(0,numStrats,numStrats)
  varStrat <- matrix(0,numStrats,numStrats)
  for (k in 1:numStrats) {
    for (j in 1:numStrats) {
      U1 <- (k-1)/numStrats+runif(numPerStrat)/numStrats
      U2 <- (j-1)/numStrats+runif(numPerStrat)/numStrats
      epsV <- qnorm(U1)
      epsU <- rho*epsV + rho2*qnorm(U2)
      VT <- V0*exp(nuTV+sigmaTV*epsV)
      UT <- U0*exp(nuTU+sigmaTU*epsU)
      dPayoff <- exp(-r*T)*pmax(0,VT-UT-K)
      meanStrat[k,j] <- mean(dPayoff)
      varStrat[k,j] <- var(dPayoff)
    }
  }
  meanTot <- mean(meanStrat)
  varTot <- mean(as.numeric(varStrat))
  alpha <- (100-c.level)/100
  half <- qnorm(1-alpha/2)*sqrt(varTot)
  ci <- c(meanTot-half,meanTot+half)
  return(list(estimate=meanTot,conf.int=ci))
}
```

FIGURE 11.4 **Pricing a spread option by double stratification.**

```
> EuSpreadNaiveMC(V0,U0,K,sigmaV,sigmaU,rho,T,r,numRepl)
$estimate
[1] 7.934347
$conf.int
[1] 7.771838 8.096855
> numStrats <- 25
> EuSpreadStrat1(V0,U0,K,sigmaV,sigmaU,rho,T,r,
        numRepl,numStrats)
$estimate
[1] 7.899219
$conf.int
[1] 7.807015 7.991424
> numStrats <- 5
```

```
> EuSpreadStrat2(V0,U0,K,sigmaV,sigmaU,rho,T,r,
        numRepl,numStrats)
$estimate
[1] 7.898201
$conf.int
[1] 7.869630 7.926772
```

The effect of stratification is rather evident. Clearly, this is a fairly trivial example, and when dealing with several random variables, full stratification is not feasible. Nevertheless, in the case of multivariate normal distributions, the aforementioned results about conditioning can be exploited to stratify with respect to the most relevant dimensions. We also note that these results are the foundation of the Brownian bridge technique. In other cases, we may lack explicit results, but rejection sampling strategies have been proposed to apply the idea.

11.2   European-style path-dependent options in the BSM world

Pricing European-style, path-dependent options is, in principle, quite easy within a Monte Carlo framework. All we need to do is generate sample paths, under the risk-neutral measure, and collect the discounted payoff. However, when prices have to be sampled at many points in time, the process can be rather inefficient. Hence, we may try to improve the accuracy/performance trade-off by applying the variance reduction strategies of Chapter 8, or sample path generation by low-discrepancy sequences. In this section we consider two cases of path-dependent options:

1. An example of barrier option, namely, a down-and-out put option, to which we apply importance sampling
2. An arithmetic average Asian call option, to which we apply control variates and path generation by scrambled Sobol sequences

11.2.1   PRICING A BARRIER OPTION

In barrier options, a specific asset price S_b is selected as a barrier value. During the life of the option, this barrier may be crossed or not, an event that may activate or cancel the option. In knock-out options, the contract is canceled if the barrier value is crossed at any time during the whole life; on the contrary, knock-in options are activated only if the barrier is crossed. The barrier S_b may be above or below the initial asset price S_0: if $S_b > S_0$, we have an up-option; if $S_b < S_0$, we have a down-option. These features may be combined with the payoffs of call and put options to define an array of barrier options. For instance, a down-and-out put option is a put option that becomes void if the asset price falls below the barrier S_b; in this case we need $S_b < S_0$ and

$S_b < K$ for the option to make sense. The rationale behind such an option is that the risk for the option writer is reduced and, therefore, it is reasonable to expect that a down-and-out put option is cheaper than a vanilla one. From the option holder's viewpoint, this means that the potential payoff is reduced; however, if you are interested in options to manage risk, and not as a speculator, this also means that you may get cheaper insurance. By the same token, other combinations of features may be devised, like up-and-out call and down-and-in put options. Actually, we do not need the ability of pricing all of them, as precise relationships between their prices may be observed. Let us consider, for instance a down-and-in put option. Holding both a down-and-out and a down-and-in put option, with the same barrier level and strike, is equivalent to holding a vanilla put option. To see this, note that the barrier is either crossed or not; hence, one of the two options will be active at maturity. Thus, we have the following parity relationship:

$$P = P_{\mathrm{di}} + P_{\mathrm{do}}, \tag{11.3}$$

where P is the price of the vanilla put, and P_{di} and P_{do} are the prices for the down-and-in and the down-and-out options, respectively. Sometimes a rebate is paid to the option holder if the barrier is crossed and option is canceled; in such a case the parity relationship above is not correct.

In principle, the barrier might be monitored continuously; in practice, periodic monitoring is applied (e.g., the price could be checked each day at the close of trading). This may affect the price, as a lower monitoring frequency makes the detection of barrier crossing less likely. As a general rule for path-dependent options, continuous-time monitoring is more amenable to analytical treatment. Indeed, exact formulas are available for certain barrier options, such as a down-and-out put option with continuous-time monitoring. The following formula, where S_0, r, σ, K, and T have the usual meaning, gives the price of this option:[2]

$$P = Ke^{-rT} \left\{ \Phi(d_4) - \Phi(d_2) - a\left[\Phi(d_7) - \Phi(d_5)\right] \right\} \\ - S_0 \left\{ \Phi(d_3) - \Phi(d_1) - b\left[\Phi(d_8) - \Phi(d_6)\right] \right\},$$

where

$$a = \left(\frac{S_b}{S_0}\right)^{-1+2r/\sigma^2}, \qquad b = \left(\frac{S_b}{S_0}\right)^{1+2r/\sigma^2},$$

---

[2]See, e.g., [17, pp. 250–251].

```
EuDOPut <- function(S0,K,Sb,T,r,sigma){
  a <- (Sb/S0)^(-1 + (2*r / sigma^2))
  b <- (Sb/S0)^(1 + (2*r / sigma^2))
  d1 <- (log(S0/K)+(r+sigma^2/2)*T)/(sigma*sqrt(T))
  d2 <- (log(S0/K)+(r-sigma^2/2)*T)/(sigma*sqrt(T))
  d3 <- (log(S0/Sb)+(r+sigma^2/2)*T)/(sigma*sqrt(T))
  d4 <- (log(S0/Sb)+(r-sigma^2/2)*T)/(sigma*sqrt(T))
  d5 <- (log(S0/Sb)-(r-sigma^2/2)*T)/(sigma*sqrt(T))
  d6 <- (log(S0/Sb)-(r+sigma^2/2)*T)/(sigma*sqrt(T))
  d7 <- (log(S0*K/Sb^2)-(r-sigma^2/2)*T)/(sigma*sqrt(T))
  d8 <- (log(S0*K/Sb^2)-(r+sigma^2/2)*T)/(sigma*sqrt(T))
  P <- K*exp(-r*T)*(pnorm(d4)-pnorm(d2)-
                  a*(pnorm(d7)-pnorm(d5))) -
      S0*(pnorm(d3)-pnorm(d1)-b*(pnorm(d8)-pnorm(d6)))
  return(P)
}
```

FIGURE 11.5 **Exact pricing for a down-and-out barrier option with continuous-time monitoring.**

and

$$d_1 = \frac{\log(S_0/K) + (r + \sigma^2/2)T}{\sigma\sqrt{T}}, \qquad d_2 = \frac{\log(S_0/K) + (r - \sigma^2/2)T}{\sigma\sqrt{T}},$$

$$d_3 = \frac{\log(S_0/S_b) + (r + \sigma^2/2)T}{\sigma\sqrt{T}}, \qquad d_4 = \frac{\log(S_0/S_b) + (r - \sigma^2/2)T}{\sigma\sqrt{T}},$$

$$d_5 = \frac{\log(S_0/S_b) - (r - \sigma^2/2)T}{\sigma\sqrt{T}}, \qquad d_6 = \frac{\log(S_0/S_b) - (r + \sigma^2/2)T}{\sigma\sqrt{T}},$$

$$d_7 = \frac{\log(S_0 K/S_b^2) - (r - \sigma^2/2)T}{\sigma\sqrt{T}}, \qquad d_8 = \frac{\log(S_0 K/S_b^2) - (r + \sigma^2/2)T}{\sigma\sqrt{T}}.$$

R code implementing these formulas is given in Fig. 11.5. When monitoring occurs in discrete time, we should expect that the price for a down-and-out option is increased, since breaching the barrier is less likely. An approximate correction has been suggested,[3] based on the idea of applying the analytical formula above with a correction on the barrier level:

$$S_b \quad \Rightarrow \quad S_b e^{\pm 0.5826 \cdot \sigma \sqrt{\delta t}}, \tag{11.4}$$

where the term 0.5826 derives from the Riemann zeta function, δt is time elapsing between two consecutive monitoring time instants, and the sign $\pm$ depends on the option type. For a down-and-out put we should choose the minus sign, as

---

[3]See [6] or [13, p. 266].

the barrier level should be lowered to reflect the reduced likelihood of crossing the barrier. Notice that this barrier option is less expensive than the corresponding vanilla; the price of the barrier option tends to that of the vanilla option if $S_b \to 0$. By lowering the barrier according to Eq. (11.4) we increase the option price. The following snapshot, where we assume daily monitoring for an option maturing in 2 months consisting of 30 days each, illustrates the difference between a vanilla and a down-and-out put, and the effect of the correction:

```
> S0 <- 50; K <- 45; Sb <- 40
> r <- 0.05; sigma <- 0.4; T <- 2/12
> numSteps <- 60; numDays <- 360
> EuVanillaPut(S0,K,T,r,sigma)
[1] 1.107108
> EuDOPut(S0,K,T,r,sigma,Sb)
[1] 0.1897568
> Sbc <- Sb*exp(-0.5826*sigma*sqrt(1/numDays))
> EuDOPut(S0,K,T,r,sigma,Sbc)
[1] 0.2420278
```

If we want to price the option using Monte Carlo simulation, we must generate sample paths like

$$\mathbf{S} = \{S_1, S_2, \ldots, S_j, \ldots, S_M\},$$

where the integer time subscript refers to time step of length δt, and $T = M\,\delta t$. The naive Monte Carlo estimator is

$$P_{\mathrm{do}} = e^{-rT}\mathrm{E}\big[I(\mathbf{S})(K - S_M)^+\big],$$

where the indicator function I is defined as

$$I(\mathbf{S}) = \begin{cases} 1, & \text{if } S_j > S_b \text{ for all } j, \\ 0, & \text{otherwise.} \end{cases}$$

By relying on the function `simvGBM` to sample a GBM process, it is easy to write R code pricing the option by Monte Carlo, as shown in Fig. 11.6. Note the use of `any` to avoid a `for` loop to inspect the path. The function returns the point estimate and the confidence interval as usual, but also the number of sample paths on which the barrier is crossed, for reasons that we will clarify shortly. With the same parameters as before we obtain

```
> EuDOPut(S0,K,T,r,sigma,Sbc)
[1] 0.2420278
> numRepl <- 50000
> set.seed(55555)
> DownOutPutNaive(S0,K,T,r,sigma,Sb,numSteps,numRepl)
$estimate
[1] 0.2437642
$conf.int
[1] 0.2368905 0.2506378
```

```
DownOutPutNaive <- function (S0,K,T,r,sigma,Sb,numSteps,
                                numRepl,c.level=0.95){
  payoff <- numeric(numRepl)
  numCrossed <- 0
  for (k in 1:numRepl){
    path <- simvGBM(S0,r,sigma,T,numSteps,1)
    if (any(path <= Sb)){
      payoff[k] <- 0
      numCrossed <- numCrossed+1
    } else
      payoff[k] <- max(0, K-path[numSteps+1])
  }
  aux <- t.test(exp(-r*T)*payoff,conf.level=c.level)
  value <- as.numeric(aux$estimate)
  ci <- as.numeric(aux$conf.int)
  return(list(estimate=value,conf.int=ci,
                numCrossed=numCrossed))
}
```

FIGURE 11.6 **Pricing a down-and-out barrier option by naive Monte Carlo.**

```
$numCrossed
[1] 7835
```

We notice that the barrier has been crossed in 7835 replications out of 50,000. If we have to price a down-and-in put option, we may price the down-and-out and use the parity relationship of Eq. (11.3), or we may adapt the code, which is very easy to do.

The choice between these alternatives may by a matter of convenience. Let us suppose that the barrier is low and that crossing it is a rare event. If we consider pricing a down-and-in option, crossing the barrier is the "interesting" event, and if it is rare, then in most replications the payoff is just zero. Hence, we may consider using importance sampling to improve the performance. In this case, it is more convenient to price the down-and-in directly, and then use parity to price the down-and-out. One possible idea is changing the drift of the asset price in such a way that crossing the barrier is more likely.[4] In this case we should decrease the drift, simulate GBM under this distorted measure, and correct the estimator by a likelihood ratio, just as we did in Section 8.6.1. Here, however, the likelihood ratio involves several random variables, and in order to figure it out we should step back and consider what we do to generate the sample path **S**. For each time step, we generate a normal variate Z_j with

---

[4]See [15].

expected value

$$\mu = \left(r - \frac{\sigma^2}{2}\right)\delta t \qquad (11.5)$$

and variance $\sigma^2 \delta t$. All of these variates are mutually independent, and the asset price path is generated by setting

$$\log S_j - \log S_{j-1} = Z_j.$$

Let $\mathbf{Z}$ be the vector of the normal variates, and let $f(\mathbf{Z})$ be its joint density. Now, let us assume that we lower the drift r by an amount b and generate normals with expected value

$$\mu - b \cdot \delta t = \mu - b_{\delta t}.$$

Note that we interpret b as an annualized drift component, with continuous compounding just like the risk-free rate r, and we have to multiply it by the time step δt. Let $g(\mathbf{Z})$ be the joint density for the normal variates generated with this modified expected value. Thanks to independence, both multivariate densities are just the product of M univariate normal densities, and their ratio can be manipulated as follows:

$$\frac{f(z_1, \ldots, z_M)}{g(z_1, \ldots, z_M)}$$

$$= \exp\left\{-\frac{1}{2}\sum_{j=1}^{M}\left(\frac{z_j - \mu}{\sigma\sqrt{\delta t}}\right)^2\right\}\exp\left\{\frac{1}{2}\sum_{j=1}^{M}\left(\frac{z_j - \mu + b_{\delta t}}{\sigma\sqrt{\delta t}}\right)^2\right\}$$

$$= \exp\left\{-\frac{1}{2\sigma^2\,\delta t}\sum_{j=1}^{M}\left[(z_j - \mu)^2 - (z_j - \mu + b_{\delta t})^2\right]\right\}$$

$$= \exp\left\{-\frac{1}{2\sigma^2\,\delta t}\sum_{j=1}^{M}\left[-2(z_j - \mu)b_{\delta t} - b_{\delta t}^2\right]\right\}$$

$$= \exp\left\{-\frac{1}{2\sigma^2\,\delta t}\left[-2b_{\delta t}\sum_{j=1}^{M}z_j + 2M\mu\,b_{\delta t} - Mb_{\delta t}^2\right]\right\}$$

$$= \exp\left\{\frac{b_{\delta t}}{\sigma^2\,\delta t}\sum_{j=1}^{M}z_j - \frac{Mb_{\delta t}}{\sigma^2}\left(r - \frac{\sigma^2}{2}\right) + \frac{Mb_{\delta t}^2}{2\sigma^2\,\delta t}\right\},$$

where in the last equality we replace the generic drift μ according to Eq. (11.5). R code implementing importance sampling for a down-and-in put option is reported in Fig. 11.7. By setting the parameter b to 0, we essentially run naive Monte Carlo, so we may just use this function to assess the variance reduction in pricing the down-and-in put option:

```
> S0 <- 50; K <- 50; Sb <- 40; r <- 0.05; sigma <- 0.4;
> T <- 5/12; numSteps <- 5*30; numDays <- 360
```

```
DownInPutIS <- function(S0,K,T,r,sigma,Sb,numSteps,
                        numRepl,b,c.level=0.95){
  # invariant quantities and vector allocation
  dt <- T/numSteps
  nudt <- (r-0.5*sigma^2)*dt
  bdt <- b*dt
  sidt <- sigma*sqrt(dt)
  numCrossed <- 0
  payoff <- numeric(numRepl)
  ISRatio <- numeric(numRepl)
  # Generate asset paths and payoffs
  for (k in 1:numRepl){
    vetZ <- nudt-bdt+sidt*rnorm(numSteps)
    logPath <- cumsum(c(log(S0), vetZ))
    path <- exp(logPath)
    if (any(path <= Sb)){
      numCrossed <- numCrossed+1
      payoff[k] <- max(0, K-path[numSteps+1])
      ISRatio[k] <-    # evaluate likelihood ratio
        exp(numSteps*bdt^2/2/sigma^2/dt +
            bdt/sigma^2/dt*sum(vetZ) -
            numSteps*bdt/sigma^2*(r - sigma^2/2))
    }
  }
  ISvet <- exp(-r*T)*payoff*ISRatio
  aux <- t.test(ISvet,conf.level=c.level)
  value <- as.numeric(aux$estimate)
  ci <- as.numeric(aux$conf.int)
  return(list(estimate=value,conf.int=ci,
              numCrossed=numCrossed))
}
```

FIGURE 11.7   **Pricing a down-and-in put option by importance sampling.**

```
> Sbc <- Sb*exp(-0.5826*sigma*sqrt(1/numDays))
> numRepl <- 50000
> set.seed(55555)
> DownInPutIS(S0,K,T,r,sigma,Sb,numSteps,numRepl,0)
$estimate
[1] 3.90588
$conf.int
[1] 3.850881 3.960879
$numCrossed
[1] 18936
> set.seed(55555)
> DownInPutIS(S0,K,T,r,sigma,Sb,numSteps,numRepl,0.8)
```

```
$estimate
[1] 3.907309
$conf.int
[1] 3.882772 3.931846
$numCrossed
[1] 40303
```

It seems that we do gain something by applying importance sampling. However, apart from the intuition that rare and interesting events should be made more likely, it is not quite clear how the drift should be selected. In Section 13.4 we will consider a more detailed analysis to justify the selection of an importance sampling measure.

11.2.2 PRICING AN ARITHMETIC AVERAGE ASIAN OPTION

Barrier options are path-dependent, but to a weak degree, since the only relevant information is whether the barrier has been crossed or not. A stronger degree of path dependency is typical of Asian options, as the payoff depends on the average asset price over the option life. Different Asian options may be devised, depending on how the average is calculated. Sampling may be carried out in discrete or (in principle) continuous time, and the average itself may be arithmetic or geometric. The discrete arithmetic average is

$$A_{\mathrm{da}} = \frac{1}{M} \sum_{j=1}^{M} S(t_j),$$

where t_j, $j = 1, \ldots, M$, are the discrete sampling times. The corresponding geometric average is

$$A_{\mathrm{dg}} = \left[\prod_{j=1}^{M} S(t_j) \right]^{1/M}.$$

If continuous sampling is adopted, we have

$$A_{\mathrm{ca}} = \frac{1}{T} \int_0^T S(t)\, dt,$$

$$A_{\mathrm{cg}} = \exp\left[\frac{1}{T} \int_0^T \log S(t)\, dt \right].$$

The chosen average A, whatever it is, defines the option payoff by playing the role of either a rate or a strike. For instance, an average rate call has a payoff given by

$$\max\{A - K, 0\},$$

whereas in an average strike call we have

$$\max\{S(T) - A, 0\}.$$

By the same token, we may define an average rate put,

$$\max\{K - A, 0\},$$

or an average strike put,

$$\max\{A - S(T), 0\}.$$

Early exercise features may also be defined in the contract. Hence, we may devise American- or Bermudan-style Asian options, even though the name does sound a bit weird.

Pricing an Asian option is, in general, a nontrivial task. The problem is not trivial not only because the payoff is more complicated than in a vanilla option, but also because when pricing the option after its inception, we must account for the average cumulated so far, and the fact that even if the sampling period δt is uniform, the time to the next price observation is different. Nevertheless, there are some analytical pricing formulas for Asian option. Consider the payoff of a discrete-time, geometric average Asian option:

$$\max\left\{\left[\prod_{j=1}^{M} S(t_j)\right]^{1/M} - K, \ 0\right\}.$$

Since the product of lognormal random variables is still lognormal, it is possible to find an analytical formula for this option, which looks like a modified Black–Scholes–Merton formula.[5] In the formula, we are pricing the option at time t, and G_t is the current geometric average, if any observation has been already collected. If so, t_m is the last time instant at which we observed the price of the underlying asset, m is the corresponding index, t_{m+1} is the time instant of the next observation, and $t_M = T$ is the option maturity:

$$P_{\mathrm{GA}} = e^{-rT}\left[e^{a+(b/2)}\Phi(x) - K\Phi\left(x - \sqrt{b}\right)\right], \qquad (11.6)$$

where

$$a = \frac{m}{M}\log(G_t) + \frac{M-m}{M}\left[\log(S_0) + \nu(t_{m+1} - t) + \frac{1}{2}\nu(T - t_{m+1})\right],$$

$$b = \frac{(M-m)^2}{M^2}\sigma^2(t_{m+1} - t) + \frac{\sigma^2(T - t_{m+1})}{6M^2}(M - m)\big[2(M - m) - 1\big],$$

$$\nu = r - \frac{1}{2}\sigma^2, \qquad x = \frac{a - \log(K) + b}{\sqrt{b}}.$$

The formula gets considerably simplified if we just consider the option price at its inception, i.e., at time $t = 0$. In such a case $m = 0$, and the resulting R implementation is illustrated in Fig. 11.8.

In the case of an arithmetic average option, since the sum of lognormals is not lognormal, we have to use numerical methods. Pricing a European-style

---

```
EuAsianGeometric <- function(S0,K,T,r,sigma,numSamples){
  dT <- T/numSamples
  nu <- r-sigma^2/2
  a <- log(S0)+nu*dT+0.5*nu*(T-dT)
  b <- sigma^2*dT+sigma^2*(T-dT)*
            (2*numSamples-1)/6/numSamples
  x <- (a-log(K)+b)/sqrt(b)
  value <- exp(-r*T)*
            (exp(a+b/2)*pnorm(x)-K*pnorm(x-sqrt(b)))
  return(value)
}
```

FIGURE 11.8 **Exact pricing of a geometric average Asian option.**

```
EuAsianNaive <- function(S0,K,T,r,sigma,numSamples,
                         numRepl,c.level=0.95){
  payoff <- numeric(numRepl)
  for (k in 1:numRepl){
    path <- simvGBM(S0,r,sigma,T,numSamples,1)
    payoff[k] <- max(0, mean(path[-1])-K)
  }
  aux <- t.test(exp(-r*T)*payoff,conf.level=c.level)
  value=as.numeric(aux$estimate)
  ci=as.numeric(aux$conf.int)
  return(list(estimate=value,conf.int=ci))
}
```

FIGURE 11.9 **Naive Monte Carlo method for a European-style Asian option.**

Asian option by Monte Carlo is quite easy in principle, since we have just to generate sample paths and take an average. In Fig. 11.9 we show the R code for a discrete, arithmetic average rate call option:

```
> S0 <- 50; K <- 50; r <- 0.05; sigma <- 0.4
> T <- 1; numSamples <- 12; numRepl <- 50000
> EuVanillaCall(S0,K,T,r,sigma)
[1] 9.011476
> set.seed(55555)
> EuAsianNaive(S0,K,T,r,sigma,numSamples,numRepl)
$estimate
[1] 5.436376
$conf.int
[1] 5.359236 5.513516
```

It is interesting to note that the Asian option is cheaper than the corresponding vanilla, as there is less volatility in the average price than in the price at maturity. This explains why Asian options may provide cheaper insurance when multiple time instants are relevant for a hedger. In order to improve the quality of estimates, we may pursue variance reduction strategies, most notably control variates, or adopt path generation by low-discrepancy sequences.

11.2.2.1 Control variates for an arithmetic average Asian option

We know from Section 8.3 that, in order to apply this kind of variance reduction strategy, we need a suitable control variate, i.e., a random variable, with a known expected value, which is correlated with the option payoff. In the BSM framework, the following possibilities can be considered:

- Using the expected value of the arithmetic average, which is known in the case of geometric Brownian motion. Let us recall the formula

$$\sum_{i=1}^{M} \alpha^i = \frac{\alpha(1 - \alpha^M)}{1 - \alpha}.$$

The expected value of the arithmetic average of stock prices is (under the risk-neutral measure):

$$E[A] = E\left[\frac{1}{M}\sum_{j=1}^{M} S(t_j)\right] = \frac{1}{M}\sum_{j=1}^{M} E[S(j\,\delta t)]$$

$$= \frac{1}{M}\sum_{j=1}^{M} S(0)e^{rj\,\delta t} = \frac{S(0)}{M}\sum_{j=1}^{M}[e^{r\,\delta t}]^j = \frac{S(0)}{M}\frac{e^{r\,\delta t}(1 - e^{rM\,\delta t})}{1 - e^{r\,\delta t}}.$$

This control variate does capture an essential feature of the option payoff, but not its nonlinearity.
- Using a vanilla call option, whose price is given by the BSM formula. This control variate, unlike the previous one, does capture nonlinearity in the option payoff, but it disregards the sample path.
- Using the geometric average Asian, whose price is given in Eq. (11.6). This control variate is definitely the most sophisticated and it embodies very deep knowledge, even though the kind of average is different.

The R function code in Fig. 11.10 is a generic function to implement control variates for the arithmetic average Asian option. The input argument `SampleCV` is actually a function that returns the value of the control variate for a given sample path; `expCV` is the expected value of the control variate. The script of Fig. 11.11 shows how to use the generic function to implement the three proposed strategies and produced the following (edited) output:

```
EuAsianCV <- function(S0,K,T,r,sigma,numSamples,numRepl,
                   SampleCV, expCV,numPilot,c.level=0.95){
  # precompute discount factor
  df <- exp(-r*T)
  # pilot replications to set control parameter
  cVarsP <- numeric(numPilot)
  arithPrices <- numeric(numPilot)
  for (k in 1:numPilot){
    path <- simvGBM(S0,r,sigma,T,numSamples,1)
    cVarsP[k] <- SampleCV(path)
    arithPrices[k] <- df*max(0,mean(path[-1]) - K)
  }
  covar <- cov(cVarsP, arithPrices)
  cstar <- -covar/var(cVarsP)
  # MC run
  CVestimate <- numeric(numRepl)
  for (k in 1:numRepl){
    path <- simvGBM(S0,r,sigma,T,numSamples,1)
    cVar <- SampleCV(path)
    arithPrice <- df*max(0,mean(path[-1]) - K)
    CVestimate[k] <- arithPrice + cstar*(cVar-expCV)
  }
  aux <- t.test(CVestimate,conf.level=c.level)
  value=as.numeric(aux$estimate)
  ci=as.numeric(aux$conf.int)
  return(list(estimate=value,conf.int=ci))
}
```

FIGURE 11.10 **A generic function to price an Asian option by control variates.**

```
> EuAsianNaive(S0,K,T,r,0,sigma,numSamples,numRepl)
$estimate
[1] 5.436376
$conf.int
[1] 5.359236 5.513516
> # CV is average price
> EuAsianCV(S0,K,T,r,0,sigma,numSamples,numRepl,
+          SampleCV1,expCV1,numPilot)
$estimate
[1] 5.396385
$conf.int
[1] 5.364076 5.428693
> # CV is vanilla call
> EuAsianCV(S0,K,T,r,0,sigma,numSamples,numRepl,
+          SampleCV2,expCV2,numPilot)
$estimate
```

```
S0 <- 50; K <- 50; r <- 0.05; sigma <- 0.4
T <- 1; numSamples <- 12; numRepl <- 50000
set.seed(55555)
EuAsianNaive(S0,K,T,r,0,sigma,numSamples,numRepl)


numPilot <- 5000; numRepl <- 45000
# CV is average price
alpha <- exp(r*T/numSamples)
expCV1 <- S0*alpha*(1-alpha^numSamples)/
          (numSamples*(1-alpha))
SampleCV1 <- function(path) mean(path[-1])
set.seed(55555)
EuAsianCV(S0,K,T,r,0,sigma,numSamples,numRepl,
          SampleCV1,expCV1,numPilot)
# CV is vanilla call
expCV2 <- EuVanillaCall(S0,K,T,r,sigma)
SampleCV2 <- function(path) exp(-r*T)*
  max(path[length(path)]-K,0)
set.seed(55555)
EuAsianCV(S0,K,T,r,0,sigma,numSamples,numRepl,
          SampleCV2,expCV2,numPilot)
# CV is geometric
SampleCV3 <- function(path) exp(-r*T)*
  max(0,(prod(path[-1]))^(1/numSamples) - K)
expCV3 <- EuAsianGeometric(S0,K,T,r,sigma,numSamples)
set.seed(55555)
EuAsianCV(S0,K,T,r,0,sigma,numSamples,numRepl,
          SampleCV3,expCV3,numPilot)
```

FIGURE 11.11 **A script to compare the three control variates.**

```
[1] 5.425655
$conf.int
[1] 5.383384 5.467925
> # CV is geometric
> EuAsianCV(S0,K,T,r,0,sigma,numSamples,numRepl,
+           SampleCV3,expCV3,numPilot)
$estimate
[1] 5.394813
$conf.int
[1] 5.390427 5.399199
```

We clearly see that the geometric average option does an excellent job, even though the two alternatives also reduce variance a bit with respect to naive Monte Carlo. This is easily explained by checking the correlation between the naive Monte Carlo estimator and the three control variates. This can be done

```
EuAsianCompareCorr <- function(S0,K,T,r,sigma,numSamples,
                                numRepl){
  df <- exp(-r*T)
  averagePrices <- numeric(numRepl)
  vanillaPrices <- numeric(numRepl)
  geomPrices <- numeric(numRepl)
  arithPrices <- numeric(numRepl)
  for (k in 1:numRepl){
    path <- simvGBM(S0,r,sigma,T,numSamples,1)
    averagePrices[k] <- mean(path[-1])
    vanillaPrices[k] <- df*max(0,path[numSamples+1] - K)
    geomPrices[k] <- df*max(0,
                      (prod(path[-1]))^(1/numSamples) - K)
    arithPrices[k] <- df*max(0,mean(path[-1]) - K)
  }
  cat('Correlations between arithmetic Asian and\n')
  aux <- cor(averagePrices, arithPrices)
  cat('   average stock price ', aux, '\n')
  aux <- cor(vanillaPrices, arithPrices)
  cat('   vanilla call ', aux, '\n')
  aux <- cor(geomPrices, arithPrices)
  cat('   geometric Asian ', aux, '\n')
}
```

FIGURE 11.12 **Comparing the correlation of the three control variates.**

by using the function of Fig. 11.12, which yields the following estimates of correlation:

```
> set.seed(55555)
> EuAsianCompareCorr(S0,K,T,r,sigma,numSamples,50000)
Correlations between arithmetic Asian and
   average stock price  0.9179336
   vanilla call  0.8543479
   geometric Asian  0.9985351
```

Indeed, there is an extremely strong correlation between the arithmetic and geometric average payoffs, which explains why this control variates does such a good job.

11.2.2.2 Using low-discrepancy sequences

Armed with the function lowDiscrGBM to generate sample paths of a GBM process (see Fig. 9.9), it is easy to price the arithmetic average Asian option using low-discrepancy sequences. The code is given in Fig. 11.13, and we may check its performance on the same example that we have used with control vari-

```
EuAsianLOWD <- function(S0,K,T,r,sigma,numSamples,
                        numRepl,type="sobol",scrambling=0){
  paths <- lowDiscrGBM(S0,r,sigma,T,numSamples,
                       numRepl,type,scrambling)
  ave <- rowMeans(paths[,-1])
  payoff <- pmax(0, ave-K)
  return(exp(-r*T)*mean(payoff))
}
```

FIGURE 11.13 **Using low-discrepancy sequences to price the arithmetic average Asian option.**

ates. We may consider the value provided by the most powerful control variate, 5.394813, as a quite accurate estimate. In the following snapshot we compare the performance of Halton and Sobol sequences with different scramblings:

```
> S0 <- 50; K <- 50; r <- 0.05; sigma <- 0.4
> T <- 1; numSamples <- 12; numRepl <- 10000
> EuAsianLOWD(S0,K,T,r,sigma,numSamples,
+             numRepl,"halton")
[1] 5.341912
> EuAsianLOWD(S0,K,T,r,sigma,numSamples,
+             numRepl,"sobol",0)
[1] 5.413302
> EuAsianLOWD(S0,K,T,r,sigma,numSamples,
+             numRepl,"sobol",1)
[1] 5.402433
> EuAsianLOWD(S0,K,T,r,sigma,numSamples,
+             numRepl,"sobol",2)
[1] 5.399809
```

We see that, with a limited sample size, scrambled Sobol sequences provide us with a satisfactory result, whereas Halton sequences are a bit less effective. See the related discussion in Section 9.5.

11.3 Pricing options with early exercise features

In this section we illustrate the basic issues of pricing derivatives with early exercise features, such as American- and Bermudan-style options. Strictly speaking, since Monte Carlo methods rely on time discretization, we can only price the latter kind of options. Anyway, the price of a Bermudan-style option can be considered as useful approximation for the corresponding American-style option, and it can be refined by Richardson extrapolation. The price of this kind of derivative is related to the following optimization problem:[6]

$$\max_{\tau} E^{\mathbb{Q}} \big[e^{-r\tau} f(S_\tau) \big] , \qquad (11.7)$$

where the function $f(\cdot)$ is the option payoff, the expectation is taken under a risk-neutral measure $\mathbb{Q}$, and τ is a stopping time. Essentially, a stopping time is a random variable representing the time at which the option is exercised. This random variable is adapted to the available information, which is the sample path observed so far; in other words, the stopping time is associated with a nonanticipative exercise policy. The early exercise decision should be made by comparing the intrinsic value of the option, i.e., the payoff obtained by exercising the option immediately, and the value of continuation, i.e., the value of keeping the option alive. This has the effect of partitioning the state space in a continuation and an exercise region; see Fig. 10.8. Finding the optimal exercise policy is not trivial in general. When dealing with low-dimensional problems, alternative approaches can be applied, like binomial lattices and finite differences, which determine the optimal exercise policy by going backward in time. These methods are not feasible for high-dimensional and possibly path-dependent problems, which is the typical range of problems well-suited to Monte Carlo methods, in the case of European-style options. Indeed, years ago it was believed that Monte Carlo methods were not suitable to price options featuring early exercise opportunities. The situation has changed to a certain degree, thanks to the application of an array of approximation methods related to stochastic optimization, which we have introduced in Chapter 10:

Simulation-based optimization. One strategy is to specify a decision rule depending on a set of parameters, and solve a simulation-based optimization problem based on a set of sample paths. For instance, one might decide to exercise the option when it is enough in-the-money, i.e., when the intrinsic value exceeds a time-varying threshold. We may define the threshold by a set of parameters that can be optimized by the approaches of Sections 10.4. This approach may work fairly well in some cases, but it suffers from definite limitations when the shape of the early exercise region is complicated and disconnected.

Scenario trees. Another approach relies on a tree representation of uncertainty. As we have seen in Section 10.5, scenario trees are the standard way to represent uncertainty in stochastic programming with recourse, as they allow to deal with nonanticipative decisions. Trees are discretized stochastic processes, just like binomial lattices, but they do not recombine. Hence, the bad side of the coin is that their size tends to explode exponentially with respect to the number of time steps, but the upside is that they can deal with path-dependent and high-dimensional problems, provided that the set of early exercise opportunities is limited.

---

[6]See Section 10.2.5.

Approximate dynamic programming. Finally, we may rely on approximate dynamic programming methods, most notably regression-based approaches, to learn the value function; see Section 10.8. The value function is, in this case, the value of the continuation. With respect to full-fledged stochastic optimization approaches, in option pricing we do not face one of the three curses of dimensionality, since the decision variable is pretty simple: exercise or continue.

These three categories are not exhaustive, since there are alternatives, such as the stochastic mesh method. Furthermore, the boundaries among them need not be clear-cut. For instance, learning a policy by simulation can be interpreted as a form of approximate dynamic programming, focused on the policy itself rather than the value function determining the policy implicitly. Relying on the value function, in principle, allows for more flexibility, and in the following we will describe in detail only examples of the two latter approaches. However, first we have to investigate bias issues. Indeed, in Section 7.1.1 we have illustrated the pricing of a chooser option, which involves an optimal decision, and we discovered that a naive Monte Carlo approach may not work at all because of induced bias.

11.3.1 SOURCES OF BIAS IN PRICING OPTIONS WITH EARLY EXERCISE FEATURES

When pricing European-style options, we usually take for granted that our estimator is unbiased and concentrate on reducing its variance. When dealing with early exercise features, the picture is not so simple. To begin with, there are two clear sources of low bias:

- If we have to price an American-style option and discretize time to generate sample paths, we are restricting the exercise opportunities, which reduces the value of the option.
- When making early exercise decisions, we typically use suboptimal rules obtained by an approximate solution of an optimization problem. Hence, this also contributes to reduce the estimate of the option value.

On the other hand, however there is also a potential source of high bias. When making decisions based on a scenario tree, like the one illustrated in Fig. 11.14, we are actually using more information in the model than we actually have in the real world, as only a set of sampled scenarios is used, ruling out other realizations of the random variables involved. As a more formal clue, let us consider the problem of estimating

$$\max \{\alpha, \mathrm{E}[X]\},$$

where α is a given number, playing the role of the intrinsic value, and X is a random variable, whose expectation plays the role of the continuation value; we estimate the maximum on the basis of a sample $X_1, \ldots, X_m$ and the sample

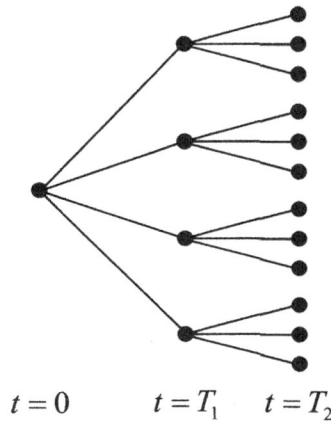

FIGURE 11.14 **Scenario tree to price a Bermudan-style option.**

mean $\bar{X}$. It is easy to see that estimator $\max\left\{\alpha, \bar{X}\right\}$ is biased high:[7]

$$E\left[\max\left\{\alpha, \bar{X}\right\}\right] \geq \max\left\{\alpha, E\left[\bar{X}\right]\right\} = \max\left\{\alpha, E[X]\right\}.$$

In general, the different sources of bias interact in unclear ways, possibly resulting in an undefined situation. It is quite recommended to run the simulation in such a way that we have a clear bias, in one way or another. The scenario tree approach, that we illustrate in Section 11.3.2, yields both a low- and a high-biased estimator, which can be merged into a valid confidence interval. The value-function-based approach that we illustrate in Section 11.3.3 is a potentially more powerful approach, but its bias is unclear. What we can do is learn an approximate policy, using a set of replications, and then evaluate the option by an independent set of replications, which yields a low-biased estimator. There are other sophisticated strategies to obtain upper bounds, but they require a deeper background on stochastic processes.[8]

11.3.2 THE SCENARIO TREE APPROACH

Pricing European-style options can be accomplished by the generation of a bunch of independent scenarios, sharing only the initial state. When dealing with early exercise features, we need more care, as we cannot make any exercise decision on the basis of the future states visited along each individual sample path. As we have seen in Section 10.5, a scenario tree can be generated in order to enforce nonanticipativity of decisions in stochastic programming

---

[7]This is a consequence of Jensen's inequality: $E[f(\mathbf{X})] \geq f(E[\mathbf{X}])$ for a random vector $\mathbf{X}$ and a convex function $f(\cdot)$. See [9, Chapter 8] for an excellent discussion of bias issues in pricing options with early exercise.

[8]See, e.g., [10].

```
sampleBGMTree <- function(S0,drift,sigma,T,b,numSteps){
  # preallocate memory
  numNodes <- sum(b^(1:(numSteps-1)))
  tree <- numeric(numNodes+1)
  # precompute invariant quantities once
  dt <- T/numSteps
  muT = (drift - 0.5*sigma^2)*dt
  sigmaT = sigma*sqrt(dt)
  # fill tree
  tree[1] <- S0
  for(i in 0:(numNodes - b^(numSteps-1))){
    tree[(i*b+1):((i+1)*b)+1] <- tree[i+1]*e
                           xp(rnorm(b,muT,sigmaT))
  }
  return(tree)
}
```

FIGURE 11.15 **Sampling a scenario tree for option pricing.**

models. Broadie and Glasserman (BG) have proposed a pricing approach that relies on such a structure.[9] The R function in Fig. 11.15 shows how a tree can be built for a trivial one-dimensional GBM process. The approach can be easily extended to multidimensional and more realistic models. A possible disadvantage of trees is that they suffer from an exponential explosion in their size when one has to branch at many time instants. Let b denote the branching factor, i.e., the number of nodes branched for each node in the tree; for the sake of simplicity, we assume that b is constant along the tree. If we branch $b = 100$ nodes from each node at each time layer in the structure, we have 1 node at time $t = 0$, 100 nodes at time $t = T_1$, 100^2 nodes at time $t = T_2$, and so on. Indeed, the BG tree approach is best suited to derivatives with limited time epochs for early exercise. An advantage, however, is that the idea works very well also for high-dimensional derivatives. A further, and possibly even more significant advantage, is that the authors have proposed an approach to find both high- and low-biased estimators.

To understand the approach, it is important to choose a sensible notation.[10] We generate the tree starting from the initial state X_0. Then:

- From the initial node X_0 we generate b successor nodes $X_1^1, \ldots, X_1^b$ at time T_1.
- From each of these nodes X_1^i, $i = 1, \ldots, b$, we branch b successors at time T_2, denoted by $X_2^{i1}, \ldots, X_2^{ib}$.

---

[9]See [5].
[10]The notation is based on [9, pp. 432–434].

- The process is repeated recursively to generate the successors at time T_3, $X_3^{i_1 i_2 1}, \ldots, X_3^{i_1 i_2 b}$, up to option expiration at time T_m.

We observe that the subscript for each node is associated with its time period, whereas the superscript is associated with the path. Indeed, this approach can be applied to path-dependent options, unlike binomial lattices. Since we are sampling the tree by straightforward Monte Carlo, all branches have the same conditional probability, $1/b$; alternative scenario generation procedures, like Gaussian quadrature, would assign different probabilities to successor nodes. The early exercise decision when in state x at time t has to be made by a comparison between the intrinsic value $h_t(x)$ and the discounted expectation of the value at future states:

$$V_t(x) = \max \left\{ h_t(x), \ e^{-r\,\delta t} \cdot \mathrm{E}^{\mathbb{Q}}\big[V_{t+1}(X_{t+1}) \mid X_t = x\big] \right\}, \qquad (11.8)$$

where we assume for simplicity that the length of the time steps is always δt. For a vanilla put, the intrinsic value is $h_t(x) = \max\{K - x, 0\}$, where the state x is the current asset price S_t and K is the asset price; clearly, we do not care about negative intrinsic values. This is a recursive dynamic programming equation defining the option value at each state. At the terminal nodes of the tree, the exercise decision is trivial, since there is no value in waiting and the form of the option payoff is given. Going backward, we may estimate the option value by replacing the expectation by a sample mean, as successor nodes are equiprobable:

$$\widehat{V}_t^{i_1 \cdots i_t} = \max \left\{ h_t\big(X_t^{i_1 \cdots i_t}\big), \ e^{-r\,\delta t} \cdot \frac{1}{b} \sum_{k=1}^{b} \widehat{V}_{t+1}^{i_1 \cdots i_t, k} \right\}.$$

For the reasons that we have outlined in Section 11.3.1, this estimator is biased high for each node of the tree. In building this estimator, the same information is used twice: first, to decide when to exercise early, and then to estimate the value of this decision. In order come up with a corresponding low-biased estimator, we should break this link and use a sample to learn the rule, and then another independent sample to estimate its value. Since this requires sampling more and more nodes, the BG approach relies on a common trick in statistics. Given a node $(i_1 i_2 \cdots i_t)$ at time t, we set apart a single successor node $(i_1 i_2 \cdots i_t k)$ in turn, $k = 1, \ldots, b$, and use the remaining $b-1$ nodes to find a low-biased estimate $\hat{\theta}_{tk}^{i_1 i_2 \cdots i_t}$ of the option value at $(i_1 i_2 \cdots i_t)$. The estimate is low-biased because, as we show below in detail, we use the other $b-1$ successor nodes to decide whether we exercise or not, but we assess the value of the decision on the successor node that was set apart. This results in b estimates, which are then averaged to find the low-biased estimate $\hat{v}_t^{i_1 i_2 \cdots i_t}$. Note the use of a lowercase v to distinguish this estimate from the high-biased value estimate, denoted by uppercase V. In detail, given node $i_1 i_2 \cdots i_t$ at time t, we set

$$
\hat{\theta}_{tk}^{i_1 i_2 \cdots i_t} =
\begin{cases}
h_t(X_t^{i_1 i_2 \cdots i_t}), & \text{if } \dfrac{e^{-r\,\delta t}}{b-1} \displaystyle\sum_{j=1, j \neq k}^{b} \hat{v}_{t+1}^{i_1 i_2 \cdots i_t, j} \leq h\left(X_t^{i_1 i_2 \cdots i_t}\right), \\[3mm]
e^{-r\,\delta t} \cdot \hat{v}_{t+1}^{i_1 i_2 \cdots i_t k}, & \text{otherwise.}
\end{cases}
$$

Then, all of the above quantities are averaged and yield the low-biased estimate

$$
\hat{v}_t^{i_1 i_2 \cdots i_t} = \sum_{k=1}^{b} \hat{\theta}_{tk}^{i_1 i_2 \cdots i_t}.
$$

These low-biased estimates are initialized at the last time layer of the tree by setting them to option payoff. The R code of Fig. 11.16 shows how to generate lower and upper bounds based on a tree generated by the function `sampleBGMTree`.

11.3.3 THE REGRESSION-BASED APPROACH

In Sections 10.7 and 10.8 we have illustrated different strategies for the numerical solution of the recursive equation of stochastic dynamic programming by approximating value functions. In particular, approximate dynamic programming (ADP) lends itself to pricing options with early exercise features by Monte Carlo sampling. In this section we describe an approach due to Longstaff and Schwartz,[11] which should be interpreted as a way to approximate the value function of dynamic programming by a linear regression against a set of basis functions. Since we approximate the value function, what we expect is a suboptimal solution; furthermore, time is discretized. Hence, we should expect some low bias in our estimate of price. Actually, to find an accurate estimate with a well-defined bias, it is recommended to use the procedure that we outline below to learn first a set of approximate value functions, which embody a strategy for early exercise decisions; then, a more extensive set of replications should be used to assess the value of the decision rules. This is not needed when ADP is used in other optimization problems whose main output is the set of decision variables, defined by a suitable strategy; here, however, it is the value of the objective function that we are interested in.

For the sake of simplicity, we will just consider a vanilla American put option on a single non-dividend-paying stock. Clearly, the approach makes sense in more complex settings. As usual with Monte Carlo simulation, we generate sample paths $(S_0, S_1, \ldots, S_t, \ldots, S_T)$, where we use t as a discrete time index, leaving the discretization step δt implicit. If we denote by $h_t(S_t)$ the intrinsic value[12] of the option at time t, the dynamic programming recursion for the value function $V_t(S_t)$ is

---

[11] See [14].

[12] Again, we identify the intrinsic value with the value of immediate exercise.

```
getboundsBG <- function(tree,b,numSteps,T,r,payoffFun){
  # This function computes the upper and lower bound on
  # a Bermudan-style option based on a single BG tree
  # It should be called repeatedly in order to build a
  # confidence interval; the two bounds are collected
  # into a vector c(lbound,ubound)

  # compute payoff on the last time layer
  # (nodes from beginLast to numNodes)
  numNodes <- length(tree)
  beginLast <- numNodes - b^(numSteps-1)
  payoffValues <- payoffFun(tree[beginLast:numNodes])
  # evaluate discount factor
  disc = exp(-r*T/numSteps)
  # initialize tree of upper bounds
  nodesToGo <- beginLast-1
  ubounds <- numeric(numNodes)
  ubounds[beginLast:numNodes] = payoffValues
  # trace back the tree and compute upper bounds
  for(i in nodesToGo:0){
    continValue <- disc*mean(ubounds[i*b+1:b+1])
    intrinsicValue <- f(tree[i+1])
    ubounds[i+1] <- max(continValue,intrinsicValue)
  }
  # now do the same with lower bounds:
  # initialize and trace back
  lbounds <- numeric(numNodes)
  lbounds[beginLast:numNodes] = payoffValues
  aux = numeric(b)
  for(i in nodesToGo:0){
    intrinsicValue <- payoffFun(tree[i+1])
    for(j in 1:b){
      aux[j] <- disc*mean(lbounds[i*b+(1:b)[-j]+1])
      aux[j] <- ifelse(intrinsicValue>aux[j],
                intrinsicValue, lbounds[i*b+j+1])
    }
    lbounds[i+1] <- mean(aux)
  }
  return(c(lbounds[1], ubounds[1]))
}
```

FIGURE 11.16 **Pricing using the Broadie–Glasserman tree.**

$$V_t(S_t) = \max\left\{ h_t(S_t),\, \mathrm{E}_t^{\mathbb{Q}}\!\left[e^{-r\,\delta t}\, V_{t+1}(S_{t+1}) \big| S_t \right] \right\}. \qquad (11.9)$$

The expectation is taken under a risk-neutral measure $\mathbb{Q}$ and is conditional on the current asset price S_t. In the case of a vanilla American put, we have $h_t(S_t) = \max\{K - S_t, 0\}$. Having to cope with continuous states is the only difficulty we have here, as time is discretized and the set of control actions is finite: Either you exercise immediately the option, or you cross your fingers and continue. Equation (11.9) is identical to Eq. (11.8). The difference between the regression-based and the scenario tree approaches lies in the scenario generation process: Here we do not branch a tree, which suffers from an exponential increase in complexity, but we generate independent sample paths, like we do with European-style options. Then, however, we must find a way to enforce nonanticipativity. It is important to realize that we *cannot* make the exercise decision along individual sample paths; if we are at a given point of a sample path generated by Monte Carlo sampling, we cannot exploit knowledge of future prices along that path, as this would imply clairvoyance.[13] What we can do is use our set of scenarios to build an approximation of the conditional expectation in Eq. (11.9), for some choice of basis functions $\psi_k(S_t)$, $k = 1, \ldots, K$. The simplest choice we can think of is regressing the conditional expectation against a basis of monomials: $\psi_1(S) = 1$, $\psi_2(S) = S$, $\psi_3(S) = S^2$, etc. In practice, orthogonal polynomials can also be used, as well as well as problem-specific basis functions capturing important features of each state. Note that we are using the same set of basis functions for each time instant, but the coefficients in the linear combination do depend on time:

$$\mathrm{E}_t^{\mathbb{Q}}\!\left[e^{-r\,\delta t} V_{t+1}(S_{t+1}) \big| S_t \right] \approx \sum_{k=1}^{K} \alpha_{kt} S_t^{k-1}.$$

Since the coefficients α_{kt} are not associated with specific sample paths, decisions are nonanticipative. The coefficients α_{kt} can be found by linear regression, going backward in time as customary with dynamic programming; note that the approximation is nonlinear in S_t, but it is linear in terms of the coefficients.

In order to illustrate the method, we should start from the last time period. Assume that we have generated N sample paths, and let us denote by S_{ti} the price at time t along sample path $i = 1, \ldots, N$. At option expiration, the value function is just the option payoff:

$$V_T(S_{Ti}) = \max\{K - S_{Ti}, 0\}, \qquad i = 1, \ldots, N.$$

These values can be used, in a sense, as the Y values in a linear regression, where the X values are the prices at time $T - 1$. More precisely, we estimate

---

[13]This point was also emphasized when we discussed the role of nonanticipativity in multistage stochastic programming.

the following regression model for the continuation value:

$$e^{-r\,\delta t} \max\left\{K - S_{Ti}, 0\right\} = \sum_{k=1}^{K} \alpha_{k,M-1} S_{T-1,i}^{k-1} + e_i, \qquad i = 1, \ldots, N,$$

where e_i is the residual for each sample path. We may find the weights $\alpha_{k,T-1}$ by the usual least-squares approach, minimizing the sum of squared residuals. Note that we are considering the discounted payoff, so that we may then compare it directly against the intrinsic value. Furthermore, we are using the state at time $T - 1$ as the regressor. As we clarify later, we should interpret this as a *post-decision* state, a fundamental concept in approximate dynamic programming that we have discussed in Section 10.8.2.

In the regression above, we have considered all of the generated sample paths. Actually, it is sometimes suggested to consider only the subset of sample paths for which we have a decision to make, i.e., the subset of sample paths in which the option is in the money at time $T - 1$. In fact, if the option is not in the money, we have no reason to exercise; using only the sample paths for which the option is in the money is called the "moneyness" criterion and it may improve the performance of the overall approach.[14] Denoting this subset by $\mathcal{I}_{T-1}$ and assuming $K = 3$, we would have to solve the following least-squares problem:

$$\min_{} \quad \sum_{i \in \mathcal{I}_{T-1}} e_i^2$$

$$\text{s.t.} \quad \alpha_{1,M-1} + \alpha_{2,M-1} S_{T-1,i} + \alpha_{3,M-1} S_{T-1,i}^2 + e_i$$
$$= e^{-r\,\delta t} \max\{K - S_{Ti}, 0\}, \qquad i \in \mathcal{I}_{T-1}. \qquad (11.10)$$

The output of this optimization problem is a set of coefficients in the approximation of the continuation value. Note that the weights are linked to time periods, and not to sample paths. Using the same approximation for each sample path in $\mathcal{I}_{T-1}$, we may decide if we exercise or not. We stress again that this is how nonanticipativity is enforced in the regression-based approach, without the need to resort to scenario trees.

We should pause and illustrate what we have seen so far by a little numerical example. We will use the same example as the original reference, where the eight sample paths given in Table 11.1 are considered for a vanilla American put with strike price $K = 1.1$. For each sample path, we also have a set of cash flows at expiration; cash flows are positive where the option is in the money. Cash flows are discounted back to time $t = 2$ and used for the first linear regression. Assuming a risk free rate of 6% per period, the discount factor is $e^{-0.06} = 0.94176$. The data for the regression are given in Table 11.2; X corresponds to current underlying asset price and Y corresponds to discounted cash

---

[14]The criterion was suggested in the original paper by Longstaff and Schwartz, and we follow the idea, even though some authors have argued against its use.

Table 11.1 Sample path and cash flows at option expiration for a vanilla American put

Path	$t = 0$	$t = 1$	$t = 2$	$t = 3$	Path	$t = 1$	$t = 2$	$t = 3$
1	1.00	1.09	1.08	1.34	1	–	–	.00
2	1.00	1.16	1.26	1.54	2	–	–	.00
3	1.00	1.22	1.07	1.03	3	–	–	.07
4	1.00	0.93	0.97	0.92	4	–	–	.18
5	1.00	1.11	1.56	1.52	5	–	–	.00
6	1.00	0.76	0.77	0.90	6	–	–	.20
7	1.00	0.92	0.84	1.01	7	–	–	.09
8	1.00	0.88	1.22	1.34	8	–	–	.00

Table 11.2 Regression data for time $t = 2$

Path	Y	X
1	.00 × .94176	1.08
2	–	–
3	.07 × .94176	1.07
4	.18 × .94176	0.97
5	–	–
6	.20 × .94176	0.77
7	.09 × .94176	0.84
8	–	–

flows in the future. We see that only the sample paths in which the option is in the money at time $t = 2$ are used. The following approximation is obtained:

$$\mathrm{E}[Y \mid X] \approx -1.070 + 2.983X - 1.813X^2.$$

Now, based on this approximation, we may compare at time $t = 2$ the intrinsic value and the continuation value. This is carried out in Table 11.3. Given the exercise decisions, we update the cash flow matrix. Note that the exercise decision does not exploit knowledge of the future. Consider, for instance, sample path 4: we exercise, earning \$0.13; along that sample path, later we would regret our decision, because we could have earned \$0.18 at time $t = 3$. This is what nonanticipativity is all about. We should also note that on some paths we exercise at time $t = 2$, and this is reflected by the updated cash flow matrix in the table.

The process is repeated going backward in time. To carry out the regression, we must consider the cash flows on each path, resulting from the early exercise decisions. Assume that we are at time step t, and consider path i. For each sample path i, there is an exercise time t_e^*, which we might set conven-

Table 11.3 **Comparing intrinsic and continuation value at time $t = 2$, and resulting cash flow matrix**

Path	Exercise	Continue		Path	$t = 1$	$t = 2$	$t = 3$
1	.02	.0369		1	–	.00	.00
2	–	–		2	–	.00	.00
3	.03	.0461		3	–	.00	.07
4	.13	.1176		4	–	.13	.00
5	–	–		5	–	.00	.00
6	.33	.1520		6	–	.33	.00
7	.26	.1565		7	–	.26	.00
8	–	–		8	–	.00	.00

tionally to $T + 1$ if the option will never be exercised in the future. Then the regression problem (11.10) should be rewritten, for the generic time period t, as:

$$\min \quad \sum_{i \in \mathcal{I}_t} e_i^2$$

$$\text{s.t.} \quad \alpha_{1t} + \alpha_{2t} S_{ti} + \alpha_{3t} S_{ti}^2 + e_i \tag{11.11}$$

$$= \begin{cases} e^{-r(t_e^* - t)\,\delta t} \max\{K - S_{t_e^*, i}\,, 0\}, & \text{if } t_e^* \leq T, \\ 0, & \text{if } t_e^* = T + 1, \end{cases} \quad i \in \mathcal{I}_t.$$

Since there can be at most one exercise time for each path, it may be the case that after comparing the intrinsic value with the continuation value on a path, the exercise time t_e^* is reset to a previous period. Stepping back to time $t = 1$, we have the regression data of Table 11.4. The discount factor $e^{-2 \cdot 0.06} = 0.88692$ is applied on paths 1 and 8. Since the cash flow there is zero, the discount factor is irrelevant, but we prefer using this to point out that we are discounting cash flows from time period $t = 3$; if we had a positive cash flow at $t = 3$ and zero cash flow at $t = 2$, this is the discount factor we should use. Least squares yield the approximation:

$$E[Y \mid X] \approx 2.038 - 3.335X + 1.356X^2.$$

This approximation may seem unreasonable, as we expect smaller payoffs for larger asset prices, yet the highest power of the polynomial has a positive coefficient here. It can be verified that, for the range of X values we are considering, the function is in fact decreasing. Based on this approximation of the continuation value, we obtain the exercise decisions illustrated in Table 11.5. Discounting all cash flows back to time $t = 0$ and averaging over the eight sample paths, we get an estimate of the continuation value of \$0.1144, which is larger than the intrinsic value \$0.1; hence, the option should not be exercised immediately.

Table 11.4 **Regression data for time $t = 1$**

Path	Y	X
1	$.00 \times .88692$	1.09
2	–	–
3	–	–
4	$.13 \times .94176$	0.93
5	–	–
6	$.33 \times .94176$	0.76
7	$.26 \times .94176$	0.92
8	$.00 \times .88692$	0.88

Table 11.5 **Comparing intrinsic and continuation value at time $t = 1$, and resulting cash flow matrix**

Path	Exercise	Continue
1	.01	.0139
2	–	–
3	–	–
4	.17	.1092
5	–	–
6	.34	.2866
7	.18	.1175
8	.22	.1533

Path	$t = 1$	$t = 2$	$t = 3$
1	.00	.00	.00
2	.00	.00	.00
3	.00	.00	.07
4	.17	.00	.00
5	.00	.00	.00
6	.34	.00	.00
7	.18	.00	.00
8	.22	.00	.00

```
genericLS <- function(paths,T,r,payoffFun,basisFuns){
  # get number of paths, exercise opportunities, etc.
  size <-dim(paths)
  numPaths <-size[1]
  numSteps <-size[2]-1 # paths include initial state
  S0 <- paths[1,1] # first column holds initial price
  paths <- paths[,-1] # get rid of first column
  dt <-T/numSteps # time step
  df <-exp(-r*dt*(1:numSteps)) # discount factors
  numBasis <-length(basisFuns) # number of basis functions
  #
  cashFlows <- sapply(paths[,numSteps],payoffFun)
  exerciseTime <- numSteps*rep(1,numPaths)
  for (step in seq(from=numSteps-1, to=1, by=-1)){
    intrinsicValue <- sapply(paths[,step],payoffFun)
    inMoney <- which(intrinsicValue > 0)
    xData <- paths[inMoney,step]
    regrMat <- matrix(nrow=length(xData), ncol=numBasis)
    for (k in 1:numBasis) regrMat[,k] <-
                              sapply(xData,basisFuns[[k]])
    yData <- cashFlows[inMoney]*df[exerciseTime[inMoney]-step]
    alpha <- as.numeric(lm.fit(regrMat,yData)$coefficients)
    continuationValue <- regrMat%*%alpha
    index <- which(intrinsicValue[inMoney]>continuationValue)
    exercisePaths <- inMoney[index]
    cashFlows[exercisePaths] <- intrinsicValue[inMoney[index]]
    exerciseTime[exercisePaths] <- step
  }
  price <- max(payoffFun(S0),mean(cashFlows*df[exerciseTime]))
  return(price)
}
```

FIGURE 11.17 **Regression-based American-style option pricing.**

Figure 11.17 shows R code to implement this regression-based approach in general. The function is modular in that it takes scenarios generated outside by another function. Figure 11.18 shows a script to check the toy example we have considered with the generic function. We insist again on the fact that a safer approach would first learn the exercise policy, on the basis of a limited number of replications. Then, using a brand new and possibly much larger set of replications, we should find the low-biased estimator.

Finally, it is important to interpret the regression-based approach within the framework of approximate dynamic programming. Unlike other numerical approaches for stochastic dynamic programming, which we have considered in Section 10.7, we did not consider the expectation of the value functions. The

```
paths <- cbind(1, c(1.09,1.16,1.22,.93,1.11,.76,.92,.88),
                  c(1.08,1.26,1.07,.97,1.56,.77,.84,1.22),
                  c(1.34,1.54,1.03,.92,1.52,.9,1.01,1.34)))
putF <- function(x) max(0,1.1-x)
r <- 0.06
T <- 3
b1 <- function(x) rep(1,length(x))
b2 <- function(x) x
b3 <- function(x) x^2
basisFuns <- list(b1,b2,b3)
source("LSRegression.R")
genericLS(paths,T,r,putF,basisFuns)
```

FIGURE 11.18 A script to check the example of Longstaff and Schwartz.

expectation is directly built into the conditional expectation, as linear regression yields a function of S_t:

$$\widetilde{V}_t(S_t) = \mathrm{E}_t^{\mathbb{Q}}\big[e^{-r\,\delta t}V_{t+1}(S_{t+1})\big|\,S_t\big].$$

In this approach, there is no need to tackle the expectation of $V_{t+1}(S_{t+1})$ by, say, Gaussian quadrature. As we have seen in Section 10.8, this corresponds to the concept of value function for the post-decision state variable. In this case the post-decision state variable is just the current asset price, which in practice is the same as the pre-decision state variable, but it should be interpreted as the state *after the decision to continue*. The point is that our decision is whether we exercise or not, and this does not influence the state variable, unlike other dynamic programming models. Indeed, linear regression yields the value of the continuation, which is then compared against the value of immediate exercise to find the option value.

We close the section by noting that in some ADP approaches, a recursive approach to linear regression is adopted, whereby the coefficients are incrementally updated by considering one sample path at time. Here we are using a batch regression, whereby all sample paths are generated at once and we run a linear regression per time step on all of them. The relative merits of the two strategies may depend on the specific problem at hand.

11.4 A look outside the BSM world: Equity options under the Heston model

Option pricing within the BSM world is only a first step to grasp the remarkable potential of Monte Carlo methods for financial engineering. We have considered some more general models than geometric Brownian motion in Chapter 3, such as the Heston stochastic volatility model and mean-reverting square-root diffusions for short rates. Path generation mechanisms for such processes have been introduced in Chapter 6. All of these tools may be integrated to relax the restrictive assumptions of the BSM formula and price options in a more general and, hopefully, more realistic setting. We should not forget that in doing so we may lose market completeness. Hence, as we have pointed out in Section 3.9.3, we may have to face a nontrivial model calibration task. In this and the next section we do not consider calibration, but we just illustrate the flexibility of Monte Carlo methods, which indeed are often (not always) the only viable pricing strategy.

In Section 3.7.5 we have considered the Heston stochastic volatility model, reported below for convenience, under the risk-neutral measure:

$$dS_t = rS_t\,dt + \sqrt{V_t}S_t\,dW_t^1,$$
$$dV_t = \alpha\big(\bar{V} - V_t\big)\,dt + \xi\sqrt{V_t}\,dW_t^2.$$

The model integrates a GBM with nonconstant volatility and a square-root diffusion modeling squared volatility, where $\bar{V}$ is a long-term value, α measures the speed of reversion to the mean, and ξ is the volatility of the square-root diffusion. Different assumptions can be made about the instantaneous correlation ρ of the two driving Wiener processes W_t^1 and W_t^2. Suppose that we are interested in pricing a vanilla call option. Unlike the case of GBM, we cannot sample the price S_T at maturity directly, but we have to go through a whole sample path generation to keep the discretization error under control. A straightforward approach to discretize the above equations is the Euler scheme

$$S_{t+\delta t} = S_t(1 + r\,\delta t) + S_t\sqrt{V_t\,\delta t}\,\epsilon_t^1,$$
$$V_{t+\delta t} = V_t + \alpha\big(\bar{V} - V_t\big)\,\delta t + \xi\sqrt{V_t\,\delta t}\,\epsilon_t^2,$$

where ϵ_t^1 and ϵ_t^2 are standard normals with correlation ρ. Since the Euler discretization does not guarantee non-negativity, we may heuristically patch the above expressions by taking the maximum between the result and 0, as shown in the code of Fig. 11.19. Here is one sample run of the function, and the reader is invited to try and play with it, in order to see the effect of the different parameters:

```
> CallStochVol(S0,K,T,r,V0,alpha,Vbar,xi,rho,numRepl,numSteps)
$estimate
[1] 5.754051
$conf.int
[1] 5.566677 5.941425
```

Actually, the Heston model allows for some semianalytical solutions in simple cases, but the Monte Carlo code can be adapted to more complicated options. As we mentioned, we have to generate a whole sample path, with a corresponding increase in computational effort with respect to the GBM case. Furthermore, since we are using a plain Euler scheme, the time step δt must be small enough

```
CallStochVol <- function(S0,K,T,r,V0,alpha,Vbar,xi,
                         rho,numRepl,numSteps,c.level=0.95){
  # invariant quantities
  dt <- T/numSteps
  alphadt <- alpha*dt
  xidt <- xi*sqrt(dt)
  oneplusrdt <- 1+r*dt
  rho2 <- sqrt(1-rho^2)
  VPath <- numeric(numSteps+1)
  VPath[1] <- V0
  SPath <- numeric(numSteps+1)
  SPath[1] <- S0
  payoff <- numeric(numRepl)
  for (k in 1:numRepl){
    eps1 <- rnorm(numSteps)
    eps2 <- rho*eps1+rho2*rnorm(numSteps)
    for (t in 1:numSteps){
      VPath[t+1] <- max(0,VPath[t]+alphadt*(Vbar-VPath[t])+
        xidt*sqrt(VPath[t])*eps1[t])
      SPath[t+1] <- max(0,SPath[t]*(oneplusrdt+
                              sqrt(VPath[t]*dt)*eps2[t]))
    }
    payoff[k] <- max(0, SPath[numSteps+1]-K)
  }
  aux <- t.test(exp(-r*T)*payoff,conf.level=c.level)
  value <- as.numeric(aux$estimate)
  ci <- as.numeric(aux$conf.int)
  return(list(estimate=value,conf.int=ci))
}

S0 <- 50; K <- 50; T <- 1; r <- 0.03; V0 <- 0.07
alpha <- 1; Vbar <- 0.05; xi <- 0.1; rho <- 0.5
numRepl <- 10000; numSteps <- 50
set.seed(55555)
CallStochVol(S0,K,T,r,V0,alpha,Vbar,xi,rho,numRepl,numSteps)
```

FIGURE 11.19 **Pricing a vanilla call option under the Heston model.**

to ensure accuracy. However, this additional cost may be not quite relevant if
we want to price more exotic, possibly path-dependent, options. Asian options
feature a payoff depending on the average underlying asset price. Another class
of options, called lookback options, feature a payoff depending on the maximum or the minimum observed price. Consider, for instance, a lookback call
option, which is characterized by the payoff

$$\max_{t=t_1,t_2,\ldots t_n} S_t - S_T,$$

where $t_1, t_2, \ldots, t_n \equiv T$ is the collection of discrete time instants at which the price is monitored. It is very easy to adapt the code of Fig. 11.19 to price this option.

11.5 Pricing interest rate derivatives

In the first sections of this chapter we have used the risk-neutral valuation principle to price options on stocks, under the assumption of a constant risk-free rate. Using this principle, the price at time t of a path-independent, European-style option maturing in T can be written as

$$V(S_t, t) = e^{-r(T-t)} \mathrm{E}^{\mathbb{Q}}[f(S_T) \,|\, S_t], \qquad (11.12)$$

where $f(S_T)$ is the payoff at maturity, depending on the stock price S_T, $\mathbb{Q}$ is the pricing measure, and the expectation is conditional on the price S_t of the underlying asset in t. If we step into the domain of interest rate derivatives, there are substantial complications.

- The first issue is that modeling interest rates is an intrinsically more difficult endeavor than modeling stock prices, since we should model a whole term structure, rather than a single stochastic process. Here we will skip this difficulty by adopting single-factor models depending only on the short-term (instantaneous) rate, like the Vasicek and the Cox–Ingersoll–Ross (CIR) models that we have mentioned in Chapter 3. These models rely on stochastic differential equations that are slightly more complicated than geometric Brownian motion, but we have seen in Section 6.3 how to generate sample paths for them.

- Since we are dealing with stochastic interest rates, we cannot take the discount factor outside the expectation, as is done in Eq. (11.12). Furthermore, since we are dealing with a short rate, we should account for the whole path of the process r_t, even if the option payoff is path-independent. The risk-neutral pricing approach leads here to a formula like

$$V(r_t, t) = \mathrm{E}^{\mathbb{Q}}\left[\exp\left(-\int_t^T r_\tau \, d\tau \right) f(r_T) \,\bigg|\, r_t \right]. \qquad (11.13)$$

- Another quite tricky point is that, as we have pointed out in Section 3.9.3, the market here is incomplete. The implication is that there is not a unique measure $\mathbb{Q}$ for pricing, but several ones. This means that we have to calibrate the short-rate model against the prices of quoted derivatives. In this chapter, we assume that such calibration has already been done, and when we use a model like

$$dr_t = \gamma\,(\bar{r} - r_t)\,dt + \sigma\,dW_t,$$

we assume that the parameters define the risk-neutral measure for pricing purposes.

In order to calibrate an interest rate model, we need quoted prices of interest rate derivatives, and the simplest such assets are, in fact, bonds. This may sound odd, as bond pricing given a term structure of interest rates is an easy task, but the point here is that we want to *estimate* a risk-neutral model for pricing purposes, accounting for the stochastic character of the interest rates. In fact, bonds are interest rate derivatives in the sense that they are assets whose value depends on that underlying risk factor. Since a coupon-bearing bond can be priced as a portfolio of zero-coupon bonds, the first interest rate derivative that we should price is a zero-coupon bond. In such a case, the option payoff is just 1, if we assume a bond with \$1 face value. Hence, the price at time t of a zero-coupon bond maturing at time T, when the short rate in t is r_t, can be expressed as

$$Z(r_t, t; T) = \mathrm{E}^{\mathbb{Q}}\left[\exp\left(-\int_t^T r_\tau \, d\tau\right)\bigg| r_t\right]. \tag{11.14}$$

This expression is fairly simple, and indeed there are analytical formulas to price zero-coupon bonds under both Vasicek and CIR models. From a pedagogical viewpoint, though, it is quite useful to find those prices by Monte Carlo simulation in order to build skills that can be put to good use when dealing with more complex (and interesting) cases. From Eq. (11.14) we immediately see that to use Monte Carlo in this setting, we should generate sample paths of the short rate process and then approximate the integral by a sum:

$$\int_t^T r_\tau \, d\tau \approx \sum_{k=0}^{m-1} r_{t_k} \, \delta t,$$

where δt is the discretization step, and r_{t_k} is the short rate at the discretized time instants t_k, $k = 0, \ldots, m$, such that $t_0 = t$ and $t_m = T$. Note that in the sum we multiply by δt the rate observed at the *beginning* of each time slice.

The next step in the learning process is to price a relatively simple option, namely, a call option on a zero-coupon bond. The risk-neutral principle must be applied exactly in the same way, but we have to realize that there are two maturities:

- The option maturity, T_O, i.e., the time at which the option can be exercised, by purchasing a zero-coupon bond at the strike price K.
- The bond maturity, T_B; clearly, for the option to make sense, we must have $T_O < T_B$.

Therefore, the option payoff at T_O is

$$\max\{Z(r_{T_O}, T_O; T_B) - K, 0\}. \tag{11.15}$$

11.5.1   PRICING BONDS AND BOND OPTIONS UNDER THE VASICEK MODEL

The price at time t of a zero-coupon bond maturing in T, with face value \$1, under the Vasicek short-rate model

$$dr_t = \gamma \left(\bar{r} - r_t\right) dt + \sigma \, dW_t,$$

can be expressed as follows:

$$Z(r_t, t; T) = e^{A(t;T) - B(t;T)r_t},$$

where

$$B(t;T) = \frac{1}{\gamma}\left[1 - e^{-\gamma(T-t)}\right], \tag{11.16}$$

$$A(t;T) = \left[B(t;T) - (T-t)\right]\left(\bar{r} - \frac{\sigma^2}{2\gamma^2}\right) - \frac{\sigma^2 B(t;T)^2}{4\gamma}. \tag{11.17}$$

We insist once again that the parameters of the Vasicek model are *not* the parameters of the real world, which could be estimated by analyzing time series of short rates, but those in the risk-neutral world. These formulas can be proved by solving the partial differential equation that we have given in Eq. (3.125),[15] and are easily implemented R, as shown in Fig. 11.20.

To price the bond using Monte Carlo, we rely on the function simVasicek, listed in Fig. 6.12, to generate sample paths. In Fig. 11.20, we also show code to price the zero-coupon bond using random sampling, and a snapshot to compare the results. Running the code yields the following output:

```
> ZCBVasicek(r0,rbar,gamma,sigma,T,F)
[1] 97.26989
> ZCBVasicekMC(r0,rbar,gamma,sigma,T,mumRepl,numSteps,F)
$estimate
[1] 97.23904
$conf.int
[1] 97.14941 97.32867
```

We see that the results are fairly reasonable. The price at time 0 of a European-style call option maturing at time T_O on a zero-coupon bond maturing in T_B is, under the Vasicek model:

$$Z(0, r_0; T_B)\Phi(d_1) - KZ(0, r_0; T_O)\Phi(d_2), \tag{11.18}$$

---

[15] See, e.g., [16].

```
# Exact ZC bond price under Vasicek
ZCBVasicek <- function(r0,rbar,gamma,sigma,T,F=100){
  B <- (1-exp(-gamma*T))/gamma
  A <- (B-T)*(rbar-sigma^2/(2*gamma^2))-sigma^2*B^2/(4*gamma)
  return(F*exp(A-B*r0))
}


# ZCB price under Vasicek with MC
ZCBVasicekMC <- function(r0,rbar,gamma,sigma,T,mumRepl,
                         numSteps,F=100,c.level=0.95){
  paths <- simVasicek(r0,rbar,gamma,sigma,T,numSteps,numRepl)
  dt <- T/numSteps
  # approximate integral with sum (paths are on rows!)
  sums <- rowSums(paths[,1:numSteps]*dt)
  prices <- F*exp(-sums)
  aux <- t.test(prices,conf.level=c.level)
  value <- as.numeric(aux$estimate)
  ci <- as.numeric(aux$conf.int)
  return(list(estimate=value,conf.int=ci))
}


# Compare exact ZCB price under Vasicek with MC
r0 <- 0.02; rbar <- 0.06; gamma <- 0.45; sigma <- 0.03
T <- 1; F <- 100; numRepl <- 1000; numSteps <- 200
ZCBVasicek(r0,rbar,gamma,sigma,T,F)
ZCBVasicekMC(r0,rbar,gamma,sigma,T,mumRepl,numSteps,F)
```

FIGURE 11.20 **Pricing a zero-coupon bond under the Vasicek model.**

where r_0 is the current value of the short rate, $\Phi(\cdot)$ is the CDF of the standard normal distribution, and

$$d_1 = \frac{1}{S(T_O)} \log\left[\frac{Z(0,r_0;T_B)}{KZ(0,r_0;T_O)}\right] + \frac{S(T_0)}{2}, \qquad (11.19)$$

$$d_2 = d_1 - S(T_0), \qquad (11.20)$$

$$S(T_O) = B(T_O,T_B)\sqrt{\frac{\sigma^2}{2\gamma}\left(1 - e^{-2\gamma T_O}\right)}. \qquad (11.21)$$

To interpret this formula, it is quite useful to note its deep similarity with the Black–Scholes–Merton price of a vanilla call option on a stock, given in Eq. (3.108), and note the following:

- In Eq. (11.18), $Z(0,r_0;T_B)$ should be interpreted as the price of the underlying asset, i.e., the bond maturing in T_B, whereas $Z(0,r_0;T_O)$ should be interpreted as a discount factor from the option maturity T_0 to time 0.

- The terms d_1 and d_2 in Eqs. (11.19) and (11.20) look much like the similar terms in the BSM formula.
- The term $\mathcal{S}(T_O)$, where $B(T_O, T_B)$ is just the function given in Eq. (11.16), plays the role of a volatility.

It is also useful to note that short-rates under the Vasicek model, which relies on an Ornstein–Uhlenbeck process, are normally distributed, and that the price of a zero-coupon bond is an exponential of these rates. Hence, the bond price is lognormally distributed, just like stock prices under the BSM model, and this is the essential reason behind the observed similarity.

In Fig. 11.21 we give R code to evaluate the exact call price, and a function to estimate it by straightforward Monte Carlo. We are cheating a bit there, as you may notice that to evaluate the payoff we are using the exact bond price. Otherwise, we should simulate sample paths of r_t from $t = 0$ to $t = T_O$ to evaluate the discount factor, and then up to $r = T_B$ to evaluate the bond price. We use a little short-cut to avoid this nightmare, but the reader will immediately appreciate the complexity of pricing a really difficult interest rate derivative.[16]

Running the code, we again find sensible results:

```
> ZCBCallVasicek(r0,rbar,gamma,sigma,TB,K,TO,F)
[1] 3.190405
> ZCBCallVasicekMC(r0,rbar,gamma,sigma,TB,K,TO,
                   numRepl,numSteps,F)
$estimate
[1] 3.192955
$conf.int
[1] 3.140366 3.245544
```

11.5.2 PRICING A ZERO-COUPON BOND UNDER THE CIR MODEL

We have noted in Chapter 6 that, since the CIR model relies on a square-root diffusion,

$$dr_t = \gamma(\bar{r} - r_t)\, dt + \sqrt{\alpha r_t}\, dW_t,$$

the short rates are no longer normal. Nevertheless, a zero-coupon bond can be priced analytically under the CIR model, even though the formulas are more complicated than those for the Vasicek model:

$$Z(r_t, t; T) = e^{A(t;T) - B(t;T)r_t},$$

---

[16]Suitable changes of measure are in fact used to price interest rate derivatives with Monte Carlo, but this requires deeper concepts that are beyond the scope of this book.

```
# Exact ZC bond call option price under Vasicek
ZCBCallVasicek <- function(r0,rbar,gamma,sigma,TB,
                           K,TO,F=100){
  Tmat <- TB-TO
  Kn <- K/F
  B <- (1-exp(-gamma*Tmat))/gamma
  ZTB <- ZCBVasicek(r0,rbar,gamma,sigma,TB,1)
  ZTO <- ZCBVasicek(r0,rbar,gamma,sigma,TO,1)
  SZ <- B*sqrt(sigma^2*(1-exp(-2*gamma*TO))/2/gamma)
  d1 <- log(ZTB/Kn/ZTO)/SZ + SZ/2
  d2 <- d1 - SZ
  unitPrice <- ZTB*pnorm(d1)-Kn*ZTO*pnorm(d2)
  return(F*unitPrice)
}
# Monte Carlo for ZC bond call option price under Vasicek
ZCBCallVasicekMC <- function(r0,rbar,gamma,sigma,TB,K,
                     TO,mumRepl,numSteps,F=100,c.level=0.95){
  # find discount factors and rates at option maturities
  dt <- TO/numSteps
  paths <- simVasicek(r0,rbar,gamma,sigma,TO,
                      numSteps,numRepl)
  sums <- rowSums(paths[,1:numSteps]*dt)
  df <- exp(-sums)
  # find bond prices for each path
  Tmat <- TB-TO
  bondPrices <- ZCBVasicek(paths[,numSteps+1],rbar,
                           gamma,sigma,Tmat,F)
  # find discounted payoff
  dPayoff <- df*pmax(0, bondPrices-K)
  aux <- t.test(dPayoff,conf.level=c.level)
  value <- as.numeric(aux$estimate)
  ci <- as.numeric(aux$conf.int)
  return(list(estimate=value,conf.int=ci))
}

F <- 100; K <- 80; TO <- 1; TB <- 5; rbar <- 0.06
gamma <- 0.46; sigma <- 0.025; r0 <- 0.016
numRepl <- 10000; numSteps <- 200
set.seed(55555)
ZCBCallVasicek(r0,rbar,gamma,sigma,TB,K,TO,F)
ZCBCallVasicekMC(r0,rbar,gamma,sigma,TB,K,TO,
                 numRepl,numSteps,F)
```

FIGURE 11.21 **Pricing a bond option under the Vasicek model.**

where

$$B(t;T) = \frac{2(e^{\psi(T-t)} - 1)}{(\gamma + \psi)(e^{\psi(T-t)} - 1) + 2\psi},$$

$$A(t;T) = \frac{2\bar{r}\gamma}{\alpha} \log \left[\frac{2\psi e^{(\psi+\gamma)\frac{T-t}{2}}}{(\gamma + \psi)(e^{\psi(T-t)} - 1) + 2\psi} \right],$$

$$\psi = \sqrt{\gamma^2 + 2\alpha}.$$

In Fig. 11.22 we show R code that implements these formulas, as well as a Monte Carlo pricer relying on the function `simCIR` of Fig. 6.14 to generate sample paths of the square-root diffusion. As usual, we also verify the sensibility of the results:

```
> ZCBCIR(r0,rbar,gamma,alpha,T,F)
[1] 97.2596
> ZCBCIRMC(r0,rbar,gamma,alpha,T,numRepl,numSteps,F)
$estimate
[1] 97.26934
$conf.int
[1] 97.25589 97.28279
```

For further reading

- A friendly introduction to numerical methods for option pricing can be found in [4].

- The most comprehensive reference on Monte Carlo methods for option pricing is [9].

- Other books that include interesting discussions of specific points are [11] and [12].

- An early survey can be found in [2], which was later updated in [3].

- An interesting collection of papers is proposed in [8].

- The background on the interest rate derivatives that we have discussed in Section 11.5 can be found, e.g., in [16].

References

1 T. Björk. *Arbitrage Theory in Continuous Time* (2nd ed.). Oxford University Press, Oxford, 2004.

2 P. Boyle. Options: A Monte Carlo approach. *Journal of Financial Economics*, 4:323–338, 1977.

3 P. Boyle, M. Broadie, and P. Glasserman. Monte Carlo methods for security pricing. *Journal of Economics Dynamics and Control*, 21:1267–1321, 1997.

```
# Exact ZC bond price under CIR
ZCBCIR <- function(r0,rbar,gamma,alpha,T,F=100){
  psi <- sqrt(gamma^2+2*alpha)
  aux <- (gamma+psi)*(exp(psi*T)-1)+2*psi
  B <- 2*(exp(psi*T)-1)/aux
  A <- 2*rbar*gamma/alpha*log(2*psi*exp((psi+gamma)*T/2)/aux)
  return(F*exp(A-B*r0))
}

# ZCB price under CIR with MC
ZCBCIRMC <- function(r0,rbar,gamma,alpha,T,mumRepl,
                        numSteps,F=100,c.level=0.95){
  paths <- simCIR(r0,rbar,gamma,alpha,T,numSteps,numRepl)
  dt <- T/numSteps
  # approximate integral with sum (paths are on rows!)
  sums <- rowSums(paths[,1:numSteps]*dt)
  prices <- F*exp(-sums)
  aux <- t.test(prices,conf.level=c.level)
  value <- as.numeric(aux$estimate)
  ci <- as.numeric(aux$conf.int)
  return(list(estimate=value,conf.int=ci))
}

# Compare exact ZCB price under CIR with MC
r0 <- 0.02
rbar <- 0.06
gamma <- 0.45
alpha <- 0.03^2
T <- 1
F <- 100
numRepl <- 1000
numSteps <- 200
ZCBCIR(r0,rbar,gamma,alpha,T,F)
ZCBCIRMC(r0,rbar,gamma,alpha,T,mumRepl,numSteps,F)
```

FIGURE 11.22 **Pricing a zero-coupon bond under the CIR model.**

4 P. Brandimarte. *Numerical Methods in Finance and Economics: A MATLAB-Based Introduction* (2nd ed.). Wiley, Hoboken, NJ, 2006.

5 M. Broadie and P. Glasserman. Pricing American-style securities using simulation. *Journal of Economic Dynamics and Control*, 21:1323–1352, 1997.

6 M. Broadie, P. Glasserman, and S.G. Kou. A continuity correction for discrete barrier options. *Mathematical Finance*, 7:325–349, 1997.

7 L. Clewlow and C. Strickland. *Implementing Derivatives Models*. Wiley,

Chichester, West Sussex, England, 1998.

8   B. Dupire, editor. *Monte Carlo. Methodologies and Applications for Pricing and Risk Management*. Risk Books, London, 1998.

9   P. Glasserman. *Monte Carlo Methods in Financial Engineering*. Springer, New York, 2004.

10   M.B. Haugh and L. Kogan. Pricing American options: A duality approach. *Operations Research*, 52:258–270, 2004.

11   P. Jaeckel. *Monte Carlo Methods in Finance*. Wiley, Chichester, 2002.

12   R. Korn, E. Korn, and G. Kroisandt. *Monte Carlo Methods and Models in Finance and Insurance*. CRC Press, Boca Raton, FL, 2010.

13   Y.K. Kwok. *Mathematical Models of Financial Derivatives*. Springer, Berlin, 1998.

14   F.A. Longstaff and E.S. Schwartz. Valuing American options by simulation: A simple least-squares approach. *Review of Financial Studies*, 14:113–147, 2001.

15   S.M. Ross and J.G. Shanthikumar. Monotonicity in volatility and efficient simulation. *Probability in the Engineering and Informational Sciences*, 14:317–326, 2000.

16   P. Veronesi. *Fixed Income Securities: Valuation, Risk, and Risk Management*. Wiley, Hoboken, NJ, 2010.

17   P. Wilmott. *Quantitative Finance (vols. I and II)*. Wiley, Chichester, West Sussex, England, 2000.

## Chapter Twelve

Sensitivity Estimation

The aim of most Monte Carlo simulations is to estimate the expected value of a function of several random variables. On the one hand, this function depends on the sample path, which we may associate with a random event ω; informally, we may think of ω as a sequence of (pseudo)random numbers. On the other hand, the function depends on some relevant parameters as well. If, for the sake of simplicity, we focus on a single parameter α, we may denote the function implemented by the simulation program as $f(\alpha, \omega)$. When taking the expected value, we obtain another function

$$g(\alpha) \equiv \mathrm{E}_{\omega}\big[f(\alpha, \omega)\big], \tag{12.1}$$

depending on α alone. Monte Carlo is a way to find an estimate $\hat{g}(\alpha)$. In many practical settings, we are also interested in the sensitivity of $g(\cdot)$ with respect to its argument α. Formally, we would like to find

$$\frac{dg}{d\alpha} \equiv \frac{d\mathrm{E}_{\omega}[f(\alpha, \omega)]}{d\alpha}. \tag{12.2}$$

The standard example in financial engineering is the computation of the option greeks, which measure the sensitivity of option prices with respect to parameters like the current price of the underlying asset, volatility, etc. More generally, we might need the sensitivity with respect to parameters that are the decision variables of a stochastic optimization problem. If the objective function depends on a vector $\alpha \in \mathbb{R}^{p}$, certain stochastic optimization algorithms need an estimate of the gradient

$$\nabla g(\alpha) = \left[\frac{\partial g(\alpha)}{\partial \alpha_1}, \frac{\partial g(\alpha)}{\partial \alpha_2}, \dots, \frac{\partial g(\alpha)}{\partial \alpha_p}\right]^{\mathsf{T}}.$$

In other cases, we might also want to find the Hessian matrix of second-order derivatives. Option gamma, for instance, is the second-order partial derivative of option price with respect to the underlying asset price. To keep it simple, we will consider first the case of the first-order derivatives, referring to option greeks of a plain vanilla, European-style option within the Black–Scholes–Merton world. There, we may also afford the luxury to compare our estimates with the exact result to gain confidence with methods and to see the relevant issues. To make things a tad more interesting, later we also deal with an Asian option.[1]

It is important to realize that even such a simple problem may hide some potential trouble. In fact, we do not really know $g(\alpha)$, and the only thing we are able to do is to sample realizations of the random variable $Y(\alpha) = f(\alpha, \omega)$. Then, we might consider the following "limit of finite differences":

$$\lim_{h \to 0} \frac{Y(\alpha + h) - Y(\alpha)}{h}, \qquad (12.3)$$

and estimate the expected value

$$\mathrm{E}\left[\lim_{h \to 0} \frac{Y(\alpha + h) - Y(\alpha)}{h}\right] \qquad (12.4)$$

by its sample mean. However, this raises a few issues:

1. In the limit of Eq. (12.3) we have differences of *random variables*. Which kind of limit should we consider? What is the role of ω?

2. What we want is actually the derivative of an expectation, but in Eq. (12.4) we are taking expectation of a "derivative." However, to begin with, this is not quite a standard derivative, since it involves random variables. Even if we disregard this issue, how can we be sure that the expectation and the limit commute?

To start appreciating the involved issues and gain a deeper understanding, we first consider a simple approach based on finite differences in Section 12.1. Then, we move on to more refined approaches that aim at estimating sensitivities using information provided by the same sample path that we use to estimate the expected value of the function itself, in our case the option price. In Section 12.2 we consider an approach based on pathwise derivatives, where the parameter is regarded as an attribute of the function itself. On the contrary, in the likelihood ratio (or score) method, which is dealt with in Section 12.3, the parameter is regarded as an attribute of the probability density, i.e., it is a property of the random variables, rather than the function.

We know from Chapter 11 that the price of a vanilla option is a function,

$$g(S_0, K, T, r, \sigma), \qquad (12.5)$$

depending on current underlying asset price S_0, strike price K, time to maturity T, risk-free interest rate r, and volatility σ. The aim of this chapter is not to give a full picture of Monte Carlo estimation procedures for the full set of option greeks. We just want to use simple examples to introduce the aforementioned approaches. Hence, we will only consider the essential greek, i.e., the option delta:

$$\Delta \equiv \frac{\partial g}{\partial S_0}.$$

---

[1] See Chapter 11 for an introduction to path-dependent options.

```
VanillaCallDelta <- function(S0,K,T,r,sigma){
  d1 <- (log(S0/K)+(r+sigma^2/2)*T)/(sqrt(T)*sigma)
  return(pnorm(d1))
}
```

FIGURE 12.1 **Exact Δ for a vanilla BSM option.**

This will suffice to appreciate the essential issues, which apply, e.g., to the estimation of vega as well. For an alternative approach to sensitivity estimation, we refer the reader to Section 10.4.2, where we deal with metamodeling approaches to simulation-based optimization; these strategies, too, can be used to estimate the gradient and the Hessian matrix of a function depending on random variables.

12.1 Estimating option greeks by finite differences

Using the Black–Scholes–Merton formula, it is not too difficult to prove that the option delta for a European-style vanilla call is

$$\Phi(d_1), \qquad d_1 = \frac{\log(S_0/K) + (r + \sigma^2/2)\, T}{\sigma\sqrt{T}}, \tag{12.6}$$

where $\Phi(\cdot)$ is the CDF of a standard normal random variable. This formula is easily implemented as shown in Fig. 12.1. Since the task may not be this simple with exotic options, we are interested in Monte Carlo estimates of greeks. Estimating the option delta by random sampling should be a straightforward business. After all, the derivative with respect to S_0 is just the limit of an increment ratio, for suitably small h:

$$\Delta \approx \frac{g(S_0 + h) - g(S_0)}{h},$$

where in $g(\cdot)$ we only point out the dependence on the initial underlying asset price. If we sample the discounted option payoff starting from $S_0 + h$ and S_0, and take the sample mean of this increment ratio, the job should be done. This simple idea is reflected in the code of Fig. 12.2. The code also returns a confidence interval, and the figure includes a script to set values and check how well the function performs against the exact result:

```
> DeltaOK
[1] 0.583839
> OutDeltaMCNaive
$estimate
[1] 0.8262371
```

```
VanillaDeltaMCNaive <- function(S0,K,T,r,sigma,dS,numRepl,
                                c.level=0.95){
  nuT <- (r - 0.5*sigma^2)*T
  sigmaT <- sigma*sqrt(T)
  payoff1 <- pmax(0,S0*exp(rnorm(numRepl,nuT,sigmaT))-K)
  payoff2 <- pmax(0,(S0+dS)*
             exp(rnorm(numRepl, nuT, sigmaT))-K)
  SampleDiff <- exp(-r*T)*(payoff2 - payoff1)/dS
  aux <- t.test(SampleDiff,conf.level=c.level)
  value <- as.numeric(aux$estimate)
  ci <- as.numeric(aux$conf.int)
  return(list(estimate=value,conf.int=ci))
}

S0 <- 50
K <- 55
T <- 1
r <- 0.1
sigma <-0.4
numRepl <- 50000
dS <- 0.5
set.seed(55555)
DeltaOK <- VanillaCallDelta(S0,K,T,r,sigma)
OutDeltaMCNaive <- VanillaDeltaMCNaive(S0,K,T,r,sigma,
                                        dS,numRepl)
```

FIGURE 12.2 **Straightforward approach to estimate a vanilla call delta.**

```
$conf.int
[1] 0.4567431 1.1957310
```

To put it mildly, the result is extremely disappointing. The point estimate is quite far from the exact value, which is something that in practical cases we would not know. What we would know, however, is that such a wide confidence interval is far from acceptable. It is so large that it includes nonsensical values, as option delta cannot be larger than 1 for this kind of option: From a financial viewpoint, Δ is the number of stock shares that a call option writer should hold in order to hedge option risk, according to the BSM model; clearly, there is no reason to hedge by using more than one stock share per option. From a mathematical viewpoint, Eq. (12.6) shows that Δ is given as a probability, which cannot be larger than 1. Therefore, something is definitely wrong.

A first issue is related to a (deliberately) missing point in Eq. (12.3). There, we did not clarify the role of the event ω, i.e., the underlying sequence of random numbers. To be more precise, what we wrote is something like

$$\Delta \approx \mathrm{E}\left[\frac{f(S_0 + h, \omega_1) - f(S_0, \omega_2)}{h}\right],$$

where $f(S_0, \omega)$ is the discounted option payoff starting from S_0 and using the sample path ω, which in this case boils down to a single observation of a standard normal variable used to generate a lognormal price. The trouble is that we have used *independent* sample paths in the evaluation. From a theoretical point of view, this does not look quite right when compared against the definitions in Eqs. (12.1) and (12.2). There, we have a function $f(\cdot, \cdot)$ depending on two arguments, α, which we perturb, and ω, the sample path, which we use in the expectation. Here, however, we have two such sample paths. From a practical point of view, we know from Chapter 8 that when we consider differences of random variables, variance can be reduced by common random numbers. Indeed, we should use the *same* sequence of random variables in both terms of the difference. Otherwise, variance (noise) will overwhelm the actual signal in the difference, especially when h is small.

There is also another issue, related to finite difference methods in numerical analysis, which are often used to solve partial differential equations,[2] including those involved in option pricing. People in the field know that sometimes an alternative finite difference should be adopted to improve accuracy. The above finite difference is called a *forward difference*. The following one is called a *central* difference,

$$\Delta \approx \mathrm{E}\left[\frac{f(S_0 + h, \omega) - f(S_0 - h, \omega)}{2h}\right],$$

and it is used in the code of Fig. 12.3, where we save the sequence of standard normals in order to use them twice, for both payoffs. Now, the results are definitely more reasonable:

```
> set.seed(55555)
> OutDeltaMC <- VanillaDeltaMC(S0,K,T,r,sigma,dS,numRepl)
> DeltaOK
[1] 0.583839
> OutDeltaMC
$estimate
[1] 0.5810456
$conf.int
[1] 0.5747994 0.5872918
```

It seems that by a simple trick we have solved all of our problems. However, we should dig a bit deeper to sharpen our understanding of the use of finite differences. In particular, we should investigate the potential bias of this kind of estimator. To set the problem within the correct framework, let us consider

---

[2]See, e.g., [2, Chapters 5 and 9].

```
VanillaDeltaMC <- function(S0,K,T,r,sigma,dS,numRepl,
                           c.level=0.95){
  nuT <- (r - 0.5*sigma^2)*T
  sigmaT <- sigma*sqrt(T)
  S1 <- S0 - dS
  S2 <- S0 + dS
  Veps <- rnorm(numRepl)
  payoff1 <- pmax(0,S1*exp(nuT+sigmaT*Veps)-K)
  payoff2 <- pmax(0,S2*exp(nuT+sigmaT*Veps)-K)
  SampleDiff <- exp(-r*T)*(payoff2 - payoff1)/(2*dS)
  aux <- t.test(SampleDiff,conf.level=c.level)
  value <- as.numeric(aux$estimate)
  ci <- as.numeric(aux$conf.int)
  return(list(estimate=value,conf.int=ci))
}
```

FIGURE 12.3 Improving the estimate of the option Δ by common random numbers and central differences.

the forward difference estimator

$$\hat{\Delta}_F = \frac{f(S_0 + h, \omega) - f(S_0, \omega)}{h}, \tag{12.7}$$

and the central difference estimator

$$\hat{\Delta}_C = \frac{f(S_0 + h, \omega) - f(S_0 - h, \omega)}{2h}. \tag{12.8}$$

Here we point out the dependence on a common stream of random numbers explicitly, as we have seen that this does play a role in terms of variance. However, this does not play any role in terms of expectation. If we take the expectation of the forward difference estimator (12.7), we find

$$E\left[\hat{\Delta}_F\right] = \frac{g(S_0 + h) - g(S_0)}{h}. \tag{12.9}$$

Assuming that $g(\cdot)$ is suitably differentiable, we may write the Taylor expansion:

$$g(S_0 + h) = g(S_0) + hg'(S_0) + \frac{1}{2}h^2 g''(S_0) + o(h^2). \tag{12.10}$$

Rearranging the equation, we find the bias in the estimator:

$$E\left[\hat{\Delta}_F - g'(S_0)\right] = \frac{1}{2}h g''(S_0) + o(h). \tag{12.11}$$

Let us compare this with the central difference estimator. In this case, it is also useful to consider the Taylor expansion:

$$g(S_0 - h) = g(S_0) - hg'(S_0) + \frac{1}{2}h^2 g''(S_0) + o(h^2). \tag{12.12}$$

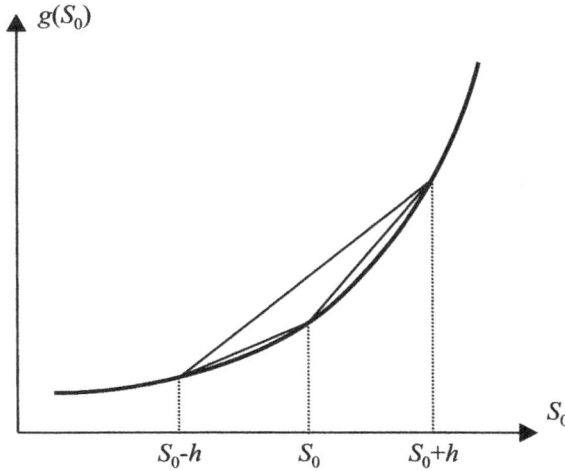

FIGURE 12.4 **Comparing finite differences.**

Subtracting Eq. (12.12) from Eq. (12.10) and rearranging, we find

$$E\left[\hat{\Delta}_C - g'(S_0)\right] = o(h), \tag{12.13}$$

since the second-order derivatives now cancel each other. This bias is of lower order with respect to Eq. (12.11). We might also consider a backward difference estimator, but Fig. 12.4 shows intuitively why the central difference estimator might be the best one.

As a more interesting example, let us consider the delta of an arithmetic average, European-style, Asian option, with payoff:

$$\max\left\{\bar{S} - K, 0\right\}, \qquad \bar{S} = \frac{1}{T}\sum_{t=1}^{T} S_t. \tag{12.14}$$

For the sake of simplicity, we consider only the option at its inception, i.e., when it is written, and we also assume that the time intervals at which the underlying asset price is sampled along the option's life to evaluate the average price are of uniform length. The code in Fig. 12.5 is based on central finite differences and common random numbers. In this case, we cannot compare what we find numerically against an exact, analytical result. One possibility for a check is to use a simple first-order metamodel, which actually boils down to pricing the option in the neighborhood of S_0, using a high-quality approach, and then fit a linear regression model. The slope of the regression line provides us with another estimate of delta. In Chapter 11 we have seen that an accurate estimate of the price of the arithmetic average Asian option is obtained by using the corresponding geometric average option as a control variate. The script of Fig. 12.6 yields the following output, which is somewhat reassuring:

```
AsianDeltaCentral <- function(S0,K,r,T,sigma,dS,numRepl,
                              numSamples,c.level=0.95){
  df <- exp(-r*T)
  dt <- T/numSamples
  nuT <- (r-sigma^2/2)*dt
  sigmaT <- sqrt(dt)*sigma
  diffs <- numeric(numRepl)
  for (k in 1:numRepl){
    # store normals for CRN
    normVet <- rnorm(numSamples,nuT,sigmaT)
    logIncrements1 <- c(log(S0-ds),normVet)
    path1 <- exp(cumsum(logIncrements1))
    logIncrements2 <- c(log(S0+ds),normVet)
    path2 <- exp(cumsum(logIncrements2))
    payoff1 <- max(0,mean(path1[-1]) - K)
    payoff2 <- max(0,mean(path2[-1]) - K)
    diffs[k] <- df*(payoff2 - payoff1)/(2*ds)
  }
  aux <- t.test(diffs,conf.level=c.level)
  value <- as.numeric(aux$estimate)
  ci <- as.numeric(aux$conf.int)
  return(list(estimate=value,conf.int=ci))
}
```

FIGURE 12.5 **Code to estimate Δ of an Asian option by central differences.**

```
> delta
[1] 0.5742706
> OutAsianDeltaCentral
$estimate
[1] 0.5727809
$conf.int
[1] 0.5675191 0.5780428
```

What is less pleasing here is that we need a significant computational effort, since we have to run simulations starting from $(S_0 - h)$, S_0, and $(S_0 + h)$, if we want to get both the option price and the delta by central differences. It would be nice to run just one set of replications for pricing, and then squeeze out additional information to find the greeks. Indeed, this can be done by the two approaches we describe in the following two sections:

1. The pathwise derivative approach, based on perturbation analysis
2. The likelihood ratio approach, also known as score function approach

```
S0 <- 90
vetS <- seq(from=88, to=92, by=0.5)
K <- 90
T <-  1
r <- 0.05
sigma <-0.4
numSamples <- 12
numRepl <- 45000
numPilot <- 5000; N <- length(vetS)
prices <- numeric(N)
set.seed(55555)
for (i in 1:N)
   prices[i] <- EuAsianCVGeom(vetS[i],K,T,r,0,sigma,
               numSamples,numRepl,numPilot)$estimate
mod <- lm(prices~vetS)
p <- as.numeric(mod$coeff)
delta <- p[2]
ds <- 0.5
set.seed(55555)
numReplTot <- numRepl+numPilot
OutAsianDeltaCentral <- AsianDeltaCentral(S0,K,r,T,
               sigma,dS,numReplTot,numSamples)
delta
OutAsianDeltaCentral
```

FIGURE 12.6 **Checking the estimate Δ of an Asian option against a regression approach.**

12.2 Estimating option greeks by pathwise derivatives

Let us consider again the function $f(\alpha, \omega)$; if we fix ω, i.e., if we consider a given sample path, and the function is suitably differentiable, we may take the derivative

$$\frac{\partial f}{\partial \alpha}\bigg|_{\omega} = \lim_{h \to 0} \frac{f(\alpha + h, \omega) - f(\alpha, \omega)}{h}.$$

Since this is a random variable, depending on ω, we may consider its expected value

$$\mathrm{E}\left[\frac{\partial f}{\partial \alpha}\bigg|_{\omega}\right]. \tag{12.15}$$

Actually, we should wonder whether the inner limit exists. To see the relevance of the question, consider the payoff of a vanilla call option. We know that there is a point of nondifferentiability for $S_T = K$, which is potentially troublesome when we differentiate with respect to $\alpha = S_0$. Nevertheless, since this is a set of measure zero, we may argue that the limit exists with probability 1 and hope for the best. As we have already pointed out, however, what we are interested in is not the expectation of the limit, but the limit of an expectation:[3]

$$g'(\alpha) = \frac{d\mathrm{E}[f(\alpha,\omega)]}{d\alpha}. \tag{12.16}$$

Comparing Eqs. (12.15) and (12.16) we see that if expectation and differentiation commute, we may use $\partial f / \partial \alpha$ as an unbiased estimator of $g'(\alpha)$. However, this cannot be taken for granted in general. Since we fix a sample path ω and work on the dependence with respect to α, this approach is called *pathwise differentiation*.

It is possible to give a sound theoretical treatment of the above issues, which is outside the introductory scope of this book.[4] However, the best way to get the essentials is by a concrete example. Let us consider a vanilla call option, with discounted payoff

$$e^{-rT} \max\{S_T - K, 0\}, \tag{12.17}$$

where

$$S_T = S_0 e^{(r-\sigma^2/2)T+\sigma\sqrt{T}\epsilon}, \tag{12.18}$$

and $\epsilon \sim \mathrm{N}(0,1)$ is the essence of the sample path ω. We may regard the discounted payoff as our function $f(\alpha,\omega)$, which, in this specific case, can be rewritten as

$$f(S_0, \epsilon).$$

Using the chain rule for differentiation, for a fixed sample path, we have

$$\frac{\partial f}{\partial S_0} = \frac{\partial f}{\partial S_T} \frac{\partial S_T}{\partial S_0}.$$

The second derivative is easy to find by considering Eq. (12.18) for a fixed ϵ:

$$\frac{\partial S_T}{\partial S_0} = e^{(r-\sigma^2/2)T+\sigma\sqrt{T}\epsilon} = \frac{S_T}{S_0}.$$

The first derivative is a bit more problematic, as the option payoff is kinky at the strike price K. Nevertheless, we have already observed that the trouble arises at a set of measure zero. At other points, we have

$$\frac{d}{dx} \max\{x - K, 0\} = \begin{cases} 0, & \text{if } x < K, \\ 1, & \text{if } x > K. \end{cases}$$

When taking the expectation in Eq. (12.15), we may disregard the only critical point and conclude

$$\frac{\partial f}{\partial S_T} = e^{-rT} \mathbf{1}_{\{S_T > K\}},$$

---

[3] In this case we have an ordinary derivative, as the function $g(\alpha) = \mathrm{E}[f(\alpha,\omega)]$ only depends on α whereas in Eq. (12.15) we have a partial derivative. In many books, only ordinary derivatives are used, but we prefer to insist clearly on the dependence on ω.

[4] See, e.g., [4, p. 393].

```
VanillaDeltaMCPath <- function(S0,K,T,r,sigma,numRepl,
                               c.level=0.95){
  nuT <- (r - 0.5*sigma^2)*T
  sigmaT <- sigma*sqrt(T)
  VLogn <- exp(nuT+sigmaT*rnorm(numRepl))
  SampleDelta <- exp(-r*T) * VLogn * (S0*VLogn > K)
  aux <- t.test(SampleDelta,conf.level=c.level)
  value <- as.numeric(aux$estimate)
  ci <- as.numeric(aux$conf.int)
  return(list(estimate=value,conf.int=ci))
}
```

FIGURE 12.7 **Estimating the option Δ by a pathwise estimator.**

where $\mathbf{1}_{\{\cdot\}}$ is the usual indicator function for a given event. Putting everything together, we obtain the following estimator of Δ:

$$e^{-rT}\frac{S_T}{S_0}\mathbf{1}_{\{S_T>K\}}.$$

It is easy to implement the idea in R, as shown in Fig. 12.7. We may check the estimate against the exact result, in the same setting as the previous section:

```
> DeltaOK
[1] 0.583839
> set.seed(55555)
> OutDeltaPath <- VanillaDeltaMCPath(S0,K,T,r,sigma,numRepl)
> OutDeltaPath
$estimate
[1] 0.5813664
$conf.int
[1] 0.5750999 0.5876330
```

We see that the approach indeed works in this case, but it is quite useful to see a counterexample.

Example 12.1 Delta of a digital call

The digital call pays $1 when the underlying asset price at maturity is above the strike. Hence, the discounted payoff of a digital option is

$$e^{-rT}\mathbf{1}_{\{S_T>K\}}. \tag{12.19}$$

This payoff is not only nondifferentiable at $S_T = K$, but also discontinuous. However, this is again a single point. The trouble is that

elsewhere the derivative of this discounted payoff is just zero, as it is
a piecewise constant function of S_T, which is in turn a function of S_0.
Hence, for the above payoff, we find

$$0 = E\left[\frac{\partial f}{\partial S_0}\right] \neq \frac{dg}{dS_0},$$

since the delta of a digital call is certainly not zero.

By comparing Eqs. (12.17) and (12.19), we immediately see that pathwise differentiation cannot be used to estimate the Γ of a vanilla call, either. In the next
section we consider an alternative approach that may overcome these issues.
Before doing so, let us consider another example of pathwise differentiation,
the delta of an Asian option. By discounting the payoff in Eq. (12.14), we are
led to the function

$$f(S_0, \omega) = e^{-rT} \max\left\{\bar{S} - K, 0\right\}, \qquad \bar{S} = \frac{1}{T}\sum_{t=1}^{T} S_t, \qquad (12.20)$$

where ω is a sequence of T independent standard normals ϵ_t, and the dependence on S_0 is implicit in the arithmetic average $\bar{S}$. Keeping the sample path
$\omega = \{\epsilon_t\}$ fixed, we follow the same drill as above:

$$\frac{\partial f}{\partial S_0} = \frac{\partial f}{\partial \bar{S}}\frac{\partial \bar{S}}{\partial S_0} = e^{-rT}\mathbf{1}_{\{\bar{S} > K\}}\frac{\partial \bar{S}}{\partial S_0}.$$

For the last derivative, we may expand the sum:

$$\frac{\partial \bar{S}}{\partial S_0} = \frac{1}{T}\sum_{t=1}^{T}\frac{\partial S_t}{\partial S_0} = \frac{1}{T}\sum_{t=1}^{T}\frac{S_t}{S_0} = \frac{\bar{S}}{S_0},$$

where we again keep the normals fixed. Therefore, the pathwise estimator for
the Asian delta boils down to

$$\frac{\partial f}{\partial S_0} = e^{-rT}\mathbf{1}_{\{\bar{S} > K\}}\frac{\bar{S}}{S_0},$$

which is readily implemented as shown in Fig. 12.8. The results we obtain by
applying this function are in line with those obtained by finite differences:

```
> S0 <- 90
> K <- 90
> T <- 1
> r <- 0.05
> sigma <-0.4
```

```
AsianDeltaPath <- function(S0,K,r,T,sigma,numRepl,
                             numSamples,c.level=0.95){
  df <- exp(-r*T)
  dt <- T/numSamples
  nuT <- (r-sigma^2/2)*dt
  sigmaT <- sqrt(dt)*sigma
  sampleDiff <- numeric(numRepl)
  for (k in 1:numRepl){
    # store normals
    normVet <- rnorm(numSamples,nuT,sigmaT)
    logIncrements <- c(log(S0),normVet)
    path <- exp(cumsum(logIncrements))
    meanS <- mean(path[-1])
    sampleDiff[k] <- df*(meanS > K)*meanS/S0
  }
  aux <- t.test(sampleDiff,conf.level=c.level)
  value <- as.numeric(aux$estimate)
  ci <- as.numeric(aux$conf.int)
  return(list(estimate=value,conf.int=ci))
}
```

FIGURE 12.8 Estimating Δ for an arithmetic average Asian option by a pathwise estimator.

```
> numSamples <- 12
> numRepl <- 50000
> set.seed(55555)
> AsianDeltaPath(S0,K,r,T,sigma,numRepl,numSamples)
$estimate
[1] 0.5726421
$conf.int
[1] 0.5673608 0.5779234
```

12.3 Estimating option greeks by the likelihood ratio method

The likelihood ratio method, also known as score function approach, casts the problem of sensitivity estimation in a quite different way with respect to pathwise differentiation. Pathwise differentiation revolves around the dependence of a function with respect to a parameter of interest. Since the functions we deal with in financial engineering are often nondifferentiable or even discontinuous option payoffs, this may raise some trouble, as we have seen in Example 12.1. On the contrary, in the likelihood ratio approach, the parameter is associated with the probability density of the underlying random variables. Hence, it is convenient to consider a function $f(\cdot)$ depending on a vector of random variables $\mathbf{X} = [X_1, \ldots, X_n]^\mathsf{T}$, and write

$$g(\alpha) = \mathrm{E}_\alpha\big[f(X_1, \ldots, X_n)\big] = \int_{\mathbb{R}^n} f(\mathbf{x}) h_\alpha(\mathbf{x})\, d\mathbf{x}, \qquad (12.21)$$

where $h_\alpha(\mathbf{x})$ is the joint density of the random vector $\mathbf{X}$, depending on the parameter α. If integration and differentiation commute, we may write

$$g'(\alpha) = \frac{d\mathrm{E}_\alpha[f(\mathbf{X})]}{d\alpha} = \int_{\mathbb{R}^n} f(\mathbf{x}) \frac{\partial h_\alpha(\mathbf{x})}{\partial \alpha}\, d\mathbf{x}. \qquad (12.22)$$

Assuming that the interchange is valid, we may use essentially the same trick we used in variance reduction by importance sampling. If we introduce the shorthand

$$\dot{h}_\alpha(\mathbf{x}) \equiv \frac{\partial h_\alpha(\mathbf{x})}{\partial \alpha}$$

to streamline notation a bit, we may rewrite Eq. (12.22) as

$$g'(\alpha) = \int_{\mathbb{R}^n} f(\mathbf{x}) \frac{\dot{h}_\alpha(\mathbf{x})}{h_\alpha(\mathbf{x})} h_\alpha(\mathbf{x})\, d\mathbf{x} = \mathrm{E}_\alpha\left[f(\mathbf{x}) \frac{\dot{h}_\alpha(\mathbf{x})}{h_\alpha(\mathbf{x})}\right]. \qquad (12.23)$$

Note that, unlike $\dot{h}_\alpha(\mathbf{x})$, $h_\alpha(\mathbf{x})$ is a proper density, and that the ratio

$$\frac{\dot{h}_\alpha(\mathbf{x})}{h_\alpha(\mathbf{x})}$$

is similar to a likelihood ratio, which we already met when dealing with importance sampling, and from which the name of the method is derived. Apparently, this approach is less natural than pathwise differentiation but, as we shall see shortly, it is actually simple to recast the problem in such a way that the parameter of interest is included in the density. The big advantage of this idea is that its validity relies on differentiability of probability densities, which are typically smooth, unlike option payoffs.

To illustrate the method, we use the two examples we are familiar with, the delta of vanilla and Asian call options. In the first case, the key point is to work on the density of S_T considering S_0 as a parameter. Assuming geometric Brownian motion as usual, we know from Section 3.7.4 that, under the risk-neutral measure,

$$\log\left(\frac{S_T}{S_0}\right) \sim \mathsf{N}\left(\left(r - \frac{\sigma^2}{2}\right)T,\ \sigma^2 T\right).$$

This says that S_T is lognormal, and we may write its density as[5]

$$h_{S_0}(x) = \frac{1}{x\sigma\sqrt{T}}\, \phi\big(\xi(x, S_0)\big), \qquad (12.24)$$

---

[5] See Section 3.2.3.1.

where

$$\phi(z) = \frac{1}{\sqrt{2\pi}} e^{-x^2/2}$$

is the density of a standard normal and

$$\xi(x, S_0) \equiv \frac{\log(x/S_0) - (r - \sigma^2/2)T}{\sigma\sqrt{T}}.$$

Using the chain rule, we find

$$\dot{h}_{S_0}(x) = \frac{\partial h_{S_0}(x)}{\partial S_0} = \frac{1}{x\sigma\sqrt{T}} \cdot \frac{d\phi(\xi)}{d\xi} \cdot \frac{\partial \xi(x, S_0)}{\partial S_0}$$

$$= \frac{1}{x\sigma\sqrt{T}} \cdot \left[-\xi(x, S_0)\,\phi\big(\xi(x, S_0)\big)\right] \cdot \frac{\partial \xi(x, S_0)}{\partial S_0},$$

which implies that the likelihood ratio is

$$\frac{\dot{h}_{S_0}(x)}{h_{S_0}(x)} = -\xi(x, S_0)\frac{\partial \xi(x, S_0)}{\partial S_0}$$

$$= \frac{\log(x/S_0) - (r - \sigma^2/2)T}{S_0\sigma^2 T}.$$

The sensitivity estimator is obtained by plugging $x = S_T$ and multiplying the likelihood ratio by the discounted payoff:

$$e^{-rT} \cdot \max\{S_T - K, 0\} \cdot \left[\frac{\log(S_T/S_0) - (r - \sigma^2/2)T}{S_0\sigma^2 T}\right]. \qquad (12.25)$$

We may recast the expression as a function of the standard normal ϵ that we use to sample S_T starting from S_0. By plugging Eq. (12.18) into (12.25) and simplifying the ratio, we end up with the estimator

$$e^{-rT} \max\{S_T - K, 0\} \left(\frac{\epsilon}{S_0\sigma\sqrt{T}}\right). \qquad (12.26)$$

The resulting code is displayed in Fig. 12.9. The result may be checked against the exact delta. Once again, the results are in line with the other methods:

```
> S0 <- 50
> K <- 55
> T <- 1
> r <- 0.1
> sigma <-0.4
> numRepl <- 50000
> VanillaCallDelta(S0,K,T,r,sigma)
[1] 0.583839
> set.seed(55555)
> VanillaDeltaMCLR(S0,K,T,r,sigma,numRepl)
$estimate
[1] 0.5729819
$conf.int
[1] 0.5592513 0.5867126
```

```
VanillaDeltaMCLR <- function(S0,K,T,r,sigma,numRepl,
                             c.level=0.95){
  nuT <- (r - 0.5*sigma^2)*T
  sigmaT <- sigma*sqrt(T)
  Z <- rnorm(numRepl)
  ST <- S0*exp(nuT+sigmaT*Z)
  SampleDelta <- exp(-r*T)*pmax(0,ST-K)*Z/(S0*sigmaT)
  aux <- t.test(SampleDelta,conf.level=c.level)
  value <- as.numeric(aux$estimate)
  ci <- as.numeric(aux$conf.int)
  return(list(estimate=value,conf.int=ci))
}
```

FIGURE 12.9 Estimating Δ for an arithmetic average Asian option by likelihood ratios.

The case of the arithmetic average Asian option proceeds much along the same lines, but we have to cope with a multidimensional PDF $h_{S_0}(S_1,\ldots,S_T)$. Since the underlying GBM process is Markovian, we may factor the PDF as the product of conditional densities:

$$h_{S_0}(S_1,\ldots,S_T) = h_1(S_1 \mid S_0)h_2(S_2 \mid S_1)\cdots h_T(S_T \mid S_{T-1}).$$

Each conditional density can be written just like in the case of the vanilla option:

$$h_t(S_t \mid S_{t-1}) = \frac{1}{x_t\sigma\sqrt{\delta t}}\phi(\xi(x_t \mid x_{t-1})),$$

where $\phi(\cdot)$ is again the density of a standard normal, δt is the time elapsing between two observations in the average, and

$$\xi(x_t \mid x_{t-1}) \equiv \frac{\log(x_t/x_{t-1}) - (r - \sigma^2/2)\,\delta t}{\sigma\sqrt{\delta t}}.$$

The score function gets considerably simplified by the fact that S_0 only occurs in the first conditional density. It is convenient to differentiate the logarithm of the above product, thus turning it into a sum. Using more or less the same calculations as in the vanilla call case, we find

$$\frac{\partial \log h_{S_0}(S_1,\ldots,S_T)}{\partial S_0} = \frac{\partial \log h_1(S_1 \mid S_0)}{\partial S_0} = \frac{\xi_1(S_1 \mid S_0)}{S_0\sigma\sqrt{\delta t}} = \frac{\epsilon_1}{S_0\sigma\sqrt{\delta t}},$$

where ϵ_1 is the first standard normal used for sample path generation, in the transition from S_0 to S_1. Therefore, the estimator of Δ is

$$e^{-rT}\max\{\bar{S} - K, 0\}\frac{\epsilon_1}{S_0\sigma\sqrt{\delta t}},$$

which is used in the code of Fig. 12.10. Once again, the results are in line with the other methods:

```
AsianDeltaLR <- function(S0,K,r,T,sigma,numRepl,numSamples,
                         c.level=0.95){
  df <- exp(-r*T)
  dt <- T/numSamples
  nuT <- (r-sigma^2/2)*dt
  sigmaT <- sqrt(dt)*sigma
  sampleDiff <- numeric(numRepl)
  for (k in 1:numRepl){
    # store STANDARD normals
    Veps <- rnorm(numSamples)
    logIncrements <- c(log(S0),nuT+Veps*sigmaT)
    path <- exp(cumsum(logIncrements))
    payoff <- max(0,mean(path[-1]) - K)
    sampleDiff[k] <- df*payoff*Veps[1]/(S0*sigmaT)
  }
  aux <- t.test(sampleDiff,conf.level=c.level)
  value <- as.numeric(aux$estimate)
  ci <- as.numeric(aux$conf.int)
  return(list(estimate=value,conf.int=ci))
}
```

FIGURE 12.10 **Estimating** Δ **for an arithmetic average Asian option by likelihood ratios.**

```
> S0 <- 90
> K <- 90
> T <-  1
> r <- 0.05
> sigma <-0.4
> numSamples <- 12
> numRepl <- 50000
> set.seed(55555)
> AsianDeltaLR(S0,K,r,T,sigma,numRepl,numSamples)
$estimate
[1] 0.5735931
$conf.int
[1] 0.5540027 0.5931835
```

For further reading

- A deeper, yet readable treatment of bias issues in sensitivity estimation can be found in [4, Chapter 7].
- For a rigorous treatment of perturbation analysis, which is the background of pathwise derivatives, you may have a look at [3].

- The score function method is dealt with in depth in [8], even though the treatment is aimed at discrete-event systems. For discrete-time systems, see, e.g., [5].
- The interplay between simulation and optimization, where sensitivity analysis does play a role, is covered in [7] or [9], where you may also find information about stochastic approximation and stochastic gradient methods. The metamodeling approach is treated in depth by [6].
- Another high-level textbook reference for the topics of this chapter is [1, Chapters 7–8].

References

1 S. Asmussen and P.W. Glynn. *Stochastic Simulation: Algorithms and Analysis*. Springer, New York, 2010.

2 P. Brandimarte. *Numerical Methods in Finance and Economics: A MATLAB-Based Introduction* (2nd ed.). Wiley, Hoboken, NJ, 2006.

3 P. Glasserman. *Gradient Estimation via Perturbation Analysis*. Kluwer Academic, Boston, 1991.

4 P. Glasserman. *Monte Carlo Methods in Financial Engineering*. Springer, New York, 2004.

5 P.W. Glynn. Likelihood ratio gradient estimation for stochastic systems. *Communications of the ACM*, 33:75–84, 1990.

6 J.P.C. Kleijnen. *Design and Analysis of Simulation Experiments*. Springer, Berlin, 2008.

7 G.C. Pflug. *Optimization of Stochastic Models: The Interface Between Simulation and Optimization*. Kluwer Academic, Dordrecht, The Netherlands, 1996.

8 R.Y. Rubinstein and A. Shapiro. *Discrete Event Systems: Sensitivity Analysis and Stochastic Optimization via the Score Function Method*. Wiley, New York, 1993.

9 J.C. Spall. *Introduction to Stochastic Search and Optimization: Estimation, Simulation, and Control*. Wiley, Hoboken, NJ, 2003.

## Chapter Thirteen

Risk Measurement and Management

In this chapter we deal with a fundamental key issue in finance, i.e., risk. Actually, risk is a multidimensional concept, and there is a long list of risk factors that are relevant in finance. Risk factors directly related to financial markets are the first that come to mind, such as volatility in stock prices, interest rates, and foreign exchange rates, but they do not exhaust the list at all. For instance, after the subprime mortgage crisis, everyone is well aware of the role of credit risk, and the increasing interplay between financial markets and energy/commodity markets contributes to complicate the overall picture. In fact, the impact of derivative markets on oil and commodity (most notably, food) prices is controversial. The growth of derivative markets casts some doubts on seemingly obvious causal links: If one thinks about interest rate derivatives, it may seem natural assume that the value of those derivatives depends on prevailing conditions on money and capital markets. However, if one realizes the sheer volume of derivatives traded on regulated exchanges and over the counter, some doubts may arise about tails wagging dogs.

The above risk factors are often addressed by Monte Carlo methods, but we should mention that there are other risk factors that are less amenable to statistical approaches. The increasing role of information/communication technologies and globalization has introduced other forms of risk, such as operational risk, political risk, and regulatory risk. Ensuring business continuity in the face of operational disruptions is quite relevant to any financial firm, but it involves a different kind of risk analysis, and it is certainly difficult to attach probability distributions to political risk factors. Despite these difficulties, it is fundamental not to miss the interplay between financial and nonfinancial stakeholders. In times of general concern about the nature of financial speculation and the abuse of derivatives, it is refreshing to keep in mind that risk affects not only financial institutions, but also nonfinancial stakeholders, like firms using derivatives for hedging purposes. A global player is typically exposed, among other things, to foreign exchange risk. Choosing the right hedging instrument (e.g., forward/futures contracts, vanilla or exotic options) requires an assessment of the amount of exposure that should be covered. In such a context, things are made difficult by volume risk, which is related to how good or bad the business is going to be. Given this level of complexity, it is clear that we cannot

591

adequately address the full set of risk dimensions in a single chapter. We will only deal with a limited subset of the above risk factors, through a few selected examples aimed at illustrating the potential of Monte Carlo applications for risk measurement and management.

Last but not least, we should remark that risk management is sometimes confused with risk measurement. It is certainly true that a sensible way to measure risk is a necessary condition to manage risk properly. But coming up with a number gauging risk, however useful, is not the same as establishing a policy, e.g., to hedge financial or commodity risk. In the first part of the chapter we deal with risk measures. Emphasis will be placed on quantile-based risk measures, such as value-at-risk (V@R) and conditional value-at-risk (CV@R). In the second part of the chapter, we move on and deal with risk management; we also illustrate a few links with certain stochastic optimization models dealt with in Chapter 10.

We start in Section 13.1 by discussing general issues concerning risk measurement, most notably the coherence of risk measures, Then, Section 13.2 is essentially devoted to V@R, a widely used and widely debated quantile-based risk measure. First we define V@R, then we illustrate a few of its shortcomings. Despite all of the criticism concerning its use, V@R is still a relevant risk measure, and it provides us with a good opportunity to appreciate Monte Carlo simulation in action. It is not too difficult to build a crude simulation model to estimate V@R, but there are several issues that we must take into account. As we have pointed out in Section 7.3, estimating quantiles raises some nontrivial statistical issues. Therefore, in Section 13.3 we estimate shortfall probabilities, from which other risk measures can be estimated. Furthermore, when a portfolio includes derivatives, it may be rather expensive to reprice them several times. Thus, we explore the possibility of using option greeks to speed up the computation. Another thorny issue is that in risk measurement we are concerned with extreme bad tails, where rare events are relevant. This may require the adoption of suitable variance reduction strategies, as we illustrate in Section 13.4. In the second half of the chapter we move from risk measurement to risk management. In Section 13.5 we take a brief detour to discuss stochastic programming models involving risk measures; in particular, we illustrate the difficulty of dealing with V@R, whereas the problem of CV@R optimization has a surprisingly simple solution, at least in principle. In Section 13.6 we illustrate how Monte Carlo methods can be used to assess the merits of a risk management policy. As a simple but instructive example, we consider delta hedging of a vanilla call option and compare it against a simpler stop-loss strategy. As expected, continuous-time delta hedging yields the Black–Scholes–Merton price for the option, assuming very idealized, not quite realistic, market conditions. Nevertheless, a simulation model is flexible and allows us to analyze the impact of additional features, like transaction costs, stochastic volatilities, and modeling errors. Finally, in Section 13.7 we discuss the interplay between financial and nonfinancial risk factors in a foreign exchange risk management problem.

13.1 What is a risk measure?

When setting up a portfolio of financial assets, stocks or bonds, or when writing an option, risk is the first and foremost concern. This is evident in basic portfolio management models in the Markowitz style, where expected return is traded off against variance. We have introduced this model in Section 3.1.1, and here we just recall that, if we consider a universe of n assets with random rate of return R_i, $i = 1, \ldots, n$, and if $\mathbf{w} \in \mathbb{R}^n$ is the vector collecting portfolio weights w_i, then the variance of the portfolio return is

$$\sigma_p^2 = \mathrm{Var}\left(\sum_{i=1}^n w_i R_i \right) = \sum_{i=1}^n \sum_{j=1}^n w_i \sigma_{ij} w_j = \mathbf{w}^{\mathsf{T}} \Sigma \mathbf{w},$$

where the matrix $\Sigma \in \mathbb{R}^{n,n}$ collects the covariances $\sigma_{ij} = \mathrm{Cov}(R_i, R_j)$ of the asset returns. Typically, the actual risk measure considered is the standard deviation σ_p, which has the same unit of measurement of expected return; the reason for minimizing variance is purely computational, as this is equivalent to minimizing standard deviation and results in an easy quadratic programming problem. Clearly, standard deviation can be considered as a risk measure: the smaller, the better. Standard deviation does capture the dispersion of a probability distribution, but is it really a good risk measure?

Before answering the question, we should actually clarify what a risk measure is. Standard deviation maps a random variable into a real number. Hence, we may define a risk measure as a function $\rho(X)$ mapping a random variable X into $\mathbb{R}$. According to our convenience, the random variable X may represent the value of a portfolio, or profit, or loss. Furthermore, when we consider loss, we may take the current wealth as a reference point, or its expected value in the future. Whatever the modeling choice is, a larger value of $\rho(X)$ is associated with a riskier portfolio. By the way, standard deviation of portfolio return R_p is just the same as the standard deviation of portfolio loss $L_p = -R_p$. Indeed, standard deviation is symmetric, as it penalizes both the upside and the downside potential of a portfolio. Hence, it may be a suitable measure if we assume a symmetric distribution for asset returns; it is a bit less so, if we include assets with skewed, i.e., asymmetric returns. Even if we accept the hypothesis that stock returns are, e.g., normally distributed, and this is not quite the case, it is certainly not true that the same property holds for derivatives written on those assets, because of the nonlinearity of the payoff functions, which induces a nonlinearity in option values before maturity as well.

Given the limitations of standard deviation, we should look somewhere else to find alternative risk measures, but which ones make sense? What are, in principle, the desirable features of a risk measure? A plausible list of desiderata derives from intuitive properties that a risk measure should enjoy. Such a list has been compiled and leads to the definition of *coherent* risk measure. Note that in the following statement we assume that X is related to portfolio value or profit (the larger, the better); if we interpret X as loss, the properties can be easily adjusted by flipping some inequalities and changing a few signs:

- **Normalization.** If the random variable is $X \equiv 0$, it is reasonable to set $\rho(0) = 0$; if you do not hold any portfolio, you are not exposed to any risk.

- **Monotonicity.** If $X_1 \leq X_2$, with probability 1, then $\rho(X_1) \geq \rho(X_2)$. In plain English, if the value of portfolio 1 is never larger than the value of portfolio 2, then portfolio 1 is at least as risky as portfolio 2.[1]

- **Translation invariance.** If we add a fixed amount to the portfolio, this will reduce risk: $\rho(X + \alpha) = \rho(X) - \alpha$.

- **Positive homogeneity.** Intuitively, if we double the amount invested in a portfolio, we double risk. Formally: $\rho(\alpha X) = \alpha \rho(X)$, for $\alpha \geq 0$.

- **Subadditivity.** Diversification is expected to decrease risk; at the very least, diversification cannot increase risk. Hence, it makes sense to assume that the risk of the sum of two random variables should not exceed the sum of the respective risks: $\rho(X + Y) \leq \rho(X) + \rho(Y)$.

The two last conditions may be combined into a convexity condition:

$$\rho\big(\lambda X + (1 - \lambda)Y\big) \leq \lambda\rho(X) + (1 - \lambda)\rho(Y) \qquad \forall \lambda \in [0, 1].$$

Convexity is a quite important feature when tackling optimization problems involving risk measures, as convex optimization problems are relatively easy to solve. We are dealing here only with a single-period problem; tackling multiperiod problems may complicate the matter further, introducing issues related to time consistency, which we do not consider here.[2]

These are the theoretical requirements of a risk measure, but what about the practical ones? Clearly, a risk measure should not be overly difficult to compute. Unfortunately, computational effort may be an issue, if we deal with financial derivatives whose pricing itself requires intensive computation. Another requirement is that it should be easily communicated to top management. A statistically motivated measure, characterizing a feature of a probability distribution, may be fine for the initiated, but a risk measure expressed in hard monetary terms can be easier to grasp. This is what led to the development of value-at-risk, which we describe in the next section. A further advantage of such a measure is that it sets all different kind of assets on a common ground. For instance, duration gives a measure of interest rate risk of bonds, and option greeks, like delta and gamma, tell something about the risk of derivatives, but it is important to find a measure summarizing all risk contributions, irrespective of the nature of the different positions.

Before closing the section, we should at least mention that in financial economics there is another approach to deal with risk aversion and decision under

---

[1] Since we are comparing random variables, the inequality should be qualified as holding almost surely, i.e., for all of the possible outcomes, with the exception of a set of measure zero. The unfamiliar reader may consider this as a technicality.

[2] The essence of time consistency of a multiperiod risk measure is that if a portfolio is riskier than another portfolio at time horizon τ, then it is riskier at time horizons $t < \tau$ as well. See, e.g., [2].

risk, based on expected utility theory. Utility functions, despite all of their limitations and the criticisms surrounding them, could be used in principle, but there are at least two related difficulties:

- It is difficult to elicit the utility function of a decision maker.
- If we are dealing with a portfolio of a mutual fund, whose risk aversion should we measure? The fund manager? The client? And how can we aggregate the risk aversion of several clients?

The last observation clearly points out that we need a hopefully *objective* risk measure, related to the risk in the portfolio, rather than the subjective attitude of a decision maker toward risk. This is necessary if risk measures are to be used as a tool by regulators and managers.

13.2 Quantile-based risk measures: Value-at-risk

The trouble with standard deviation as a risk measure is that it does not focus on the bad tail of the profit/loss distribution. Hence, in order to account for asymmetry, it is natural to look for risk measures related to quantiles. The best-known such measure is value-at-risk. It is common to denote value-at-risk as VaR, where the last capital letter should avoid confusion with variance.[3] We follow here the less ambiguous notational style of V@R. Informally, V@R aims at measuring the maximum portfolio loss one could suffer, over a given time horizon, within a given confidence level. Technically speaking, it is a quantile of the probability distribution of loss. Let L_T be a random variable representing the loss of a portfolio over a holding period of length T; note that a negative value of loss corresponds to a profit. Then, V@R at confidence level $1 - \alpha$ can be defined as the smallest number $\text{V@R}_{1-\alpha}$ such that

$$P\{L_T \leq \text{V@R}_{1-\alpha}\} \geq 1 - \alpha. \tag{13.1}$$

This definition is also valid for discrete probability distributions. If L_T is a continuous random variable and its CDF is invertible, we may rewrite Eq. (13.1) as

$$P\{L_T \leq \text{V@R}_{1-\alpha}\} = 1 - \alpha. \tag{13.2}$$

In the following discussion we will mostly assume for simplicity that loss is a continuous random variable, unless a discrete one is explicitly involved. For instance, if we set $\alpha = 0.05$, we obtain V@R at 95%. The probability that the loss exceeds V@R is α.[4]

Actually, there are two possible definitions of V@R, depending on the reference wealth that we use in defining loss. Let W_0 be the initial portfolio wealth.

---

[3]Arguably, the lowercase letter in the middle should also avoid confusion with VAR, which usually refers to vector autoregressive models.

[4]We stick to the statistical convention that α is the small area associated with the tail, but sometimes the opposite notation is adopted.

If R_T is the random (rate of) return over the holding period, the future wealth is

$$W_T = W_0(1 + R_T).$$

Its expected value is

$$\mathrm{E}[W_T] = W_0(1 + \mu),$$

where μ is the expected return. The absolute loss over the holding period is related to the initial wealth:

$$L_T^a = W_0 - W_T = -W_0 R_T.$$

The quantile of absolute loss at level $1 - \alpha$ is the *absolute* V@R at that confidence level. We find the *relative* V@R if we take the expected future wealth as the reference in defining loss:

$$L_T^r = \mathrm{E}[W_T] - W_T = W_0(\mu - R_T).$$

If we work with a short time period, say, one day, drift is dominated by volatility;[5] hence, the expected return is essentially zero and the two definitions boil down to the same thing. Since certain bank regulations require the use of a risk measure in order to set aside enough cash to be able to cover *short-term* losses, in this section we consider absolute V@R and drop the superscript from the definition of loss. Nevertheless, we should keep in mind that relative V@R may be more relevant to longer term risk, as the one faced by a pension fund. Furthermore, even if the underlying risk factors do not change at all, the sheer passage of time does have an effect on bonds and derivatives.

The loss L_T is a random variable depending on a set of risk factors, possibly through complicated transformations linked to option pricing. To illustrate V@R using a simple example, let us consider a portfolio of stock shares and assume that the risk factor is portfolio return itself. If the holding period return has a continuous distribution, Eq. (13.2) implies that loss will exceed V@R with a low probability,

$$\mathrm{P}\left\{L_T \geq \mathrm{V@R}_{1-\alpha}\right\} = \alpha,$$

where we may indifferently write "$\geq$" or "$>$," since the distribution involved is continuous. This can be rewritten as follows:

$$\begin{aligned}
\mathrm{P}\left\{L_T \geq \mathrm{V@R}_{1-\alpha}\right\} &= \mathrm{P}\left\{-W_0 R_T \geq \mathrm{V@R}_{1-\alpha}\right\} \\
&= \mathrm{P}\left\{R_T \leq -\frac{\mathrm{V@R}_{1-\alpha}}{W_0}\right\} \\
&= \mathrm{P}\{R_T \leq r_\alpha\} = \alpha, \tag{13.3}
\end{aligned}$$

---

[5]Intuitively, *drift* is related to expected return, and *volatility* is related to standard deviation. On a short time interval of length δt, since drift scales linearly with δt, whereas volatility is proportional to $\sqrt{\delta t}$, drift goes to zero more rapidly than does volatility. See the square-root rule, which we considered in Section 3.7.1.

where we have defined

$$r_\alpha \equiv -\frac{\text{V@R}_{1-\alpha}}{W_0}.$$

The return r_α will be negative in most practical cases and is just the quantile at level α of the distribution of portfolio return, i.e., the worst-case return with confidence level α. Thus, absolute V@R is

$$\text{V@R} = -W_0 r_\alpha, \qquad (13.4)$$

whereas relative V@R is

$$\text{V@R} = -W_0(r_\alpha - \mu). \qquad (13.5)$$

Computing V@R is very easy if we assume that the rate of return is normally distributed with

$$\text{E}[R_T] \approx 0, \qquad \text{Var}(R_T) = \sigma^2,$$

in which case there is no difference between absolute and relative V@R, and loss $L_T = -W_0 R_T$ is normal as well, $L_T \sim \text{N}(0, W_0^2 \sigma^2)$. We may take advantage of the symmetry of the normal distribution, as the critical return r_α is, in absolute value, equal to the quantile $r_{1-\alpha}$. Then, to compute $\text{V@R}_{1-\alpha}$, we may use the familiar standardization/destandardization drill for normal variables:

$$\text{P}\{L_T \leq \text{V@R}_{1-\alpha}\} = \text{P}\left\{\frac{L_T - 0}{W_0 \sigma} \leq \frac{\text{V@R}_{1-\alpha} - 0}{W_0 \sigma}\right\} = \text{P}\left\{Z \leq \frac{\text{V@R}_{1-\alpha}}{W_0 \sigma}\right\},$$

where Z is a standard normal variable. We have just to find the standard quantile $z_{1-\alpha}$ and set

$$\text{V@R}_{1-\alpha} = z_{1-\alpha} \sigma W_0.$$

▣ Example 13.1 Elementary V@R calculation

> We have invested \$100,000 in Quacko Corporation stock shares, whose daily volatility is 2%. Then, V@R at 95% level is
>
> $$\$100,000 \times 0.02 \times z_{0.95} = \$100,000 \times 0.02 \times 1.6449 = \$3289.71.$$
>
> We are "95% sure" that we will not lose more than \$3289.71 in one day. V@R at 99% level is
>
> $$\$100,000 \times 0.02 \times z_{0.99} = \$100,000 \times 0.02 \times 2.3263 = \$4652.70.$$
>
> Clearly, increasing the confidence level by 4% has a significant effect, since we are working on the tail of the distribution.

The assumption of normality of returns can be dangerous, as the normal distribution has a relatively low kurtosis; alternative distributions have been proposed, featuring fatter tails, in order to better account for tail risk, which is what we are concerned about in risk management. Nevertheless, the calculation based on the normal distribution is so simple and appealing that it is tempting to use it even when we should rely on more realistic models. In practice, we are not interested in V@R for a single asset, but in V@R for a whole portfolio. Again, the normality assumption streamlines our task considerably.

Example 13.2 V@R in multiple dimensions

Suppose that we hold a portfolio of two assets. The portfolio weights are $w_1 = \frac{2}{3}$ and $w_2 = \frac{1}{3}$, respectively. We also assume that the returns of the two assets have a jointly normal distribution; the two daily volatilities are $\sigma_1 = 2\%$ and $\sigma_2 = 1\%$, respectively, and the correlation is $\rho = 0.7$. Let the time horizon be $T = 10$ days; despite this, we assume again that expected holding period return is zero. To obtain portfolio risk, we first compute the variance of the holding period return:

$$\sigma_p^2 = \begin{bmatrix} w_1 & w_2 \end{bmatrix} \begin{bmatrix} \sigma_1^2 T & \rho\sigma_1\sigma_2 T \\ \rho\sigma_1\sigma_2 T & \sigma_2^2 T \end{bmatrix} \begin{bmatrix} w_1 \\ w_2 \end{bmatrix} = 0.0025111.$$

Hence, $\sigma_p = 0.05011$. If the overall portfolio value is \$10 million, and the required confidence level is 99%, we obtain

$$\text{V@R}_{0.99} = z_{0.99} \cdot \sigma_p \cdot W_0 = 2.3263 \times 0.05011 \times 10^7 = \$1,165,709.$$

Once again, we stress that the calculations in Example 13.2 are quite simple (probably *too* simple) since they rely on a few rather critical assumptions. To begin with, we have scaled volatility over time using the square-root law, which assumes independence of returns over time. If you are not willing to believe in the efficient market hypothesis, this may be an uncomfortable assumption. Then, we have taken advantage of the analytical tractability of the normal distribution, and the fact the return of stock shares was the only risk factor involved. Other risk factors may be involved, such as inflation and interest rates, and the portfolio can include derivatives, whose value is a complicated function of underlying asset prices. Even if we assume that the underlying risk factors are normally distributed, the portfolio value may be a nonlinear function of them, and the analytical tractability of the normal distribution is lost. Needless to say, Monte Carlo methods play a relevant role in estimating V@R in such cases, as we will illustrate later. However, we should also mention that a completely different route may be taken, based on historical V@R. So far, we

5%

5%

V@R   Loss

V@R   Loss

(a)

(b)

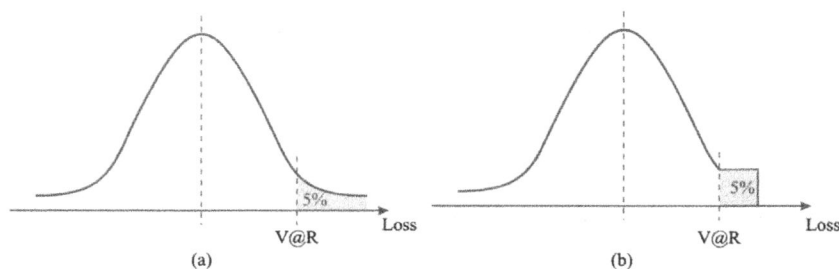

FIGURE 13.1 **Value-at-risk can be the same in quite different situations.**

have relied on a *parametric* approach, based on a theoretical, not necessarily normal, probability distribution. One advantage of the normal distribution is that it simplifies the task of characterizing the joint distribution of returns, since we need only a correlation matrix. However, we know that correlations may not fully capture dependence between random variables, and this is especially true under the stress conditions that are relevant in risk management. The copula framework of Section 3.4 may be adopted to improve accuracy. Alternatively, rather than assuming a specific joint distribution, we may rely on a *nonparametric* approach based on historical data. The advantage of historical data is that they should naturally capture dependence. Hence, we may combine them, according to *bootstrapping* procedures, to generate future scenarios and estimate V@R by historical simulation.

Whatever approach we use for its computation, V@R is not free from some fundamental flaws, which depend on its definition as a quantile. We should be well aware of them, especially when using sophisticated computational tools that may lure us into a false sense of security. For instance, a quantile cannot distinguish between different tail shapes. Consider the two loss densities in Fig. 13.1. In Fig. 13.1(a) we observe a normal loss distribution and its 95% V@R, which is just its quantile at probability level 95%; the area of the right tail is 5%. In Fig. 13.1(b) we observe a sort of truncated distribution, obtained by appending a uniform tail to a normal density, which accounts for 5% of the total probability. By construction, V@R is the same in both cases, since the areas of the right tails are identical. However, we should not associate the same measure of risk with the two distributions. In the case of the normal distribution there is no upper bound to loss; in the second case, there is a clearly defined worst-case loss. Whether the risk for density (a) is larger than density (b) or not, it depends on how we measure risk exactly; the point is that V@R does not indicate any difference between them. In order to discriminate between the two cases, we may consider the expected value of loss *conditional* on being on the right (bad) tail of the loss distribution. This conditional expectation yields the midpoint of the uniform tail in the truncated density; the conditional expected value may be larger in the normal case, because of its unbounded support. This observation has led to the definition of alternative risk measures, such as *conditional value-*

at-risk (CV@R), which is the expected value of loss, conditional on being to the right of V@R.

Risk measures like V@R or CV@R could also be used in portfolio optimization, by solving mathematical programs with the same structure as problem (10.12), where variance is replaced by such measures. The resulting optimization problem can be rather difficult. In particular, it may lack the convexity properties that are so important in optimization. It turns out that minimizing V@R, when uncertainty is modeled by a finite set of scenarios (which may be useful to capture complex distributions and dependencies among asset prices), is a nasty nonconvex problem, whereas minimizing CV@R is, from a computational viewpoint, easier as it yields a convex optimization problem.[6]

There is one last issue with V@R that deserves mention. Intuitively, risk is reduced by diversification. This should be reflected by any risk measure $\rho(\cdot)$ we consider and, as we have seen before, the subadditivity condition is needed to express such a requirement. The following counterexample is often used to show that V@R lacks this property.

■ Example 13.3 **V@R is not subadditive**

Let us consider two corporate bonds, A and B, whose issuers may default with probability 4%. Say that, in the case of default, we lose the full face value, \$100 (in practice, we might partially recover the face value of the bond). Let us compute the V@R of each bond with confidence level 95%. Since loss has a discrete distribution in this example, we should use the more general definition of V@R provided by Eq. (13.1). The probability of default is 4%, and $1 - 0.04 = 0.96 > 0.95$; therefore, we find

$$V@R(A) = V@R(B) = V@R(A) + V@R(B) = \$0.$$

Now what happens if we hold both bonds and assume independent defaults? We will suffer:

- A loss of \$0, with probability $0.96^2 = 0.9216$
- A loss of \$100, with probability $2 \times 0.96 \times 0.04 = 0.0768$
- A loss of \$200, with probability $0.04^2 = 0.0016$

Now the probability of losing \$0 is smaller than 95%, and

$$P\{L_T \leq 100\} = 0.9216 + 0.0768 > 0.95.$$

Hence, with that confidence level, $V@R(A+B) = 100 > V@R(A) + V@R(B)$, which means that risk, as measured by V@R, may be increased by diversification.

---

[6]We illustrate this later in Section 13.5.

The counterexample shows that V@R lacks one of the fundamental properties of a coherent risk measure. We should mention that if we restrict our attention to specific classes of distributions, such as the normal, V@R is subadditive. Nevertheless, the above considerations suggest the opportunity of introducing other risk measures, like CV@R. It can be shown that CV@R is a coherent risk measure.

13.3 Issues in Monte Carlo estimation of V@R

Using Monte Carlo methods to estimate V@R is, in principle, quite straightforward. Let $V(\mathbf{S}(t), t)$ denote the value of a portfolio at time t, depending on a vector of risk factors $\mathbf{S}(t) \in \mathbb{R}^n$, with components $S_i(t)$, $i = 1, \ldots, n$. The risk factors could be the underlying asset prices in a portfolio including derivatives, but we might also consider interest rates or exchange rates. After a time step of length δt, the risk factors are changed by a shock

$$\delta S_i \equiv S_i(t + \delta t) - S_i(t), \qquad i = 1, \ldots, n.$$

To streamline notation, we will use $S_i \equiv S_i(t)$, $i = 1, \ldots, n$, to denote the current value of the risk factors, and $\delta_i = \delta S_i$ to denote their shocks; these variables are collected into vectors $\mathbf{S}$ and $\boldsymbol{\delta}$, respectively. If we assume a joint distribution for $\boldsymbol{\delta}$, Monte Carlo estimation of V@R is straightforward, in principle:

1. We sample independent observations of the shocks $\boldsymbol{\delta}$.
2. We reprice each asset included in the portfolio for the new values of the risk factors, and we assess the corresponding loss

$$L = V(\mathbf{S}, t) - V(\mathbf{S} + \boldsymbol{\delta}, t + \delta t).$$

3. Given a set of loss observations, we estimate the required quantile.

A quick and dirty way to estimate the quantile would be to sort, say, 1000 observations of loss in decreasing order, and to report the one in position 50 if, for instance, we want V@R at 95% confidence level. For a slightly more careful discussion of quantile estimation, see Section 7.3, where we also underline the related difficulties. An alternative approach, proposed in [6, 7], is to fix a suitable loss threshold x and to estimate the probability of losing more than this value, i.e., the *shortfall probability*:

$$P\{L > x\}. \tag{13.6}$$

Given M replications, $k = 1, \ldots, k$ we can estimate the above probability as

$$\frac{1}{M} \sum_{k=1}^{M} \mathbf{1}_{\{L_k > x\}}, \tag{13.7}$$

where $\mathbf{1}_{\{\cdot\}}$ is the indicator function for the relevant event. Then, we may build confidence intervals for this probability based on parameter estimation for the binomial distribution (using, e.g., the R function `bino.fit`). Furthermore, on the basis of the probability estimates corresponding to a few selected values of the threshold x, we may build an empirical CDF, which enables us to come up with an estimate of V@R. Given the role of shortfall probabilities, in the following we will only be concerned with them.

Unfortunately, the first two steps in the above procedure are not quite trivial, either. Specifying a sensible distribution for multiple factors, especially under stress conditions, is far from trivial. Even if we pretend, as we do in the rest of this section, that a multivariate normal is a sensible model for the underlying risk factors, repricing each derivative contract in a real life portfolio may be a daunting task, possibly itself requiring expensive Monte Carlo runs. Using option greeks, it is possible to develop an approximation which may help in estimating V@R. If we assume that we own a portfolio of derivatives and that the only risk factors are the underlying asset prices, we might rely on a delta–gamma approximation to develop linear or quadratic models.[7] Let us collect the first-order sensitivities into the vector

$$\boldsymbol{\Delta} \equiv [\Delta_1, \ldots, \Delta_n]^{\mathsf{T}}, \qquad \Delta_i \equiv \frac{\partial V}{\partial S_i}. \tag{13.8}$$

We should note that Δ_i should not be interpreted as the delta of a single option in the portfolio; rather, it is the sensitivity of the overall portfolio to a single risk factor. The portfolio may consist of long or short positions in any number of vanilla call or put options, or even exotic derivatives. Thus, Δ_i will be, in general, a linear combination of option deltas. Then, loss can be approximated by the linear model

$$L = -\delta V \approx -\sum_{i=1}^{n} \Delta_i \cdot \delta_i = -\boldsymbol{\Delta}^{\mathsf{T}} \boldsymbol{\delta}. \tag{13.9}$$

A similar line of reasoning may be pursued for bonds, using their durations. Clearly, L boils down to a normal random variable, if the underlying shocks are jointly normal. This makes computations quite simple, but possibly oversimplified. In fact, we are missing one point which may be quite relevant for both options and bonds, the passage of time: In general, long option positions lose value, which is measured by the option theta, which is usually negative. Let Θ be the portfolio theta, which may be positive or negative, depending on the sign of the positions. As we pointed out in the previous section, there are cases in which δt is very small and the effect of time can be ignored; for the sake of generality, here we do not neglect this issue. Furthermore, a linear approximation may be inaccurate, and we may include second-order sensitivities, related to option gammas, as a remedy. If we collect the second-order sensitivities into

---

[7]To be precise, if we assume that the risk factors are just the underlying asset prices, we may rely on the standard formulas of Section 3.9.2 for option greeks. Otherwise, option sensitivities must be determined, and estimated, by a careful analysis.

the symmetric matrix

$$\mathbf{\Gamma} \in \mathbb{R}^{n,n}, \quad \Gamma_{ij} = \frac{\partial^2 V}{\partial S_i \, \partial S_j}, \qquad i,j = 1, \ldots, n,$$

the resulting second-order approximation is

$$L \approx -\frac{\partial V}{\partial t}\, \delta t - \sum_{i=1}^{n} \Delta_i \cdot \delta_i - \frac{1}{2} \sum_{i=1}^{n} \sum_{j=1}^{n} \frac{\partial^2 V}{\partial S_i \, \partial S_j} \, \delta_i \, \delta_j,$$

$$= -\Theta \, \delta t - \mathbf{\Delta}^{\mathsf{T}} \boldsymbol{\delta} - \frac{1}{2} \boldsymbol{\delta}^{\mathsf{T}} \mathbf{\Gamma} \boldsymbol{\delta}. \tag{13.10}$$

The sensitivity with respect to time plays the role of a constant, but the quadratic term makes the overall approximation non-normal, even when $\boldsymbol{\delta}$ is jointly normal.

To check the quality of delta–gamma approximations, let us compare the shortfall probability estimates obtained by:

1. A straightforward Monte Carlo procedure based on full repricing of a portfolio of derivatives. The R code is displayed in Fig. 13.2. Note that shocks in the value of underlying assets are generated according to normal distribution, rather than from a lognormal, to ease the comparison with the approach that we describe in the next section. The probability measure we use, of course, is the real one, as risk-neural measures are relevant for pricing purposes only.

2. The corresponding estimate by a delta–gamma approximation, shown in Fig. 13.3. This function relies on R code to evaluate greeks; see Fig. 3.51.

Here and in the following section we aim at replicating an experiment carried out in [7]. In that paper, the shortfall probability $P\{L > x\}$ is estimated, among other things, for a portfolio consisting of a short position in ten calls and five puts, all written on ten stock shares with similar characteristics; note that the number of calls and puts is negative. The input data are shown in Fig. 13.4, where we give a script to compare the two estimates. In the output lists `out1` and `out2` we also include vectors of losses corresponding to each scenario. Note that, by resetting the state of random generators, we are comparing estimates based on the same 3000 scenarios. It is interesting to produce a scatterplot of the true loss against the delta–gamma estimates and to check their correlation:

```
> plot(out1$loss,out2$loss,pch=20)
> cor(out1$loss,out2$loss)
[1] 0.9979958
```

The correlation between losses is quite high and indeed the plot in Fig. 13.5 shows that the approximation looks fairly good. The two point estimates also look in fairly good agreement:

```
> out1$estimate
[1] 0.3206667
```

```
ProbShortfallCrudeMC<-function(S0,K,T,mu,r,sigma,numStocks,
          numCalls,numPuts,dt,numRepl,thresh,c.level=0.95){
  # evaluate initial portfolio value
  Call0 <- EuVanillaCall(S0,K,T,r,sigma)
  Put0 <- EuVanillaPut(S0,K,T,r,sigma)
  Value0 <- numStocks*(numCalls*Call0+numPuts*Put0)
  # check if loss is larger than threshold
  numAbove <- 0
  loss <- numeric(numRepl)
  for (k in 1:numRepl){
    # scenarios are generated according to normal
    # distribution for comparison purposes
    newS <- S0*(1+rnorm(numStocks,mu*dt,sigma*sqrt(dt)))
    newCall <- EuVanillaCall(newS,K,T-dt,r,sigma)
    newPut <- EuVanillaPut(newS,K,T-dt,r,sigma)
    newValue <- numCalls*sum(newCall)+numPuts*sum(newPut)
    loss[k] <- Value0-newValue
    if (loss[k]>thresh) numAbove <- numAbove+1
  }
  aux <- binom.test(numAbove,numRepl,conf.level=c.level)
  value <- as.numeric(aux$estimate)
  ci <- as.numeric(aux$conf.int)
  return(list(estimate=value,conf.int=ci,loss=loss))
}
```

R programming notes:

1. The function takes a portfolio of call and put options written on a number numStocks of stocks, whose prices are assumed independent, and evaluate the probability of shortfall over a time interval dt with respect to the threshold thresh.

2. The vectors numCalls and numPuts specify the number of options of each type on each underlying asset.

3. The functions EuVanillaCall and EuVanillaPut to price options were discussed in Section 3.9.2.

4. Note the use of binom.test to build the point estimate and the confidence interval.

5. The code also returns a vector consisting of the loss for each scenario.

FIGURE 13.2 **Estimating a shortfall probability by straightforward Monte Carlo for a portfolio of call and put options.**

```
ProbShortfallDG<-function(S0,K,T,mu,r,sigma,numStocks,
        numCalls,numPuts,dt,numRepl,thresh,c.level=0.95){
  # evaluate greeks
  DeltaCall <- BSMdelta(S0,K,T,r,sigma)
  DeltaPut <- DeltaCall-1
  Gamma <- BSMgamma(S0,K,T,r,sigma)
  ThetaCall <- BSMthetacall(S0,K,T,r,sigma)
  ThetaPut <- BSMthetaput(S0,K,T,r,sigma)
  DeltaVet <-
          rep(numCalls*DeltaCall+numPuts*DeltaPut,numStocks)
  GammaMat <- diag((numCalls+numPuts)*Gamma,numStocks)
  ThetaTot<-numStocks*(numCalls*ThetaCall+numPuts*ThetaPut)
  # check if loss is larger than threshold
  numAbove <- 0
  loss <- numeric(numRepl)
  for (k in 1:numRepl){
    # scenarios are generated according to normal
    # distribution for comparison purposes
    dS <- S0*rnorm(numStocks,mu*dt,sigma*sqrt(dt))
    loss[k] <- -(ThetaTot*dt+crossprod(dS,DeltaVet)+
                0.5*dS%*%GammaMat%*%dS)
    if ( loss[k] > thresh) numAbove<-numAbove+1
  }
  aux <- binom.test(numAbove,numRepl,conf.level=c.level)
  value <- as.numeric(aux$estimate)
  ci <- as.numeric(aux$conf.int)
  return(list(estimate=value,conf.int=ci,loss=loss))
}
```

FIGURE 13.3 **Estimating a shortfall probability by a delta–gamma approximation.**

```
> out2$estimate
[1] 0.33
```

A more careful check shows that there are some problems, though:

```
> cor(out1$loss>50,out2$loss>50)
[1] 0.9470662
> cor(out1$loss>150,out2$loss>150)
[1] 0.9334365
> cor(out1$loss>250,out2$loss>250)
[1] 0.9035132
```

Here we are comparing indicator variables detecting when the loss threshold is exceeded. These correlations get smaller and smaller when the threshold is increased. This seems to suggest that the quality of the delta–gamma approximation tends to deteriorate under extreme scenarios. Hence, we may be

```
S0 <- 100
r <- 0.05
mu <- 0.05
sigma <- 0.3
T <- 0.5
K <- 100
numStocks <- 10
numCalls <- -10
numPuts <- -5
numRepl <- 3000
dt <- 0.1
thresh <- 50
set.seed(55555)
out1 <- ProbShortfallCrudeMC(S0,K,T,mu,r,sigma,numStocks,
                        numCalls,numPuts,dt,numRepl,thresh)
set.seed(55555)
out2 <- ProbShortfallDG(S0,K,T,mu,r,sigma,numStocks,
                        numCalls,numPuts,dt,numRepl,thresh)
```

FIGURE 13.4   **Comparing two estimates of shortfall probability.**

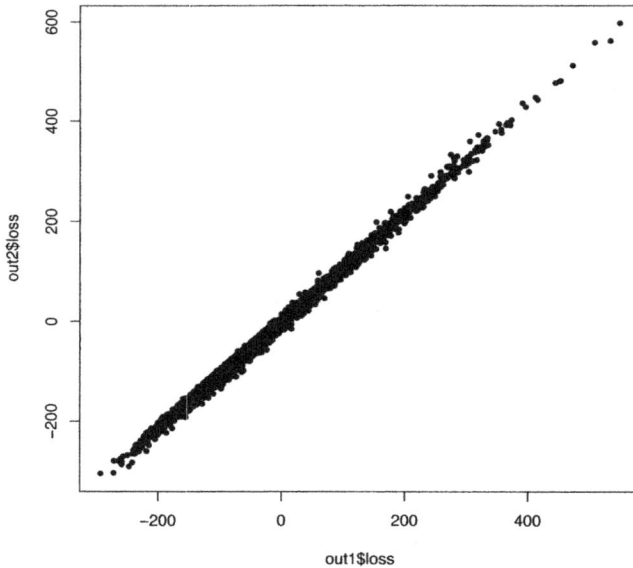

FIGURE 13.5   **A scatterplot of true loss against the delta–gamma approximation.**

forced to resort to more sophisticated approximations, when available, or to full option repricing. Nevertheless, this kind of approximation may help improving the efficiency of Monte Carlo methods, as we illustrate in the next section.

13.4   Variance reduction methods for V@R

In this section we consider variance reduction methods, like importance sampling, to measure the risk of a portfolio of derivatives. The approach we describe was originally proposed in [7], and it is an excellent illustration of clever variance reduction strategies, taking advantage of the delta–gamma approximation we have discussed in the previous section. The treatment follows [6, Chapter 9], where more details can be found, as well as an extension to heavy-tailed distributions. As in the previous section, the objective is to estimate the probability $P\{L > x\}$ that loss L exceeds a given threshold x. It would be convenient to approximate loss by the quadratic function of Eq. (13.10), but, as we have seen, this may be an inaccurate approximation for market scenarios under stress. Furthermore, risk measurement has to cope with tail risk, which implies that many replications might be actually useless, when exceeding the loss threshold is a rare event. As we have seen in Chapter 8, this is a typical case for importance sampling and, possibly, other variance reduction strategies. The main issue with importance sampling is how to find a suitable change of probability measure. To this end, we may take advantage of the quadratic delta–gamma approximation.

Let us assume, for the sake of simplicity, that $\delta \sim N(0, \Sigma)$, where δ is the vector collecting the changes δS_i in the underlying risk factors. As we have seen in Section 5.5.2, sampling such a multivariate normal may be accomplished by choosing a matrix C such that $CC^T = \Sigma$, generating a vector of independent standard normals $Z \sim N(0, I)$, and then computing

$$\delta = CZ. \tag{13.11}$$

Given the selected matrix C, the delta–gamma quadratic approximation of Eq. (13.10) can be rewritten as

$$L \approx a + b^T Z - \frac{1}{2} Z^T (C^T \Gamma C) Z \equiv Q, \tag{13.12}$$

where $a = -\Theta \, \delta t$ and $b = -C^T \Delta$. Standard choices for the matrix C are the lower triangular Cholesky factor or the (symmetric) square root of the covariance matrix. However, a different choice can be advantageous. In fact, the quadratic form is a function of standard normals, but it is not normal itself, as it involves squares Z_j^2 and products $Z_i Z_j$ of standard normals. If we could get rid of these products, the quadratic form would be a much simpler function of standard normals, paving the way for useful analysis. This would be the case if the Hessian matrix $C^T \Gamma C$ in Eq. (13.12) were diagonal. In order to find a matrix C such that $CC^T = \Sigma$ and the resulting Hessian is diagonal, we must find a way to transform the more familiar factorizations of the covariance matrix Σ. Let the square matrix $\mathbf{G}$ be one such factorization, i.e., a matrix such that $\mathbf{GG}^{\mathsf{T}} = \boldsymbol{\Sigma}$, like the lower triangular Cholesky factor of $\boldsymbol{\Sigma}$. Since the matrix $-\frac{1}{2}\mathbf{G}^{\mathsf{T}}\boldsymbol{\Gamma}\mathbf{G}$ is symmetric, it can be diagonalized:

$$-\frac{1}{2}\mathbf{G}^{\mathsf{T}}\boldsymbol{\Gamma}\mathbf{G} = \mathbf{U}\boldsymbol{\Lambda}\mathbf{U}^{\mathsf{T}}, \tag{13.13}$$

where $\mathbf{U}$ is an orthogonal matrix (i.e, $\mathbf{U}^{\mathsf{T}}\mathbf{U} = \mathbf{UU}^{\mathsf{T}} = \mathbf{I}$, where $\mathbf{I}$ is the identity matrix) collecting the unit eigenvectors of $-\frac{1}{2}\mathbf{G}^{\mathsf{T}}\boldsymbol{\Gamma}\mathbf{G}$. The diagonal matrix $\boldsymbol{\Lambda}$ consists of the corresponding eigenvalues:

$$\boldsymbol{\Lambda} = \begin{bmatrix} \lambda_1 & & & \\ & \lambda_2 & & \\ & & \ddots & \\ & & & \lambda_n \end{bmatrix}.$$

Note that we are not diagonalizing a covariance matrix and that these eigenvalues might also be negative. Now, let us consider the matrix

$$\mathbf{C} = \mathbf{GU}. \tag{13.14}$$

This matrix meets our two requirements:

1. It factors the covariance matrix:

$$\mathbf{CC}^{\mathsf{T}} = \mathbf{GUU}^{\mathsf{T}}\mathbf{G}^{\mathsf{T}} = \mathbf{GG}^{\mathsf{T}} = \boldsymbol{\Sigma}.$$

2. The resulting Hessian matrix is diagonal:

$$-\frac{1}{2}\mathbf{C}^{\mathsf{T}}\boldsymbol{\Gamma}\mathbf{C} = -\frac{1}{2}\mathbf{U}^{\mathsf{T}}\mathbf{G}^{\mathsf{T}}\boldsymbol{\Gamma}\mathbf{GU} = \mathbf{U}^{\mathsf{T}}\mathbf{U}\boldsymbol{\Lambda}\mathbf{U}^{\mathsf{T}}\mathbf{U} = \boldsymbol{\Lambda}.$$

Therefore, if we sample the factor shocks using the matrix $\mathbf{C}$ of Eq. (13.14), we may rewrite the quadratic approximation Q in Eq. (13.12) as

$$Q = a + \mathbf{b}^{\mathsf{T}}\mathbf{Z} + \mathbf{Z}^{\mathsf{T}}\boldsymbol{\Lambda}\mathbf{Z}$$
$$= a + \sum_{i=1}^{n}(b_i Z_i + \lambda_i Z_i^2). \tag{13.15}$$

This form of Q includes squared standard normals, which is a bit annoying, but at least it is the sum of independent terms. We may analyze its distribution by evaluating its characteristic function, or we may just complete the square,

$$b_i Z_i + \lambda_i Z_i^2 = \lambda_i \left(Z_i + \frac{b_i}{2\lambda_i} \right)^2 - \frac{b_i^2}{4\lambda_i},$$

which hints at a noncentral chi-square random variable. In fact, we recall from Section 3.2.3.2 that the chi-square distribution is obtained as a sum of squares of independent standard normals. In the expression above the standard normals Z_i are shifted and there are additional transformations, but the resulting distribution can be analyzed by using moment and cumulant generating functions (see Section 8.6.2). We refrain from doing so, as this requires a deeper background, but we give the result of the analysis. The cumulant generating function for Q is

$$\psi(\theta) \equiv \log \mathrm{E}\left[e^{\theta Q}\right]$$
$$= a\theta + \frac{1}{2}\sum_{i=1}^{n}\left[\frac{\theta^2 b_i^2}{1 - 2\theta\lambda_i} - \log(1 - 2\theta\lambda_i)\right] \qquad (13.16)$$

On the basis of this knowledge, we might even compute the exact probability $\mathrm{P}\{Q > x\}$, which immediately suggests a control variate estimator to improve variance in Eq. (13.7):

$$\frac{1}{M}\sum_{k=1}^{M}\mathbf{1}_{\{L_k > x\}} - c\left(\frac{1}{M}\sum_{k=1}^{M}\mathbf{1}_{\{Q_k > x\}} - \mathrm{P}\{Q > x\}\right). \qquad (13.17)$$

A potential issue with this approach is that the correlation between L and Q may be weak just where it matters most, when loss is high, as we have seen in the previous section; hence, we will not pursue this approach here.[8] Intuition suggests that importance sampling should work better in this specific case, and that we should sample in such a way that large losses, exceeding the threshold x, are more likely. This means that we should change the distribution of the normals Z_i. If we look again at Eq. (13.15), we observe immediately that we may increase loss as follows:

- When $b_i > 0$, we should have a positive Z_i as well, which suggests increasing the expected value of Z_i from 0 to a positive number.
- When $b_i < 0$, we should have a negative Z_i as well, which suggests decreasing the expected value of Z_i from 0 to a negative number.
- When $\lambda_i > 0$, since it multiplies Z_i^2, we could increase loss by increasing the variance of Z_i.

This intuition is made more precise in [7] by resorting to exponential tilting. In Section 8.6.3 we have seen that:

1. Exponential tilting transforms a normal variable X into another normal variable.
2. The likelihood ratio is given by $\exp\left\{-\theta X + \psi(\theta)\right\}$, where $\theta > 0$ is the tilting parameter and $\psi(\theta)$ is the cumulant generating function (see Eq. 8.24).

Now the problem is how to choose a suitable random variable playing the role of X in the likelihood ratio. Such a variable is the loss Q in the delta–gamma

---

[8]See [6, Section 9.2].

approximation of Eq. (13.15). If we used Q, the application of exponential tilting to the estimation of the shortfall probability would involve the importance sampling estimator $e^{-\theta Q + \psi(\theta)} \mathbf{1}_{\{Q > x\}}$, which yields

$$P\{Q > x\} = E_\theta \left[e^{-\theta Q + \psi(\theta)} \mathbf{1}_{\{Q > x\}} \right], \tag{13.18}$$

where $E_\theta[\cdot]$ denotes expectation under the importance sampling measure parameterized by the tilting coefficient θ. Note that, for $\theta > 0$, the tilted measure gives more weight to large losses, and the likelihood ratio in Eq. (13.18) compensates for that. To better see why we may hope to reduce variance, let us consider the second-order moment of the estimator:

$$E_\theta \left\{ \left[e^{-\theta Q + \psi(\theta)} \mathbf{1}_{\{Q > x\}} \right]^2 \right\} = E_\theta \left\{ e^{-2\theta Q + 2\psi(\theta)} \mathbf{1}_{\{Q > x\}} \right\} \tag{13.19}$$

$$= E \left\{ e^{-\theta Q + \psi(\theta)} \mathbf{1}_{\{Q > x\}} \right\} \tag{13.20}$$

$$\leq E \left\{ e^{-\theta x + \psi(\theta)} \mathbf{1}_{\{Q > x\}} \right\} \tag{13.21}$$

$$\leq e^{-\theta x + \psi(\theta)}, \tag{13.22}$$

where:

- The equality in Eq. (13.19) follows from familiar properties of the exponential function and the fact that $1^2 = 1$.
- We move back to the original probability measure in Eq. (13.20), which requires multiplying the estimator by the inverse of the likelihood ratio.
- The inequality in Eq. (13.21) follows from the fact that if $Q > x$, then $e^{-\theta Q} < e^{-\theta x}$ for $\theta > 0$.
- The inequality in Eq. (13.22) follows from the fact that $\mathbf{1}_{\{Q > x\}} \leq 1$.

Clearly, this bound is reduced by increasing θ. However, since Q may be a poor replacement of the true loss L, we should apply the idea as follows:

$$P\{L > x\} = E_\theta \left[e^{-\theta Q + \psi(\theta)} \mathbf{1}_{\{L > x\}} \right]. \tag{13.23}$$

Again, if we consider the second-order moment of this estimator (be sure to notice the change of measure in the expectation),

$$E \left[e^{-\theta Q + \psi(\theta)} \mathbf{1}_{\{L > x\}} \right], \tag{13.24}$$

we see that it is small if $\theta > 0$ and if Q is large when the event $\{L > x\}$ occurs. The correlation between Q and L may not be as large as we wish, but it should be enough for the idea to work. However, now we face the problem of sampling Q and L under the tilted measure, which may not be trivial in general. If we assume that the shocks δ are normally distributed, what we actually do is sample a vector of independent standard normals $\mathbf{Z}$ and set $\delta =$

CZ, where the matrix C is given by the diagonalization process that we have described before. These shocks may be used to reprice the portfolio exactly, rather than by the quadratic approximation. Then, the estimator is actually a function of the normal variables Z, and a key issue is what their distribution is under the tilted measure. We have seen in Example 8.10 a simple case in which by tilting a normal, we end up with another normal. This nice result, as shown in the original paper [7], holds here as well. Under the tilted measure with parameter θ, the distribution of the driving normals Z is still normal, $Z \sim N\big(\mu(\theta), \Sigma(\theta)\big)$, where the covariance matrix is diagonal. The components of $\mu(\theta)$ and the diagonal entries of $\Sigma(\theta)$ are related to the n eigenvalues λ_i of the Hessian matrix of the quadratic approximation:

$$\mu_i(\theta) = \frac{\theta b_i}{1 - 2\lambda_i \theta}, \qquad \sigma_i^2(\theta) = \frac{1}{1 - 2\lambda_i \theta}, \qquad i = 1, \dots, n. \qquad (13.25)$$

By the way, these results agree with our intuitive discussion about how the probability of a large loss can be increased under the importance sampling measure: The sign of the expected value of each normal variable depends on the sign of the corresponding coefficient b_i, and its variance is increased if the corresponding eigenvalue λ_i is positive. A quick look at these expressions shows that we have constraints on the tilting parameter:

$$1 - 2\lambda_i \theta > 0 \quad \Rightarrow \quad \theta < \frac{1}{2\lambda_i}, \qquad \forall \lambda_i > 0. \qquad (13.26)$$

These constraints are actually required to ensure the existence of the cumulant generating function in Eq. (13.16). Furthermore, we need some clue about a good setting of θ. Equation (13.22) provides us with an upper bound on the second-order moment of the estimator. This lower bound is minimized by minimizing the argument of the exponential in the expression of the upper bound,

$$\psi(\theta) - \theta x.$$

Taking advantage of the convexity property of the cumulant generating function, we enforce the first-order optimality condition and find the minimum by solving the equation

$$\psi'(\theta) = x. \qquad (13.27)$$

The reader may notice that this is related to the result of equation (8.26).

What we have achieved may be summarized by the following procedure:

Step 1. Calculate the sensitivities (greeks) of the option portfolio.

Step 2. Using diagonalization of the Hessian matrix, find the parameters a, b_i, and λ_i in the quadratic approximation Q of Eq. (13.15).

Step 3. Find the optimal tilting parameter θ by solving Eq. (13.16).

Step 4. For each replication:

- Sample Z from a multivariate normal with parameters given by Eq. (13.25).

```
psi <- function(tilt,a,b,lambda){
  # auxiliary function to compute the psi function
  # see Eq. (9.5) in Glasserman
  aux <- 1-2*tilt*lambda
  out <- a*tilt + 0.5*sum((tilt*b)^2/aux-log(aux))
  return(out)
}

psiPrime <- function(tilt,a,b,lambda){
  # auxiliary function to compute the derivative of
  # the psi function
  aux <- 1-2*tilt*lambda
  out <- a + sum(tilt*b^2*(1-tilt*lambda)/(aux^2) +
                 lambda/aux)
  return(out)
}

findTilt <- function(thresh,a,b,lambda){
  if (any(lambda > 0))
    tiltMax <- 1/(2*max(lambda))
  else
    tiltMax <- 1000 # some BIG number
  fun <- function(x) psiPrime(x,a,b,lambda)-thresh
  tiltStar <- uniroot(fun,c(0,tiltMax))$root
  return(tiltStar)
}
```

FIGURE 13.6 **Some auxiliary functions to implement importance sampling for estimation of a shortfall probability.**

- Evaluate the delta–gamma approximation Q on the basis of Eq. (13.15), as well as the likelihood ratio $e^{-\theta Q+\psi(\theta)}$.
- Evaluate the shock $\boldsymbol{\delta} = \mathbf{C}\mathbf{Z}$ on the risk factors $\mathbf{S}$, as in Eq. (13.11), reprice the portfolio to find $V(\mathbf{S} + \boldsymbol{\delta}, t + \delta t)$, and evaluate the loss L.
- Calculate the importance sampling estimator

$$e^{-\theta Q+\psi(\theta)}\mathbf{1}_{\{L>x\}}.$$

Step 5. Return the average of the estimator.

To accomplish all of the above in the simple example involving vanilla call and put options (see Fig. 13.4), we use again the functions to evaluate option greeks, but we also need some auxiliary functions depicted in Fig. 13.6.

- The function `psi` evaluates the cumulant generating function for a given value of the tilting parameter `tilt`.

- The function `psiPrime` evaluates the derivative $\psi'(\theta)$.
- The function `findTilt` finds the optimal value of θ by solving Eq. (13.27). To this end, we use the function `uniroot` to find a positive root bounded by Eq. (13.26).

Armed with these functions, we may implement the procedure as shown in Fig. 13.7. Note that we are assuming normal shocks on the underlying asset prices, which is not quite consistent with lognormality of prices in the GBM model. We should generate normals Z_i and compute their exponential, but using a simple Taylor expansion we see

$$e^{Z_i} \approx 1 + Z_i.$$

Thus, the approximation is not too bad for small shocks. Unfortunately, it is large shocks that we are interested in, and we make them larger using exponential tilting. Indeed, we should change our definition of the greeks, in order to express the sensitivity of option prices to the underlying normal shocks, rather than asset prices. We avoid doing so for the sake of simplicity. Anyway, we compare results for functions based on the same approximation; thus, the comparison should be fair. In order to do so for the toy example, we use the script in Fig. 13.8, which produces the following output:

```
estimate MC = 0.03497667
estimate IS = 0.03534474
percentage error crude MC   = 3.194342 %
percentage error imp. samp. = 1.079004 %
```

In order to avoid issues with the non-normality of the estimator, rather than producing a confidence interval using one sequence of replications, we run each function 100 times, producing 100 averages of replications. Then, we build a confidence interval based on these averages and evaluate a percentage error. In fact, we do see a significant reduction in variance when we use importance sampling. Clearly, this is just meant as an illustration and not as a conclusive evidence. We also mention that the variance reduction becomes really impressive when importance sampling is integrated with stratification; see [6].

13.5 Mean–risk models in stochastic programming

Estimating V@R, or any other risk measure, does not necessarily mean that we are actually *managing* risk. To this end, we need to set up a strategy, or a decision model. For instance, the classical mean–variance portfolio optimization model is a decision support for managers interested in trading off expected return against variance/standard deviation. Since we have other risk measures at our disposal, it is natural to generalize the idea by replacing variance with, say, V@R or CV@R. This leads to mean–risk decision models. In this section we are not concerned with the appropriateness of each risk measure, but with the computational viability of each class of models. Minimizing variance subject

```
ProbShortfallIS <- function(S0,K,T,mu,r,sigma,numStocks,
                   numCalls,numPuts,dt,numRepl,thresh){
  # evaluate initial portfolio value
  Call0 <- EuVanillaCall(S0,K,T,r,sigma)
  Put0 <- EuVanillaPut(S0,K,T,r,sigma)
  Value0 <- numStocks*(numCalls*Call0+numPuts*Put0)
  # evaluate greeks
  DeltaCall <- BSMdelta(S0,K,T,r,sigma)
  DeltaPut <- DeltaCall-1
  Gamma <- BSMgamma(S0,K,T,r,sigma)
  ThetaCall <- BSMthetacall(S0,K,T,r,sigma)
  ThetaPut <- BSMthetaput(S0,K,T,r,sigma)
  DeltaVet<-rep(numCalls*DeltaCall+numPuts*DeltaPut,numStocks)
  GammaMat<-diag((numCalls+numPuts)*Gamma,numStocks)
  ThetaTot<-numStocks*(numCalls*ThetaCall+numPuts*ThetaPut)
  # covariance matrix is trivial here
  varDeltaS <- S0^2*sigma^2*dt
  sigmaMat <- diag(varDeltaS,numStocks)
  # find a, b, and lambdas in quadratic approximation
  G <- t(chol(sigmaMat)) # upper triangle
  hessian <- -0.5 * G %*% GammaMat %*% t(G)
  aux <- eigen(hessian);   lambda <- aux$values
  U <- aux$vectors; C <- G %*% U
  a <- -dt*ThetaTot; b <- -C %*% DeltaVet
  # find optimal tilting parameter
  tilt <- findTilt(thresh,a,b,lambda)
  # find parameters of tilted normals
  muTheta <- tilt*b/(1-2*lambda*tilt)
  sigmaTheta <- 1/sqrt((1-2*lambda*tilt))
  psiVal <- psi(tilt,a,b,lambda)
  # Now sample scenarios
  numAbove <- 0
  estimate <- numeric(numRepl)
  for (k in 1:numRepl){
    Z <- rnorm(numStocks,muTheta,sigmaTheta)
    Q <- a+sum(b*Z+lambda*Z^2)
    likRatio <- exp(-tilt*Q+psiVal)
    newS <- S0 + muTheta + C %*% Z
    newCall <- EuVanillaCall(newS,K,T-dt,r,sigma)
    newPut <- EuVanillaPut(newS,K,T-dt,r,sigma)
    newValue <- numCalls*sum(newCall)+numPuts*sum(newPut)
    loss <- Value0-newValue
    estimate[k] <- likRatio * (loss > thresh)
  }
  return(mean(estimate))
}
```

FIGURE 13.7 **Estimating a shortfall probability by importance sampling.**

```
S0 <- 100; r <- 0.05; mu <- 0.05; sigma <- 0.3
T <- 0.5; K <- 100; numStocks <- 10
numCalls <- -10; numPuts <- -5
numRepl <- 3000
dt <- 0.1; thresh <- 250
set.seed(55555)
N <- 100
outMC <- numeric(N)
outIS <- numeric(N)
for (k in 1:N){
  aux <- ProbShortfallCrudeMC(S0,K,T,mu,r,sigma,
             numStocks,numCalls,numPuts,dt,numRepl,thresh)
  outMC[k] <- aux$estimate
  outIS[k] <- ProbShortfallIS(S0,K,T,mu,r,sigma,numStocks,
                numCalls,numPuts,dt,numRepl,thresh)
}
# get estimates and confidence intervals
estMC <- t.test(outMC)
estIS <- t.test(outIS)
percMC<-(estMC$conf.int[2]-estMC$conf.int[1])/estMC$estimate
percIS<-(estIS$conf.int[2]-estIS$conf.int[1])/estIS$estimate
cat("estimate MC =",estMC$estimate,"\n")
cat("estimate IS =",estIS$estimate,"\n")
cat("percentage error crude MC   =",percMC*100,"%\n")
cat("percentage error imp. samp. =",percIS*100,"%\n")
```

FIGURE 13.8 **Comparing crude Monte Carlo and importance sampling.**

to linear constraints, including a lower bound on expected return, is a straightforward convex quadratic programming model, because we may express variance as an explicit function of portfolio weights. In some specific cases, also minimizing V@R and CV@R may be quite simple. However, in general, we have to use the tools of stochastic programming to generate scenarios and solve the corresponding optimization problem. Needless to say, scenario generation methods, including Monte Carlo, play a key role here.[9]

Unfortunately, the minimization of V@R is not a convex optimization problem in general. One reason is its lack of subadditivity. Another reason is the interplay with Monte Carlo sampling. To see why, let us consider two assets, with sampled returns R_i^s, where $i = 1, 2$ refers to assets and s is the scenario index. For a given scenario, the portfolio return is linear function of portfolio weights w and $1 - w$:

$$R_p^s = wR_1^s + (1 - w)R_2^s.$$

---

[9]See Section 10.5.3.

```
library(MASS) # this package includes mvrnorm
s1 <- 0.4
s2 <- 0.3
rho <- 0.6
Sigma <- matrix(c(s1^2,rho*s1*s2,rho*s1*s2,s2^2), nrow=2)
mu <- c(0.12,0.10)

plotVaR <- function(numScen,alpha){
  rets <- mvrnorm(n=numScen,mu=mu,Sigma=Sigma)
  w <- seq(from=0,to=1,by=0.001)
  n <- length(w)
  VaR <- numeric(n)
  for (i in 1:n){
    portRets <- w[i]*rets[,1]+(1-w[i])*rets[,2]
    sortedRets <- sort(portRets)
    # use naive estimation of quantiles, based on
    # sorted ordered statistics
    VaR[i] <- -sortedRets[alpha*numScen]
  }
  plot(w,VaR,type='l')
}

set.seed(55555)
par(mfrow=c(2,2))
plotVaR(100,0.05)
plotVaR(100,0.05)
plotVaR(100,0.05)
plotVaR(100,0.05)
par(mfrow=c(1,1))
plotVaR(10000,0.05)
```

FIGURE 13.9 **Plotting V@R as a function of portfolio weights: The case of two assets and discretized scenarios.**

If we see V@R as a function of w, what we obtain is the envelope of linear affine functions. To see why, consider the worst-case return as a function of w:

$$R_p^{\mathrm{WC}}(w) = \min_s \left\{ w R_1^s + (1-w) R_2^s \right\}.$$

Thus, we find the lower envelope of a family of affine functions, which is clearly a continuous piecewise linear function, but nondifferentiable and not necessarily convex. Actually, we should select the worst case at confidence level α, but this does not change the nature of the resulting function. To visualize a picture and further investigate the issue, let us try the code in Fig. 13.9. Here we consider two assets with normally distributed returns. Of course, by using the concepts of Section 13.2, we may evaluate V@R as a function of w explicitly,

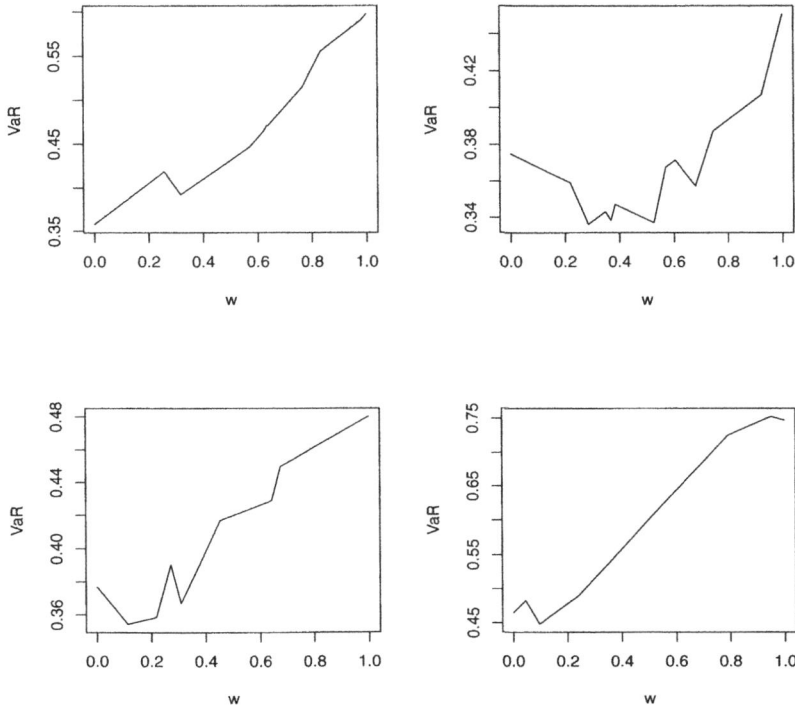

FIGURE 13.10 **Plotting V@R in a two-asset portfolio, for normally distributed returns: the case of 100 random scenarios.**

but let us consider a sample of randomly generated returns. In order to estimate V@R, we use the naive quantile estimation approach of Eq. (7.9). If we generate 100 scenarios and plot the V@R at 95% level, by repeating the random sampling four times, we obtain the plots in Fig. 13.10. There, we notice:

- There is quite some sampling variability; this is not quite surprising, as there is a significant volatility in the distributions of return, and we are collecting a statistic affected by tail behavior.

- The resulting function is nonconvex and nondifferentiable: an optimization nightmare.

If we increase the sample size to 10,000, the plot looks more reasonable, as we see in Fig. 13.11. As a check, if we set portfolio weight to $w = 0$ and $w = 1$, V@R should be as follows:

```
> qnorm(0.95)*0.4-0.12
[1] 0.5379415
```

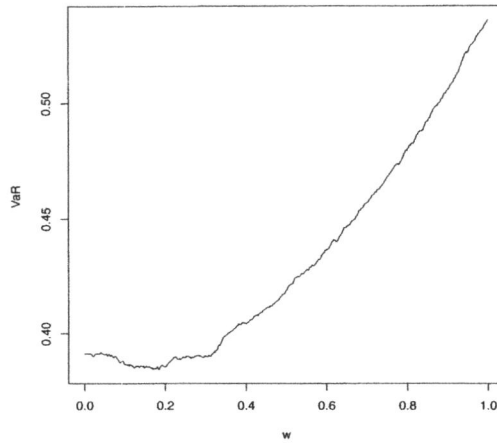

FIGURE 13.11 Plotting V@R in a two-asset portfolio, for normally distributed returns: the case of 10000 random scenarios.

```
> qnorm(0.95)*0.3-0.10
[1] 0.3934561
```

This is in good agreement with the picture, and we see that, by increasing the number of scenarios, we get closer to a smooth and convex function. Indeed, V@R is subadditive for a multivariate normal distribution, but, in a general case, anything can happen.

Since CV@R looks like a complication of V@R, it seems reasonable to expect that it is even a more difficult beast to tame. On the contrary, CV@R is much better behaved, which may be not quite surprising after all, since CV@R is a coherent risk measure and V@R is not. More surprisingly, minimizing CV@R may even lead to a (stochastic) linear programming model formulation.[10] Let $f(\mathbf{x}, \mathbf{Y})$ be a loss or cost function, depending on a vector of decision variables $\mathbf{x}$ and a vector of random variables $\mathbf{Y}$ with joint density $g_{\mathbf{Y}}(\mathbf{y})$, and consider function $F_{1-\alpha}(\mathbf{x}, \zeta)$ defined as

$$F_{1-\alpha}(\mathbf{x}, \zeta) = \zeta + \frac{1}{\alpha} \int \left[f(\mathbf{x}, \mathbf{y}) - \zeta \right]^{+} g_{\mathbf{Y}}(\mathbf{y}) \, d\mathbf{y},$$

where $[z]^{+} \equiv \max\{z, 0\}$, and $\zeta \in \mathbb{R}$ is an auxiliary variable. It can be shown that minimization of CV@R, at confidence level $1 - \alpha$, is accomplished by the minimization of $F_{1-\alpha}(\mathbf{x}, \zeta)$ with respect to its arguments. Furthermore, the resulting value of the auxiliary variable ζ turns out to be the corresponding V@R.

---

[10]In this section, we rely on results from Rockafellar and Uryasev [9, 10], which we take for granted, thereby cutting a few corners.

If we discretize the distribution of $\mathbf{Y}$ by a set $\mathcal{S}$ of scenarios, characterized by realizations $\mathbf{y}^s$ and probabilities π^s, $s \in \mathcal{S}$, we may recast the problem as a stochastic programming problem:

$$\min \quad \zeta + \frac{1}{\alpha} \sum_{s \in \mathcal{S}} \pi^s z^s$$

$$\text{s.t.} \quad z^s \geq f(\mathbf{x}, \mathbf{y}^s) - \zeta \qquad \forall s \in \mathcal{S},$$

$$z^s \geq 0 \qquad \forall s \in \mathcal{S}.$$

Furthermore, if the loss $f(\mathbf{x}, \mathbf{y}^s)$ in scenario s is a linear function, and the same applies to the additional constraints depending on the specific model, the minimization of CV@R boils down to the solution of a linear programming model. This important result should be tempered by the difficulty in getting a quantile-based estimate right using a limited number of scenarios. See, e.g., [5] for some critical remarks on the coherence of risk measures and their estimates.

13.6 Simulating delta hedging strategies

In this section we illustrate how Monte Carlo simulation may be used to check the effectiveness of hedging strategies. As an illustrative example, we consider delta hedging in its simplest form. We know from Chapter 3 that the price of a call option written on a non-dividend-paying stock is essentially the cost of a delta hedging strategy and that the continuous-time hedging strategy requires holding an amount Δ of the underlying asset. To summarize the essence of the reasoning, we have a portfolio whose value is a function $\Pi(S)$ of the underlying asset price S. To hedge against changes in S, we may hold an amount h of the underlying asset. Then, the overall value of the portfolio is

$$\Pi(S) + hS.$$

If we look for first-order immunization, we set the first-order derivative to zero:

$$\frac{\partial \Pi}{\partial S} + h \frac{\partial S}{\partial S} = \Delta_\Pi + h = 0 \quad \Rightarrow \quad h = -\Delta_\Pi.$$

Delta hedging is a commonly mentioned risk management strategy, but it is open to criticism:

- We have a perfect hedge for infinitesimal perturbations, rather than a good hedge for larger shocks.
- We should rebalance the hedge frequently, incurring transaction costs.
- Delta hedging as a pricing argument in the BSM model neglects not only transaction costs, but stochastic volatilities, jumps, market impact, herding behavior, etc.

In a word, naive delta hedging is subject to modeling errors. Hence, we should probably at least check its effectiveness and robustness by simulation experiments in a realistic setting. Needless to say, this is a job for Monte Carlo methods. For illustration purposes we consider hedging a vanilla call option in the BSM world. Clearly, in real life the net factor exposure due to a whole portfolio of options should be considered.

It may also be instructive to compare delta hedging against another strategy, the stop-loss strategy.[11] Stop loss is simpler than delta hedging and model-free. The idea is that we should have a covered position (hold one share) when the option is in the money, and a naked position (hold no share) when it is out of the money. So, we should buy a share when the asset price goes above the strike price K, and we should sell it when it goes below. Ideally, if we buy at K and sell at K, the cash flows cancel each other, assuming that we disregard the time value of money. This strategy makes intuitive sense, but it is not that trivial to analyze in continuous time.[12] Nevertheless, it is easy to evaluate its performance in discrete time by Monte Carlo simulation. The problem with executing the strategy in discrete time is that we cannot really buy or sell at the strike price: We buy at a price larger than K, when we detect that the price went above that critical value, and we sell at a price which is slightly lower. Hence, we cannot claim that cash flows due to buying and selling at K cancel each other. So, even without considering transaction costs, which would affect delta hedging as well, we see a potential trouble with the stop-loss strategy.

An R function to estimate the average cost of a stop-loss strategy is given in Fig. 13.12. The function receives the matrix `paths` of sample paths, possibly generated as we have seen in Chapter 6, by the function `simvGBM`. Note that in this case, unlike option pricing, the real drift `mu` must be used in the simulation, as we check hedging in the real world, and not under the risk-neutral measure. Note that the true number of steps (time intervals) is one less the number of columns in matrix `paths`, which includes the initial price. If we need to buy shares of the underlying stock, we may need to borrow money, which should be taken into account. But, since we assume deterministic and constant interest rates, we will not account for borrowed money, since we can simply record cash flows from trading and discount them back to time $t = 0$, having precomputed discount factors in the vector `discountFactors`. We use a state variable, `covered`, to detect when we cross the strike price going up or down. Since cash flow is negative when we buy, and positive when we sell, the option "price" is evaluated as the average total discounted cash flow, with a change in sign. We should also pay attention to what happens at maturity: If the option is in the money, the option holder will exercise her right and we will also earn the strike price, which should be included in the cash flow stream.

Since vectorizing R code is often beneficial, we also show a vectorized version of this code in Fig. 13.13. The main trick here is using a vector `OldPrice`,

---

[11] See [8, pp. 300–302].
[12] See [4].

```
stopLoss <- function(paths,K,r,T){
  # paths is a matrix
  numRepl <- dim(paths)[1]
  numSteps <- dim(paths)[2]
  numSteps <-  numSteps - 1 # true number of steps
  cost <- numeric(numRepl)
  dt <- T/numSteps
  discountFactors <- exp(-r*dt*(0:numSteps))
  for (k in 1:numRepl){
    # clear cash flow vector
    cashFlows <- numeric(numSteps+1)
    if (paths[k,1] >= K){ # should I cover at beginning?
      covered <- TRUE
      cashFlows[1] <- -paths[k,1]
    }
    else
      covered <- FALSE
    # now go on until end of horizon
    for (t in 2:(numSteps+1)){
      if (covered  && (paths[k,t] < K)){
        # Sell
        covered <- FALSE
        cashFlows[t] <- paths[k,t]
      }
      else if (!covered && (paths[k,t] > K)){
        # Buy
        covered <- TRUE
        cashFlows[t] <- -paths[k,t]
      }
    }
    # what happens at maturity?
    if (paths[k,numSteps + 1] >= K) # Option is exercised
      cashFlows[numSteps + 1] <- cashFlows[numSteps + 1] + K
    # at the end of each replication get discounted cost
    cost[k] <- -crossprod(discountFactors,cashFlows)
  } # end of main loop
  return(list(mean=mean(cost),sd=sd(cost)))
}
```

FIGURE 13.12 **Evaluating the cost of a stop-loss hedging strategy.**

which is essentially a shifted copy of paths, to spot where the price crosses the critical level, going up or down. The time instants at which we go up are recorded in vector upTimes, where we have a negative cash flow; a similar consideration applies to downTimes.

```
stopLossV <- function(paths,K,r,T){
  # paths is a matrix
  numRepl <- dim(paths)[1]
  numSteps <- dim(paths)[2]
  numSteps <-  numSteps - 1 # true number of steps
  cost <- numeric(numRepl)
  dt <- T/numSteps
  discountFactors <- exp(-r*dt*(0:numSteps))
  cashFlows <- matrix(0, nrow=numRepl, ncol=numSteps+1);
  oldPrice <- c(numeric(numRepl), paths[,1:numSteps])
  upTimes <- which((oldPrice < K) & (paths >= K))
  downTimes <- which((oldPrice >= K) & (paths < K))
  cashFlows[upTimes] <- -paths[upTimes]
  cashFlows[downTimes] <- paths[downTimes]
  exPaths <- which(paths[,numSteps+1] >= K)
  cashFlows[exPaths,numSteps+1] <-
            cashFlows[exPaths,numSteps+1] + K
  cost = -cashFlows %*% discountFactors
  return(list(mean=mean(cost),sd=sd(as.numeric(cost))))
}
```

FIGURE 13.13 **Vectorized code for the stop-loss hedging strategy.**

Now let us check if the two functions are actually consistent, i.e., if they yield the same results, and whether there is any advantage in vectorization:

```
> S0 <- 50
> K <- 52
> mu <- 0.1;
> sigma <- 0.4;
> r <- 0.05;
> T <- 5/12;
> numRepl <- 100000
> numSteps <- 10
> set.seed(55555)
> paths <- simvGBM(S0,mu,sigma,T,numSteps,numRepl)
> system.time(cost <- stopLoss(paths,K,r,T))
   user   system elapsed
   4.93     0.00     4.93
> system.time(costV <- stopLossV(paths,K,r,T))
   user   system elapsed
   0.20     0.00     0.21
> cost$mean
[1] 4.833226
> costV$mean
[1] 4.833226
```

```
deltaHedging <- function(paths,K,sigma,r,T){
  # paths is a matrix
  numRepl <- dim(paths)[1]
  numSteps <- dim(paths)[2]
  numSteps <-  numSteps - 1 # true number of steps
  cost <- numeric(numRepl)
  dt <- T/numSteps
  cashFlows <- numeric(numSteps+1)
  discountFactors <- exp(-r*dt*(0:numSteps))
  for (k in 1:numRepl){
    path <- paths[k,]
    position <- 0 # initial stock holding is zero
    # get vector of deltas at each point on the sample path
    deltas <- BSMdelta(path[1:numSteps],K,
                       T-(0:(numSteps-1))*dt,r,sigma);
    for (j in 1:numSteps){
      # delta hedge along the sample path
      cashFlows[j] <- (position - deltas[j])*path[j]
      position <- deltas[j]
    }
    # check if option is exercised at maturity
    # and liquidate position
    if (path[numSteps+1] > K)
      cashFlows[numSteps+1] <-
                      K-(1-position)*path[numSteps+1]
    else
      cashFlows[numSteps+1] <- position*path[numSteps+1]
    # at the end of each replication get discounted cost
    cost[k] <- -crossprod(discountFactors,cashFlows)
  } # end of main loop
  return(list(mean=mean(cost),sd=sd(cost)))
}
```

FIGURE 13.14 **Evaluating the performance of delta hedging.**

Indeed, using the function system.time we observe that there is a significant difference in CPU time. Now we should compare the cost of the stop-loss strategy against the cost of delta hedging, as well as the theoretical option price. A code to estimate the average cost of delta hedging is displayed in Fig. 13.14. The code is similar to the stop-loss strategy, but it is not vectorized. The only vectorization we have done is in calling BSMdelta once to get the option Δ for each point on the sample path. Note that Δ must be computed using the current asset price and the current time to maturity. The current position in the stock is updated given the new Δ, generating positive or negative cash flows that are discounted back to time $t = 0$.

```
S0 <- 50
K <- 52
mu <- 0.1;
sigma <- 0.4;
r <- 0.05;
T <- 5/12;
numRepl <- 100000
numSteps <- 10
# exact call price
C <- EuVanillaCall(S0,K,T,r,sigma,q=0)
# try 10 steps
set.seed(55555)
paths <- simvGBM(S0,mu,sigma,T,numSteps,numRepl)
SL <- stopLossV(paths,K,r,T)
DC <- deltaHedging(paths,K,sigma,r,T)
cat("true price = ", C, "\n")
cat("StopLoss 10 steps: mean =", SL$mean,
    "sd =", SL$sd, "\n")
cat("DeltaHedging 10 steps: mean =", DC$mean,
    "sd =", DC$sd, "\n")
# try 100 steps
numSteps <- 100
set.seed(55555)
paths <- simvGBM(S0,mu,sigma,T,numSteps,numRepl)
SL <- stopLossV(paths,K,r,T)
DC <- deltaHedging(paths,K,sigma,r,T)
cat("StopLoss 100 steps: mean =", SL$mean,
    "sd =", SL$sd, "\n")
cat("DeltaHedging 100 steps: mean =", DC$mean,
    "sd =", DC$sd, "\n")
```

FIGURE 13.15 **A script to compare hedging strategies.**

Figure 13.15 displays a script to compare performances of the two hedging strategies, in terms of both expected value and standard deviation of the present value of hedging cost. In the first pair of runs, we are only using 10 hedging steps, which has an impact on hedging errors; in the second pair this is increased to 100 steps. We also compute the exact option price using the BSM formula, for comparison purposes. By running the script, we obtain

```
true price =  4.732837
StopLoss 10 steps: mean =  4.833226 sd =  4.542682
DeltaHedging 10 steps: mean =  4.729912 sd =  1.380229
StopLoss 100 steps: mean =  4.807347 sd =  4.074755
DeltaHedging 100 steps: mean =  4.734969 sd =  0.4529393
```

We may observe the following:

- The average cost of the stop-loss strategy is larger than the average cost of delta hedging.
- What is more significant, however, is the standard deviation of the hedging cost. This is converging to zero with delta hedging, whose expected cost converges to the BSM option price.

Indeed, the price of an option is related to its hedging cost. In the idealized BSM world, the market is complete and we may hedge perfectly the option risk. In other words, we are replicating it exactly. Of course, the introduction of stochastic volatility, transaction costs, modeling errors, etc., make hedging less than perfect. As we mentioned, delta hedging has been often criticized as an over-simplified, possibly misleading, approach. The important message is that by Monte Carlo methods we may assess all of the above effects, for this and other hedging policies.

13.7 The interplay of financial and nonfinancial risks

Let us consider a firm that is subject to currency exchange risk, but also to *volume* risk.[13] To make the problem as simple as possible, let us assume that the current exchange rate is 1.0, expressed in any currency exchange ratio you like. There is considerable uncertainty on the exchange rate at some future time T, and let us assume that it is uniformly distributed between 0.7 and 1.3. The firm has to buy a significant amount of the foreign currency for its activity, say, 50,000. If the rate increases, the firm will have to pay much more for that amount of currency; if, on the contrary, the rate decreases, that will be good news. Let us say that the base case is when the rate stays at the current level. Considering the two extreme scenarios, if the rate goes up and turns out to be 1.3, the firm will incur a loss given by

$$50,000 \times (1.3 - 1.0) = 15,000.$$

On the contrary, it will have a corresponding gain if the rate turns out to be 0.7. Let us assume that the firm can hedge using two types of derivative:

- A long position in a forward contract, where we assume that the forward price is just 1.0.[14]
- A long position in at-the-money call options. Unlike forward contracts, this requires an upfront payment. Let us assume that the price of each option is 0.1; in other words, we have to pay 10% of the nominal amount we hedge, in the domestic currency.

---

[13] For a real life example of such a situation, please refer to the following business case, which has been an inspiration for the section: M.A. Desai, A. Sjoman, and V. Dessain. *Hedging Currency Risks at AIFS*. Case no. 9-205-026, Harvard Business School Publishing.

[14] Therefore, we are assuming that there is no difference between the interest rates in the two currencies.

The decision that the firm has to make is twofold:

- They have do decide how much to hedge.
- They have to decide the hedge mix, i.e., how much to hedge with forward contracts vs. options.

Hence, if we only consider the scenarios $(0.7, 1.0, 1.3)$, the corresponding outcomes would be

- $(15,000, 0, -15,000)$ with 0% hedging.
- $(0, 0, 0)$ with 100% hedging using forward contracts.
- $(-5000, -5000, 10,000)$ with 100% hedging using options; note that in this case we have to pay 5000 for the options, which expire worthless in the first two outcomes, whereas we have to subtract the option price from the option payoff in the third case.

With no hedging there is considerable uncertainty, which is completely eliminated by forward contracts. With options, we limit the downside a bit, while retaining some upside. The choice is a matter of risk aversion, as well as probabilities, assuming that we are willing to associate probabilities with each event.

In many practical settings, however, there is still another complication: volume risk. We may not really know precisely the amount we have to hedge. Say that 50,000 is the volume in the nominal scenario, i.e., the base case, but the amount we need to hedge is uniformly distributed between 10,000 and 90,000. For the sake of simplicity, we assume that the two sources of risk are independent but, clearly, Monte Carlo methods may be applied in more complicated and realistic settings. To see why volume risk is relevant, let us check the outcome resulting from 100% hedging with forward contracts in the following scenario: exchange rate is 0.7 and required volume is 10,000. Now we have to buy (and arguably to sell immediately) 40,000 useless units of foreign currency, with a loss of

$$40,000 \times (1.0 - 0.7) = 12,000.$$

Apart from the loss, we are here in a probably bad business setting, since the reduced requirements is likely to come from lost business. With 100% coverage using call options, which give the right but not the obligation to buy, the loss would be just the foregone option premium, 5000.

Deciding the best course of action is by no means a trivial task, even in this idealized setting. To support the decision makers, it would be nice to have a picture of the profit/loss scenarios as a function of the decision variables. Then we might also want to plot a given risk measure. All of these important tasks can be accomplished by Monte Carlo sampling.

For further reading

- An elementary introduction to risk aversion and related issues can be found in [3].

- The original treatment of coherent risk measures can be found in [1].
- For Monte Carlo applications to credit risk, see [6, Chapter 9].
- The reformulation of CV@R in a way suitable for stochastic optimization was originally proposed in [9, 10].

References

1 P. Artzner, F. Delbaen, J.-M. Eber, and D. Heath. Coherent measures of risk. *Mathematical Finance*, 9:203–228, 1999.

2 P. Artzner, F. Delbaen, J.-M. Eber, D. Heath, and H. Ku. Coherent multi-period risk adjusted values and Bellman's principle. *Annals of Operations Research*, 152:5–22, 2007.

3 P. Brandimarte. *Quantitative Methods: An Introduction for Business Management*. Wiley, Hoboken, NJ, 2011.

4 P.P. Carr and R.A. Jarrow. The stop-loss start-gain paradox and option valuation: a new decomposition into intrinsic and time value. *Review of Financial Studies*, 3:469–492, 1990.

5 F.J. Fabozzi and R. Tunaru. On risk management problems related to a coherence property. *Quantitative Finance*, 6:75–81, 2006.

6 P. Glasserman. *Monte Carlo Methods in Financial Engineering*. Springer, New York, 2004.

7 P. Glasserman, P. Heidelberger, and P. Shahabuddin. Variance reduction techniques for estimating value-at-risk. *Management Science*, 46:1349–1364, 2000.

8 J.C. Hull. *Options, Futures, and Other Derivatives* (8th ed.). Prentice Hall, Upper Saddle River, NJ, 2011.

9 R.T. Rockafellar and S. Uryasev. Optimization of conditional value-at-risk. *The Journal of Risk*, 2:21–41, 2000.

10 R.T. Rockafellar and S. Uryasev. Conditional value-at-risk for general loss distributions. *Journal of Banking and Finance*, 26:1443–1471, 2002.

Markov Chain Monte Carlo and Bayesian Statistics

We have introduced Bayesian parameter estimation in Section 4.6, as a possible way to overcome some limitations of orthodox statistics. The essence of the approach can be summarized as follows:

$$\text{posterior} \propto \text{prior} \times \text{likelihood},$$

where the prior collects information, possibly of subjective nature, that we have about a set of parameters before observing new data; the likelihood measures how likely is what we observe, on the basis of our current knowledge or belief; and the posterior merges the two above pieces of information in order to provide us with an updated picture. This is an informal restatement of Eq. (4.12), which we rewrite below for convenience:

$$p_n(\theta_1, \ldots, \theta_q \,|\, x_1, \ldots, x_n) \propto f_n(x_1, \ldots, x_n \,|\, \theta_1, \ldots, \theta_q) \cdot p(\theta_1, \ldots, \theta_q). \quad (14.1)$$

The parameters θ_j, $j = 1, \ldots, q$, are regarded as random variables within the Bayesian framework, and $p(\theta_1, \ldots, \theta_q)$ is their prior. The prior can be a probability density function (PDF) or a probability mass function (PMF), or a mixed object, depending on the nature of the parameters involved. For the sake of simplicity, in this chapter we will not deal systematically with both cases, and we will refer to either one according to convenience. We have a set of independent observations, i.e., realizations of a random variable taking values x_i, $i = 1, \ldots, n$. Note that here we are considering n observations of a scalar random variable, whose distribution depends on q parameters; the concepts can be generalized to multivariate distributions. We know from maximum-likelihood estimation that an important role is played by the likelihood function $f_n(x_1, \ldots, x_n \,|\, \theta_1, \ldots, \theta_q)$, for given parameters. Since we are considering independent observations, the likelihood is the product of PDF values corresponding to each observation x_i. Bayes' theorem yields the posterior density $p_n(\theta_1, \ldots, \theta_q \,|\, x_1, \ldots, x_n)$, summarizing our knowledge about the parameters, merging the possibly subjective prior and the empirical evidence.

Armed with the posterior density, we may proceed to find estimates of the parameters $\theta_1, \ldots, \theta_q$, e.g., by finding their expected values. We may also find credible intervals, which are the Bayesian counterpart of confidence intervals, and test hypotheses about parameters. All of this requires knowledge of the posterior density, but there is a missing piece in Eq. (14.1). There, we state that the posterior is *proportional* to the product of prior and likelihood, but to fully specify the posterior density, we should also find a normalization constant B such that the PDF integrates to 1. Hence, we should divide the posterior by

$$B = \int \cdots \int f_n(x_1, \ldots, x_n \,|\, \theta_1, \ldots, \theta_q)\, p(\theta_1, \ldots, \theta_q)\, d\theta_1 \cdots d\theta_q.$$

Unless there is some special structure to be exploited, the calculation of this multidimensional integral is difficult. As we have seen in Section 4.6, there are nice cases in which the form of the posterior is easy to find. If we deal with observations from a normal distribution and the prior is normal, then the posterior is normal too, and figuring out the normalizing constant is easy. When the posterior belongs to the same family of distributions as the prior, for a given kind of likelihood, we speak of *conjugate* distributions; the prior is also called the conjugate prior. In less lucky cases, one possibility to find the normalization constant, needless to say, is the numerical evaluation of the multidimensional integral by Monte Carlo methods. However, there is a possibly better idea, if what we need is just the ability to *sample* from the posterior density. As it turns out, there are ways to sample from a density which is known up to a constant, and in fact we have already met one: acceptance–rejection.

In Section 14.1 we show that we may apply acceptance–rejection sampling when we miss a normalization constant, but the resulting approach need not be the most efficient. There is a class of sampling approaches, collectively known as Markov chain Monte Carlo (MCMC) methods, that can be used. The idea, as the name suggests, is to simulate a Markov chain, whose stationary distribution is the posterior density. MCMC methods can be introduced as a tool for computational Bayesian statistics, but they have a wider application range, as they can be used as random sample generators for difficult cases that cannot be tackled using standard approaches. We review a few basic facts about the stationary distribution of Markov chains Section 14.2. Then, we outline the Metropolis–Hastings algorithm and some of its variations in Section 14.3. Finally, in Section 14.4 we show how the simulated annealing algorithm, a stochastic search algorithm for both global and combinatorial optimization that we introduced in Section 10.3, can be cast and analyzed within the MCMC framework.

14.1   Acceptance–rejection sampling in Bayesian statistics

Acceptance–rejection sampling is a general purpose approach to sample from a distribution with a given PDF or PMF, whereas the inverse transform method works with the CDF. As we have seen in Section 5.4, given a target density $f(\cdot)$, we choose an instrumental density $g(\cdot)$ and proceed as follows:

   1. Generate $Y \sim g$.

2. Generate $U \sim \mathsf{U}(0, 1)$, independent of Y.

3. If

$$U \leq \frac{f(Y)}{cg(Y)},$$

where c is an upper bound on the ratio f/g, accept and return $X = Y$; otherwise, reject and repeat the procedure.

If you look at the proof of the validity of the method, in Section 5.4, you will notice that what really matters is that c is an upper bound on the ratio of f/g, and we may also work with an unscaled density $f(\cdot)$, i.e., a density known up to a constant. In the Bayesian framework, we have an unscaled posterior

$$f_n(\mathbf{x} \mid \boldsymbol{\theta}) \, p(\boldsymbol{\theta}),$$

where $\mathbf{x}$ is the vector of (given) observations and $\boldsymbol{\theta}$ the vector of (random) parameters. To sample a vector of parameters from the unscaled posterior, we have to find a candidate density $g(\boldsymbol{\theta})$ and an upper bound M on the ratio:

$$\frac{f_n(\mathbf{x} \mid \boldsymbol{\theta}) \, p(\boldsymbol{\theta})}{g(\boldsymbol{\theta})} \leq M.$$

Then, we sample a candidate value $\tilde{\boldsymbol{\theta}}$ randomly from $g(\cdot)$, and we accept it with probability

$$\frac{f_n\big(\mathbf{x} \mid \tilde{\boldsymbol{\theta}}\big) \, p(\tilde{\boldsymbol{\theta}})}{M g(\tilde{\boldsymbol{\theta}})}.$$

Improvements to this basic strategy have been proposed, but they do not solve one basic difficulty: the number of iterations that are needed to accept a sampled value can be large. We know that this is given by M; hence, the closer the instrumental density to the target, the better. If we are working with a distribution with bounded support, like a beta distribution, we may use a uniform as the instrumental density. However, the choice of the target is more difficult if the target has unbounded support, in which case the candidate should be a heavy-tailed distribution. Unfortunately, when we are dealing with multiple parameters, acceptance–rejection may require the generation of many proposals before one is accepted. Because of this practical limitation, alternative methods, described in the following, have been proposed. Nevertheless, acceptance–rejection is a viable approach for the case of a single parameter.

14.2   An introduction to Markov chains

Before we introduce MCMC, we need to refresh some essential concepts pertaining to discrete-time Markov chains. MCMC methods are indirect sampling strategies, based on the simulation of a Markov process, whose long-run state density is the target density. We have already met Markov processes in the book,

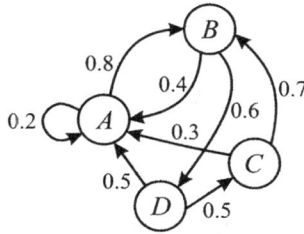

FIGURE 14.1   **A discrete-state Markov chain.**

since processes described by Itô stochastic differential equations are Markovian. Here, though, we focus on their long-term state distribution. Within a Bayesian framework, the state of the stochastic process is related to the value of the parameters $\boldsymbol{\theta}$. When the parameters are real numbers, we have to deal with a continuous-state process. However, the necessary intuition to understand MCMC methods can be built on the basis of the much simpler discrete-state case, which is referred to as Markov chain. A discrete-time Markov chain is a process with a state variable

$$X_t, \qquad t = 0, 1, 2, 3, \ldots,$$

taking values on a discrete set, which may be identified with the set of integer numbers, or a subset of it, or a finite discrete set (e.g., the credit rating of a bond). We may represent the process as a graph, which is where the name "chain" comes from. A simple example is shown in Fig. 14.1. There, the state space is the set $\{A, B, C, D\}$, and nodes correspond to states. Directed arcs are labeled by transition probabilities. For instance, if we are in state C, there is a probability 0.7 to be in state B at the next step, and a probability 0.3 to be in state A. Note that these probabilities depend only on the current state, and not on the whole past history. Indeed, the essential property of Markov processes is related to their limited memory: The only information that is needed to characterize the future system evolution is the current state. Formally, we may express this in terms of the conditional transition probability

$$\begin{aligned}
\mathrm{P}\{X_{t+1} = i_{t+1} \mid X_t = i_t, X_{t-1} = i_{t-1}, X_{t-2} = i_{t-2}, \ldots\} \\
= \mathrm{P}\{X_{t+1} = i_{t+1} \mid X_t = i_t\}.
\end{aligned}$$

Hence, we just need to know the one-step transition probabilities to characterize the chain. A further simplification occurs if the process is stationary, i.e., if the transition probabilities do not depend on time. In such a case, we may describe the chain in terms of the transition probabilities

$$P_{ij} \equiv \mathrm{P}\{X_{t+1} = j \mid X_t = i\},$$

which may be collected into the one-step transition probability matrix $\mathbf{P}$. For the chain in Fig. 14.1, we have

$$
\mathbf{P} = \begin{bmatrix} 0.2 & 0.8 & 0 & 0 \\ 0.4 & 0 & 0 & 0.6 \\ 0.3 & 0.7 & 0 & 0 \\ 0.5 & 0 & 0.5 & 0 \end{bmatrix}.
$$

Such a matrix has rather peculiar properties, related to the fact that after a transition we must land somewhere within the state space, and therefore each and every row adds up to 1:

$$
\sum_{j=1}^{N} P_{ij} = 1,
$$

where for the sake of simplicity we assume a finite state space of size N. Let $\pi_0 \in \mathbb{R}^N$ denote the initial state distribution at step 0, i.e.,

$$
\pi_{i,0} = \mathrm{P}\{X_0 = i\}.
$$

The total probability theorem implies that

$$
\mathrm{P}\{X_1 = j\} = \sum_{i=1}^{N} \mathrm{P}\{X_1 = j \mid X_0 = i\} \cdot \mathrm{P}\{X_0 = i\},
$$

which can be rewritten as

$$
\pi_1^\mathsf{T} = \pi_0^\mathsf{T} \mathbf{P},
$$

where $\pi_1 \in \mathbb{R}^N$ is a vector with components $\pi_{i,1} = \mathrm{P}\{X_1 = i\}$, $i = 1, \ldots, N$. Note how, given the shape of the transition probability matrix, we write the relationship in terms of premultiplication by a *row* vector. Thanks to the Markov property, we may easily compute multiple step transition probabilities:

$$
\mathrm{P}\{X_2 = j \mid X_0 = i\} = \sum_{l=1}^{N} \mathrm{P}\{X_2 = j, X_1 = l \mid X_0 = i\}
$$

$$
= \sum_{l=1}^{N} \mathrm{P}\{X_2 = j \mid X_1 = l, X_0 = i\} \cdot \mathrm{P}\{X_1 = l \mid X_0 = i\}
$$

$$
= \sum_{l=1}^{N} \mathrm{P}\{X_2 = j \mid X_1 = l\} \cdot \mathrm{P}\{X_1 = l \mid X_0 = i\}.
$$

Here, in the first line we have used the fact that the event $\{X_2 = j\}$ can be expressed as the union of the disjoint events $\{X_2 = j, X_1 = i\}$; then we have used conditioning and the Markov property. Since the final expression can be interpreted as an entry in the row-by-column matrix product, the two-step transition probability matrix can be expressed as

$$
\mathbf{P}_2 = \mathbf{P}\mathbf{P} = \mathbf{P}^2.
$$

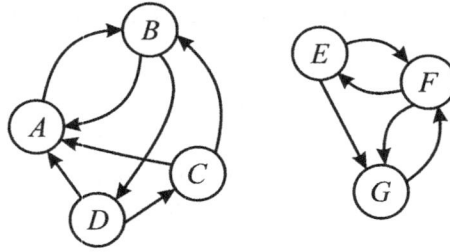

FIGURE 14.2 **A nonirreducible periodic Markov chain.**

By the same token, if we consider an arbitrary number m of steps,

$$\mathbf{P}_m = \mathbf{P}^m.$$

Now imagine that we start with distribution $\boldsymbol{\pi}_0$ and we let the system run for a long time. What happens to the limit

$$\lim_{m \to \infty} \boldsymbol{\pi}_0^{\mathsf{T}} \mathbf{P}^m \ ?$$

We might guess that, after such a long time, the memory of the initial state is lost, and that the system reaches a steady-state (or long-run) distribution $\boldsymbol{\pi}$. We might also guess that this distribution should be an equilibrium distribution characterized by the system of linear equations

$$\boldsymbol{\pi}^{\mathsf{T}} = \boldsymbol{\pi}^{\mathsf{T}} \mathbf{P}. \tag{14.2}$$

This system can be interpreted by saying that if we start with the state distribution $\boldsymbol{\pi}$, the next state has the same distribution. Hence, $\boldsymbol{\pi}$ is an eigenvector of $\mathbf{P}$ associated with the eigenvalue 1. To be precise, since we premultiply $\mathbf{P}$ by a row vector, we should use the term *left* eigenvector. To make sense as a probability distribution, this should be a unit (normalized) and non-negative eigenvector. It turns out that, in fact, we may find a non-negative eigenvector, provided that some conditions are satisfied for $\mathbf{P}$. Indeed, we cannot take for granted that a stationary distribution exists, independent from the initial state, as a few things may go wrong:

- The state space could be, in some sense, disconnected, as shown in Fig. 14.2. Clearly, in such a case, if we start in one region, we will stay there and the memory of the initial state cannot be lost completely. We rule out this case by considering only *irreducible* chains.
- The chain could be periodic, as the one shown in Fig. 14.3. As we see, despite some uncertainty between states C and D, states A, B, and E will be repeated every four steps. This kind of behavior rules out the existence of a stationary, long-run distribution. Hence, we have to refer to *aperiodic* chains.

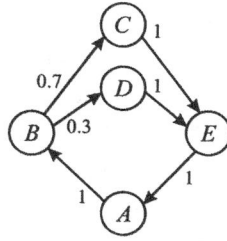

FIGURE 14.3 **A periodic Markov chain.**

We will refrain from giving a formal definition of irreducible and aperiodic chains, as intuition is all we need for our limited purposes.

Another very useful way to interpret the equilibrium condition of Eq. (14.2) can be obtained by focusing on a single equation,

$$\pi_i = \sum_{j=1}^{N} \pi_j P_{ji},$$

which gives the equilibrium probability π_i of state i, and by rewriting it as follows:

$$\pi_i = \pi_i \cdot 1 = \pi_i \sum_{j=1}^{N} P_{ij} = \sum_{j=1}^{N} \pi_i P_{ij} = \sum_{j=1}^{N} \pi_j P_{ji}.$$

The last equality can be interpreted intuitively in terms of equilibrium of "probability flows:"

- The term $\sum_{j=1}^{N} \pi_i P_{ij}$ is the probability flow out of state i to states $j = 1, \ldots, N$ (there is some flow from i to itself if the transition from i to itself is possible).

- The term $\sum_{j=1}^{N} \pi_j P_{ji}$ is the probability flow from states $j = 1, \ldots, N$ into state i.

Thus, equating the two flows for every state i establishes a *global* balance condition, which is associated with equilibrium. For some Markov chains, a stronger equilibrium condition is satisfied, corresponding to a *detailed*, rather than global balance:

$$\pi_i P_{ij} = \pi_j P_{ji}, \qquad \forall i \neq j. \tag{14.3}$$

This condition enforces a probability flow equilibrium for any pair of states, and it is easy to see that if we can find a vector $\boldsymbol{\pi}$, whose components add up to 1, satisfying Eq. (14.3), then this vector will also satisfy Eq. (14.2). To prove this, it is sufficient to sum the detailed balance over states i:

$$\sum_{i=1}^{N} \pi_i P_{ij} = \pi_j \sum_{i=1}^{N} P_{ji} = \pi_j, \qquad \forall j.$$

The detailed balance conditions are quite interesting for a few reasons:

- The resulting chain is time reversible, i.e., we cannot distinguish sample paths going forward and backward in time.

- This structure facilitates the solution of the equilibrium equations, leading to nice structures, such as product form queueing networks.

These concepts are beyond the scope of the book, but from our viewpoint, detailed balance conditions are useful when we have to go the other way around: Given a target equilibrium distribution, how can we build a Markov chain that has that equilibrium distribution? This is what we need in MCMC methods applied to Bayesian statistics, where the target equilibrium distribution is the posterior, with the additional twist that the posterior is known up to a multiplicative constant. This would be difficult in general, but, by taking advantage of detailed balance equations, it can be easily accomplished, as we show in the next section.

14.3   The Metropolis–Hastings algorithm

In Bayesian inference, we need to sample from the posterior density, which is hard to find computationally, with the exception of conjugate priors. A computational breakthrough was achieved by realizing that we may create a discrete-time Markov process, whose states are values of the parameters of interest, such that its long-run distribution can be shaped according to our needs. In other words, we need to define a state space based on the values of the parameters, and compute suitable transition probabilities. This is a daunting task in general, but it turns out that we may take advantage of the simple structure of time-reversible Markov chains, whose transition probabilities satisfy the detailed balance equations (14.3).

The original idea was devised in 1953 and is known as the Metropolis–Hastings algorithm.[1] The intuition is as follows. We want to find a Markov chain with steady-state probabilities π_i, for discrete states $i = 1, \ldots, N$ (the idea can be easily adapted to continuous distributions). To this end, we need suitable transition probabilities P_{ij}, satisfying the detailed balance equations. A first point is that whatever P_{ii} we choose, it will do the job, as the trivial balance condition

$$\pi_i P_{ii} = \pi_i P_{ii}$$

is clearly satisfied. Unfortunately, in general, the equilibrium condition does not hold. For instance, if

$$\pi_i P_{ij} < \pi_j P_{ji},$$

then we should restore the balance by rejecting some of the transitions from j to i, whereas all of the transitions from i to j can be accepted. A similar

[1] See [10] for an early reference. The idea was refined and improved in [6].

consideration applies if $\pi_i P_{ij} > \pi_j P_{ji}$. The trick can be done, for any desired steady-state distribution π and any transition matrix $\mathbf{P}$, by adjusting the transition probabilities using a sort of acceptance–rejection strategy. For each pair of states i and j, we adjust the transition probability multiplying it by an acceptance probability

$$\alpha_{ij} = \min\left\{\frac{\pi_j P_{ji}}{\pi_i P_{ij}}, 1\right\}. \qquad (14.4)$$

Hence, whenever we are in state i and a candidate transition from i to j is randomly selected in the simulation, it is actually accepted with probability α_{ij}, which implicitly defines an adjusted probability matrix $\widetilde{\mathbf{P}}$, where $\widetilde{P}_{ij} = \alpha_{ij} P_{ij}$, for $i \neq j$, and

$$\widetilde{P}_{ii} = 1 - \sum_{j \neq i} \widetilde{P}_{ij}.$$

It is not difficult to show that the pair consisting of π and $\widetilde{\mathbf{P}}$ satisfies the detailed balance conditions, since:

$$\pi_i \widetilde{P}_{ij} = \pi_i \cdot \min\left\{\frac{\pi_j P_{ji}}{\pi_i P_{ij}}, 1\right\} \cdot P_{ij} = \min\left\{\pi_j P_{ji}, \pi_i P_{ij}\right\},$$

$$\pi_j \widetilde{P}_{ji} = \pi_j \cdot \min\left\{\frac{\pi_i P_{ij}}{\pi_j P_{ji}}, 1\right\} \cdot P_{ji} = \min\left\{\pi_i P_{ij}, \pi_j P_{ji}\right\},$$

which together imply $\quad \pi_i \widetilde{P}_{ij} = \pi_j \widetilde{P}_{ji}, \quad \forall i, j.$

Therefore, we can easily build a Markov chain with the desired long-run distribution by adjusting any candidate transition matrix $\mathbf{P}$. Now, the reader should be a bit puzzled, for a couple of reasons:

1. In computational Bayesian statistics, we do not really know the posterior. Hence, how can we select it as the long-run distribution? A look at Eq. (14.4) shows where the trick is: Only the *ratio* of the probabilities is involved; therefore, even if they are known up to a scaling constant, we can still simulate the chain, and sample from the underlying long-run distribution.

2. It may seem very weird that, whatever $\mathbf{P}$ we choose, it will work. Indeed, so far we did not say anything about efficiency. Not all transition matrices are equally efficient, and only with a suitable choice, we may hope to devise a reasonably fast algorithm.

The Metropolis–Hastings algorithm takes advantage of the above ideas:

- Given a target distribution π, possibly known up to a normalization constant, we select a candidate transition matrix $\mathbf{P}$.
- We simulate the resulting Markov chain for a suitably long time horizon.
- The resulting state sequence X_t is a random sample from the target distribution.

An important point to bear in mind is that, unlike the random sampling approaches that we have described so far in the book, the sample produced by this procedure is *not* independent, since the visited states are correlated. This must be taken into due account when performing output analysis, which may require the batching method that we have discussed in Section 7.1. Furthermore, since the target density is achieved in the long run, we should discard the initial portion of the sample path, which is called the *burn-in* phase.

We have discussed the Metropolis–Hastings algorithm in a discrete-state setting, but when the parameter to be estimated is continuous, we should simulate a continuous-state Markov process. We will not analyze in detail this extension, as it would not provide us with much additional insight.[2] The major change is that densities and integrals are involved, rather than probability mass functions and sums. Since the probability of a single point within a continuous state is always zero, we should work with subsets within the state space, over which we integrate the density. Hence, the equilibrium condition of Eq. (14.2) is replaced by the following one:

$$\int_B \pi(\mathbf{x})\,d\mathbf{x} = \int \pi(\mathbf{x})P(\mathbf{x}, B)\,d\mathbf{x}, \qquad (14.5)$$

for any (measurable) subset B of the state space. Here, $\pi(\mathbf{x})$ is the equilibrium density, so that the integral on the left-hand side is the long-run probability of being in set B. The role of the transition matrix is played by the transition kernel function $P(\mathbf{x}, B)$, which gives the probability of moving from state $\mathbf{x}$ to set B. The interpretation of Eq. (14.5) is the same as Eq. (14.2).

Now, in order to be more concrete, let us apply the above ideas to a Bayesian inference for a single-parameter problem. The state space in the Markov process is the set of possible values of parameter θ and, given a vector of observations with value $\mathbf{x}$, we want to sample from the posterior density $p_n(\theta \mid \mathbf{x})$. To this aim, we have to define a transition kernel satisfying the detailed balance equation, which in practical terms requires the definition of a transition density $\tilde{q}(\theta, \theta')$ that satisfies the detailed balance condition

$$p_n(\theta \mid \mathbf{x})\tilde{q}(\theta, \theta') = p_n(\theta' \mid \mathbf{x})\tilde{q}(\theta', \theta), \qquad \forall \theta, \theta'.$$

By integrating the transition density $\tilde{q}(\theta, \theta')$ for θ' in a set B, we obtain the corresponding transition kernel, $P(\theta, B)$. The transition density is used to move from state θ to another state θ' by random sampling.[3] In order to satisfy the balance condition, we use the Metropolis–Hastings trick: Start with any transition density $q(\theta, \theta')$, and enforce the balance condition by multiplying it by

$$\alpha(\theta, \theta') = \min\left\{1, \frac{p_n(\theta' \mid \mathbf{x})q(\theta', \theta)}{p_n(\theta \mid \mathbf{x})q(\theta, \theta')}\right\}. \qquad (14.6)$$

---

[2] See, e.g., [4, pp. 129–130] for a readable proof.

[3] Some authors use the notation $q(\theta' \mid \theta)$ to clarify the role of the two arguments.

Once again, we notice that we only need the posterior up to a constant, as whatever constant we multiply the posterior by, it gets canceled in the above ratio. The algorithm runs as follows:

1. Choose an initial state θ_0 and repeat the following for $k = 1, \ldots, M$:

    (a) Select a candidate θ' by sampling from the transition density $q(\theta_{k-1}, \theta')$.

    (b) Accept the candidate with probability $\alpha(\theta_{k-1}, \theta')$; if accepted, set $\theta_k = \theta'$; otherwise, set $\theta_k = \theta_{k-1}$.

The random acceptance mechanism is simulated by sampling a uniform variable $U \sim \mathsf{U}(0, 1)$ and accepting the candidate if $U \leq \alpha(\theta_{k-1}, \theta')$.

Last but not least, to get the whole thing going we need a last piece of the puzzle: the transition density $q(\theta, \theta')$. There are two basic possibilities:

1. The *random walk* candidate density, whereby we choose a density function $h(\cdot)$, symmetric with respect to the origin, and set

$$q(\theta, \theta') = h(\theta' - \theta).$$

With this choice, we have $q(\theta, \theta') = q(\theta', \theta)$, because of symmetry. Hence, Eq. (14.6) simplifies to

$$\alpha(\theta, \theta') = \min \left\{ 1, \frac{p_n(\theta' \mid \mathbf{x})}{p_n(\theta \mid \mathbf{x})} \right\}. \tag{14.7}$$

The implication is that if we move from state θ to a state θ' with a larger (unscaled) posterior, the move is always accepted, as suggested by intuition. The acceptance probability is less than 1 when we move "downhill."

2. The *independent* density, whereby we choose the next candidate state according to a PDF

$$q(\theta, \theta') = h(\theta'),$$

which does not depend on the current state θ. With such a choice, Eq. (14.6) can be rewritten as

$$\alpha(\theta, \theta') = \min \left\{ 1, \frac{p_n(\theta' \mid \mathbf{x})}{p_n(\theta \mid \mathbf{x})} \times \frac{h(\theta)}{h(\theta')} \right\}. \tag{14.8}$$

Here, the acceptance probability depends on the ratio of the posterior densities, but also on the ratio of the candidate densities. The former ratio tends to favor uphill moves in terms of the posterior; the latter one tends to favor moves toward the tails of the candidate density, where $h(\theta')$ is small. This feature may induce a better exploration behavior.

Finding a good candidate density requires finding a satisfactory trade-off between two requirements. On the one hand, we want to explore the state space, which means that, for a given θ, we should be free to move outside its close neighborhood. On the other hand, we do not want to generate candidates that are too likely to be rejected. As the reader can imagine, striking a satisfactory balance may take some experimentation. Another issue is the potential bias introduced by the arbitrary choice of θ_0. As we have mentioned, this issue may be overcome by allowing a burn-in, i.e., by letting the system run for a while, before starting the collection of relevant statistics.

▣ Example 14.1 Implementing random walk MCMC in R

Let us implement the above ideas in R, using a mixture of two normals as the target density. We assume that the posterior is proportional to

$$20 \times \left\{ 0.7 \times \frac{1}{3} \exp\left[-\frac{(\theta - 1)^2}{2 \times 3^2} \right] + 0.3 \times \frac{1}{2} \exp\left[-\frac{(\theta - 10)^2}{2 \times 2^2} \right] \right\},$$

which is not a proper density, as we multiply it by 20, rather than dividing by $\sqrt{2\pi}$. As the candidate density, we use a random walk transition density $N(0, 3)$. We first run a 1000 step burn-in phase, where no statistics are collected, starting from some arbitrary initial state θ_0; then we generate a sample of size 50,000, whose histogram is plotted against the true density. The R code implementing a naive version of this random walk Metropolis–Hastings algorithm is displayed in Fig. 14.4, and it produces the histogram plotted in Fig. 14.5. As we see, there is a good agreement between the true density and the sampled one.

14.3.1 THE GIBBS SAMPLER

The Metropolis–Hastings algorithm is the starting point for several types of samplers, most notably the Gibbs sampler, which may be useful when we are dealing with multiple parameters, i.e., with a Markov chain featuring a vector state variable. Consider a vector of parameters

$$\boldsymbol{\theta} = \left[\theta_1, \theta_2, \dots, \theta_m \right]^{\mathsf{T}}.$$

Also let us denote by $\boldsymbol{\theta}_{-j}$ the vector of parameters with θ_j removed:

$$\boldsymbol{\theta}_{-j} = \left[\theta_1, \theta_2, \dots, \theta_{j-1}, \theta_j, \dots, \theta_m \right]^{\mathsf{T}}.$$

Since sampling directly from the multivariate distribution can be difficult, we may consider sampling each parameter in turn, while keeping the remaining ones fixed. This means that we should use a transition kernel based on a density $q(\theta_j, \theta'_j \,|\, \boldsymbol{\theta}_{-j})$ The idea can also be applied to blocks of parameters, rather than single ones, and is known as blockwise Metropolis–Hastings algorithm.

```
# define mixture of two normals
muA <- 1; sigmaA <- 3; muB <- 10; sigmaB <- 2; p <- 0.7
# define the unscaled target density
target <- function(theta) 20*(p*exp(-(theta-muA)^2/
            sigmaA^2/2)/sigmaA
  + (1-p)*exp(-(theta-muB)^2/sigmaB^2/2)/sigmaB)
# set candidate density
muQ <- 0; sigmaQ <- 3
# set starting value
theta0 <- (muA+muB)/2
burninSteps <- 1000; numSteps <- 50000
set.seed(55555)
# generate random variates, including BurnIn
Zb <- rnorm(burninSteps, muQ, sigmaQ)
Ub <- runif(burninSteps)
Z <- rnorm(numSteps, muQ, sigmaQ)
U <- runif(numSteps)
# first Burn In
oldTheta <- theta0
for (i in 1:burninSteps){
  tryTheta <- oldTheta + (muQ+Zb[i]*sigmaQ)
  alpha <- min(1,target(tryTheta)/target(oldTheta))
  if (Ub[i] <= alpha) # accept
    oldTheta <- tryTheta
}
# now collect info about target sample
targetSample <- numeric(numSteps)
for (i in 1:numSteps){
  tryTheta <- oldTheta + (muQ+Z[i]*sigmaQ)
  alpha <- min(1,target(tryTheta)/target(oldTheta))
  if (U[i] <= alpha)# accept
    oldTheta <- tryTheta
  targetSample[i] <- oldTheta
}
# plot actual density and histogram
theta<-seq(from=min(muA-3*sigmaA, muB-3*sigmaB),
            to=max(muA+3*sigmaA, muB+3*sigmaB),by=0.001)
trueF <- p*dnorm(theta, muA, sigmaA)+
                (1-p)*dnorm(theta,muB,sigmaB)
h<-hist(targetSample, plot = FALSE, nclass = 50)
ymax<-max(c(h$density,trueF))*1.05
hist(targetSample, prob = TRUE, col = "light blue",
        xlim = range(theta), ylim = c(0,ymax),
        main = "Sample from target density",
    nclass = 50, xlab = 'x', ylab = 'Density')
lines(theta, trueF)
```

FIGURE 14.4 **An illustration of the Metropolis–Hastings algorithm.**

Sample from target density

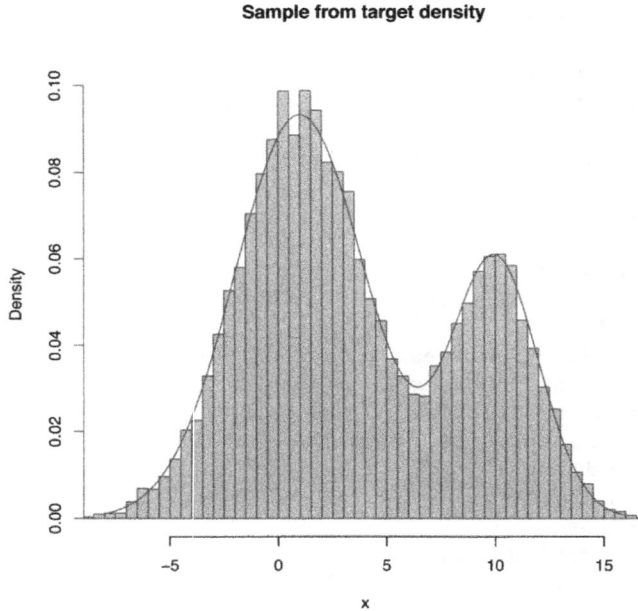

FIGURE 14.5 **Histogram produced by the sampler of Fig. 14.4.**

The Gibbs sampler is a specific case of blockwise Metropolis–Hastings algorithm, where we use the exact conditional density as the candidate density. One possible version of the algorithm is as follows:

- Let $f(\boldsymbol{\theta})$ be the joint density and $f_j(\theta_j \mid \boldsymbol{\theta}_{-j})$ be the conditional density for component $j = 1, \ldots, m$.
- Given the current state $\boldsymbol{\theta}^{(k)}$ at step k, we generate the next state $\boldsymbol{\theta}^{(k+1)}$ iteratively:
  - Sample $\theta_1^{(k+1)} \sim f_1(\theta_1 \mid \theta_2^{(k)}, \ldots, \theta_m^{(k)})$.
  - Sample $\theta_2^{(k+1)} \sim f_2(\theta_2 \mid \theta_1^{(k+1)}, \theta_3^{(k)}, \ldots, \theta_m^{(k)})$.
  - ...
  - Sample $\theta_m^{(k+1)} \sim f_m(\theta_m \mid \theta_1^{(k+1)}, \theta_2^{(k+1)}, \ldots, \theta_{m-1}^{(k+1)})$.

Thus, we see that Gibbs sampling is an iterative sampling strategy that should be contrasted with one-step sampling from a multivariate distribution. Furthermore, since we are using the true conditional density as the candidate, there is no need to check acceptance, as the acceptance probability is always 1. An alternative version of the Gibbs sampler, rather than cycling systematically through each variable, selects one dimension at a time randomly for sampling. The approach makes sense when conditional distributions are available, which is the case, e.g., for hierarchical models.

■ Example 14.2 **An illustration of Gibbs sampling**

As a very simple illustration, let us implement a Gibbs sampler for a bivariate normal distribution

$$\begin{bmatrix} X_1 \\ X_2 \end{bmatrix} \sim \mathsf{N}\left(\begin{bmatrix} 0 \\ 0 \end{bmatrix}, \begin{bmatrix} 1 & \rho \\ \rho & 1 \end{bmatrix} \right).$$

We know from Section 3.3.4.1 that the two conditional distributions are:

$$X_1 \,|\, X_2 = x_2 \sim \mathsf{N}(\rho x_2, (1 - \rho^2)),$$
$$X_2 \,|\, X_1 = x_1 \sim \mathsf{N}(\rho x_1, (1 - \rho^2)).$$

The code in Fig. 14.6 implements the corresponding Gibbs sampler. The function receives the sample size, the length of the burn-in period, and the correlation. The output is used by the script to produce the scatterplot matrix, with histograms on the diagonal, shown in Fig. 14.7. The result is in fact what we would expect from such a bivariate normal distribution.

14.4 A re-examination of simulated annealing

The Metropolis–Hastings algorithm should ring some bells for readers familiar with simulated annealing, which we described in Section 10.3.2. In both algorithms we generate a candidate state, which is accepted with some probability. With the random walk candidate density, the acceptance probability in the Metropolis–Hastings algorithm is given by Eq. (14.7); uphill moves are always accepted, whereas the acceptance of downhill moves is random. When applying simulated annealing to a minimization problem,

$$f^* \equiv \min_{\mathbf{x} \in S} f(\mathbf{x}),$$

the roles of uphill and downhill moves are swapped:

- Given the current solution $\bar{\mathbf{x}}$, with cost $f(\bar{\mathbf{x}})$, an alternative (candidate) solution $\mathbf{x}^\circ$ is sampled in the neighborhood of $\bar{\mathbf{x}}$.

```
gibbsNormal <- function(n,burnIn,rho){
  # precompute invariant quantity
  r2 <- sqrt(1 - rho^2)
  x1 <- 0; x2 <- 0
  # run burn-in
  for (i in 1:burnIn){
    x1 <- rnorm(1,rho*x2,r2)
    x2 <- rnorm(1,rho*x1,r2)}
  mat <- matrix(nrow=n,ncol=2)
  for (i in 1:n) {
    x1 <- rnorm(1,rho*x2,r2)
    x2 <- rnorm(1,rho*x1,r2)
    mat[i,] <- c(x1,x2)}
  return(mat)
}

set.seed(55555)
out <- gibbsNormal(10000,500,0.85)
library(car)
scatterplotMatrix(out,diagonal="histogram",
                  smooth=FALSE,reg.line=FALSE,pch=20)
```

FIGURE 14.6 **A simple Gibbs sampler.**

- The candidate is accepted with probability

$$\min\left\{1, \exp\left(-\frac{\Delta f}{T}\right)\right\}, \tag{14.9}$$

where $\Delta f = f(\mathbf{x}^\circ) - f(\bar{\mathbf{x}})$ and T is a control parameter playing the role of a temperature.

The acceptance probability is 1 if $\Delta f < 0$, i.e., the candidate solution improves the current one; otherwise, we have an uphill move, which is accepted with a probability depending on T. When T is large, it is relatively easy to accept an uphill move, whereas the system tends to favor only improvements when the temperature is low.

The similarity with the Metropolis–Hastings algorithm is in fact a deep one, as simulated annealing can be regarded as a MCMC optimization method, whose aim is to sample from a distribution concentrated around the global optimizers. Let us define the set of global optimizers

$$\mathcal{M} \equiv \{\mathbf{x} \in S : f(\mathbf{x}) = f^*\}.$$

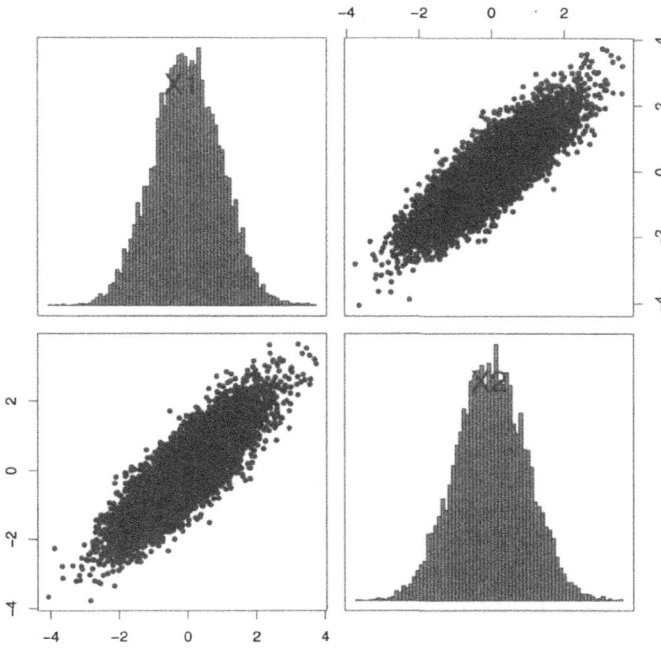

FIGURE 14.7 **Scatterplot matrix, with histograms on the diagonal, produced by the Gibbs sampler of Fig. 14.6.**

The optimal solution need not be unique, so let us denote the cardinality of this set by $|\mathcal{M}|$. Consider the target PMF:

$$\pi_\lambda(\mathbf{x}) = \frac{e^{-\lambda f(\mathbf{x})}}{\displaystyle\sum_{\mathbf{x}' \in S} e^{-\lambda f(\mathbf{x}')}} \tag{14.10}$$

for a given parameter $\lambda > 0$. Note that we may speak of a PMF if the set S is finite or enumerable, as it happens in combinatorial optimization problems; we assume that this is the optimization problem we are dealing with, rather than a continuous and nonconvex problem. Note that this PMF favors low-cost solutions, for which the negative exponential is larger. In fact, when $\lambda \to +\infty$, the above PMF tends to

$$\frac{\mathbf{1}_\mathcal{M}(\mathbf{x})}{|\mathcal{M}|},$$

where the indicator function $\mathbf{1}_\mathcal{M}(\mathbf{x})$ is 1 if $\mathbf{x}$ is optimal, zero otherwise. To see this, it is sufficient to multiply both numerator and denominator in Eq. (14.10) by $e^{-\lambda f^*}$, which yields:

$$
\begin{aligned}
\pi_\lambda(\mathbf{x}) &= \frac{e^{-\lambda[f(\mathbf{x})-f^*]}}{\displaystyle\sum_{\mathbf{x}'\in S} e^{-\lambda[f(\mathbf{x}')-f^*]}} \\[2mm]
&= \frac{e^{-\lambda[f(\mathbf{x})-f^*]}}{\displaystyle\sum_{\mathbf{x}'\in\mathcal{M}} e^{-\lambda[f(\mathbf{x}')-f^*]} + \sum_{\mathbf{x}'\in S\setminus\mathcal{M}} e^{-\lambda[f(\mathbf{x}')-f^*]}} \\[2mm]
&= \frac{e^{-\lambda[f(\mathbf{x})-f^*]}}{|\mathcal{M}| + \displaystyle\sum_{\mathbf{x}'\in S\setminus\mathcal{M}} e^{-\lambda[f(\mathbf{x}')-f^*]}}
\end{aligned}
$$

where \ denotes set difference. By letting $\lambda \to +\infty$, we obtain a very desirable distribution assigning the same positive probability to optimal solutions, and zero to nonoptimal solutions.

Now we have to define a candidate density to generate an alternative candidate solution $\mathbf{x}^\circ$. A simple approach is to sample randomly the neighborhood $\mathcal{N}(\bar{\mathbf{x}})$ of the current solution; the candidate PMF assigns a uniform probability, $1/|\mathcal{N}(\bar{\mathbf{x}})|$, with each solution in that neighborhood. Then, given the desired target PMF, the acceptance probability is

$$
\min\left\{1, \frac{e^{-\lambda f(\mathbf{x}^\circ)}/|\mathcal{N}(\mathbf{x}^\circ)|}{e^{-\lambda f(\bar{\mathbf{x}})}/|\mathcal{N}(\bar{\mathbf{x}})|}\right\}.
$$

If the neighborhood structure is the same for each solution, then the above expression boils down to

$$
\min\left\{1, e^{-\lambda[f(\mathbf{x}^\circ)-f(\bar{\mathbf{x}})]}\right\},
$$

which is just Eq. (14.9), provided that we substitute $\lambda = 1/T$. This is a reassuring result, as it suggests a convergence guarantee of simulated annealing to a globally optimal solution. However, we need to find a suitable cooling schedule for the temperature. Furthermore, a cynical reader might argue that, if the set of feasible solutions is finite, any complete enumeration algorithm will find an optimal solution in *finite* time, rather than in the long run. The objection does not necessarily apply if we are considering a continuous nonconvex optimization model, possibly arising from a difficult model calibration problem.

For further reading

- Bayesian statistics can be a challenging topic, but friendly introductions are provided in [3] and [9]. These books also offer R code.
- A quite readable introduction to computational Bayesian statistics can be found in [4], where variations of acceptance–rejection strategies are also

discussed. Among other things, applications to Bayesian linear regression are discussed.

- MCMC methods are not only aimed at Bayesian statistics; a generic treatment can be found in [11, Chapter 10].

- Some further applications of MCMC are discussed in [8, Chapter 6], which includes MATLAB code.

- Readers interested in financial applications might want to learn about the link between Bayesian statistics and portfolio management. The Black–Litterman model, originally introduced in [1], may be interpreted as a Bayesian approach; see also [7].

- The role of Bayesian methods in risk management is illustrated in [2].

- There are a few books on Bayesian econometrics, too; a recent one is [5].

References

1   F. Black and R. Litterman. Global portfolio optimization. *Financial Analysts Journal*, 48:28–43, 1992.

2   K. Böecker, editor. *Rethinking Risk Measurement and Reporting: Volumes I and II*. Risk Books, London, 2010.

3   W.M. Bolstad. *Introduction to Bayesian Statistics* (2nd ed.). Wiley, Hoboken, NJ, 2007.

4   W.M. Bolstad. *Understanding Computational Bayesian Statistics*. Wiley, Hoboken, NJ, 2010.

5   E. Greenberg. *Introduction to Bayesian Econometrics* (2nd ed.). Cambridge University Press, Cambridge, 2012.

6   W.K. Hastings. Monte Carlo sampling methods using Markov chains. *Biometrika*, 89:731–743, 1970.

7   G. He and R. Litterman. The intuition behind Black–Litterman model portfolios. Technical report, *Investment Management Research*, Goldman & Sachs Company, 1999. A more recent version can be downloaded from http://www.ssrn.org.

8   D.P. Kroese, T. Taimre, and Z.I. Botev. *Handbook of Monte Carlo Methods*. Wiley, Hoboken, NJ, 2011.

9   J.K. Kruschke. *Doing Bayesian Data Analysis: A Tutorial with R and BUGS*. Academic Press, Burlington, MA, 2011.

10   N.A. Metropolis, A.W. Rosenbluth, M.N. Rosenbluth, A.H. Teller, and E. Teller. Equations of state calculations by fast computing machines. *Journal of Chemical Physics*, 21:1087–1092, 1953.

11   S. Ross. *Simulation* (2nd ed.). Academic Press, San Diego, CA, 1997.

Index

absolute loss, 596
acceptance–rejection, 255, 269, 278, 282, 346, 380, 396, 630, 637
ACF, *see* autocorrelation, function
action (in dynamic programming), 486
additive model, 102, 161
ADP, *see* dynamic programming, approximate
antithetic sampling, 342, 361
APT, *see* arbitrage pricing theory
AR, *see* autoregressive
arbitrage
    opportunity, 194
    pricing theory, 192
arc (in a network), 470
ARCH, 158, 178
    model, 139
ARIMA, 139, 155, 237
ARMA, 151
asset pricing, 23
asset–liability management, 455
association, linear, 122
autocorrelation, 139
    function, 139, 143, 329
        sample, 140
    partial, *see* partial autocorrelation
autocorrelogram, 140, 144
autocovariance, 139
autoregressive
    conditionally heteroskedastic, *see* ARCH
    integrated moving-average, *see* ARIMA
    model, 139

    moving-average, *see* ARMA
    process, 8, 147

backshift operator, 154
balance condition for Markov chains
    detailed, 635
    global, 635
bandwidth, 110
barrier option, *see* option, barrier
basis
    function, 431, 489, 509, 552, 554
    of a lattice, 61
batch method, 329
batching, 32
Bayes' theorem, 242, 629
Bayesian
    computational statistics, 253
    estimation, 242
    parameter estimation, 629
    statistics, 106, 241
    view, 35
bequest, utility from, 476
Bernoulli
    random variable, 88, 106, 193, 196, 336
    trial, 243
beta
    distribution, 105, 244, 269
    function, 105, 244, 271
    variate generation, 272
bias, 292, 552, 577
big-M constraint, 423
binomial
    coefficient, 90
    distribution, 88, 106, 243
    lattice, 67, 89

model, 89
random variable, 89, 336
black swan, 35
Black–Scholes–Merton
    formula, 42, 196, 198, 330,
        575
    partial differential equation,
        197
    world, 573
Boltzmann constant, 436
bond, 205, 602
bootstrap, 110
bootstrapping, 599
    bond yields, 430
Box–Jenkins model, 139, 155
Box–Muller method, 277, 346,
    395
branching factor, 550
Brownian bridge, 127, 304, 364,
    399
BSM, *see* Black–Scholes–Merton
Buffon's needle, 5
burn-in, 638

càdlàg function, 95
capital asset pricing model, 192,
    424
CAPM, *see* capital asset pric-
    ing model
cardinality constraint, 424
Cartesian product, 53
CDF, *see* cumulative distribu-
    tion function
central limit theorem, 101, 336
    for quantiles, 337
chain rule, 159
chance-constrained model, 449
chi-square
    distribution, 102, 108, 356
        noncentral, 103, 310, 312,
            608
    random variable, 354
    test, 228, 229
    variable, 282
Cholesky

factor, 125, 278, 283, 527,
    607
factorization, 302
CIR, *see* Cox–Ingersoll–Ross
code vectorization, 620
coefficient of determination, 234
coefficient of variation, 214
collocation
    method, 489, 509
    node, 489
combinatorial optimization, 645
common random numbers, 348,
    447, 577
complete market, 203
concave
    function, 413
    optimization problem, 414
    problem, 414
concordance, 122, 134
conditional
    density, 370
    expectation, 198, 353, 554
    Monte Carlo, 354
    probability, 454
    value-at-risk, 80, 599, 618,
        see also value-at-risk, con-
        ditional
    variance, 354
conditioning, 353
confidence interval, 31, 57, 211,
    239, 255, 300, 326, 344,
    629
    half-length, 334
confidence level, 240, 595
conjugate family (of priors), 246
conjugate prior, 630
consumption–saving problem, 29,
    81, 493
continuation
    region, 429
    value, 428
continuous-time model, 84
continuously compounded
    interest rate, 84
contour plot, 23
control variable, 13

control variate, 219, 349, 532, 542, 579, 609

convex

function, 410

polyhedral, 410

strictly, 410

hull, 445

optimization, 594

optimization problem, 22, 414, 600

set, 408

convexity, 408, 594, 600

cooling schedule, 437

copula, 129, 132, 283, 599

t, 133

estimation, 211

Gaussian, 133

normal, 133

product, 133

Student's t, 283

theory, 38, 118

correlation, 32, 110, 343, 349, 350

coefficient, 119

instantaneous, 177, 526

negative, 343

Pearson's, 119, 134

sample, 218

testing significance, 218

cost-to-go, 472

counting process, 17, 95

covariance, 118, 139, 189, 593

matrix, 79, 278, 417, 527, 607

sample, 181

coverage (in confidence intervals), 214

covered position, 620

Cox–Ingersoll–Ross model, 177, 203, 310

Cramér–Rao bound, 221

credible interval, 629

cross-entropy method, 377, 520

cross-sectional

dependence, 112, 116

dependency, 110

model, 77

cumulant generating function, 374, 609

cumulative distribution function, 42, 87, 267

empirical, 337

joint, 113

curse of dimensionality, 30, 405, 477

CV@R, *see* value-at-risk, conditional

decision rules, 405

decision variable, 407

binary, 423

logical, 422

semicontinuous, 423

default, 600

degrees of freedom, 102, 104, 105, 127

delta hedging, 619

density

candidate, 272

instrumental, 272

target, 272

derivative

financial, 192

over-the-counter, 25

derivative-free method, 406

destandardization, 125

deterministic equivalent, 451

detrending, 142

diagonal model, 191

differencing, 151

second-order, 155

differential equation, 13, 85

digital shift, 394

dimensionality reduction, 178

direction number, 389

discount factor

subjective, 83

discrepancy, 379, 381

star, 381

discrete-event model, 16

discrete-time model, 10

discretization error, 292

distribution, *see also specific distribution*
    cumulative function, 199
    beta, 244
    bimodal, 110
    binomial, 243
    conditional, 305
    cumulative function
      joint, 113
    discrete empirical, 268, 503
    elliptical, 124
    lognormal, 53
    multivariate normal, 278
    right-skewed, 92
    skewed, 214
    uniform, 256
disturbance, 13, 474
diversification, 594
DP, *see* dynamic programming
drift, 14, 68, 85, 158, 173, 175, 620
duration, 602
dynamic programming, 53, 405, 453, 469, 551
    approximate, 405, 506, 548, 552
    numerical, 405, 480
    stochastic, 28

early exercise, 546
    boundary, 429
efficient market hypothesis, 77, 110, 154, 160, 424, 598
eigenvalue, 280, 608, 634
eigenvector, 182, 608, 634
elliptical distribution, 127
EMH, *see* efficient market hypothesis
empirical distribution, 108
endogenous uncertainty, 35
envelope theorem, 491
epsilon-greedy policy, 517
equilibrium condition, 635
Erlang
    distribution, 108, 282
    random variable, 94

error
    absolute, 334
    discretization, 292
    relative, 335
    sampling, 291
estimation error, 381
estimator
    biased, 32
    consistent, 222
    controlled, 349
    efficient unbiased, 221
    high-biased, 549, 550
    low-biased, 550
    point, 221
    stratified, 357
    unbiased, 212, 221, 350
Euler discretization scheme, 14, 291, 561
Euler–Maruyama discretization scheme, *see* Euler
excess return, 82
exclusive OR, 265
exogenous uncertainty, 35
expected value, 8
    of a function, 41
explicit scheme, 295
exploitation, 435, 517
exploration, 435, 517
exponential
    distribution, 92, 95, 107, 214, 224, 268
      memoryless property, 93
    random variable, 17, 316
    variable, 282
exponential tilting, 374, 609
exponential twisting, *see* exponential tilting

F distribution, 105
factor analysis, 179
factor model, 112, 189
factorial, 90, 106
factorization, Cholesky, 278
fat solution, 448
fat tail, 100, 104, 237
Faure sequence, 387

feasible
    region, 407
    set, 407
feedback policy, 475
Feynman–Kač formula, 198
financial asset, 7
finite difference, 348, 575
    central, 577
    forward, 577
firefly algorithm, 442
frequentist probability, 35
Fubini theorem, 53
function
    2-increasing, 131
    grounded, 130
    indicator, 41
    piecewise linear, 193
functional equation, 472
fundamental pricing equation,
    204

gamma
    distribution, 95, 107, 282
    function, 106
        and factorials, 106
    random variable, 103
GARCH, 139, 158, 178, 315
Gauss–Hermite
    quadrature, 52
Gaussian
    copula, 133
    distribution, 98
    process, 85, 290
    quadrature, 4, 28, 48, 57,
        405, 458, 467, 492, 496,
        551, *see also* quadrature,
        Gaussian
    white noise, 142
GBM, *see* geometric Brownian
    motion
generalized autoregressive con-
    ditionally heteroskedastic
    model, *see* GARCH
generalized feedback shift reg-
    isters, 265
generalized inverse, 337

genetic algorithm, 26, 406, 439
geometric
    distribution, 90
    random variable, 275
geometric Brownian motion, 9,
    14, 67, 86, 173, 196, 292,
    308, 332, 397, 496, 502,
    542, 586
    bidimensional, 526
    multidimensional, 301
GFSR, *see* generalized feedback
    shift registers
Gibbs sampler, 640
global optimization, 380
goodness of fit, 103
    test, 260
graph optimization, 470
Gray code, 392
grounded function, 130

half-length (of a confidence in-
    terval), 59, 334
Halton sequence, 60, 382
Hardy–Krause variation, 394
hedging strategy, 619
Hessian matrix, 607
Heston model, 178, 315
heteroskedasticity, 158
homoskedasticity, 158
hypercube, 129
hypothesis
    alternative, 214
    null, 214

i.i.d. (independent and identi-
    cally distributed), 142, 160
implicit scheme, 296
importance sampling, 41, 364,
    532, 607
income uncertainty, 493
incomplete market, 203, 431
increment
    stationary and independent,
        98, 164, 290
independence, 5
    vs. lack of correlation, 121

independent increment, 164
indicator function, 8, 336, 370,
        535, 583, 601
inferential statistics, 211
inner product, 49
integer programming, 519
integral
        multidimensional, 4, 630
        Riemann, 167
        stochastic, 167
integrating factor, 310
integration error, 45
interest rate
        continuously compounded,
        13
        instantaneous, 309
        modeling, 308
interior point methods, 416
interpolation, 43, 482
        piecewise linear, 108
intrinsic value, 428
inverse problem, 431
inverse transform method, 109,
        133, 267, 269, 278, 346,
        362, 380, 396, 468
Itô
        isometry, 311
        lemma, 170, 196, 203, 294,
        297, 310
        process, 167, 291
        stochastic differential equa-
        tion, 290, 632
        stochastic integral, 169

Jacobian
        determinant, 277
        matrix, 491
Jarque–Bera test, 233
Jensen's inequality, 102, 549
joint
        probability density function,
        5, 114
        probability mass function,
        114
jump, 92
jump–diffusion process, 319

Kendall's tau, 134
kernel density, 315
kernel-based estimation, 109
Koksma–Hlawka theorem, 393
Kolmogorov–Smirnov
        measure of fit, 381
        test, 231
Korobov point set, 64, 394
kurtosis, 52, 69, 100, 171, 233,
        374, 598
        excess, 104

labor income, 82
lack of memory (exponential dis-
        tribution), 92
Lagrange multiplier, 183, 359,
        406
Lagrange polynomial, 45
Lagrangian function, 183, 359
large numbers
        strong law of, 56
Latin hypercube sampling, 376
lattice, 61
        binomial, 67
        calibration, 68
        generator, 61
        projection regular, 65
        rank-r rule, 64
        structure in LCG, 261, 278
law of iterated expectations, 353
law of one price, 195
LCG, *see* linear congruential gen-
        erator
least-squares, 555
        nonlinear problem, 431
Levenberg–Marquardt method,
        432
Lévy flight, 442
Lévy–Itô decomposition, 319
Lévy process, 18, 178, 317
likelihood
        function, 26, 223, 242, 629
        log-function, 224
        method of maximum, 223
likelihood ratio, 364, 371, 375,
        609

derivative estimation, 580
    method, 585
linear congruential generator, 256,
    379
linear programming, 22, 415,
    619
    canonical form, 416
    mixed integer, 417
    standard form, 416
linear regression, 233, 341, 350,
    447, 510, 552, 554, 579
    probabilistic view, 137
linearity, 43
local improvement, 433
local maxima, 411
local search, 433
    best-improving, 434
    first-improving, 434
log-likelihood function, 224
log-return, 158
logarithm, 224
lognormal
    distribution, 14, 67, 101,
        161, 175, 176
    random variable, 9, 83, 540
longitudinal
    dependence, 110, 112
    model, 77
loss function, 463
low-discrepancy sequence, 4, 28,
    60, 178, 253, 306, 405,
    458, 545
    scrambled, 61
LP, see linear programming
lurking variable, 150

margin, 130
marginal, 114
    cumulative distribution func-
        tion, 114
    density, 114
    distribution, 112
    probability density function,
        114
market
    complete, 203

incomplete, 203
market portfolio, 190
Markov
    chain, 13, 486, 630
        aperiodic, 634
        continuous-time, 18, 19
        discrete time, 502
        discrete-time, 493, 631
        irreducible, 634
        long-run distribution, 634
        time reversible, 636
    decision process, 484
    process, 13, 631
Markov chain Monte Carlo, 253,
    630
Markovian
    dynamic system, 471
    process, 476, 588
Markowitz portfolio model, 77,
    79, 593
mathematical programming, 408
matrix
    positive definite, 280
    positive semidefinite, 140,
        417
    transposition, 62
maximization problem, 224, 407
maximum-likelihood
    estimation, 223, 406, 430,
        629
    estimator, 226
    method, 237
    properties of estimator, 226
MCMC, see Markov chain Monte
    Carlo
MDP, see Markov decision pro-
    cess
mean reversion, 86, 177, 309
mean reverting process, 154
mean–risk decision model, 613
mean–variance
    efficient portfolio, 79
    portfolio optimization, 77,
        421, 613
median, 98

memoryless property (exponential distribution), 19, 93
Mersenne
    number, 266
    twister generator, 266
metaheuristic, 433
metamodel, 446, 579
metamodeling, 575
method of batches, 219
method of moments, 221, 222
Metropolis–Hastings algorithm, 636
    blockwise, 640
MILP, *see* linear programming, mixed integer
Milstein discretization scheme, 292
MLE, *see* maximum-likelihood estimator
mod (math operator), 257
mode, 92, 98
model calibration, 23, 205, 406, 430, 519, 564
model identification, 138, 139, 237
modulus (in random number generation), 257
moment
    first-order, 80, 139
    sample, 222
    second-order, 80, 139
moment generating function, 101, 171, 373
    normal distribution, 374
moment matching, 48, 67, 89, 377, 458
monomial basis, 554
monotonicity, 594
monotonicity (requirement for variance reduction), 344
Monte Carlo integration, 56
Monte Carlo simulation
    as an integration problem, 255
    structure of, 254
moving-average model, 139

moving-average process, 142
multiplicative model, 102, 160
multiplier (in random number generation), 257

naked position, 620
neighborhood structure, 433
Nelder–Mead method, 445
network optimization, 470
newsvendor problem, 419, 425
Newton's method, 491
Newton–Cotes
    quadrature formula, 54
node (in a network), 470
node (in a quadrature formula), 43, 46
nonanticipativity, 331, 547, 554
noncentrality parameter, 103
nonconvex
    optimization, 23, 380, 406
    optimization problem, 22, 427
    set, 423
nonlinear least squares, 26
nonlinear programming, 228, 359, 417, 519
nonparametric test, 228
nonstationary process, 151
normal
    bivariate distribution, 126
    copula, 133
    distribution, 98, 214, 630
        mixture of, 640
        standard, 31, 98
    moment generating function, 374
    multivariate conditional distribution, 126
    multivariate variable, 124
    random variable, 9, 49
    standard
        distribution, 104, 199
    standard random variable, 395, 575
    standard variable, 42
    variate, 346

numerical integration, 28, *see* quadrature
numerical optimization, 228

objective function, 407
OLS, *see* ordinary least squares
optimal control, continuous-time, 407
optimality equation, 476
optimization
    combinatorial, 645
    constrained problem, 408
    finite-dimensional, 22
    model, 205
    nonconvex, 645
    unconstrained problem, 408
    under uncertainty, 444
option, 192
    American-style, 9, 29, 32, 193, 237, 292, 331, 334, 428, 507, 546
    as-you-like-it, 330
    Asian, 574
        arithmetic average, 539
        geometric, 540
        geometric average, 539
    Asian-style, 68
    barrier, 532
    Bermudan-style, 546
    call, 192
    chooser, 330
    continuation value, 428
    deeply out-of-the-money, 370
    delta, 201, 574, 602
    digital call, 583
    down-and-in put, 533
    down-and-out put, 532
    European-style, 9, 41, 193, 573
    exchange, 526
    gamma, 201
    greek, 201, 573, 594, 602
    holder, 192
    in-the-money, 370, 428, 555, 620
    intrinsic value, 428

lookback, 315
lookback call, 562
multidimensional, 526
out-of-the-money, 620
path dependent, 68
path-dependent, 532
pricing, 361
sensitivity, 201, 348, 349
spread, 526
theta, 201, 602
vanilla call, 346, 361
vega, 201
writer, 192
order statistics, 108, 232, 240, 337
ordinary least squares, 233
Ornstein–Uhlenbeck process, 86, 177, 310
orthodox statistics, 213, 239
orthogonal
    matrix, 180
    polynomial, 49, 554
overfitting, 112
overflow, 90

PACF, *see* partial autocorrelation function
panel
    data, 112
    model, 77
parameter estimation, 138, 406
    Bayesian, 629
parity
    for barrier options, 533
    put–call, 199
partial autocorrelation function, 148
    sample, 150
partial differential equation, 197
particle swarm optimization, 26, 406, 441
passive portfolio management, 423
pathwise derivative estimation, 580
pathwise differentiation, 582

PCA, *see* principal component analysis

PDE, *see* partial differential equation

PDF, *see* probability density function

penalty function, 434

periodicity (in random number generators), 259

permanent shock, 82

perturbation analysis, 447, 580

PMF, *see* probability mass function

point estimate, 255

point estimator, 239

Poisson
  distribution, 92
  process, 17, 19, 95, 176, 268, 319
    compound, 17, 98, 178, 316
    inhomogeneous, 17, 98
    random variable, 95, 319

polar
  coordinate, 277
  rejection, 278, 396

policy iteration, 493

polyhedral set, 409

polyhedron, 409

polynomial, 411
  interpolation, 43
  primitive, 390

polynomial regression, 234

portfolio
  benchmark, 425
  compression, 423
  tracking, 423, 424

position
  covered, 620
  naked, 620

positive homogeneity, 594

post-decision
  value function, 512

posterior
  density, 629
  distribution, 242

predictor-corrector method, 292

prime number, 382, 383

principal component, 181

principal component analysis, 179, 457

prior
  density, 629
  distribution, 241

probability
  density function, 5, 70, 87
    joint, 113
  mass function, 87
    marginal, 114
  measure, 364
  risk-neutral, 196

process
  adapted, 167
  standard Wiener, 163
  Wiener
    nondifferentiability of, 166

product rule, 54

projection regular (lattice), 65

pseudorandom, *see also* random number, 56, 254, 343, 573
  number generator, 63
  state, 6

PSO, *see* particle swarm optimization

put–call parity, 199, 330

Pythagorean theorem, 183

Q-factor, 514

Q-learning, 514

QP, *see* quadratic programming

Q-Q plot, 229

quadratic form, convex, 417

quadratic programming, 80, 417, 421, 593, 615

quadrature
  formula, 43
    closed, 44
    composite, 46
    Gauss–Laguerre, 51
    Gauss–Legendre, 51
    Newton–Cotes, 46
    open, 44

order, 45
    Gauss–Hermite, 49
    Gaussian, 48
    product rule, 54
    trapezoidal rule, 46
quantile, 31, 34, 276, 595
    estimated, 337
    estimation, 601
    of normal distribution, 99
quantum PSO, 442
quasi–Monte Carlo, 380
quasi–random number, 254

Radon–Nikodym derivative, 364
random number, 254, *see also*
    pseudorandom
random number generator, 256
    combined, 265
    generalized feedback shift
    registers, 265
    Mersenne twister, 266
    multiple recursive, 265
    multiplicative, 259
    separate stream, 264
    Tausworthe, 265
    Wichman–Hill, 265
random sample, 87
random shift, 394
random shock, 142
random variable, *see also spe-*
    cific random variable
    continuous, 87
    discrete, 87
    independent, 114
    jointly continuous, 113
    lognormal, 9
    normal, 9, 49
    standard normal, 276
    uniform, 5
random walk, 83, 151
rank correlation, 134
Rastrigin function, 23, 437
rate, 17
    continuously compounded,
    194
    risk-free, 194

recourse
    function, 451, 479
    matrix, 449
    simple, 450
    variable, 449
rectangle rule, 44
recursive equation, 472
recursive functional equation, 30
rejection region, 232
rejection sampling, *see* acceptance–
    rejection
replicating portfolio, 194
replication, 326
residual, 555
response surface, 446
return
    continuously compounded,
    175
    rate of, 79
Richardson extrapolation, 292,
    546
risk
    aversion, 78, 405, 427, 448
    idiosyncratic, 190
    management, 76, 136
    measure, 78, 593
        coherent, 593
    specific, 190
    systematic factor, 190
risk-neutral
    measure, 370, 554, 586
    option pricing, 76
    pricing, 192, 196, 289
    probability measure, 41, 192

SACF, *see* autocorrelation func-
    tion, sample
sample
    correlation, 218
    covariance, 218
    mean, 19, 56, 211
    partial autocorrelation func-
    tion, 150
    path, 15, 573
    variance, 211
sampling vs. simulation, 9

scale parameter, 108
scatterplot, 120
scenario, 454, 600
  extreme, 111
  generation, 67, 492
  tree, 405, 547, 549
score function, 580, 585
scrambling
  Faure–Tezuka, 395
  of a low-discrepancy sequence, 395
  Owen, 395
SDE, *see* stochastic differential equation
seasonality, 140
seed, 256, 258
  setting in random number generators, 258
semiannual compounding, 84
semicontinuous variable, 423
sensitivity estimation, 349
shape parameter, 108
Shapiro–Wilk test, 232
shift (in random number generation), 257
short rate, 309
short selling, 193, 195
shortest path problem, 470
shortfall probability, 8, 601
simplex, 445
  algorithm, 416
  method, 406
  search, 445
Simpson's rule, 47
simulated annealing, 26, 435, 630, 643
  generalized, 439
simulation-based optimization, 405, 547, 575
skewness, 52, 69, 100, 233
  negative, 158
Sklar's theorem, 132
small world PSO, 442
Sobol sequence, 388, 469, 532
SP, *see* stochastic programming

SPACF, *see* partial autocorrelation function,sample
Spearman's rho, 134
spline, 431
  cubic, 482, 495
  natural, 496
square-root diffusion, 104, 177, 312, 315, 561
square-root matrix, 125, 607
square-root rule, 158, 162, 596
stability
  in sample, 406, 461
  out of sample, 406, 461
stable distribution, 101
standard deviation, 593
  of return, 421
standard normal distribution, 70
standard Wiener processes, 397
standardization, 125
star discrepancy, 60
state transition
  equation, 11
  function, 264, 474
state variable, 11
  post-decision, 511, 560
stationary and independent increments, 178
stationary increment, 164
steepest descent, 447
stochastic calculus, 14, 77, 196
stochastic differential, 163
stochastic differential equation, 8, 38, 84, 159, 162
stochastic dynamic optimization, 32
stochastic dynamic programming, 28
stochastic integral, 163
stochastic optimization, 11, 22, 42, 331, 418, 573
stochastic process, 85, 162
  mean reverting, 154
  nonstationary process, 151
  predictable, 82
  stationary, 632
  weakly stationary, 139

stochastic programming, 405, 547, 615
    multistage with recourse, 28
    two-stage, 450
    with chance constraints, 449
    with recourse, 448, 449
stochastic search, 26
stochastic volatility, 315
stop-loss hedging, 620
stopping time, 428, 547
strategic oscillation, 435
stratification, 127, 306, 468
stratified sampling, 357
stream, in random number generators, 264
strike price, 192
Student's t
    distribution, 104, 212, 237
    multivariate distribution, 127
    random variable, 354
subadditivity, 594, 600, 615
subjective
    knowledge, 240
    probability, 35
sufficient statistics, 226
surface plot, 23

tabu search, 435
tail dependence, 136
tail risk, 598
Taylor expansion, 171, 293, 578
t distribution
    multivariate, 283
term structure of interest rates, 188, 430
TEV, *see* tracking error variance
tilting parameter, 376
time bucket, 10
time consistency, 594
time series
    analysis, 77
    model, 8, 112, 237
total probability theorem, 114, 242, 633
tracking error variance, 425

tracking portfolio, 424
transaction cost, 455
transient phase, 32, 237
transition
    density, 638
    kernel, 638
    matrix, 633
    probability, 632
        matrix, 486
transitory shock, 82
translation invariance, 594
transposition, 79

uncorrelated variables, 120
uniform
    random variable, 5, 87, 319
unit hypercube, 44, 59, 60, 255, 380
unk-unks (unknown unknowns), 35
utility
    expected, 456
    function, 456
    quadratic, 427
utility function, 78, 83, 595
    logarithmic, 499
    power, 499

value function, 30, 472
value-at-risk, 34, 80, 156, 162, 336, 376, 594, 595
    absolute, 596
    conditional, 422, *see also* conditional, value-at-risk, 599
    historical, 598
    relative, 596
Van der Corput sequence, 382
VAR, *see* vector autoregressive model
V@R, *see* value-at-risk
variance, 156, 593
    of a sum of random variables, 119
variance reduction, 28, 33, 58, 253, 405, 458, 607
Vasicek model, 177, 203, 309

vector autoregressive model, 156
vector transposition, 62
volatility, 14, 68, 85, 158, 173,
     175
     cluster, 157
     stochastic, 157, 178
volume risk, 625

weakly stationary process, 139
wealth, 7
Weierstrass theorem, 45
weight function, 48
weights (in a quadrature formula),
     43
white noise, 142
     Gaussian, 142
whitening, 125
Wichman–Hill generator, 265
Wiener process, 14, 85, 163, 197,
     304
     bidimensional, 528
     correlated, 177, 302
     differential, 166
     generalized, 167, 291, 317
     lack of differentiability, 165
     standard, 163, 290, 318

yield curve, 430

zero-coupon bond, 564

Wiley Handbooks in
FINANCIAL ENGINEERING AND ECONOMETRICS

Advisory Editor
Ruey S. Tsay
The University of Chicago Booth School of Business, USA

The dynamic and interaction between financial markets around the world have changed dramatically under economic globalization. In addition, advances in communication and data collection have changed the way information is processed and used. In this new era, financial instruments have become increasingly sophisticated and their impacts are far-reaching. The recent financial (credit) crisis is a vivid example of the new challenges we face and continue to face in this information age. Analytical skills and ability to extract useful information from mass data, to comprehend the complexity of financial instruments, and to assess the financial risk involved become a necessity for economists, financial managers, and risk management professionals. To master such skills and ability, knowledge from computer science, economics, finance, mathematics and statistics is essential. As such, financial engineering is cross-disciplinary, and its theory and applications advance rapidly.

The goal of this Handbook Series is to provide a one-stop source for students, researchers, and practitioners to learn the knowledge and analytical skills they need to face today's challenges in financial markets. The Series intends to introduce systematically recent developments in different areas of financial engineering and econometrics. The coverage will be broad and thorough with balance in theory and applications. Each volume will be edited by leading researchers and practitioners in the area and covers state-of-the-art methods and theory of the selected topic.

Published Wiley Handbooks in Financial Engineering and Econometrics

Bauwens, Hafner, and Laurent · *Handbook of Volatility Models and Their Applications*
Brandimarte · *Handbook in Monte Carlo Simulation: Applications in Financial Engineering, Risk Management, and Economics*
Chan and Wong · *Handbook of Financial Risk Management: Simulations and Case Studies*
James, Marsh, and Sarno · *Handbook of Exchange Rates*
Viens, Mariani, and Florescu · *Handbook of Modeling High-Frequency Data in Finance*
Szylar · *Handbook of Market Risk*

Forthcoming Wiley Handbooks in Financial Engineering and Econometrics

Bali and Engle · *Handbook of Asset Pricing*
Chacko · *Handbook of Credit and Interest Rate Derivatives*

Cruz, Peters, and Shevchenko · *Handbook of Operational Risk*

Florescu, Mariani, Stanley, and Viens · *Handbook of High-Frequency Trading and Modeling in Finance*

Jacquier · *Handbook of Econometric Methods for Finance: Bayesian and Classical Perspecitves*

Longin · *Handbook of Extreme Value Theory and Its Applications to Finance and Insurance*

Starer · *Handbook of Equity Portfolio Management: Theory and Practice*

Veronesi · *Handbook of Fixed-Income Securities*

Printed and bound by CPI Group (UK) Ltd, Croydon, CR0 4YY

04/04/2024

14479345-0001